T0328309

BEHAVIOUR IN OUR BONES

BEHAVIOUR IN OUR BONES

How Human Behaviour Influences Skeletal Morphology

Edited by

CARA STELLA HIRST
Institute of Archaeology, University College London, London, United Kingdom

REBECCA J. GILMOUR
Department of Sociology & Anthropology, Mount Royal University, Calgary, AB, Canada

KIMBERLY A. PLOMP
School of Archaeology, University of the Philippines, Diliman, Quezon City, Manila, Philippines

FRANCISCA ALVES CARDOSO
LABOH—Laboratory of Biological Anthropology and Human Osteology,
CRIA—Center for Research in Anthropology, NOVA University of Lisbon—School of Social Sciences
and Humanities (NOVA FCSH), Lisbon, Portugal

Cranfield Forensic Institute, Cranfield University, Defence Academy of the United Kingdom, Shrivenham,
United Kingdom

ELSEVIER

Elsevier
Radarweg 29, PO Box 211, 1000 AE Amsterdam, Netherlands
The Boulevard, Langford Lane, Kidlington, Oxford OX5 1GB, United Kingdom
50 Hampshire Street, 5th Floor, Cambridge, MA 02139, United States

Notices

Knowledge and best practice in this field are constantly changing. As new research and experience broaden our understanding, changes in research methods, professional practices, or medical treatment may become necessary.

Practitioners and researchers must always rely on their own experience and knowledge in evaluating and using any information, methods, compounds, or experiments described herein. In using such information or methods they should be mindful of their own safety and the safety of others, including parties for whom they have a professional responsibility.

To the fullest extent of the law, neither the Publisher nor the authors, contributors, or editors, assume any liability for any injury and/or damage to persons or property as a matter of products liability, negligence or otherwise, or from any use or operation of any methods, products, instructions, or ideas contained in the material herein.

ISBN: 978-0-12-821383-4

For information on all Elsevier publications
visit our website at https://www.elsevier.com/books-and-journals

Publisher: Stacy Masucci
Acquisitions Editor: Ana Claudia A. Garcia
Editorial Project Manager: Susan E. Ikeda
Production Project Manager: Fahmida Sultana
Cover Designer: Matthew Limbert

Typeset by STRAIVE, India

Contents

7. Behaviour and the bones of the thorax and spine

Kimberly A. Plomp

8. Human behaviour and the pelvis

Sarah-Louise Decrausaz and Natalie Laudicina

9. Horse riding and the lower limbs

William Berthon, Christèle Baillif-Ducros, Matthew Fuka, and Ksenija Djukic

10. Locomotion and the foot and ankle

Kimberleigh A. Tommy and Meir M. Barak

11. Injury, disease, and recovery: Skeletal adaptations to immobility and impairment

Rebecca J. Gilmour, Liina Mansukoski, and Sarah Schrader

12. Acting on what we have learned and moving forward with skeletal behaviour

Rebecca J. Gilmour, Kimberly A. Plomp, and Francisca Alves Cardoso

Contributors

Francisca Alves Cardoso LABOH—Laboratory of Biological Anthropology and Human Osteology, CRIA—Center for Research in Anthropology, NOVA University of Lisbon—School of Social Sciences and Humanities (NOVA FCSH), Lisbon, Portugal; Cranfield Forensic Institute, Cranfield University, Defence Academy of the United Kingdom, Shrivenham, United Kingdom

Christèle Baillif-Ducros French National Institute for Preventive Archaeological Research (INRAP), Great East Region, Châlons-en-Champagne; UMR 6273 CRAHAM (Michel de Boüard Centre, Centre for Archaeological and Historical Research in Ancient and Medieval Times), CNRS/University of Caen Normandy, Caen, France

Meir M. Barak Department of Veterinary Biomedical Sciences, College of Veterinary Medicine, Long Island University, Brookville, NY, United States

William Berthon Chair of Biological Anthropology Paul Broca, École Pratique des Hautes Études (EPHE), PSL University, Paris, France; Department of Biological Anthropology, University of Szeged, Szeged, Hungary

Sarah-Louise Decrausaz Department of Anthropology, University of Victoria, Victoria, BC, Canada

Lily J.D. DeMars Department of Anthropology, Pennsylvania State University, University Park, State College, PA, United States

Ksenija Djukic Centre of Bone Biology, Faculty of Medicine, University of Belgrade, Belgrade, Serbia

Christopher J. Dunmore Skeletal Biology Research Centre, School of Anthropology and Conservation, University of Kent, Canterbury, United Kingdom

Matthew Fuka Cultural Resource Analysts, Inc., Richmond, VA, United States

Aaron Gasparik Institute of Archaeology, University College London, London, United Kingdom

Rebecca J. Gilmour Department of Sociology & Anthropology, Mount Royal University, Calgary, AB, Canada

Cara Stella Hirst Institute of Archaeology, University College London, London, United Kingdom

Fotios Alexandros Karakostis DFG (*Deutsche Forschungsgemeinschaft*) Center for Advanced Studies "Words, Bones, Genes, Tools," Eberhard Karls University of Tübingen, Tübingen, Germany

Natalie Laudicina Department of Biomedical Sciences, Grand Valley State University, Allendale, MI, United States

Timo van Leeuwen Department of Development and Regeneration, KU Leuven, KULAK, Kortrijk, Belgium

Szu-Ching Lu Laboratory for Innovation in Autism, University of Strathclyde, Glasgow, United Kingdom

Liina Mansukoski Department of Health Sciences, University of York, Heslington, York, United Kingdom

Lumila Paula Menéndez Konrad Lorenz Institute for Evolution and Cognition Research, Klosterneuburg, Austria; Department Anthropology of the Americas, University of Bonn, Bonn, Germany

Justyna J. Miszkiewicz School of Archaeology and Anthropology, Australian National University, Canberra, ACT; School of Social Science, University of Queensland, Brisbane, QLD, Australia

Kimberly A. Plomp School of Archaeology, University of the Philippines, Diliman, Quezon City, Manila, Philippines

Tomos Proffitt Technological Primate Research Group, Max Planck Institute for Evolutionary Anthropology, Leipzig, Germany

Jessica Ryan-Despraz Laboratory of Prehistoric Archaeology and Anthropology, Department F.-A. Forel for Environmental and Aquatic Sciences, Section of Earth and Environmental Sciences, University of Geneva, Geneva, Switzerland

Sarah Schrader Faculty of Archaeology, Leiden University, Leiden, Netherlands

Kimberleigh A. Tommy Human Variation and Identification Research Unit, School of Anatomical Sciences, University of the Witwatersrand, Johannesburg, South Africa

Nicole Torres-Tamayo School of Life and Health Sciences, University of Roehampton, London, United Kingdom

Suzy White School of Biological Sciences, University of Reading, Reading; Department of Anthropology, University College London, London, United Kingdom

Foreword

Bone has long been understood as a dynamic tissue that is sensitive to environmental influences during development and that remodels throughout adult life. It has now been 130 years since Julius Wolff first observed that the structure and orientation of trabecular bone corresponded with the principal axes of mechanical loading of the skeleton, which came to be known as Wolff's Law (Wolff, 1892). Throughout the 20th century, the promise of interpreting behaviour from the skeleton remained of great interest to anthropologists, but progress in the field was limited by two factors: (i) the underlying complexity of functional adaptation in the skeleton and (ii) the fact that most studies relied on relatively simple linear Euclidean measurements of the skeleton which are, at best, peripherally influenced by activity (Ruff, 1987). These challenges made functional interpretations of skeletal variation controversial, and early attempts to interpret behavior from the skeleton were often critiqued as telling 'stories from the skeleton' rather than being based in clear and demonstrated functional relationships (Jurmain, 1999). Methodological improvements since the 1980s and 1990s, including the application of CT scanning to the interpretation of long bone cross-sectional geometry, have fueled new research, led to theoretical shifts, and provided new confidence in our ability to read 'behaviour' from the skeleton (Ruff, 2019).

Functional morphology and, in the broadest sense, the entire field of bioarchaeology are dedicated to the interpretation of behaviour from the skeleton (Larsen, 2015).

While bioarchaeological research employs methodologies that range from the study of indirect influences of sociocultural variation on human growth, development, health, and diseases, there is a subset of research dedicated to the direct study of the influences of habitual activity on observable variation in the skeleton. This is a point of intersection with functional morphology that also includes the functional relevance of skeletal variation at among fossil and extant species. This book is about the direct mechanical influences on the skeleton that result from variation in habitual activity. It builds upon the long history of study of the relationship between mechanical loading and skeletal form that started with Julius Wolff, but it demonstrates how researchers are now able to overcome the limitations that restricted our ability to interpret behaviour from the skeleton for so long.

This volume represents a collection of papers from researchers who are pushing the boundaries of functional morphology by applying new methods, better imaging, more integrative and interdisciplinary analyses, and new theoretical perspectives to long-standing questions in functional morphology. The chapters consider regions of the skeleton as functional complexes and demonstrate not only what we currently know about behavioural influences on variation in those regions but also how novel research approaches have been employed to overcome the limitations of earlier work in the field. The research that has pushed the discipline forward includes a range of methods and approaches that are applied in the chapters.

They include high-resolution approaches to imaging of the skeleton such as microCT or surface scanning, which allow researchers to accurately capture and compare 3D geometric and structural variation in skeletal tissue. New experimental methodologies have allowed researchers to investigate the relationship between mechanical function and anatomical form in living humans, which allows us to better interpret variation in the past. The broadening of comparative morphology between species and human populations enables us to better understand the relationship between skeletal form and function. Histological analyses of bone tissue and analyses of entheseal changes have a long history of application in skeletal biomechanics, but new methods and more integrative research in both areas are extending our knowledge of localized tissue-level responses to broader patterns of activity. An additional area where new approaches are paying dividends is through the application of functional morphology to better understand palaeopathology. Here, researchers have demonstrated that interpretations of mechanical loading can both help explain susceptibility to skeletal pathology and inform us about disability, treatment, and recovery from illness and trauma in the past. Each of these methods is now being commonly framed and interpreted within broad sociocultural and ecological contexts that allow for more nuanced interpretations of the relationship between habitual behaviour and observable variation in the skeleton.

The authors of each chapter in this volume are researchers who have been instrumental in the development of these new methods and approaches, and they are all currently shaping the future of the discipline. This book is a crucial milestone in the long history of our understanding of the relationship between the form of the human skeleton and behaviour during the lifespan. The advancements presented here provide not only a clear articulation of current research in functional morphology of the human skeleton but also a wealth of ideas for future research. The interdisciplinary approaches described in these pages represent a generational shift in the available approaches to the interpretation of behaviour from the skeleton and provide a template for the types of research that will advance the discipline in the future.

Jay T. Stock
Department of Anthropology, Western University, London, ON, Canada

References

Jurmain, R. (1999). *Stories from the skeleton: Behavioral reconstruction in human osteology*. Gordon and Breach.

Larsen, C. S. (2015). *Bioarchaeology: Interpreting behavior from the human skeleton*. Cambridge: Cambridge University Press.

Ruff, C. (1987). Sexual dimorphism in human lower limb bone structure: Relationship to subsistence strategy and sexual division of labor. *Journal of Human Evolution, 16,* 391–416.

Ruff, C. B. (2019). Biomechanical analyses of archaeological human skeletons. In A. Grauer, & M. Anne Katzenberg (Eds.), *Biological anthropology of the human skeleton* (pp. 183–206). Hoboken, NJ, USA: John Wiley & Sons, Inc.

Wolff, J. (1892). *Das Gesetz der Transformation der Knochen*. Berlin: A. Hirchwild.

1

Skeletons in action: Inferring behaviour from our bones

Kimberly A. Plomp[a], Rebecca J. Gilmour[b], and Francisca Alves Cardoso[c,d]

[a]School of Archaeology, University of the Philippines, Diliman, Quezon City, Manila, Philippines, [b]Department of Sociology & Anthropology, Mount Royal University, Calgary, AB, Canada, [c]LABOH—Laboratory of Biological Anthropology and Human Osteology, CRIA—Center for Research in Anthropology, NOVA University of Lisbon—School of Social Sciences and Humanities (NOVA FCSH), Lisbon, Portugal, [d]Cranfield Forensic Institute, Cranfield University, Defence Academy of the United Kingdom, Shrivenham, United Kingdom

Human bodies record and store vast amounts of information about the way we move, what we eat, where we live, and our experiences of health and social circumstances. Nearly everything we do is reflective of our 'behaviour', that is, "the potential and expressed capacity for physical, mental, and social activity during the phases of human life" (Bornstein et al., 2020, para. 1), and our bodies preserve evidence for much of this. Our actions and interactions, all of which are individually, socially, and culturally influenced, can be embodied in our physical self. The everyday physical actions we perform, such as walking, running, lifting, throwing, as well as the environments we live in and the tools we use, will all subject our bodies to stresses and forces that can leave visually observable markers or traits on bone—skeletal evidence for our active behaviours. Beyond the physical, our sociocultural experiences can also leave biological indicators, especially in the case of injury and experiences of inequality—skeletal evidence for our social and cultural behaviours. Through comprehensive biological anthropological analyses of behaviour preserved in skeletal remains, we can gain insight into the depth of the physical and social lives of past people and populations.

By investigating variations, either healthy or pathological, in the human body, biological anthropologists create a window through which they can visualise and interpret aspects of an individual's behaviour, from their diet, to their health, movements and migrations, cultural practices, medical interventions, and adaptability to changing environments (Grauer, 2012; Larsen, 2015). Along with studies involving modern and/or living humans, biological anthropologists specialising in past humans and their fossil relatives, (i.e., bioarchaeologists and palaeoanthropologists) also ask these questions using evidence preserved in skeletonised, mummified, and fossilised remains. Inquiry into human behaviour, a topic that itself is very broad, underpins many biological anthropology questions, encompassing wide reaching topics, such as 'when did human ancestors start to walk bipedally?', to more focused areas like 'what effect do/did certain medical interventions have on pathological conditions?'

Earlier books have extensively and successfully investigated the topic of human behaviour as it is preserved in the skeleton. Jurmain's (1999) book, *Stories from the skeleton: Behavioural reconstruction in human osteology*, organises its chapters around osteoarthritis, entheseal changes, trauma, and predicts bone geometry as an area for future development. Larsen's (2015) book, *Bioarchaeology interpreting behaviour from the human skeleton*, builds on many of these topics through discussions of palaeopathology and musculoskeletal modifications (e.g. entheseal changes and osteoarthritis). A large segment of Larsen's (2015) book is dedicated to sections on skeletal geometry and mechanical loading of the postcranial and cranial skeleton. Larsen (2015) also introduces other important methodological approaches that encourage the reader to consider how isotopes, palaeodemography, biological distance analyses, and growth and development (as it relates to stress) can help us understand the behaviour of people in the past. Other, more recent, volumes are dedicated to in-depth discovery of specific methodological approaches used to infer behaviour, occasionally focussing on specific geographic or temporal contexts. For example, Ruff (2018) expands on how biomechanical approaches contribute to activity and body reconstructions across Europe, and Schrader (2019) applies osteological indicators of behaviour, including entheseal changes and osteoarthritis, with dietary isotopes to describe everyday life along the ancient Nile Valley. The present book brings much of this research together, synthesising how biological anthropologists are currently working to tell detailed behavioural stories about particular skeletal regions and/or activities, using a range of theoretical and methodological approaches. *Behaviour in our bones* compiles and summarises this current research, while introducing it to readers new to this area of study and updating others to its growing possibilities.

There are numerous ways to study and understand human behaviour. (Bio)archaeologists, for example, have previously analysed material culture, trade routes, and even used ratios of stable isotopes in bones and teeth to determine diet and migrations. This book focuses on the indicators of behaviour that are directly visible on bone and as variations in bone morphology. These include typical sources of evidence that are often used by biological anthropologists to infer behaviour and activity, such as cross-sectional geometry and trabecular microstructure studies to interpret loading environments, as well as evidence for musculoskeletal stress through careful analyses of entheseal changes, all of which are discussed in detail in the forthcoming chapters. However, because human activities are so varied, this book also takes a broad definition of 'skeletal indicators of behaviour' and includes a diverse range of evidence, such as pathological changes to the bones, marks left on bone through trauma, intentional or unintentional deformation of bone morphologies, and evolutionary adaptations. The use of a broad range of evidence should inspire biological anthropologists to take creative approaches to thinking about evidence for behaviour, and, therefore, gain a deeper understanding of both physical actions and social trends throughout human history.

Behaviour in our bones merges a range of topics and approaches, weaving together clinical and archaeological literature to demonstrate how human skeletons preserve evidence for many different types of behaviour and how biological anthropologists can read this evidence to identify and better understand human activities. The information covered in each chapter is not restricted to archaeological skeletal evidence, but also explores clinical, soft tissue, and fossil evidence, making this book relevant to a range of biological anthropologists. In essence, the aim of this book is to provide an open scaffolding where the authors of each chapter examine, analyse, critique, and discuss the research in their own area of expertise that attempts to infer behaviour from the human skeleton. There is no limitation on time or geographic area, allowing the authors to examine bioarchaeological and related research from all parts of the globe, spanning human history, including our fossil relatives. Each chapter tackles a different skeletal region of the body, and the contents are organised to work down the body from head to toe, to discuss how biological anthropologists have attempted to, successfully or not, interpret evidence of behaviour on bones. Through exploration of each chapter, readers will encounter the diversity in methodological approaches, theoretical perspectives, interpretations, and descriptions of actual bioarchaeological studies where they have been applied to investigate behaviour in the human skeleton. Each chapter will introduce new activities and assist readers in discovering a variety of behaviours preserved in our skeleton, including modes of locomotion (Chapters 7–10), the use of tools (Chapters 5 and 6), responses to pathology and social circumstances (Chapters 7 and 11), as well as the anatomical and evolutionary explanations underpinning activity (Chapters 2, 3, 4, and 8). The chapters are diverse in their use and discussion of methodological approaches. For example, readers interested in entheseal changes and joint disease can consult Chapters 4, 5, 6, 9, and 10, those interested in the functional adaptation of cortical and trabecular bone should seek Chapters 2, 6, and 11, while others concerned with broad investigations of shape and geometric morphometrics may explore Chapters 3, 7, 8, and 9. With this book, we aim for readers to be inspired to imagine how different methods and ways of interpreting information might be applied throughout the skeleton and to further investigations into other activities, above and beyond those outlined in this book.

To prepare readers and provide a basic understanding of how bone adapts and responds to activity, the book begins with a detailed discussion of *how* we can identify evidence of behaviour in bone. In Chapter 2, *'Bone biology and microscopic changes in response to behaviour'*, authors Lily DeMars, Nicole Torres-Tamayo, Cara Hirst, and Justyna Miszkiewicz aim to outline the key principles of skeletal biology, microstructure, and biomechanics. They provide in-depth discussion of how biological anthropologists, including bioarchaeologists, have attempted to reconstruct human behaviour in the past using two-dimensional (2D) histological and three-dimensional (3D) microcomputed tomography (micro-CT) on both cortical and trabecular bone. The aim of this chapter is to provide the reader with the background knowledge and understanding of bone biology that will be important to contextualise the methods and interpret the evidence of behaviour presented in the remaining chapters of the book.

Since the book covers all regions of the skeleton, we will begin at the top, with Chapter 3, *'Biosocial Complexity and the Skull'* by Suzanna White and Lumila Menéndez. In this chapter, the authors provide a thorough overview of the anatomy and development of the human skull and outline the various research trajectories that can use evidence on the cranium to infer behaviour in the past. The breadth of topics they cover encompasses human evolution, population histories, transitions in subsistence practices and climate, communication, and cultural practices. The authors expertly integrate this diverse range of topics to highlight the biosocial complexity of the human skull as each region is shaped by an individual's biology, ecology, agency, and their social and cultural environments.

Moving down the body, in Chapter 4, *'Activity and the shoulder: From soft tissues to bare bones'*, Francisca Alves Cardoso and Aaron Gasparik critically evaluate the use of entheses and degenerative pathological changes often found in the shoulder as indicators of activity, activity patterns, and to a certain degree 'occupation' in the past. They assess the shoulder joint and its function for humans, introducing how it relates to movement, activity, entheseal changes, and changes to the joint. Through exploration of many bioarchaeological studies, they compare the frequencies of entheseal and degenerative changes in the shoulder and highlight how exploring activity with a focus on just the shoulder analysis is rare, as the majority of the studies explore activity in the entirety of the skeleton. They also conclude that bioarchaeological studies tend to agree that entheseal and degenerative changes of the shoulders, or any joint, correlate more closely with increasing age than habitual activity. Most importantly, although general trends in skeletal changes may provide some evidence of movement and action, Alves Cardoso and Gasparik caution against using the presence of shoulder entheseal changes and pathologies to argue for specific activities, such as throwing or rowing.

Chapter 5, *'Archery and the arm'* by Jessica Ryan-Despraz, provides a detailed assessment of archery in the past from a kinesiological and an osteological perspective. It accomplishes this by critically evaluating the basic biomechanics related to archery and outlining common bone changes potentially linked to archery in past populations, such as degenerative joint disease, entheseal changes, and cross-sectional bone geometry. These features are integrated with the types of injuries reported in modern competitive archers to provide a clearer approach to interpreting lesions in the archaeological record. This chapter provides a preliminary framework to analyse osteological collections to identify the practice of specialised archery in the past, an activity that arguably had an immeasurable impact on human evolution and countless human societies.

From the arm, we move to the hand and wrist in Chapter 6, *'Tool use and the hand'* by Christopher J. Dunmore, Timo van Leeuwen, Alexandros Karakostis, Szu-Ching Lu, and Tomos Proffitt, which provides an overview of how behaviour is evidenced in the hand. Through detailed descriptions of skeletal evidence for grip types (e.g. power and precision) and their relationship with (stone) tool-related activities, manual manipulation, and dexterity, the authors demonstrate comparative and quantifiable evidence for technological complexity as it relates to behaviour in the hand among humans, other hominins, and primates. This chapter includes a detailed discussion of methodological improvements in the study of hand bones, and specifically describes the use of varied and integrated methods to elucidate soft tissues and loading in the archaeological record (e.g. enthesophytes [VERA], geometric morphometrics, cross-sectional geometry, and trabecular bone morphology). This chapter covers a lot of ground, but their contextualised review allows the authors to suggest that through the investigation of behaviour using hands, biological anthropologists can see potential lifestyle indicators among different hominin species.

In Chapter 7, *"Behaviour and the bones of the thorax and spine"*, Kimberly Plomp explores how human behaviours have affected and can be observed in our ribs, vertebrae, and sternum. Our thorax and spine are involved in a huge variety of functions, from basic respiration to incurring stresses associated with our mode of locomotion. Plomp reviews how these topics have been investigated in clinical and evolutionary anthropology contexts, especially reviewing evidence for relationships among the spine and bipedality in humans and other hominins. Importantly demonstrating how social behaviours can become incorporated in the skeleton, Plomp also takes time to review how evidence for pathological conditions (especially trauma) can provide evidence for social circumstances. Her section on the physical and morphological changes to the human torso that are associated with the cultural practice of corseting acts especially underscores how social interpretations, in this case pertaining to fashion and perceptions of beauty, can be drawn from behaviour preserved in skeletal remains.

Next, Sarah-Louise Decrausaz and Natalie Laudicina discuss the human hip in Chapter 8, 'Human behaviour and the pelvis'. The authors present a wide reaching overview of the types of behaviour that bioarchaeologists try to infer from the pelvis when researchers use approaches from bioarchaeology, palaeoanthropology, obstetric and gynaecological medicine, comparative anatomy and evolution, and public health. They begin by deconstructing the complex evolutionary relationship between human pelvic shape, locomotion, and childbirth. They also critically assess the use of skeletal indicators, such as parturition scarring, to determine if a female has given birth, and discuss possible scenarios where bioarchaeological evidence could indicate the loss of life due to childbirth. Moving from childbirth, they also provide a detailed overview of a number of different 'everyday' behaviours that can leave interpretable evidence on the bones of the pelvis, such as corsetry and horseback riding, which together can provide unique insight into the daily lives of past peoples.

Chapter 9, 'Horse riding and the lower limbs' by William Berthon, Christèle Baillif-Ducros, Matthew Fuka, and Ksenija Djukic, takes a more focused and detailed look at how horseback riding, a critical activity in human history, can leave evidence on the skeletal lower limbs. Horseback riding had a boundless impact on ancient societies, affecting our subsistence strategies, dispersal, and warfare. This chapter weaves together biological anthropology with archaeology, human anatomy, and sports medicine to understand how horse riding affects the skeleton and how we can identify those indicators to infer the practice of horse riding in the past. Importantly, they also provide a critical assessment of some methodological approaches to this subject and suggest future research trajectories to help us more reliably identify skeletal changes related to horse riding.

And finally, we reach Chapter 10 where Kimberleigh Tommy and Meir Barak discuss 'Locomotion and the foot'. This chapter aligns well with Chapters 7 and 8 in that the human foot and ankle has many important adaptations that allow us to walk on two legs. The authors discuss how comparative studies of animal models helped us understand how the bones of the lower leg and foot respond to mechanical loading imposed by locomotor and postural changes and how this understanding can enable bioarchaeologists and palaeoanthropologists to infer locomotor behaviour from skeletal remains of the foot. Bipedalism is a defining trait of our lineage, the hominins, and so, being able to identify bipedal adaptations in the foot and ankle, as well as the spine (Chapter 7) and pelvis (Chapter 8), allows us to identify hominin taxa in the record. In addition, the authors also integrate a similar approach to investigate variation in locomotor behaviour, such as gait and loading, within and between human populations, including highly mobile foragers and more industrialised communities.

In the last full Chapter 11, *'Injury, disease, and recovery: Skeletal adaptations to immobility and impairment'*, the authors, Rebecca J. Gilmour, Liina Mansukoski, and Sarah Schrader, change focus and discuss how pathological conditions and their functional consequences, such as impairment, adaptation, and recovery, can be identified in the human skeleton. Evidence of use or disuse after disease or trauma on a skeleton can provide invaluable insight into not only the pathological conditions, but also, importantly, what life may have been like after the illness and/or injury. The authors integrate research from clinical medicine and animal studies to develop a framework to identify individual phases and features of bone change that bioarchaeologists can recognise and use to infer behaviour after illness or trauma, and provide detailed discussions of case studies where this has been done in bioarchaeology. Importantly, they end their chapter with suggestions for future research directions for bioarchaeologists.

From these chapter synopses it is clear that biological anthropological research into human behaviour is wide and varied. This has also translated into researcher perspectives in the application and interpretation of various techniques. Differences in perspectives are inherent in the field and not possible to unify in this book, especially as this book aims to introduce readers to the range of current biological anthropological perspectives on behaviour. We end the book with a brief synopsis of where we stand as biological anthropologists and archaeologists in our ability to infer behaviour in the past through evidence left on human skeletons and outline some potential trajectories and suggestions for future research to improve this field. We encourage readers to explore methodological parallels among the chapters and investigate each author group's diverse applications of and perspectives on interpretations. While not all methodological approaches for identifying behaviour are addressed in every chapter, and not every chapter takes the same approach in the application of each method, we hope that from this compilation, readers will draw inspiration for their own work and imagine creative new ways that these existing methods may be applied in other skeletal regions and to interpret other types of human behaviours as they are preserved in our bones.

References

Bornstein, M. H., Kagan, J., & Lerner, R. M. (2020, November 5). *Human behaviour. Encyclopedia Britannica*. https://www.britannica.com/topic/human-behavior.

Grauer, A. L. (Ed.). (2012). *A companion to paleopathology*. Chichester: Blackwell Publishing.

Jurmain, R. (1999). *Stories from the skeleton: Behavioral reconstruction in human osteology*. London: Taylor and Francis.

Larsen, C. S. (Ed.). (2015). *Bioarchaeology: Interpreting behavior from the human skeleton*. Cambridge: Cambridge University Press.

Ruff, C. B. (Ed.). (2018). *Skeletal variation and adaptation in Europeans: Upper Paleolithic to the twentieth century*. Hoboken, NJ: John Wiley & Sons.

Schrader, S. (2019). *Activity, diet and social practice: Addressing everyday life in human skeletal remains*. Cham, Switzerland: Springer.

2

Bone biology and microscopic changes in response to behaviour

Lily J.D. DeMars[a], Nicole Torres-Tamayo[b],
Cara Stella Hirst[c], and Justyna J. Miszkiewicz[d,e]

[a]Department of Anthropology, Pennsylvania State University, University Park, State College, PA, United States, [b]School of Life and Health Sciences, University of Roehampton, London, United Kingdom, [c]Institute of Archaeology, University College London, London, United Kingdom, [d]School of Archaeology and Anthropology, Australian National University, Canberra, ACT, Australia, [e]School of Social Science, University of Queensland, Brisbane, QLD, Australia

2.1 Introduction

Analyses of behaviour within biological anthropology have traditionally relied on the examination of skeletal anatomy through morphological or morphometric techniques (Meyer et al., 2011). This has largely been the case because noninvasive approaches are suitable for studying irreplaceable anthropological remains. Generally, such methodologies have allowed biological anthropologists to infer past behaviour by visually evaluating muscle attachment and insertion morphology (entheseal changes) (Villotte and Knüsel, 2013), taking external bone measurements to quantify shape and size variation (Stock and Shaw, 2007), and applying radiography and computed tomography to assess internal bone structure (Lieberman et al., 2004; Ruff and Hayes, 1983; Stock and Shaw, 2007). Over the last three decades, an increase in the use of microscopic methods to address questions of biological anthropological significance, such as understanding variation in primate locomotion and past human lifestyles, has allowed researchers to focus their questions on internal bone (re)modelling dynamics (Crowder and Stout, 2011; Fajardo et al., 2007; Gross et al., 2014; Miszkiewicz and Mahoney, 2017; Miszkiewicz et al., 2022; Pfeiffer et al., 2019; Pitfield et al., 2019; Ryan and Shaw, 2015; Stout and Lueck, 1995). As such, microscopic techniques increasingly form a complementary analytical approach to the study of behavioural indicators underlying the exterior bone morphology. In particular, the use of two-dimensional (2D) hard tissue histology examining cortical bone matrix composition, and three-dimensional (3D) microcomputed tomography (micro-CT) to image trabecular bone microarchitecture, provides a growing body of data about the complex relationship between bone structure and behaviour (Crowder and Stout, 2011; Kivell, 2016). In this chapter, we provide an introductory overview of the fundamental principles of bone biology and relevant biomechanical theory to first outline the theoretical and micro-anatomical basis to bone form and function. Next, we discuss published examples of cortical bone histological and trabecular bone structural analyses where inferences about behaviour are based upon these principles.

2.2 Bone anatomy and cells

The skeletal system has several primary functions: support and site of attachment for muscles, protection of vital organs, metabolism of calcium and phosphate, and storage of red and yellow marrow for cell production (Carter, 1984; Hadjidakis and Androulakis, 2006; Henriksen et al., 2009; Sherman, 2012). Bone is a composite material consisting of both organic (collagen) and inorganic (hydroxyapatite crystal) matter, which allows bone to be both stiff and somewhat flexible. Additionally, all skeletal elements are made up of a combination of trabecular and cortical bone. In long bones (e.g. the femur), cortical bone forms the outer surfaces of the element, including the diaphysis, or shaft, and a thin shell that covers the joint surfaces. Trabecular, or cancellous, bone is located internally in the proximal and distal metaphyses and epiphyses (Currey, 2002). During endochondral development, bones are formed when a cartilage scaffold, or model, of a long bone is replaced by woven bone that eventually matures into lamellar bone. Woven bone is highly disorganised with randomly oriented collagen fibres, which are rapidly turned over during early childhood (Su et al., 2003). In contrast, lamellar bone consists of tightly organised sheets of collagen (Weiner et al., 1999). As a long bone forms, it transitions from primary to secondary tissue, increasingly accruing a vascularised network, and forming into a tube-like structure that stores yellow marrow in its internal cavity (Ortega et al., 2004). This medullary cavity (part of the diaphysis) is surrounded by relatively thick cortical walls that become thinner towards the ends of the bone (distal and proximal epiphyses), where the internal structure becomes filled with trabeculae. Trabecular bone is often described as having a spongy, honeycomb-like structure (Fig. 2.1).

2.2.1 Bone cells

To maintain strength, function, and mineral homeostasis, the skeleton undergoes a continuous biological process known as remodelling (Boskey and Posner, 1984; Robling et al., 2006), also often interchangeably referred to as turnover, metabolic activity, or physiology of bone (Walsh, 2015). This process is executed by a coordinated action of different bone cells—osteoblasts (bone forming), osteocytes (bone maintaining), osteoclasts (bone resorbing), and bone lining cells (quiescent osteoblasts) (Andersen et al., 2013; Florencio-Silva et al., 2015; Sims and Gooi, 2008). We will introduce these processes in this section and return to bone remodelling in Section 2.3. Osteoblasts and osteocytes differentiate from osteoprogenitor stem cells (precursors), also known as osteogenic cells, found in bone marrow.

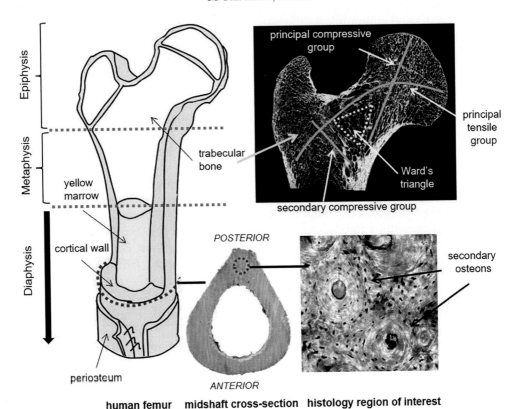

human femur   midshaft cross-section   histology region of interest

FIG. 2.1   Schematic illustration of a human femur showing different bone types and internal components of the trabecular and cortical organisation. *Images courtesy of Lily DeMars and Justyna Miszkiewicz.*

Osteoblasts are specialised bone forming cells that are derived from mesenchymal stem cells (Florencio-Silva et al., 2015; Katsimbri, 2017). Osteoblasts are located along the bone surface occupying approximately 4%–6% of its space (Florencio-Silva et al., 2015). The primary function of osteoblasts is to form new bone through the synthesis and secretion of Type I collagen, the major bone matrix protein (Katsimbri, 2017; Kenkre and Bassett, 2018). Additionally, osteoblasts mineralise newly formed initially unmineralised bone, called osteoid, through the excretion of phosphates from osteoblast-derived matrix vesicles within the osteoid (Katsimbri, 2017).

Osteocytes are the most abundant bone cell comprising 90%–95% of the total number of bone cells and have a lifespan of up to 25 years (Florencio-Silva et al., 2015; Katsimbri, 2017). They are located in lacunae within mineralised bone and are derived from osteoblasts that have undergone terminal differentiation and have been engulfed by osteoid during bone formation (Katsimbri, 2017). Osteocytes have long dendritic processes resulting in a star-shaped appearance (Katsimbri, 2017). These dendritic processes extend into canaliculi to interact with other osteocytes and osteoblasts, forming a lacunar-canalicular network (Katsimbri, 2017; Kenkre and Bassett, 2018; Prideaux et al., 2016). It is thought that osteocytes detect bone mechanical signals and communicate the need for related bone resorption or deposition to other bone cells (Katsimbri, 2017; Manolagas and Parfitt, 2010; Prideaux et al., 2016).

Osteoclasts are large multinucleated cells which are derived from mononuclear precursor cells of the macrophages/monocytes lineage, have a lifespan of approximately 12.5 days (Katsimbri, 2017), and are the only known cells capable of resorbing bone (Florencio-Silva et al., 2015; Katsimbri, 2017). They have irregular 'ruffled' borders that consist of folded layers of protein-rich cell membrane that they use to 'attach' themselves to bone surfaces within a sealing zone. A proton pump located in the membrane aids in the release of hydrogen ions to demineralise bone within a Howship's lacuna—a cavity containing osteoclasts that are actively resorbing bone.

Bone lining cells are quiescent, flat-shaped osteoblasts located on all bone surfaces (Black and Tadros, 2020). Microscopically, they appear as a thin seam of nonmineralised matrix. This thin layer of bone lining cells is thought to act as a membrane that separates bone from the interstitial fluids preventing the direct interaction between osteoclasts and bone matrix when bone resorption should not occur (Florencio-Silva et al., 2015; Wein, 2017). Bone lining cells are also involved in osteoclast differentiation (Florencio-Silva et al., 2015).

2.3 Long bone micro-anatomy, modelling, and remodelling

During the first two decades of life, bone is modelled, i.e., changes size and shape as cells add or remove bony matrix (Frost, 1994; Pearson and Lieberman, 2004). Once skeletal maturity is reached, the process of bone modelling occurs with much less frequency. However, because bone is an active, living tissue that incurs damage both during growth and throughout an individual's lifetime, remodelling occurs by removing and replacing bone in particular locations (Martin et al., 1998). In the cortex, bone vascularisation progresses from simple primary canals to secondary osteon structures that are distinctly seen in the bone matrix as having a cement line that contains concentric lamellar layers surrounding a Haversian canal (Pitfield et al., 2017) (Fig. 2.2). These primary canals, known as primary osteons, become essentially erased by secondary osteons. With age and subsequent remodelling events, secondary osteons become further erased by newly formed secondary osteons, resulting in fragmented and intact secondary osteons seen histologically (Fig. 2.2). The secondary osteons, sometimes referred to as Bone Structural Units (BSUs) are products of bone remodelling activity executed by teams of bone cells collectively known as Bone Multicellular Units (BMUs) or Bone Remodelling Compartments (BRCs) (Boivin and Meunier, 2002; Kular et al., 2012; Sherman, 2012) composed of four units: osteoclasts resorbing bone, osteoblasts building bone, which become incorporated into the secondary osteons as bone maintaining osteocytes, and bone lining cells which cover the bone surface (Hadjidakis and Androulakis, 2006; Kular et al., 2012; Martin and Sims, 2005; Ryser et al., 2009; Sims and Gooi, 2008; Wang and Seeman, 2008).

Long bones play a number of important physiological roles, including the formation of haematopoietic cells within the marrow (Gurkan and Akkus, 2008) but to a biological anthropologist, they are also useful in biomechanical studies. As bones undergo longitudinal and diametric growth, they are modelled to cope with the various normal mechanical loads applied to the skeleton every day (Robling and Stout, 2008). Modelling involves adding or resorbing bone in an imbalanced way, so that either bone resorption or deposition occurs in one

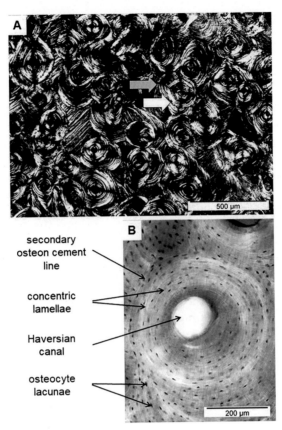

FIG. 2.2 Examples of cortical bone histology showing secondary osteons with Haversian canals under polarised (A) and transmitted (B) light in midshaft femur (A) and rib (B) cortical bone from an archaeological individual (medieval Canterbury, United Kingdom). *Grey arrow* indicates a fragmentary secondary osteon, whereas the *white arrow* indicates an intact secondary osteon. Their accumulation in a region of interest is used to estimate osteon population density in cortical bone histology samples. *Images courtesy of Justyna Miszkiewicz.*

region (Robling et al., 2006). Bone deposition predominantly occurs very close to the outer bone surface—the periosteal layer (Fig. 2.1)—simultaneously, bone resorption occurs on the endosteal surface. Ultimately, these processes lead to cortical drift, which allows the bone to change shape in response to mechanical load (Goldman et al., 2009). In comparison, remodelling (mentioned before) occurs in a balanced manner so that old bone is replaced with new bone within a relatively small area. The BMUs turn over discrete regions of bone following the sequence of activation, resorption, reversal, formation, and resting (Delaisse, 2014; Heřt et al., 1994). While cortical and trabecular bone are identical tissues in their chemical composition, they differ both macro- and microscopically (Hadjidakis and Androulakis, 2006). Approximately 80% of bone mass is comprised of cortical bone, which is dense and compact. The remaining 20% of the adult human skeleton is composed of a network of trabecular plates or spicules (Clarke, 2008). Studies have reported that cortical bone experiences a turnover rate of anywhere between 2% and 5% per year (Katsimbri, 2017; Parfitt, 2002), but trabecular bone remodels at a higher rate (e.g. 15%–25% per year) due to higher bone surface area-to-volume ratio (Choi et al., 1990; Eriksen, 2010; Huiskes et al., 2000; Martin, 2000). However, as emphasised by Parfitt (2002), the lifetime and remodelling activity of individual bones will vary across individuals and populations, so approximating remodelling rates should be done cautiously. As much as remodelling is the same process in both cortical and trabecular bone, no tunnels are formed by BMUs within the trabeculae. Instead, packets of bone are resorbed and replaced on the trabecular surfaces (Parfitt, 1984). At any one point, approximately up to 20% of bone will undergo remodelling (Parfitt, 1984). Individual BMUs are suggested to be locally controlled as they are geographically and chronologically separated from each other (Sims and Martin, 2014). Remodelling is essentially the metabolic process of living bone tissue, whereby it serves three key functions: calcium metabolism, healing pathological issues, and adaptation to mechanical strain. It is the biomechanical signature that results from adaptation to mechanical strain that is the focus of this chapter.

When the amount of resorbed bone is replaced with the same amount of new bone it is known as 'the bone balance' (Robling et al., 2006, p. 459). However, in biomechanical extremes (such as severe disuse), bone remodelling becomes out of balance (see Chapter 11). When bone experiences excessive load, its microstructure is damaged, which is known as 'microdamage' (Turner, 1998). The presence of small cracks within osteon structures activates BMU activity and begins the process of remodelling compromised tissue. Under mechanical strain in long bones, new bone tissue is typically deposited subperiosteally in a remodelling 'hot spot'— the outer third of the cortical wall (Mattheck, 1990; Robling et al., 2002). In extreme cases of overuse, this can lead to modelling where the shape of the bone changes even though skeletal maturity has been reached. In cases of severe disuse, bone removal dominates over bone deposition and is increased particularly on the endocortical and cancellous surfaces (Robling et al., 2006; Schlecht et al., 2012). Thus cases of both extreme use and disuse can change bone structure and shape. Biological anthropologists interested in behaviour usually seek to understand bone shape and structural differences that occur between individuals and populations based on variation in mechanical loading related to physical activity repertoires (many of which are discussed in chapters throughout this book). For example, intra-population studies might look for evidence of a sexual division of labour, while interpopulation studies could be interested in comparing groups of people with different subsistence strategies (e.g. hunter-gatherers vs agriculturalists). However, we also remind the reader that this approach needs to consider additional parameters that may act on bone shape and structure, including population-specific morphologies, diseases, and the effect of age and sex on bone.

Bone remodelling follows a four-stage process: resorption, reversal, formation, and termination/resting phases (Katsimbri, 2017; Kenkre and Bassett, 2018) (Fig. 2.3). The bone resorption phase involves the removal of the mineral and organic constituents of bone matrix by osteoclasts which are aided by osteoblasts (Katsimbri, 2017; Kenkre and Bassett, 2018). This phase begins with the dissemination of osteoclast progenitors from hemopoietic tissue in the bone. The osteoclast progenitors then differentiate into osteoclasts through interaction with osteoblast stromal cells (Katsimbri, 2017; Prideaux et al., 2016). Bone lining cells that prevent osteoclast activity are removed from the mineralised osteoid layer through the production of proteolytic enzymes, including matrix metalloproteinases, collagenase, and gelatinase (Katsimbri, 2017; Kenkre and Bassett, 2018; Prideaux et al., 2016). The removal of these bone lining cells allows the osteoclasts access to the underlying mineralised bone. Osteoclasts are

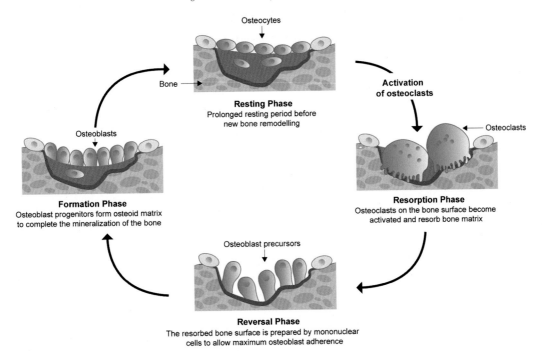

FIG. 2.3 Schematic illustration of the four phases of bone remodelling sequence here shown using trabecular bone bone surface as an example. *Reproduced from Beck-Nielsen, S. S., Greggio, N. A., & Hagenäs, L. (2021). Defining a growing and maturing skeleton and its relevance in diseases that affect skeletal growth, such as X-Linked Hypophosphataemia (XLH). International Journal of Rare Diseases and Disorders, 4:029(1), 1–13. https://doi.org/10.23937/2643-4571/1710029, under open-access Creative Commons Attribution Licence.*

then activated by osteoblasts, and the activated osteoclasts resorb the bone through the production of hydrogen ions and proteolytic ions (Kenkre and Bassett, 2018). The resorption phase is terminated when osteoclasts undergo cell death known as apoptosis (Katsimbri, 2017).

The reversal phase lasts approximately nine days and follows once the maximum depth of bone resorption has been achieved (Katsimbri, 2017; Raggatt and Partridge, 2010). It is thought that osteoclasts essentially 'leave' the bone surface and release osteogenic signals so that bone formation can begin (Abdelgawad et al., 2016; Delaisse, 2014; Katsimbri, 2017). The mechanism that arrests osteoclast activity and stimulates osteoblast activity is poorly understood, but it has been suggested that these macrophage-like cells may release factors that inhibit osteoclasts and stimulate osteoblasts (Katsimbri, 2017; Kenkre and Bassett, 2018; Raggatt and Partridge, 2010).

The bone formation phase results from the proliferation of mesenchymal cells (Kini and Nandeesh, 2012). Although the process is not yet fully understood, it is suggested that the process that attracts the osteoblasts precursor cells to the site of the defect results from a chemostatic reaction mediated by local factors produced during the resorption phase (Dempster and Raisz, 2015; Katsimbri, 2017; Raggatt and Partridge, 2010; Siddiqui and Partridge, 2016). Osteoblast precursor cells then differentiate into mature osteoblasts. The differentiation of osteoblast precursors is likely controlled by a number of bone-derived growth factors although the precise growth factors involved have not been determined (Katsimbri, 2017; Siddiqui and Partridge, 2016). The osteoblasts then secrete osteoid, an unmineralised matrix, which is 94% composed of Type 1 Collagen (Boskey and Posner, 1984; Kular et al., 2012).

The termination phase begins about 30 days after the formation of new osteoid and lasts for approximately 90 days for trabecular bone and 130 days for cortical bone (Katsimbri, 2017; Raggatt and Partridge, 2010). During the termination phase, newly formed osteoid is mineralised, which requires calcium and phosphorus in the extracellular fluid and occurs within 10–15 days (Boskey and Posner, 1984; Katsimbri, 2017). Following mineralisation, bone returns to a dormant phase when osteoblasts undergo apoptosis and are either embedded in the newly mineralised matrix or remain on the surface as bone lining cells (Boskey and Posner, 1984; Katsimbri, 2017; Raggatt and Partridge, 2010).

Bone remodelling is a complex process influenced by many extraneous factors beyond mechanical stimuli, such as systemic hormones, locally produced growth factors, and inorganic agents (ions, oxygen, free radicals) (Wozney et al., 1990). Some of these factors might be interrelated, for example, sex hormones play an important role in both stochastic bone remodelling and healing arising from pathologies. Hormones may influence bone remodelling through their action on osteoblasts, osteoclasts, or cell proliferation and differentiation; these include polypeptide, steroid, and thyroid hormones (Matsuo and Irie, 2008). Oestrogen regulates osteoclastogenesis and oestrogen deficiency increases bone remodelling (Deckers et al., 2000; Seeman, 2002). Growth factors, such as polypeptides that regulate the replication and the differentiated function of cells, are present in the bone matrix in significant quantities and can influence the growth properties and function of bone cells in vitro (Crane and Cao, 2014; Wozney et al., 1990). Additionally, several vitamins are critical for skeletal homeostasis (Kenkre and Bassett, 2018; Williamson et al., 2017). Calcium deficiency can lead to lower bone strength (Hernández-Becerra et al., 2020). Vitamin D3 deficiency can also have a detrimental effect on bone remodelling (McKenna et al., 2018). Furthermore, several metabolic and other bone conditions may also influence bone remodelling, including vitamin D deficiency, hypophosphataemic syndromes, hyperparathyroidism, renal osteodystrophy, osteoporosis, Paget's disease, osteopetrosis, diffuse idiopathic skeletal hyperostosis, and trauma (Feng and McDonald, 2011; Katsimbri, 2017; Sherman, 2012).

Age and sex are key variables impacting bone remodelling effectiveness with older individuals and menopausal females experiencing increased bone loss (Forwood and Burr, 1993; Khosla and Monroe, 2018; Lanyon, 1996). In old age, bone responsiveness to strain declines (Forwood and Burr, 1993; Lanyon, 1996). Muscle mass also decreases with age, which accounts for some of the reduction in bone mass and mineral density (Burr, 1997). Bone remodelling in response to activity is suggested to be more sensitive during childhood and adolescence providing a protective function against bone loss later in life (Pearson and Lieberman, 2004) through the construction of a 'bone bank' (accumulation of peak bone mass one reaches at the end of skeletal maturation) in earlier life that we 'withdraw' from as we age (von Scheven, 2007). Skeletal sexual dimorphism and fluctuating levels of sex steroids (oestrogens and androgens) impact bone remodelling (Khosla and Monroe, 2018). Changes can also be temporary, including reversible calcium level depletion during pregnancy and lactation in women (Kovacs and Fuleihan, 2006). The main action of oestrogens is inhibiting osteoclast activity so that prolonged bone loss does not occur (Frost, 1999a). In women experiencing menopause, the drop in oestrogen often leads to osteoporosis if not treated (Tella and Gallagher,

2014). Osteoporosis has been studied in archaeological human remains with varying levels of success (due to bone preservation and taphonomic issues), but its occurrence can be linked to behavioural mitigation in different ages and sexes (see Agarwal and Stout, 2003; Brickley et al., 2020; Miszkiewicz and Cooke, 2019; Miszkiewicz et al., 2021). While in experimental animal studies some of these factors may be controlled for, this is inherently not possible in archaeological research. Instead, an awareness of these factors and their potential influence on bone remodelling is necessary in order to properly interpret remodelling responses in archaeological remains.

2.4 Bone functional adaptation

Approximately 20%–40% of adult peak bone mass is the result of extrinsic (nongenetic) behavioural and environmental factors, including diet, climate, pathogen load, and activity level (Bridges, 1989; Larsen, 1981; Nelson et al., 2015; Robbins et al., 2018; Ruff et al., 1984; Stieglitz et al., 2016). While all of these factors are important in determining adult bone structure, the relationship between bone structure and mechanical load (often measured qualitatively where activity level is a proxy) has received significant attention within biological anthropology. As discussed in the previous section, bone structure is determined, in part, by a response to mechanical strain experienced during growth and to a lesser extent throughout an individual's lifetime. This relationship between bone structure and function is well established and commonly referred to as 'bone functional adaptation' (Frost, 1979, 1994, 1999a,b; Pearson and Lieberman, 2004; Ruff et al., 2006). Several of the theoretical principles that underlie the plasticity of bone, and have contributed to the formation of the more general principle of bone functional adaptation, include Wolff's Law (Wolff, 1892, 1986) and his trajectorial theory, the Mechanostat Theory (Frost, 1987), and the Utah Paradigm of Skeletal Physiology (Frost, 1987, 2000, 2001; Jee, 2000; Turner, 1992; Wolff, 1892, 1986).

In biological anthropology, Wolff's Law (Wolff, 1892) is often cited in behavioural research because it explains why a load imposed on a bone will induce structural adaptation. A more specific aspect of Wolff's Law about the relationship between bone structure and mechanical loading is the Trajectorial Theory which refers specifically to the orientation of trabeculae within the proximal and distal epiphyses. The architectural arrangement of bone is in line with the direction of loading so that it can withstand strain but remain light and rigid in a way that trabeculae are 'mutually perpendicular' (Doube et al., 2011; Kuo and Carter, 1991, p. 918). Although not the first to make this observation (e.g. von Meyer, 1867; Ward, 1838), Wolff popularised the concept. However, Wolff's views were developed using limited experimental settings and based on a mathematical model where bones are assumed to be solid and static (Cowin, 2011; Ruff et al., 2006). He did not incorporate the many different contextual aspects to mechanical plasticity, including genetics and environmental factors (Forwood and Burr, 1993; Pearson and Lieberman, 2004). We now understand that bones loaded in vivo respond to intermittent and dynamic, not static stress (Lanyon and Rubin, 1984).

Nevertheless, Wolff's Law is still sometimes used to describe the general ability of bones to adapt to load, though it is acknowledged that this is a complex relationship (Cowin, 2011; Martin et al., 1998; Pearson and Lieberman, 2004). Importantly, degree of anisotropy (a measurement of how oriented trabeculae are within a given volume) remains an important variable in studies of trabecular bone. And trabecular orientation is known to have an important impact on overall trabecular bone stiffness (Maquer et al., 2015; Stauber et al., 2006). The proximal end of the human femur is an excellent example of this, whereby trabeculae are visibly arranged according to stress distribution. A region of highly oriented trabeculae is visible in a coronal cross-section of the femoral head, where trabeculae are aligned supero-inferiorly from the superomedial aspect of the femoral head (where it articulates with the acetabulum of the hip) to the inferior portion of the femoral neck, where the trabecular bone meets the endocortical surface (Fig. 2.1). This organisation is thought to allow (or be the result of) the mechanical strain at the hip from walking (or running) to be transferred through the trabecular bone of the proximal femur and into the cortical bone of the diaphysis (Rudman et al., 2006).

The Mechanostat Theory was put forward by Frost (Frost, 1987; Jee, 2000) to stipulate that there is a threshold in bone physiology that determines when and if an adaptive response to load will be evoked within a bone (Frost, 1987; Jee, 2000). This idea is governed through the concept of a minimum effective strain (MES) that must be exceeded to induce a bone's transfer and interpretation of mechanical signals. Frost's (1987) strain–bone response model stipulates that high strains (suprathreshold events) lead to bone formation, and low strain levels accelerate bone removal on the endocortical and cancellous surfaces (Frost, 1987). The Mechanostat has been demonstrated experimentally using a series of in vivo models, including rabbit tibiae (Lim et al., 2007), human limbs and cross-sectional area and muscle activity (Schiessl et al., 1999), rat femora and tibiae (Umemura et al., 1997), and hip and lumbar spine strain in children (Fuchs et al., 2001; Lim et al., 2007; Schiessl et al., 1999; Umemura et al., 1997). Frost's Mechanostat (Frost, 1987) is now considered a principal component of the Utah Paradigm of Skeletal Physiology (Frost, 1987, 2000). The Utah Paradigm was finalised in the late 1990s (Frost, 2000) and clarified bone adaptation to overload or underload events (Frost, 2000). In both disuse and overuse events, bone remodelling is preceded by osteocyte apoptosis (cell death) where localised tissue can no longer be maintained (Noble et al., 2003). Robling et al. (2006) suggest cell failure is the exchange of biological nutrients supporting the canalicular network. But, bone can also enter the phase of modelling when overloaded (Turner, 1998). In summary, bones are anisotropic as they modify their structure under mechanical stimuli (Doblaré and García, 2001; Sevostianov and Kachanov, 2000). This load can be applied in various ways, including compression, tension, bending, and/or torsion (Ruff, 2008). Bones experience strain which is a longitudinal alteration per unit length (Pearson and Lieberman, 2004), and it is strain that signals bone maintenance and repair. Strain is measured by magnitude and can be of a different mode, direction, rate, frequency, distribution, and energy (Nigg and Gimston, 1999; Pearson and Lieberman, 2004).

The previously mentioned theoretical principles and the bone physiological and biological responses to mechanical stimuli are the basis from which biological anthropologists can infer mechanical loading histories when examining bone tissue in skeletal remains. Bone functional adaptation is often used by researchers as a theoretical framework that allows them to interpret variation in skeletal morphology within and between species, and to reconstruct behaviour in the archaeological and palaeontological records. High-resolution micro-CT images can be used to quantify trabecular bone structural organisation, and remodelling responses can be reconstructed and calculated from histological thin sections, both of which will be discussed in the subsequent sections of this chapter.

2.5 Bone histology and behaviour in archaeological humans

Inferring behaviour using bone histology in human remains from archaeological contexts has, to date, not been routinely undertaken, though an increasing number of studies have recently recognised the need for microscopy in their research design (Miszkiewicz and Mahoney, 2016; Pitfield et al., 2019). Histological methods rely on extracting pieces of bone from often irreplaceable specimens, processing the samples into thin sections, and then subsequent analysis under the microscope (Meyer et al., 2011; Miszkiewicz and Mahoney, 2017; Rühli et al., 2007). However, in comparison to some ancient DNA or stable isotope analysis techniques where samples are destroyed, histological examination is, arguably, less invasive whereby samples of 1 cm height will suffice as long as histological surfaces are suitably preserved (Miszkiewicz and Mahoney, 2017). Moreover, as the processed 'slice' of bone will be preserved on a glass slide and protected with a cover slip, it can be repatriated with the human remains or kept as part of osteological collections with the associated set of human remains for many decades to come. Nevertheless, it needs to be highlighted that any research design involving removal of bone for histology should be thoroughly planned and organised, following established codes of conduct, ethics, and practice to avoid unnecessary destruction of bone (Mays et al., 2013). Where this is undertaken successfully, a reconstruction of mechanical loading history can be achieved in a variety of archaeological contexts and time periods, which is the focus of this chapter section.

The usefulness of histology in reconstructing bone remodelling from archaeological bone can be attributed to a statement by Stout (1978, p. 603): 'It is time to go beyond merely reaffirming the preservation of histological structures or estimating age at death and begin the study of paleophysiology'. While bone physiology can be influenced by a series of other factors, including diet, hormones, and disease (Crowder and Stout, 2011), if the environmental context, lifestyle, and physical activity information are known for a given archaeological site and skeletal assemblage, behavioural inferences made from bone remodelling can be proposed (Crowder and Stout, 2011). This is typically done by analysing histology images in a dedicated software (e.g. ImageJ/FIJI) to extract the size, shape, and other geometric properties of Haversian canals and secondary osteons, as well as their osteocyte lacunae densities (Doube et al., 2010; Miszkiewicz, 2020; Miszkiewicz and Mahoney, 2017; Stout and Crowder, 2011; van Oers et al., 2008). These data can then be compared intra-skeletally between bones that experience different loading regimes (e.g. the femur and the rib, Skedros et al., 2013), within samples that are of known behavioural differences (e.g. socioeconomic disparities, Miszkiewicz and Mahoney, 2016), or spatially and temporally (e.g. comparing humans from different time periods and geographical locations, e.g. Pfeiffer et al., 2006; Robling and Stout, 2003). The bone histology variables typically examined in such studies are listed in Table 2.1.

Because of the micro-anatomical differences between cortical and trabecular bone described earlier in this chapter, bone histologists working within biological anthropology typically produce thin sections using cortical rather than trabecular samples. Cortical bone, particularly when sampled transversely at the midshaft of a long bone, offers a relatively wide transverse surface area for examination under the microscope (Stout and Crowder, 2011). Further, it shows different layers of bone types and widespread osteon and Haversian canal structures produced during the BMU activity of remodelling which can be neatly linked with biomechanical properties of long bones obtained through simple measurements of cortical

TABLE 2.1 Definitions of commonly quantified 2D cortical bone histology variables.

Variable	Abbreviation (Unit)	Definition
Intact osteon density	N.On	Total number of secondary osteons that show intact cement lines within a region of interest (ROI) divided by the examined area typically in mm^2 (e.g. Iwaniec et al., 1998; Miszkiewicz, 2016)
Fragmentary osteon density	N.On.Fg	Total number of fragmentary secondary osteons that show indicators or partial remodelling within a region of interest (ROI) divided by the examined area typically in mm^2 (e.g. Miszkiewicz, 2016; Schlecht et al., 2012)
Osteon population density	OPD	Sum of N.On and N.On.Fg (e.g. Miszkiewicz, 2016; Robling and Stout, 2003; Stout and Paine, 1994)
Haversian canal area	H.Ar (μm^2)	The area of a Haversian canal measured by tracing its intact borders. Typically, an average H.Ar is reported by measuring all canals visible with a ROI and dividing the sum by the total number of canals examined (e.g. Miszkiewicz and Mahoney, 2019; Pfeiffer et al., 2006)
Haversian canal diameter	H.Dm	The maximum and/or minimum (in μm) diameter of a Haversian canal. Typically, an average H.Dm is reported by measuring all canals visible with a ROI and dividing the sum by the total number of canals examined (e.g. Miszkiewicz and Mahoney, 2016; Pitfield et al., 2019)
Haversian canal circularity	H.Cr	Circularity value of a Haversian canal whereby 1 indicates a perfect circle, but < 1 indicates an irregular shape of the canal. $H.Cr = 4\pi(area/perimeter^2)$. The same principle can be applied to the circularity of a secondary osteon (On.Cr) (e.g. Keenan et al., 2017; Maggio and Franklin, 2019)
Secondary osteon area	On.Ar (μm^2)	The area of a secondary osteon measured by tracing its intact cement line border. Typically, an average On.Ar is reported by measuring all osteons visible with a ROI and dividing the sum by the total number of osteons examined (e.g. Dominguez and Agnew, 2016; Miszkiewicz, 2016; Pfeiffer et al., 2006)
Secondary osteon diameter	On.Dm (μm)	The maximum and/or minimum diameter of a secondary osteon canal. Typically, an average On.Dm is reported by measuring all osteons visible with an ROI and dividing the sum by the total number of osteons examined (e.g. Britz et al., 2009; Skedros et al., 2013)
Secondary osteon inverse of aspect ratio	On.AspR^{-1}	An elongation measure of a secondary osteon transverse shape which takes into consideration the raw Aspect Ratio (AR = major axis/minor axis) which is inverted to obtain a value from 0 (infinitely elongated) to 1 (perfectly circular). The same principle can be applied to the circularity of a Haversian canal (H.AspR^{-1}) (e.g. Hennig et al., 2015)
Osteocyte lacunae density	Ot.Dn	Total number of osteocyte lacunae counted within a ROI and divided by the area examined (typically in mm^2). The variable is a proxy for osteoblast proliferation (e.g. Bromage et al., 2009; Miszkiewicz, 2016)

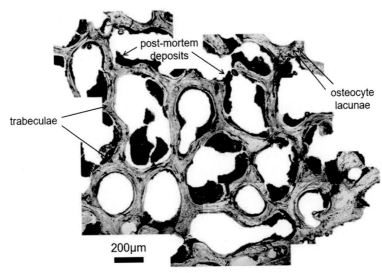

FIG. 2.4 Montage of histology images showing a transverse view of trabecular spicules in a distal tibia from an archaeological individual (medieval Canterbury, United Kingdom). *Image courtesy of Justyna Miszkiewicz.*

thickness and cross-sectional geometry (Stout et al., 2019; Stout and Crowder, 2011). On the contrary, thin sections made from trabecular bone, regardless of whether taken in a transverse or longitudinal plane, show relatively large gaps in the bone sample which simply reflect marrow spaces in between individual trabeculae (Müller et al., 1998) (Fig. 2.4). The trabeculae themselves will show osteocyte lacunae with some vascular canals, but only in well preserved or fresh samples (Müller et al., 1998). Trabecular bone histology in archaeological remains is often diagenetically compromised because of trabecular spicule thinness which makes them susceptible to postmortem agents (Miszkiewicz, 2015). Albeit, clinical and experimental research in biomechanics considers both trabecular and cortical bone in longitudinal sections from animals (e.g. Fox et al., 2007; Nordstrom et al., 2007; Yang et al., 2021). Some prior biological anthropology studies examining trabecular bone histology in ground sections have predominantly focused on disease (Kučera et al., 2017; Miszkiewicz, 2015), and age-at-death estimation (Kotting, 1977; Schranz, 1959), rather than behaviour. One notable case study of trabecular bone histology used to examine microstructural changes in mechanically atrophied bones is of a medieval male from Gruczno, Poland (Kozłowski and Piontek, 2000). Trabeculae and osteocyte lacunae densities examined histologically in this individual's metatarsals from one atrophied and one healthy lower limb were less dense in the former, demonstrating the detrimental bone effects of loss of function on limb use (Kozłowski and Piontek, 2000). Generally, 3D volumetric analyses of trabecular bone in relation to behaviour are arguably more efficient than 2D histology. Experimental research confirmed a significant agreement between serial 2D and volumetric 3D microscopic trabecular data (see Müller et al., 1998), so it makes more sense to gather 3D data nondestructively. The next Section 2.6 will discuss this further.

One of the first larger histology analyses applied to ancient human cortical bone was performed by Thompson and Gunness-Hey (1981) where adult femoral samples from the Yupik-Inupiaq Indigenous skeletal remains were examined for microstructural variation with age, cortical thickness, and bone mineral content. The samples represented St. Lawrence Island (1st century) ($n = 53$), Kodiak Island (700 BC–AD 1700) ($n = 92$), Baffin Island (19th century) ($n = 44$), and Southampton Island (19th century) ($n = 69$) and were compared to contemporary human cadavers. Based on Haversian canal and osteon area indicative of bone remodelling (among other variables), Thompson and Gunness-Hey (1981) suggested that the archaeological Yupik-Inupiaq skeletons grew more bone when compared to contemporary populations. However, the archaeological Yupik-Inupiaq had thinner cortices with more osteonal bone than the modern sample. Thompson and Gunness-Hey (1981) proposed that bone mineralisation must have played a role in these differences and linked their results to the Yupik-Inupiaq individuals leading a physically active lifestyle. However, the authors also emphasised the confounding effects of genetics and dietary factors.

Another earlier histology study where modern humans were compared to an archaeological sample was by Burr et al. (1990) who examined femur histology in 55 Pecos individuals from New Mexico (AD 14th–19th centuries). As per historical evidence, the Pecos practised a physically active lifestyle with no known evidence of dietary malnutrition. Burr et al. (1990) included sex comparisons of their femur histology data and found that the Pecos females showed smaller Haversian canals and larger secondary osteon walls when compared to contemporary females. However, Pecos males had greater densities of secondary osteons. These data for the ancient Pecos were linked with active lifestyle, but it was noted that high levels of physical activity alone did not shield both the Pecos males and females from losing bone on the endocortical surfaces.

Histological sampling has also been successfully employed in case studies that investigated behavioural signals in the limb bones in single individuals (Lazenby and Pfeiffer, 1993; Miszkiewicz et al., 2020; Snoddy et al., 2022; Walker et al., 2022). For example, Lazenby and Pfeiffer (1993) applied histology to examine cortical bone remodelling differences between the left and right femur in a AD 19th century amputee. A young male fugitive from Texas who spent his last years of life in the Middlesex County prison, Ontario, lived with a wooden prosthesis attached to his amputated leg served as a 'natural experiment' for examining bone adaptation to a modified mechanical environment (Lazenby and Pfeiffer, 1993, p. 19). Histological and gross morphological samples from this individual revealed enhanced bone remodelling of the cortical bone and expanded resorption of the endocortical space. This case study is an excellent example of the Mechanostat Theory demonstrating how a change in mechanical strain impacts bone redistribution within adult bone tissue (Frost, 1987). More recent case studies have used histology to show short-term remodelling changes at the midshaft femur in individuals with asymmetric femora due to one-sided abnormalities, such as hip joint ankylosis following injury in a male from the Metal Period, Philippines (Miszkiewicz et al., 2020) and a 19th century male from Aotearoa New Zealand who had suffered from tuberculosis-induced femoro-acetabular joint destruction (Snoddy et al., 2022). In both cases, the left femur from the affected hip joint received abnormally low mechanical load compared to the right femur, resulting in prolonged cortical bone resorption at the midshaft. The effect of long-term immobilisation of limbs on cortical bone histology has also been investigated in one archaeological case study. Walker et al. (2022) reported remarkable retention of primary bone and stunted bone remodelling in significantly atrophied humeri and femora of a young adult male from 1906 to 1523 cal BC Vietnam (also see Oxenham et al., 2009). This individual had likely lived with Klippel-Feil Syndrome Type III related paraplegia, and possibly quadriplegia, which created a minimal/absent mechanical loading environment for their postcranium. The key limitation in case studies is, of course, the small sample size, meaning the histology findings cannot be translated to population-wide implications. Some of these case studies are further discussed in Chapter 11.

Intra-skeletal comparisons of cortical bone histology within archaeological humans typically use the rib and the femur as these two bones receive different levels and types of mechanical load (Miszkiewicz et al., 2022; Robling and Stout, 2003; also see experimental evaluation in Stewart et al., 2021). For example, Robling and Stout (2003) applied this approach to examine temporal change in subsistence strategies in a Peruvian Paloma sample ($n = 48$) from the Chilca Valley (6500 BP–4700 BP). The sample represented a population characterised by sex-specific activities that decreased in intensity of labour (such as intense walking and carrying of items such as different foods) and became more sedentary through time. Robling and Stout (2003) incorporated rib and femur histology into a broader analysis that also included femoral length and body mass. Using the rib to adjust femoral data, a decrease in the densities of secondary osteons in the femur along with femur size indicated a reduction in bone remodelling through time which matched the archaeological evidence for temporal changes in activity. More recently, Miszkiewicz et al. (2022) examined bone histology in a femur, humerus, and rib of a 1720 BP male individual from the Ebon Atoll, Marshall Islands, in eastern Micronesia. Also using the rib as a control sample, the authors were able to link higher densities of relatively small vascular canals in the humerus to strenuous arm use resulting from behaviours, including gardening, food production, and fishing in the ancient Marshall Islands.

Another example of histological research and behaviour is when socioeconomic status of past societies can be inferred. A recent study by Pitfield et al. (2019) examined cortical bone from the humeri and ribs in 175 juvenile remains from across different medieval sites in the United Kingdom (Canterbury, York, Newcastle). Within the Canterbury sample, low- and high-status groups of children were compared for differences in histological indicators of adaptation to varying levels of mechanical load. Pitfield et al. (2019) reported the Canterbury children to show lower densities of secondary osteons when compared to the other sites, which was interpreted as reflecting less strenuous physical activity. Status comparisons with the Canterbury sample also showed that the bone histology of low-status children was characterised with smaller Haversian canals suggesting engagement in more physically demanding labour. This matched evidence for medieval children entering the workforce around 7 years old. An earlier study accessing bone histology from the same medieval site in Canterbury, but comparing adults of different socioeconomic status, also reported the lower status individuals to show histological characteristics indicative of hard manual labour (Miszkiewicz and Mahoney, 2016). However, the authors also noted ill health and poor nutritional status of the lower class leading to poor bone health in this population (Miszkiewicz and Mahoney, 2016).

In contrast to the above, other studies have found inconsistent links between behaviour and histological indicators of mechanically induced bone remodelling (Ericksen, 1980;

Pfeiffer et al., 2006; Richman et al., 1979). For example, Pfeiffer et al.'s (2006) examination of ribs and femora from three temporally and spatially distributed populations did not support the use of secondary osteon and Haversian canal geometric properties as behavioural markers. The authors measured these histological structures in samples dated to 18th, 19th, and 20th centuries, representing Holocene South African foragers ($n = 44$), Spitalfields (London), and St. Thomas (Canada) ($n = 40$) (Pfeiffer et al., 2006). In this study, individuals who would have likely engaged in strenuous physical activity did not show smaller osteons as hypothesised. Further, females showed opposite patterns in the size of histological structures compared to males, which was not the expectation either. Based on this, Pfeiffer et al. (2006, p. 466) cautioned that reconstructing behaviour from secondary osteon and Haversian canal size is not as straightforward as was assumed. Indeed, it is vital to emphasise that the expression of bone histology as seen under the microscope can be influenced by multiple other factors than just mechanical load. These are impossible to identify exactly in archaeological contexts, but include a combination of diet, function of hormones, disease, age, and genetic underpinning (Heaney et al., 2000). Therefore, ideally, cortical bone histology from archaeological populations should be carefully interpreted within the archaeological context of a given sample, and several alternative interpretations of the data should be proposed.

2.6 Trabecular bone structure and behaviour

Trabecular, or cancellous, bone is a highly porous network of bony plates and rods found in the epiphyses and metaphyses of long bones (e.g. the femur and humerus), short bones such as carpals and tarsals, irregular bones such as vertebral bodies, and in the internal portions of flat bones like the innominates and sternum. Studies of cortical bone cross-sectional geometry tend to dominate the bone functional adaptation literature and have provided important information about the relationship between diaphyseal shape, bending properties, and behaviour (Bridges, 1989; Macintosh et al., 2017; Ruff et al., 2015; Shaw and Stock, 2013). However, the complex structure, large surface area, and high rate of annual remodelling (15%–25%) characteristic of trabecular bone make it both interesting and potentially useful for investigating and understanding the relationship between bone structure, biomechanical loading, and behaviour (Eriksen, 2010; Kivell, 2016; Saers et al., 2016). This section discusses how trabecular bone is measured and analysed, reviews several studies that have compared trabecular bone structure between archaeological human groups with distinct subsistence strategies and inferred activity levels, and examines several limitations as well as future directions for the use of trabecular bone analysis within biological anthropology.

Trabecular bone is often measured and analysed using 3D, volumetric data obtained from micro-CT images. It is also possible to use high-resolution peripheral quantitative computed tomography (HRPQCT) to image in vivo samples, though this is not very common within anthropological literature (Best et al., 2017). Two-dimensional methods such as peripheral quantitative computed tomography, pQCT (Chirchir et al., 2015, 2017) or a combination of thick sectioning and X-ray (e.g. Agarwal et al., 2004) have also been used to assess trabecular bone structure. However, 3D methods offer various benefits over 2D approaches, one of which is their nondestructive nature; as such, this section will focus on the application of 3D methodologies to trabecular bone microstructural analyses.

Each trabecula is approximately 200 micrometres (0.2 mm) in size, and consequently most studies of human trabecular bone structure within anthropology aim to use CT scans with resolutions less than or equal to 0.05 mm in order to allow for adequate image resolution and ensure accurate measurement of structural variables (Kivell et al., 2011b; Kothari et al., 1998). After a skeletal element of interest has been scanned, all subsequent analyses are undertaken digitally. Variables that are often quantified include bone volume fraction (BV/TV), degree of anisotropy (DA), trabecular thickness (Tb.Th), trabecular separation (Tb.Sp), connectivity density (Conn.D), trabecular number (Tb.N), and bone surface density (BS/BV). Table 2.2 includes a description of each variable. Based on finite element (FE) modelling, bone volume fraction and degree of anisotropy are considered to be the most biomechanically relevant variables, accounting for approximately 87% and 10% of trabecular stiffness, respectively (Currey, 2002; Kivell, 2016; Maquer et al., 2015; Stauber et al., 2006). Variables such as trabecular thickness, separation, and number tend to be highly correlated with bone volume fraction and do not explain additional variance in trabecular stiffness within FE models, but are often included in analyses because they are useful for describing the size and shape of the trabeculae (Kivell, 2016).

In terms of quantifying the previously mentioned variables, there are two common 3D approaches: the Volume of Interest (VOI) approach and the whole bone analysis method. The VOI method is used to extract a subvolume (usually a cube or sphere) of trabecular bone from an anatomical region of interest (e.g. the femoral head) within a micro-CT dataset (DeSilva and Devlin, 2012; Doershuk et al., 2019; Fajardo and Müller, 2001; Fajardo et al., 2007; Ketcham and Ryan, 2004; Kuo et al., 2013; Ryan and Shaw, 2012, 2015; Ryan and Walker, 2010; Saers et al., 2016; Schuif et al., 2016). Multiple VOIs can also be placed within a single skeletal element in order to target multiple regions of bone hypothesised to be biomechanically and/or biologically important (Saers et al., 2019). After VOIs are extracted, bone analysis programs such as BoneJ can be used to quantify the trabecular variables of interest (Doube et al., 2010). Average values for each trabecular variable are then compared

TABLE 2.2   Definitions of commonly quantified 3D trabecular bone variables.

Variable	Abbreviation (unit)	Definition
Bone volume fraction	BV/TV	Relative trabecular bone volume to total volume within the volume of interest
Trabecular thickness	Tb.Th (mm)	The mean thickness of trabecular struts
Trabecular separation	Tb.Sp (mm)	The mean distance between adjacent trabeculae
Connectivity density	Conn. D (mm^3)	The number of trabeculae per unit volume (Odgaard and Gundersen, 1993)
Trabecular number	Tb.N (mm^{-1})	The number of trabecula per mm
Bone surface density	BS/BV (mm^{-1})	The ratio of trabecular bone surface area to total trabecular bone volume in the region of interest
Degree of anisotropy	DA	The distribution of bone in 3D space. 1-[smallest eigenvalue/largest eigenvalue] DA=1; total anisotropy, DA=0; total isotropy (Doube et al., 2010)

between groups. By using a standardised protocol, researchers intend to ensure homology between individuals within the study group because each VOI is considered to be relatively the same size (or scaled based on estimated body mass) and in the same position across individuals. However, the VOI method has several limitations. Although standardised VOI positioning protocols do help to ensure homology between individuals within studies, the location and size of VOIs can significantly alter results, making it difficult to compare research projects that use different VOI placement protocols (Kivell et al., 2011b; Lazenby et al., 2011). Additionally, large single VOIs likely do not capture localised variation within a joint structure because the analysis of each VOI results in a single average value for each variable. If a VOI encompasses a large volume of bone within a joint, variation in bone structure within that volume is averaged out.

An alternative method used to quantify and visualise trabecular bone structure utilises the entire trabecular volume of the skeletal element of interest. The whole bone analysis method is described in detail elsewhere (Gross et al., 2014), but briefly, using the software program Medtool 4.3 (Dr. Pahr Ingenieurs e.U.), the entire volume of trabecular bone is digitally isolated from the cortical bone within the dataset (Dr. Pahr Ingenieurs e.U, 2019; Gross et al., 2014). Subsequently, a series of sampling spheres (essentially small VOIs) are positioned sequentially on a grid encompassing the trabecular volume. From these, trabecular variables are measured in 3D at each node of the grid, across the entire volume of trabecular bone. A 3D tetrahedral mesh, or model, of the skeletal element is then generated and the trabecular variable results are mapped as scalar values to create colour maps of the distribution of each quantified variable. Although the whole bone analysis method allows for sophisticated quantification and visualisation of trabecular variables, it can be difficult to compare results between individuals and groups because the sampling spheres are not homologous between individuals due to variation in the size and shape of each skeletal element. Additionally, this method is considerably more computationally and time intensive. Figs. 2.5 and 2.6 show a visual representation of each trabecular bone analysis method.

The final methodological process that needs to be mentioned in relation to trabecular bone structural analysis is segmentation or the removal of any non-bone inclusions from the CT image datasets. It is not uncommon for archaeological material to have significant soil infill, or for fossil specimens to be filled with sediment of different densities, all of which must be removed prior to analysis. A detailed discussion of the various segmentation methods is beyond the scope of this chapter (Dunmore et al., 2018; Yazdani et al., 2019). It is worth noting that the quality of the image segmentations can have a major impact on the results of trabecular bone analyses regardless of which quantification method is used. If non-bone inclusions are not removed, any subsequent analysis would include both bone and non-bone materials. This would cause individuals or populations with high levels of non-bone inclusions to appear more robust than they actually are, which could result in inaccurate interpretations of the relationship between bone structure and behaviour.

Trabecular bone structure has been used in a variety of contexts to address questions within biological anthropology. Utilising a bone functional adaptation framework, researchers have undertaken experimental studies using animal models to investigate how trabecular bone models and remodels under controlled conditions. The subjects of these studies are generally split into control and experimental groups, where the behaviour of the experimental group is altered in some way that is expected to elicit a bony response (Barak et al., 2011; Carlson et al., 2008;

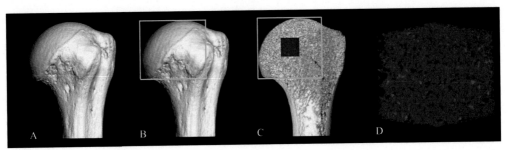

FIG. 2.5 (A) Three-dimensional volume rendering of a complete micro-CT dataset of the proximal humerus. (B) Region of interest bounding box placed around the humeral head. (C) The cubic VOI is represented in *blue*, in this example the VOI size is calculated as having dimensions of 33% of the region of interest. (D) The isolated, extracted cubic volume of interest. Note: This is a 2D visualisation of a 3D method. *Images courtesy of Lily DeMars.*

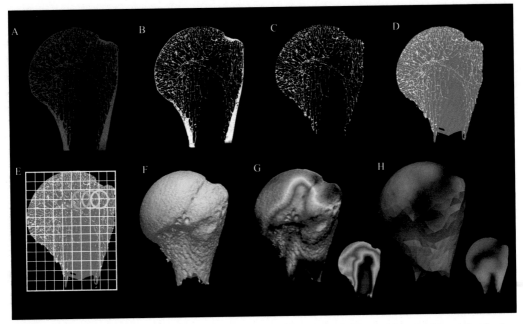

FIG. 2.6 (A) A coronal slice from a micro-CT scan of a proximal humerus. (B) The segmented image has had non-bone inclusions removed and the image had been binarised. (C) The cortical bone has been removed leaving just the isolated trabecular bone. (D) The image has been recast such that all voxels (three-dimensional pixels) representing bone, air inside the bone, and air outside the bone have different greyscale values. (E) A 3D grid has been placed over the entire volume of the isolated and recast trabecular bone dataset. Overlapping sampling spheres (represented in *orange*) are placed at each node of the grid. Only five are shown for image clarity, but sampling spheres are used to calculate trabecular properties throughout the entire 3D volume. (F) The 3D mesh, or model, of the isolated trabecular bone (cortical bone has been removed). (G) Scalar values for bone volume fraction have been mapped to the 3D mesh based on the values obtained from step E. Warm colours equate to higher bone volume fraction, cool colours indicate lower bone volume fraction. (H) Scalar values for degree of anisotropy have been mapped to the 3D mesh. *Dark purple* indicates higher degree of anisotropy or more oriented bone. Note: The small images in the bottom right corners of (G) and (H) show the same slice seen in steps A–E, demonstrating that this method is applied across the entire trabecular volume. *Images courtesy of Lily DeMars.*

Polk et al., 2008; Pontzer et al., 2006; Robbins et al., 2018; Wallace et al., 2015). Another active area of research in biological anthropology explores to what degree trabecular bone structure reflects differences in posture and locomotor modes between species of primates (Dunmore et al., 2019; Fajardo et al., 2007; Georgiou et al., 2019; Kivell et al., 2011a; Kuo et al., 2013; Ryan and Shaw, 2012; Shaw and Ryan, 2012; Sukhdeo et al., 2018; Tsegai et al., 2013, 2017). Trabecular bone structure has also been used to interpret behaviour based on fossil hominin morphology (Barak et al., 2013; Kivell et al., 2011a; Ryan et al., 2018; Skinner et al., 2015; Stephens et al., 2016; Su and Carlson, 2017).

Over the last several years, a number of studies have focused on investigating how trabecular bone structure varies within and between human groups with distinct subsistence strategies, and thus different levels of physical activity and mechanical loading. Previous bone functional adaptation-based research found that modern humans tend to have low bone mass compared to other hominins and most nonhuman primates and that this trend of reduced skeletal robusticity (in both cortical and trabecular bone) appears in the last ~ 10,000 years, and seems to coincide with reduced mobility and changes in physical activity associated with the emergence of agricultural subsistence strategies (Chirchir et al., 2015; Ruff et al., 2015). Ryan and Shaw (2015) used a single, cubic VOI positioned in the centre of the femoral head to compare trabecular bone structure between North American hunter-gatherers and agriculturalists. They found that the hunter-gatherers had significantly higher bone volume fractions than the agriculturalists and suggest that this could be due to a higher level of physical activity and mechanical loading experienced by the more mobile hunter-gatherers compared with the more sedentary agriculturalists (Ryan and Shaw, 2015). Scherf et al. (2016) used a single, spherical VOI positioned in the centre of the humeral head to compare the trabecular bone structure of Neolithic and contemporary Europeans. Trabecular bone structure between males and females within each group was also compared. The authors hypothesised that due to differences in physical workload (agricultural labour vs. office work), the Neolithic farmers would have stronger trabecular architecture than the contemporary individuals. Further, based on the results of a previous study that shows a sexual division of labour where females undertake repetitive activities with lower magnitude but higher loading frequency (Eshed et al., 2004), they hypothesised that Neolithic females would have lower bone volume fraction and more anisotropic (highly oriented) bone structure than Neolithic males. As a note, degree of anisotropy is generally associated with variation in loading direction where more anisotropic bone is thought to result from more uniform loading direction (Barak et al., 2013). In general, they found that the male and female Neolithic farmers did have more, thicker, and less separated trabeculae. Neolithic females also had more highly oriented trabecular bone in their humeral heads than Neolithic males and contemporary females, which supports their hypothesis that males and females undertook different types of manual labour.

Ryan and Shaw (2015) and Scherf et al. (2016) both employed large single VOIs and compared a single skeletal region of interest. Other studies compared single VOIs across multiple skeletal elements for each individual in their study. Saers et al. (2016) used single, cubic VOIs positioned across the joints of the lower limb (proximal femur, distal femur, proximal tibia, and distal tibia) and maximised the size of the VOI that would fit within the joint. The study compared trabecular bone structure between three human groups with high, intermediate, and low levels of inferred habitual activity but also explored how trabecular bone is organised among the joints of interest. Specifically, they examined whether or not, like cortical bone, trabecular bone structure reflects limb tapering or reduction in bone mass. The group with the highest level of mobility (the hunter-gatherers) also had the highest bone volume fraction, while the group with the lowest level of mobility (sedentary agriculturalists) had the least amount of bone. Interestingly, the differences between groups were most obvious in the proximal femur and became progressively less pronounced in the more distally positioned VOIs. The authors mention that this pattern indicates that the trabecular bone of the lower limb may be differentially sensitive to mechanical loading depending on the joint in question, and that like some cortical bone variables, trabecular bone volume fraction decreases from proximal to distal along the lower limb. Mulder et al. (2020) used single, spherical VOIs positioned across the lower limb in the same joints utilised by Saers et al. (2016) but each VOI was scaled according to body size and positioned just below the subchondral bone. The authors compared three subgroups of a single population of individuals from late medieval Cambridge, United Kingdom. The individuals in the study are broadly contemporary with each other but were excavated from either a parish church, hospital, or Augustinian friary. The authors note that although there are some differences between each of the sites, their behavioural patterns would likely not have been very different and each site can be classified as urban and largely sedentary. Despite similarities between the groups, the individuals from the parish church had lower bone volume fraction and trabecular thickness than the other two sites, which the authors suggest may be due to a difference in diet. Sexual dimorphism was also compared across sites. The authors found that trabecular thickness and connectivity density differed between males and females, but that patterns of difference between skeletal elements (femur and tibia) were inconsistent. Overall, the authors highlight the need for further understanding of the range of variation in trabecular bone structure present within *and* between human groups, noting that their intra-population study found considerably less variation in trabecular bone variables than other studies investigating interpopulation differences.

In contrast to Saers et al. (2016) and Mulder et al. (2020), Doershuk et al. (2019) utilised single, cubic VOIs of maximum size to compare trabecular bone structure between the upper (proximal humerus) and lower limbs (proximal femur) in five human groups with various levels of high, intermediate, or low inferred activity level (based on subsistence strategy). The authors predicted that in both the humerus and femur, groups with higher levels of inferred activity would have more robust trabecular bone, but that in all groups, the femur would have more robust trabecular bone than the humerus. Additionally, authors hypothesised that if there were differences between the sexes, males would have more robust trabecular bone than females. Results indicated that, as with other studies, trabecular bone structure of femur reflects an activity gradient where highly active groups have more bone than less active groups. Interestingly, in the humerus, there was no significant difference in trabecular bone structure between the groups classified as high and intermediate activity level. The authors suggest that this finding, in addition to inconsistent coefficients of variation between skeletal elements and groups, indicates localised, nonsystemic response to differences in loading. They also found, that as in other primates, the femur consistently has higher bone volume fraction and more anisotropic bone than the humerus. In terms of sex differences, in contrast with Scherf et al. (2016), there were no consistent patterns that indicated sexual dimorphism in the trabecular bone variables reported.

Another approach used by researchers investigating intra-population trabecular bone structural variation is to use multiple VOIs positioned in several skeletal elements. Saers et al. (2019) used 17 VOIs within the talus, calcaneus, and first metatarsal of four human groups classified as having relatively high or low mobility. The position of each VOI was based on biomechanical expectations of forces experienced in the foot during bipedal locomotion. In addition to testing if the trabecular bone structure of the foot follows expectations based on bone functional adaptation, the authors also assess patterns of sexual dimorphism. The large number and specific, hypothesis-driven placement of the VOIs provided the researchers a more fine-grained view of variation in trabecular bone structure across the foot compared with other VOI methods, though the resulting dataset was correspondingly large and complex. In general, they found that groups with high mobility had higher bone volume fraction, as well as thicker, fewer, and more tightly packed trabeculae. Similar to other studies discussed here, little evidence of sexual dimorphism was apparent in the trabecular bone of the foot (Saers et al., 2019). Across studies, the lack of sexual dimorphism in trabecular bone is interesting because studies of cortical bone cross-sectional geometry often report differences between males and females.

Thus far, the studies discussed have used various forms of the VOI approach to quantify trabecular bone structure. In contrast, Stephens et al. (2018) used whole bone analysis combined with multiple methods of statistical evaluation to compare trabecular bone structure in 22 bones of the hand between human groups classified as either foragers or post-Neolithic. The forager sample spans a broad timescale, including early modern human individuals, as well as 19th century foragers from Tierra del Fuego. Stephens and colleagues hypothesised that because all individuals classified as *Homo sapiens* have very similar hand morphology, patterns of trabecular bone structure would be similar between the forager and post-Neolithic groups, but may reflect some differences in patterns of hand use. Based on a bone functional adaptation framework, they also expected foragers to have generally more robust trabecular bone, consistent with greater levels of loading in the hands and upper body. Of note, both a benefit and challenge of the whole bone analysis approach is the large amount of data output as part of the analysis. As such, the summary here generalises the author's findings. The distribution of trabecular variables between the forager and post-Neolithic groups suggested that hand use was similar between the two in terms of posture and loading positions. Additionally, Stephens et al. (2018) found that the foragers (inferred to have higher levels of mechanical loading) tended to have higher bone volume fraction and lower degree of anisotropy, as seen in other studies discussed in this section. Overall, their results are consistent with a bone functional adaptation framework where higher levels of loading are associated with increased trabecular robusticity. Importantly, the whole bone analysis approach used in this study allowed researchers to investigate trabecular bone structural variation at a very fine scale across a large anatomical region, though the large volume of data can be difficult to interpret, the method allows for very detailed visualisation of variation in bone structure.

Research examining the relationship between trabecular bone structural variation and behaviour has greatly enhanced our knowledge of how bone responds to mechanical loading. Results from the studies discussed above consistently found that when comparing groups with divergent levels of physical activity, groups with higher activity levels have more robust trabecular bone structure. However, all of the previously mentioned studies also contained results that were not consistent, at some level, with their predictions. Like other bone tissues, and as outlined earlier in this chapter, trabecular bone structure is determined by many factors (e.g. disease, diet, hormones) other than adaptation to mechanical strain, therefore, it is not unexpected that the results of trabecular bone research are often complex, difficult to interpret, and somewhat inconsistent. These challenges can also be attributed, in part, to a lack of standardised methodological and analysis protocols, although this is partially due to a major push over the last several years to improve how trabecular bone is analysed and meaningfully compared between groups (DeMars et al., 2021; Dunmore et al., 2019; Stephens et al., 2018; Sukhdeo et al., 2018; Sylvester and Terhune, 2017). Trabecular bone analyses can also be expensive, time consuming, and computationally intensive. The necessity of using high-resolution scans in order to have good quality data results in larger more complex datasets that take longer both to acquire and analyse. Despite these challenges, trabecular bone research within biological anthropology is an active area of study contributing to interpretations of behaviour from the skeleton with many important questions.

2.7 Conclusions

Bone is a dynamic tissue that responds to mechanical signals by rearranging its cortical and trabecular microstructure. This knowledge has seen useful applications within biological anthropology to address a range of questions that relate to behaviour. Both contemporary and archaeological human samples retain the effect of biomechanical signals on cortical bone histology parameters, which have been used to reconstruct temporal, spatial, intra- and inter-skeletal, and population-level behaviours. In order to further successfully explore such links between behaviour and cortical bone histology, future studies should consider examining as many histological and other 3D variables as possible while controlling for the effect of age at death and sex. Ideally, large sample sizes should be studied to detect patterns in bone histological trends within well-documented archaeological contexts. While there are many questions to pursue within trabecular bone microstructure research, one especially interesting area that should be considered more is how trabecular bone structure varies in relation to other levels of bone structural organisation such as cortical bone cross-sectional geometry. A number of studies have investigated intra-individual variation in trabecular bone structure, but it would be interesting to better understand the relationship and structure of covariation between different types of bony tissues. Combining 2D and 3D analyses of cortical and trabecular bone in the future will offer complementary lines of evidence for the way human bones adapt and respond to function.

Acknowledgments

We would like to thank Cara Hirst for inviting us to contribute this chapter to the book, the reviewers and editors for their constructive feedback on the chapter content. This project was supported by the National Science Foundation Graduate Research Fellowship Program under Grant No. DGE1255832 (to LJDD) and the Australian Research Council (Grant No. DE190100068 to JJM). Any opinions, findings, conclusions, or recommendations expressed in this material are those of the authors and do not necessarily reflect the views of the National Science Foundation or the Australian Research Council.

References

Abdelgawad, M. E., Delaisse, J. M., Hinge, M., Jensen, P. R., Alnaimi, R. W., Rolighed, L., et al. (2016). Early reversal cells in adult human bone remodeling: Osteoblastic nature, catabolic functions and interactions with osteoclasts. *Histochemistry and Cell Biology, 145*(6), 603–615. https://doi.org/10.1007/s00418-016-1414-y.

Agarwal, S. C., Dumitriu, M., Tomlinson, G. A., & Grynpas, M. D. (2004). Medieval trabecular bone architecture: The influence of age, sex, and lifestyle. *American Journal of Physical Anthropology, 124*(1), 33–44. https://doi.org/10.1002/ajpa.10335.

Agarwal, S. C., & Stout, S. D. (Eds.). (2003). *Bone loss and osteoporosis: An anthropological perspective*. Kluwer Academic.

Andersen, T. L., Abdelgawad, M. E., Kristensen, H. B., Hauge, E. M., Rolighed, L., Bollerslev, J., et al. (2013). Understanding coupling between bone resorption and formation: Are reversal cells the missing link? *The American Journal of Pathology, 183*(1), 235–246. https://doi.org/10.1016/j.ajpath.2013.03.006.

Barak, M. M., Lieberman, D. E., & Hublin, J. J. (2011). A Wolff in sheep's clothing: Trabecular bone adaptation in response to changes in joint loading orientation. *Bone, 49*(6), 1141–1151. https://doi.org/10.1016/j.bone.2011.08.020.

Barak, M. M., Lieberman, D. E., Raichlen, D., Pontzer, H., Warrener, A. G., & Hublin, J. J. (2013). Trabecular evidence for a human-like gait in Australopithecus africanus. *PLoS One, 8*(11). https://doi.org/10.1371/journal.pone.0077687, e77687.

Best, A., Holt, B., Troy, K., & Hamill, J. (2017). Trabecular bone in the calcaneus of runners. *PLoS One, 12*(12). https://doi.org/10.1371/journal.pone.0188200, e0190553.

Black, J. D., & Tadros, B. J. (2020). Bone structure: From cortical to calcium. *Orthopaedics and Traumatology, 34*(3), 113–119. https://doi.org/10.1016/j.mporth.2020.03.002.

Boivin, G., & Meunier, P. J. (2002). Changes in bone remodeling rate influence the degree of mineralization of bone. *Connective Tissue Research, 43*(2–3), 535–537. https://doi.org/10.1080/03008200290000934.

Boskey, A. L., & Posner, A. S. (1984). Structure and formation of bone mineral. In *Natural and living biomaterials* (pp. 27–41). CRC Press.

Brickley, M. B., Ives, R., & Mays, S. (2020). *The bioarchaeology of metabolic bone disease*. Elsevier Academic Press.

Bridges, P. S. (1989). Changes in activities with the shift to agriculture in the Southeastern United States. *Current Anthropology, 30*(3), 385–394. https://doi.org/10.1086/203756.

Britz, H. M., Thomas, C. D. L., Clement, J. G., & Cooper, D. M. (2009). The relation of femoral osteon geometry to age, sex, height and weight. *Bone, 45*(1), 77–83. https://doi.org/10.1016/j.bone.2009.03.654.

Bromage, T. G., Lacruz, R. S., Hogg, R., Goldman, H. M., McFarlin, S. C., Warshaw, J., et al. (2009). Lamellar bone is an incremental tissue reconciling enamel rhythms, body size, and organismal life history. *Calcified Tissue International, 84*(5), 388–404. https://doi.org/10.1007/s00223-009-9221-2.

Burr, D. B. (1997). Muscle strength, bone mass, and age-related bone loss. *Journal of Bone and Mineral Research, 12*(10), 1547–1551. https://doi.org/10.1359/jbmr.1997.12.10.1547.

Burr, D. B., Ruff, C. B., & Thompson, D. D. (1990). Patterns of skeletal histologic change through time: Comparison of an archaic native american population with modern populations. *The Anatomical Record, 226*(3), 307–313. https://doi.org/10.1002/ar.1092260306.

Carlson, K. J., Lublinsky, S., & Judex, S. (2008). Do different locomotor modes during growth modulate trabecular architecture in the murine hind limb? *Integrative and Comparative Biology, 48*(3), 385–393. https://doi.org/10.1093/icb/icn066.

Carter, D. R. (1984). Mechanical loading histories and cortical bone remodeling. *Calcified Tissue International, 36*(1), S19–S24. https://doi.org/10.1007/BF02406129.

Chirchir, H., Kivell, T. L., Ruff, C. B., Hublin, J.-J., Carlson, K. J., Zipfel, B., et al. (2015). Recent origin of low trabecular bone density in modern humans. *Proceedings of the National Academy of Sciences, 112*(2), 366–371. https://doi.org/10.1073/pnas.1411696112.

Chirchir, H., Ruff, C. B., Junno, J. A., & Potts, R. (2017). Low trabecular bone density in recent sedentary modern humans. *American Journal of Physical Anthropology, 162*(3), 550–560. https://doi.org/10.1002/ajpa.23138.

Choi, K., Kuhn, J. L., Ciarelli, M. J., & Goldstein, S. A. (1990). The elastic moduli of human subchondral, trabecular, and cortical bone tissue and the size-dependency of cortical bone modulus. *Journal of Biomechanics, 23*(11), 1103–1113. https://doi.org/10.1016/0021-9290(90)90003-L.

Clarke, B. (2008). Normal bone anatomy and physiology. *Clinical Journal of the American Society of Nephrology, 3*(3), S131–S139. https://doi.org/10.2215/CJN.04151206.

Cowin, S. C. (2011). *Bone biomechanics handbook*. CRC Press.

Crane, J. L., & Cao, X. (2014). Bone marrow mesenchymal stem cells and TGF-β signaling in bone remodeling. *The Journal of Clinical Investigation, 124*(2), 466–472. https://doi.org/10.1172/JCI70050.

Crowder, C., & Stout, S. (2011). *Bone histology: An anthropological perspective*. CRC Press.

Currey, J. D. (2002). *Bones: Structure and mechanics*. Princeton University Press.

Deckers, M. M., Karperien, M., van der Bent, C., Yamashita, T., Papapoulos, S. E., & Löwik, C. W. (2000). Expression of vascular endothelial growth factors and their receptors during osteoblast differentiation. *Endocrinology, 141*(5), 1667–1674. https://doi.org/10.1210/endo.141.5.7458.

Delaisse, J.-M. (2014). The reversal phase of the bone-remodeling cycle: Cellular prerequisites for coupling resorption and formation. *BoneKEy Reports, 3*, 561. https://doi.org/10.1038/bonekey.2014.56.

DeMars, L. J. D., Stephens, N. B., Saers, J. P. P., Gordon, A., Stock, J. T., & Ryan, T. M. (2021). Using point clouds to investigate the relationship between trabecular bone phenotype and behavior: An example utilizing the human calcaneus. *American Journal of Human Biology, 33*(2). https://doi.org/10.1002/ajgh.23468, e23468.

Dempster, D. W., & Raisz, L. G. (2015). Bone physiology: Bone cells, modeling, and remodeling. In *Nutrition and bone health* (pp. 37–56). Humana Press.

DeSilva, J. M., & Devlin, M. J. (2012). A comparative study of the trabecular bony architecture of the talus in humans, non-human primates, and Australopithecus. *Journal of Human Evolution, 63*(3), 536–551. https://doi.org/10.1016/j.jhevol.2012.06.006.

Doblaré, M., & García, J. M. (2001). Application of an anisotropic bone-remodelling model based on a damage-repair theory to the analysis of the proximal femur before and after total hip replacement. *Journal of Biomechanics, 34*(9), 1157–1170. https://doi.org/10.1016/S0021-9290(01)00069-0.

Doershuk, L. J., Saers, J. P. P., Shaw, C. N., Jashashvili, T., Carlson, K. J., Stock, J. T., et al. (2019). Complex variation of trabecular bone structure in the proximal humerus and femur of five modern human populations. *American Journal of Physical Anthropology, 168*(1), 104–118. https://doi.org/10.1002/ajpa.23725.

Dominguez, V. M., & Agnew, A. M. (2016). Examination of factors potentially influencing osteon size in the human rib. *The Anatomical Record, 299*(3), 313–324. https://doi.org/10.1002/ar.23305.

Doube, M., Kłosowski, M. M., Arganda-Carreras, I., Cordelières, F. P., Dougherty, R. P., Jackson, J. S., et al. (2010). BoneJ: Free and extensible bone image analysis in ImageJ. *Bone, 47*(6), 1076–1079. https://doi.org/10.1016/j.bone.2010.08.023.

Doube, M., Kłosowski, M. M., Wiktorowicz-Conroy, A. M., Hutchinson, J. R., & Shefelbine, S. J. (2011). Trabecular bone scales allometrically in mammals and birds. *Proceedings of the Royal Society B: Biological Sciences, 278*(1721), 3067–3073. https://doi.org/10.1098/rspb.2011.0069.

Dr. Pahr Ingenieurs e.U. (2019). *Medtool. Dr. Pahr Ingenieure e.U.* http://www.drpahr.at

Dunmore, C. J., Kivell, T. L., Bardo, A., & Skinner, M. M. (2019). Metacarpal trabecular bone varies with distinct hand-positions used in hominid locomotion. *Journal of Anatomy, 235*(1), 45–66. https://doi.org/10.1111/joa.12966.

Dunmore, C. J., Wollny, G., & Skinner, M. M. (2018). MIA-Clustering: A novel method for segmentation of paleontological material. *PeerJ, 6*. https://doi.org/10.7717/peerj.4374, e4374.

Ericksen, M. F. (1980). Patterns of microscopic bone reomdeling in three aboriginal American populations. In *Early native Americans: Prehistoric demography, economy, and technology* (pp. 239–270). The Hague.

Eriksen, E. F. (2010). Cellular mechanisms of bone remodeling. *Reviews in Endocrine and Metabolic Disorders, 11*, 219–227. https://doi.org/10.1007/s11154-010-9153-1.

Eshed, V., Gopher, A., Galili, E., & Hershkovitz, I. (2004). Musculoskeletal stress markers in Natufian hunter-gathers and Neolithci farmers in the Levant: The upper limb. *American Journal of Physical Anthropology, 123*(4), 303–315. https://doi.org/10.1002/ajpa.10312.

Fajardo, R. J., & Müller, R. (2001). Three-dimensional analysis of nonhuman primate trabecular architecture using micro-computed tomography. *American Journal of Physical Anthropology, 115*(4), 327–336. https://doi.org/10.1002/ajpa.1089.

Fajardo, R., Müller, R., Ketcham, R., & Colbert, M. (2007). Nonhuman anthropoid primate femoral neck trabecular architecture and its relationship to locomotor mode. *Anatomical Record, 290*(4), 422–436. https://doi.org/10.1002/ar.20493.

Feng, X., & McDonald, J. M. (2011). Disorders of bone remodeling. *Annual Review of Pathology: Mechanisms of Disease, 6*, 121–145. https://doi.org/10.1146/annurev-pathol-011110-130203.

Florencio-Silva, R., Sasso, G. R. D. S., Sasso-Cerri, E., Simões, M. J., & Cerri, P. S. (2015). Biology of bone tissue: Structure, function, and factors that influence bone cells. *BioMed Research International*. https://doi.org/10.1155/2015/421746.

Forwood, M. R., & Burr, D. B. (1993). Physical activity and bone mass: Exercises in futility? *Bone and Mineral, 21*(2), 89–112. https://doi.org/10.1016/S0169-6009(08)80012-8.

Fox, J., Miller, M. A., Newman, M. K., Turner, C. H., Recker, R. R., & Smith, S. Y. (2007). Treatment of skeletally mature ovariectomized rhesus monkeys with PTH (1-84) for 16 months increases bone formation and density and improves trabecular architecture and biomechanical properties at the lumbar spine. *Journal of Bone and Mineral Research, 22*(2), 260–273. https://doi.org/10.1359/jbmr.061101.

Frost, H. (1979). A chondral modeling theory. *Calcified Tissue International, 28*(1), 181–200. https://doi.org/10.1007/BF02441236.

Frost, H. (1987). Bone "mass" and the "mechanostat": A proposal. *The Anatomical Record, 219*(1), 1–9. https://doi.org/10.1002/ar.1092190104.

Frost, H. (1994). Wolff's Law and bone's structural adaptations to mechanical usage: An overview for clinicians. *Angle Orthodontist, 64*(3), 175–188. https://doi.org/10.1043/0003-3219(1994)064<0175:WLABSA>2.0.CO;2.

Frost, H. M. (1999a). On the estrogen–bone relationship and postmenopausal bone loss: A new model. *Journal of Bone and Mineral Research, 14*(9), 1473–1477. https://doi.org/10.1359/jbmr.1999.14.9.1473.

Frost, H. (1999b). Joint anatomy, design, and arthroses: Insights of the Utah paradigm. *Anatomical Record, 255*(2), 162–174. https://doi.org/10.1002/(SICI)1097-0185(19990601)255:2<162::AID-AR6>3.0.CO;2-1.

Frost, H. (2000). The Utah paradigm of skeletal physiology: An overview of its insights for bone, cartilage and collagenous tissue organs. *Journal of Bone and Mineral Metabolism, 18*(6), 305–316. https://doi.org/10.1007/s007740070001.

Frost, H. (2001). From Wolff's law to the Utah paradigm: Insights about bone physiology and its clinical applications. *Anatomical Record, 262*(4), 398–419. https://doi.org/10.1002/ar.1049.

Fuchs, R. K., Bauer, J. J., & Snow, C. M. (2001). Jumping improves hip and lumbar spine bone mass in prepubescent children: A randomized controlled trial. *Journal of Bone and Mineral Research, 16*(1), 148–156. https://doi.org/10.1359/jbmr.2001.16.1.148.

Georgiou, L., Kivell, T. L., Pahr, D. H., Buck, L. T., & Skinner, M. M. (2019). Trabecular architecture of the great ape and human femoral head. *Journal of Anatomy, 234*(5), 679–693. https://doi.org/10.1111/joa.12957.

Goldman, H. M., McFarlin, S. C., Cooper, D. M. L., Thomas, C. D. L., & Clement, J. G. (2009). Ontogenetic patterning of cortical bone microstructure and geometry at the human mid-shaft femur. *Anatomical Record, 292*(1), 48–64. https://doi.org/10.1002/ar.20778.

Gross, T., Kivell, T. L., Skinner, M. M., Nguyen, N. H., & Pahr, D. H. (2014). A CT-image-based framework for the holistic analysis of cortical and trabecular bone morphology. *Palaeontologia Electronica, 17*(3;33A), 1–13. http://palaeo-electronica.org/content/2014/889-holistic-analysis-of-bone.

Gurkan, U. A., & Akkus, O. (2008). The mechanical environment of bone marrow: A review. *Annals of Biomedical Engineering, 36*(12), 1978–1991. https://doi.org/10.1007/s10439-008-9577-x.

Hadjidakis, D. J., & Androulakis, I. I. (2006). Bone remodeling. *Annals of the New York Academy of Sciences, 1092*(1), 385–396. https://doi.org/10.1196/annals.1365.035.

Heaney, R. P., Abrams, S., Dawson-Hughes, B., Looker, A., Looker, A., Marcus, R., et al. (2000). Peak bone mass. *Osteoporosis International, 11*(12), 985–1009. https://doi.org/10.1007/s001980070020.

Hennig, C., Thomas, C. D. L., Clement, J. G., & Cooper, D. M. (2015). Does 3D orientation account for variation in osteon morphology assessed by 2D histology? *Journal of Anatomy, 227*(4), 497–505. https://doi.org/10.1111/joa.12357.

Henriksen, K., Neutzsky-Wulff, A. V., Bonewald, L. F., & Karsdal, M. A. (2009). Local communication on and within bone controls bone remodeling. *Bone, 44*(6), 1026–1033. https://doi.org/10.1016/j.bone.2009.03.671.

Hernández-Becerra, E., Jímenez-Mendoza, D., Mutis-Gonzalez, N., Pineda-Gomez, P., Rojas-Molina, I., & Rodríguez-García, M. E. (2020). Calcium deficiency in diet decreases the magnesium content in bone and affects femur physicochemical properties in growing rats. *Biological Trace Element Research, 197*, 224–232. https://doi.org/10.1007/s12011-019-01989-9.

Heřt, J., Fiala, P., & Petrtýl, M. (1994). Osteon orientation of the diaphysis of the long bones in man. *Bone, 15*(3), 269–277. https://doi.org/10.1016/8756-3282(94)90288-7.

Huiskes, R., Ruimerman, R., Van Lenthe, G. H., & Janssen, J. D. (2000). Effects of mechanical forces on maintenance and adaptation of form in trabecular bone. *Nature, 405*(6787), 704–706. https://doi.org/10.1038/35015116.

Iwaniec, U. T., Crenshaw, T. D., Schoeninger, M. J., Stout, S. D., & Ericksen, M. F. (1998). Methods for improving the efficiency of estimating total osteon density in the human anterior mid-diaphyseal femur. *American Journal of Physical Anthropology, 107*(1), 13–24. https://doi.org/10.1002/(SICI)1096-8644(199809)107:1<13::AID-AJPA2>3.0.CO;2-E.

Jee, W. S. (2000). Principles in bone physiology. *Journal of Musculoskeletal & Neuronal Interactions, 1*, 11–13.

Katsimbri, P. (2017). The biology of normal bone remodelling. *European Journal of Cancer Care, 26*(6). https://doi.org/10.1111/ecc.12740, e12740.

Keenan, K. E., Mears, C. S., & Skedros, J. G. (2017). Utility of osteon circularity for determining species and interpreting load history in primates and nonprimates. *American Journal of Physical Anthropology, 162*(4), 657–681. https://doi.org/10.1002/ajpa.23154.

Kenkre, J. S., & Bassett, J. H. D. (2018). The bone remodelling cycle. *Annals of Clinical Biochemistry, 55*(3), 308–327. https://doi.org/10.1177/0004563218759371.

Ketcham, R. A., & Ryan, T. M. (2004). Quantification and visualization of anisotropy in trabecular bone. *Journal of Microscopy, 213*(2), 158–171. https://doi.org/10.1111/j.1365-2818.2004.01277.x.

Khosla, S., & Monroe, D. G. (2018). Regulation of bone metabolism by sex steroids. *Cold Spring Harbor Perspectives in Medicine, 8*(1). https://doi.org/10.1101/cshperspect.a031211, a031211.

Kini, U., & Nandeesh, B. N. (2012). Physiology of bone formation, remodeling, and metabolism. In *Radionuclide and hybrid bone imaging* (pp. 29–57). Springer.

Kivell, T. L. (2016). A review of trabecular bone functional adaptation: What have we learned from trabecular analyses in extant hominoids and what can we apply to fossils? *Journal of Anatomy, 228*(4), 569–594. https://doi.org/10.1111/joa.12446.

Kivell, T. L., Kibii, J. M., Churchill, S. E., Schmid, P., & Berger, L. R. (2011a). Australopithecus sediba hand demonstrates mosaic evolution of locomotor and manipulative abilities. *Science, 333*(6048), 1411–1417. https://doi.org/10.1126/science.1202625.

Kivell, T. L., Skinner, M. M., Lazenby, R., & Hublin, J. J. (2011b). Methodological considerations for analyzing trabecular architecture: An example from the primate hand. *Journal of Anatomy, 218*(2), 209–225. https://doi.org/10.1111/j.1469-7580.2010.01314.x.

Kothari, M., Keaveny, T. M., Lin, J. C., Newitt, D. C., Genant, H. K., & Majumdar, S. (1998). Impact of spatial resolution on the prediction of trabecular architecture parameters. *Bone, 22*(5), 437–443. https://doi.org/10.1016/S8756-3282(98)00031-3.

Kotting, D. (1977). *Trabecular involution of the proximal femur as a means of estimating age at death* (Master Thesis). Kent, OH: Kent State University.

Kovacs, C. S., & Fuleihan, G. E. H. (2006). Calcium and bone disorders during pregnancy and lactation. *Endocrinology and Metabolism Clinics, 35*(1), 21–51. https://doi.org/10.1016/j.ecl.2005.09.004.

Kozłowski, T., & Piontek, J. (2000). A case of atrophy of bones of the right lower limb of a skeleton from a medieval (12th-14th centuries) burial ground in Gruczno, Poland. *Journal of Paleopathology, 12*(1), 5–16.

Kučera, J., Rasmussen, K. L., Kameník, J., Kubešová, M., Skytte, L., Považil, C., et al. (2017). Was he murdered or was he not?—Part II: Multi-elemental analyses of hair and bone samples from Tycho Brahe and histopathology of his bones. *Archaeometry, 59*(5), 918–933. https://doi.org/10.1111/arcm.12284.

Kular, J., Tickner, J., Chim, S. M., & Xu, J. (2012). An overview of the regulation of bone remodelling at the cellular level. *Clinical Biochemistry, 45*(12), 863–873. https://doi.org/10.1016/j.clinbiochem.2012.03.021.

Kuo, A., & Carter, D. (1991). Computational methods for analyzing the structure of cancellous bone in planar sections. *Journal of Orthopaedic Research, 9*(6), 918–931. https://doi.org/10.1002/jor.1100090619.

Kuo, S., Desilva, J. M., Devlin, M. J., Mcdonald, G., & Morgan, E. F. (2013). The effect of the achilles tendon on trabecular structure in the primate calcaneus. *Anatomical Record, 296*(10), 1509–1517. https://doi.org/10.1002/ar.22739.

Lanyon, L. E. (1996). Using functional loading to influence bone mass and architecture: Objectives, mechanisms, and relationship with estrogen of the mechanically adaptive process in bone. *Bone, 18*(1), S37–S43. https://doi.org/10.1016/8756-3282(95)00378-9.

Lanyon, L. E., & Rubin, C. T. (1984). Static vs dynamic loads as an influence on bone remodelling. *Journal of Biomechanics, 7*(12), 897–905. https://doi.org/10.1016/0021-9290(84)90003-4.

Larsen, C. (1981). Functional implications of postcranial size reduction on the prehistoric Georgia coast, U.S.A. *Journal of Human Evolution, 10*(6), 489–502. https://doi.org/10.1016/S0047-2484(81)80095-4.

Lazenby, R. A., & Pfeiffer, S. K. (1993). Effects of a nineteenth century below-knee amputation and prosthesis on femoral morphology. *International Journal of Osteoarchaeology, 13*(1), 19–28. https://doi.org/10.1002/oa.1390030103.

Lazenby, R. A., Skinner, M. M., Kivell, T. L., & Hublin, J. J. (2011). Scaling VOI size in 3D μcT studies of trabecular bone: A test of the over-sampling hypothesis. *American Journal of Physical Anthropology, 144*(2), 196–203. https://doi.org/10.1002/ajpa.21385.

Lieberman, D. E., Polk, J. D., & Demes, B. (2004). Predicting long bone loading from cross-sectional geometry. *American Journal of Physical Anthropology, 123*(2), 156–171. https://doi.org/10.1002/ajpa.10316.

Lim, Y. J., Lee, B. U., Heo, S. J., Koak, J. Y., Kim, S. K., Kim, Y. S., et al. (2007). Bone response to cyclic loading with anodized implants. *Key Engineering Materials, 342*, 117–120.

Macintosh, A. A., Pinhasi, R., & Stock, J. T. (2017). Prehistoric women's manual labor exceeded that of athletes through the first 5500 years of farming in Central Europe. *Science Advances, 3*(11). https://doi.org/10.1126/sciadv.aao3893, eaao3893.

Maggio, A., & Franklin, D. (2019). Histomorphometric age estimation from the femoral cortex: A test of three methods in an Australian population. *Forensic Science International, 303*. https://doi.org/10.1016/j.forsciint.2019.109950, 109950.

Manolagas, S. C., & Parfitt, A. M. (2010). What old means to bone. *Trends in Endocrinology and Metabolism*, 21(6), 369–374. https://doi.org/10.1016/j.tem.2010.01.010.

Maquer, G., Musy, S. N., Wandel, J., Gross, T., & Zysset, P. K. (2015). Bone volume fraction and fabric anisotropy are better determinants of trabecular bone stiffness than other morphological variables. *Journal of Bone and Mineral Research*, 30(6), 1000–1008. https://doi.org/10.1002/jbmr.2437.

Martin, R. B. (2000). Toward a unifying theory of bone remodeling. *Bone*, 26(1), 1–6. https://doi.org/10.1016/S8756-3282(99)00241-0.

Martin, R. B., Burr, D. B., & Sharkey, N. A. (1998). *Skeletal tissue mechanics*. Springer.

Martin, T. J., & Sims, N. A. (2005). Osteoclast-derived activity in the coupling of bone formation to resorption. *Trends in Molecular Medicine*, 11(2), 76–81. https://doi.org/10.1016/j.molmed.2004.12.004.

Matsuo, K., & Irie, N. (2008). Osteoclast–osteoblast communication. *Archives of Biochemistry and Biophysics*, 473(2), 201–209. https://doi.org/10.1016/j.abb.2008.03.027.

Mattheck, C. (1990). Design and growth rules for biological structures and their application to engineering. *Fatigue & Fracture of Engineering Materials & Structures*, 13(5), 535–550. https://doi.org/10.1111/j.1460-2695.1990.tb00623.x.

Mays, S., Elders, J., Humphrey, L., White, W., & Marshall, P. (2013). *Science and the dead: A guideline for the destructive sampling of archaeological human remains for scientific analysis*. Advisory Panel on the Archaeology of Burials in England. English Heritage.

McKenna, M. J., Murray, B., Lonergan, R., Segurado, R., Tubridy, N., & Kilbane, M. T. (2018). Analysing the effect of multiple sclerosis on vitamin D related biochemical markers of bone remodelling. *The Journal of Steroid Biochemistry and Molecular Biology*, 177, 91–95. https://doi.org/10.1016/j.jsbmb.2017.09.002.

Meyer, C., Nicklisch, N., Held, P., Fritsch, B., & Alt, K. W. (2011). Tracing patterns of activity in the human skeleton: An overview of methods, problems, and limits of interpretation. *HOMO-Journal of Comparative Human Biology*, 62(3), 202–217. https://doi.org/10.1016/j.jchb.2011.03.003.

Miszkiewicz, J. J. (2015). Histology of a Harris line in a human distal tibia. *Journal of Bone and Mineral Metabolism*, 33(4), 462–466. https://doi.org/10.1007/s00774-014-0644-0.

Miszkiewicz, J. J. (2016). Investigating histomorphometric relationships at the human femoral midshaft in a biomechanical context. *Journal of Bone and Mineral Metabolism*, 34(2), 179–192. https://doi.org/10.1007/s00774-015-0652-8.

Miszkiewicz, J. (2020). The importance of open access software in the analysis of bone histology in biological anthropology. *Evolutionary Anthropology*, 29(4), 165–167. https://doi.org/10.1002/evan.21859.

Miszkiewicz, J. J., & Cooke, K. M. (2019). Socio-economic determinants of bone health from past to present. *Clinical Reviews in Bone and Mineral Metabolism*, 17(3), 109–122. https://doi.org/10.1007/s12018-019-09263-1.

Miszkiewicz, J., & Mahoney, P. (2016). Ancient human bone microstructure in medieval England: Comparisons between two socio-economic groups. *Anatomical Record*, 299(1), 42–59. https://doi.org/10.1002/ar.23285.

Miszkiewicz, J., & Mahoney, P. (2017). Human bone and dental history in an archaeological context. In *Imaging human remains: Another dimension* (pp. 29–43). Elsevier/Academic Press.

Miszkiewicz, J. J., & Mahoney, P. (2019). Histomorphometry and cortical robusticity of the adult human femur. *Journal of Bone and Mineral Metabolism*, 37(1), 90–104–104. https://doi.org/10.1007/s00774-017-0899-3.

Miszkiewicz, J. J., Matisoo-Smith, E. A., & Weisler, M. I. (2022). Behavior and intra-skeletal remodeling in an adult male from 1720 BP Ebon Atoll, Marshall Islands, eastern Micronesia. *The Journal of Island and Coastal Archaeology*. https://doi.org/10.1080/15564894.2020.1837305.

Miszkiewicz, J., Rider, C., Kealy, S., Vrahnas, C., Sims, N., Vongsvivut, J., et al. (2020). Asymmetric midshaft femur remodeling in an adult male with left sided hip joint ankylosis, Metal Period Nagsabaran, Philippines. *International Journal of Paleopathology*, 31, 14–22. https://doi.org/10.1016/j.ijpp.2020.07.003.

Miszkiewicz, J. J., Valentin, F., Vrahnas, C., Sims, N. A., Vongsvivut, J., Tobin, M. J., et al. (2021). Bone loss markers in the earliest Pacific Islanders. *Scientific Reports*, 11(1), 1–16. https://doi.org/10.1038/s41598-021-83264-3.

Mulder, B., Stock, J. T., Saers, J. P. P., Inskip, S. A., Cessford, C., & Robb, J. E. (2020). Intrapopulation variation in lower limb trabecular architecture. *American Journal of Physical Anthropology*, 173(1), 112–129. https://doi.org/10.1002/ajpa.24058.

Müller, R., Van Campenhout, H., Van Damme, B., Van der Perre, G., Dequeker, J., Hildebrand, T., et al. (1998). Morphometric analysis of human bone biopsies: a quantitative structural comparison of histological sections and micro-computed tomography. *Bone*, 23(1), 59–66. https://doi.org/10.1016/S8756-3282(98)00068-4.

Nelson, D., Agarwal, S., & Darga, L. (2015). Bone Health from an Evolutionary Perspective: Development in Early Human Populations. In M. Holick, & J. Nieves (Eds.), *Nutrition and bone health* (pp. 3–20). Humana Press. https://doi.org/10.1007/978-1-4939-2001-3_1.

Nigg, B. M., & Gimston, S. K. (1999). Bone: Morphology and histology. In *Biomechanics of the musculo-skeletal system* (pp. 86–91). John Wiley and Sons Inc.

Noble, B. S., Peet, N., Stevens, H. Y., Brabbs, A., Mosley, J. R., Reilly, G. C., et al. (2003). Mechanical loading: Biphasic osteocyte survival and targeting of osteoclasts for bone destruction in rat cortical bone. *American Journal of Physiology - Cell Physiology, 284*(4), C934–C943. https://doi.org/10.1152/ajpcell.00234.2002.

Nordstrom, S. M., Carleton, S. M., Carson, W. L., Eren, M., Phillips, C. L., & Vaughan, D. E. (2007). Transgenic over-expression of plasminogen activator inhibitor-1 results in age-dependent and gender-specific increases in bone strength and mineralization. *Bone, 41*(6), 995–1004. https://doi.org/10.1016/j.bone.2007.08.020.

Odgaard, A., & Gundersen, H. J. G. (1993). Quantification of connectivity in cancellous bone, with special emphasis on 3-D reconstructions. *Bone, 14*(2), 173–182. https://doi.org/10.1016/8756-3282(93)90245-6.

Ortega, N., Behonick, D. J., & Werb, Z. (2004). Matrix remodeling during endochondral ossification. *Trends in Cell Biology, 14*(2), 86–93. https://doi.org/10.1016/j.tcb.2003.12.003.

Oxenham, M. F., Tilley, L., Matsumura, H., Nguyen, L. C., Nguyen, K. T., Nguyen, K. D., et al. (2009). Paralysis and severe disability requiring intensive care in Neolithic Asia. *Anthropological Science, 117*(2), 107–112. https://doi.org/10.1537/ase.081114.

Parfitt, A. M. (1984). The cellular basis of bone remodeling: The quantum concept reexamined in light of recent advances in the cell biology of bone. *Calcified Tissue International, 36*(1), S37–S45. https://doi.org/10.1007/BF02406132.

Parfitt, A. M. (2002). Misconceptions (2): Turnover is always higher in cancellous than in cortical bone. *Bone, 30*, 807–809. https://doi.org/10.1016/S8756-3282(02)00735-4.

Pearson, O. M., & Lieberman, D. E. (2004). The aging of Wolff's "law": Ontogeny and responses to mechanical loading in cortical bone. *Yearbook of Physical Anthropology, 125*(S39), 63–99. https://doi.org/10.1002/ajpa.20155.

Pfeiffer, S., Cameron, M. E., Sealy, J., & Beresheim, A. C. (2019). Diet and adult age-at-death among mobile foragers: A synthesis of bioarcheological methods. *American Journal of Physical Anthropology, 170*(1), 131–147. https://doi.org/10.1002/ajpa.23883.

Pfeiffer, S., Crowder, C., Harrington, L., & Brown, M. (2006). Secondary osteon and Haversian canal dimensions as behavioral indicators. *American Journal of Physical Anthropology, 131*(4), 460–468. https://doi.org/10.1002/ajpa.20454.

Pitfield, R., Deter, C., & Mahoney, P. (2019). Bone histomorphometric measures of physical activity in children from medieval England. *American Journal of Physical Anthropology, 169*(4), 730–746. https://doi.org/10.1002/ajpa.23853.

Pitfield, R., Miszkiewicz, J. J., & Mahoney, P. (2017). Cortical histomorphometry of the human humerus during ontogeny. *Calcified Tissue International, 101*(2), 148–158. https://doi.org/10.1007/s00223-017-0268-1.

Polk, J. D., Blumenfeld, J., & Ahluwalia, D. (2008). Knee posture predicted from subchondral apparent density in the distal femur: An experimental validation. *Anatomical Record, 291*(3), 293–302. https://doi.org/10.1002/ar.20653.

Pontzer, H., Lieberman, D. E., Momin, E., Devlin, M. J., Polk, J. D., Hallgrímsson, B., et al. (2006). Trabecular bone in the bird knee responds with high sensitivity to changes in load orientation. *Journal of Experimental Biology, 209*(1), 57–65. https://doi.org/10.1242/jeb.01971.

Prideaux, M., Findlay, D. M., & Atkins, G. J. (2016). Osteocytes: The master cells in bone remodelling. *Current Opinion in Pharmacology, 28*, 24–30. https://doi.org/10.1016/j.coph.2016.02.003.

Raggatt, L. J., & Partridge, N. C. (2010). Cellular and molecular mechanisms of bone remodeling. *Journal of Biological Chemistry, 285*(33), 25103–25108. https://doi.org/10.1074/jbc.R109.041087.

Richman, E. A., Ortner, D. J., & Schulter-Ellis, F. P. (1979). Differences in intracortical bone remodeling in three aboriginal American populations: Possible dietary factors. *Calcified Tissue International, 28*(1), 209–214. https://doi.org/10.1007/BF02441238.

Robbins, A., Tom, C. A. T. M. B., Cosman, M. N., Moursi, C., Shipp, L., Spencer, T. M., et al. (2018). Low temperature decreases bone mass in mice: Implications for humans. *American Journal of Physical Anthropology, 167*(3), 557–568. https://doi.org/10.1002/ajpa.23684.

Robling, A., Castillo, A., & Turner, C. (2006). Biomechanical and molecular regulation of bone remodeling. *Annual Review of Biomedical Engineering, 8*, 455–498. https://doi.org/10.1146/annurev.bioeng.8.061505.095721.

Robling, A., Hinant, F., Burr, D., & Turner, C. (2002). Improved bone structure and strength after long-term mechanical loading is greatest if loading is separated into short bouts. *Journal of Bone and Mineral Research, 17*(8), 1545–1554. https://doi.org/10.1359/jbmr.2002.17.8.1545.

Robling, A. G., & Stout, S. D. (2003). Histology, geometry, and mechanical loading in past populations. In *Bone loss and osteoporosis: An anthropological perspective* (pp. 189–206). Kluwer Academic/Plenum Publishers.

Robling, A., & Stout, S. (2008). Histomorphometry of human cortical bone: applications to age estimation. In *Biological anthropology of the human skeleton* (pp. 149–173). Wiley-Liss.

Rudman, K. E., Aspden, R. M., & Meakin, J. R. (2006). Compression or tension? The stress distribution in the proximal femur. *Biomedical Engineering Online, 5*(1), 1–7. https://doi.org/10.1186/1475-925X-5-12.

Ruff, C. B. (2008). Biomechanical analyses of archaeological human skeletons. In *Biological anthropology of the human skeleton* (pp. 183–206). John Wiley and Sons, Inc.

Ruff, C., & Hayes, W. (1983). Cross-sectional geometry of Pecos Pueblo femora and tibiae—A biomechanical investigation: I. Method and general patterns of variation. *American Journal of Physical Anthropology, 60*(3), 359–381. https://doi.org/10.1002/ajpa.1330600308.

Ruff, C., Holt, B., Niskanen, M., Sladek, V., Berner, M., Garofalo, E., et al. (2015). Gradual decline in mobility with the adoption of food production in Europe. *Proceedings of the National Academy of Sciences, 112*(23), 7147–7152. https://doi.org/10.1073/pnas.1502932112.

Ruff, C., Holt, B., & Trinkaus, E. (2006). Who's Afraid of the Big Bad Wolff?: "Wolff's Law" and Bone Functional Adaptation. *American Journal of Physical Anthropology, 129*, 484–498. https://doi.org/10.1002/ajpa.

Ruff, C., Larsen, C., & Hayes, W. (1984). Structural changes in the femur with the transition to agriculture on the Georgia coast. *American Journal of Physical Anthropology, 64*(2), 125–136. https://doi.org/10.1002/ajpa.1330640205.

Rühli, F. J., Kuhn, G., Evison, R., Müller, R., & Schultz, M. (2007). Diagnostic value of micro-CT in comparison with histology in the qualitative assessment of historical human skull bone pathologies. *American Journal of Physical Anthropology, 133*(4), 1099–1111. https://doi.org/10.1002/ajpa.20611.

Ryan, T. M., Carlson, K. J., Gordon, A. D., Jablonski, N., Shaw, C. N., & Stock, J. T. (2018). Human-like hip joint loading in Australopithecus africanus and Paranthropus robustus. *Journal of Human Evolution, 121*, 12–24. https://doi.org/10.1016/j.jhevol.2018.03.008.

Ryan, T. M., & Shaw, C. N. (2012). Unique suites of trabecular bone features characterize locomotor behavior in human and non-human anthropoid primates. *PLoS One, 7*(7). https://doi.org/10.1371/journal.pone.0041037, e41037.

Ryan, T. M., & Shaw, C. N. (2015). Gracility of the modern *Homo sapiens* skeleton is the result of decreased biomechanical loading. *Proceedings of the National Academy of Sciences, 112*(2), 372–377. https://doi.org/10.1073/pnas.1418641112.

Ryan, T. M., & Walker, A. (2010). Trabecular bone structure in the humeral and femoral heads of anthropoid primates. *Anatomical Record, 293*(4), 719–729. https://doi.org/10.1002/ar.21139.

Ryser, M. D., Nigam, N., & Komarova, S. V. (2009). Mathematical modeling of spatio-temporal dynamics of a single bone multicellular unit. *Journal of Bone and Mineral Research, 24*(5), 860–870. https://doi.org/10.1359/jbmr.081229.

Saers, J. P. P., Cazorla-Bak, Y., Shaw, C. N., Stock, J. T., & Ryan, T. M. (2016). Trabecular bone structural variation throughout the human lower limb. *Journal of Human Evolution, 97*, 97–108. https://doi.org/10.1016/j.jhevol.2016.05.012.

Saers, J. P. P., Ryan, T. M., & Stock, J. T. (2019). Trabecular bone functional adaptation and sexual dimorphism in the human foot. *American Journal of Physical Anthropology, 168*(1), 154–169. https://doi.org/10.1002/ajpa.23732.

Scherf, H., Wahl, J., Hublin, J. J., & Harvati, K. (2016). Patterns of activity adaptation in humeral trabecular bone in Neolithic humans and present-day people. *American Journal of Physical Anthropology, 159*(1), 106–115. https://doi.org/10.1002/ajpa.22835.

Schiessl, H., Willnecker, J., & Niemeyer, G. T. (1999). Muscle cross-sectional area and bone cross-sectional area in the human lower leg measured with peripheral computed tomography. *Musculoskeletal Interactions, 2*, 47–52.

Schlecht, S. H., Pinto, D. C., Agnew, A. M., & Stout, S. D. (2012). Brief communication: The effects of disuse on the mechanical properties of bone: What unloading tells us about the adaptive nature of skeletal tissue. *American Journal of Physical Anthropology, 149*(4), 599–605. https://doi.org/10.1002/ajpa.22150.

Schranz, D. (1959). Age determination from the internal structure of the humerus. *American Journal of Physical Anthropology, 17*(4), 273–277. https://doi.org/10.1002/ajpa.1330170403.

Seeman, E. (2002). Pathogenesis of bone fragility in women and men. *The Lancet, 359*(9320), 1841–1850. https://doi.org/10.1016/S0140-6736(02)08706-8.

Sevostianov, I., & Kachanov, M. (2000). Impact of the porous microstructure on the overall elastic properties of the osteonal cortical bone. *Journal of Biomechanics, 33*(7), 881–888. https://doi.org/10.1016/S0021-9290(00)00031-2.

Shaw, C. N., & Ryan, T. M. (2012). Does skeletal anatomy reflect adaptation to locomotor patterns? Cortical and trabecular architecture in human and nonhuman anthropoids. *American Journal of Physical Anthropology, 147*(2), 187–200. https://doi.org/10.1002/ajpa.21635.

Shaw, C. N., & Stock, J. T. (2013). Extreme mobility in the Late Pleistocene? Comparing limb biomechanics among fossil Homo, varsity athletes and Holocene foragers. *Journal of Human Evolution, 64*(4), 242–249. https://doi.org/10.1016/j.jhevol.2013.01.004.

Sherman, K. P. (2012). Metabolic bone disease. *Orthopaedics and Traumatology*, 26(3), 220–225. https://doi. org/10.1016/j.mporth.2012.04.003.

Siddiqui, J. A., & Partridge, N. C. (2016). Physiological bone remodeling: Systemic regulation and growth factor involvement. *Physiology*, 31(3), 233–245. https://doi.org/10.1152/physiol.00061.2014.

Sims, N. A., & Martin, T. J. (2014). Coupling the activities of bone formation and resorption: A multitude of signals within the basic multicellular unit. *Bonekey Reports*, 3. https://doi.org/10.1038/bonekey.2013.215, 481.

Sims, N. A., & Gooi, J. H. (2008). Bone remodeling: Multiple cellular interactions required for coupling of bone formation and resorption. *Seminars in Cell & Developmental Biology*, 19(5), 444–451. https://doi.org/10.1016/j. semcdb.2008.07.016.

Skedros, J. G., Keenan, K. E., Williams, T. J., & Kiser, C. J. (2013). Secondary osteon size and collagen/lamellar organization ("osteon morphotypes") are not coupled, but potentially adapt independently for local strain mode or magnitude. *Journal of Structural Biology*, 181(2), 95–107.

Skinner, M. M., Stephens, N. B., Tsegai, Z. J., Foote, A. C., Nguyen, N. H., Gross, T., et al. (2015). Human-like hand use in Australopithecus africanus. *Science*, 347(6220), 395–399. https://doi.org/10.1126/science.1261735.

Snoddy, A. M. E., Miszkiewicz, J. J., Cooke, K. M., Petchey, P., & Buckley, H. (2022). Bone remodelling changes in an individual with tuberculosis induced left-sided femoro-acetabular joint destruction, from 19th century Milton, New Zealand. *Bioarchaeology International*. https://doi.org/10.5744/bi.2021.0010. In press.

Stauber, M., Rapillard, L., Van Lenthe, G. H., Zysset, P., & Müller, R. (2006). Importance of individual rods and plates in the assessment of bone quality and their contribution to bone stiffness. *Journal of Bone and Mineral Research*, 21(4), 86–595. https://doi.org/10.1359/jbmr.060102.

Stephens, N. B., Kivell, T. L., Gross, T., Pahr, D. H., Lazenby, R. A., Hublin, J. J., et al. (2016). Trabecular architecture in the thumb of Pan and Homo: Implications for investigating hand use, loading, and hand preference in the fossil record. *American Journal of Physical Anthropology*, 161(4), 603–619. https://doi.org/10.1002/ajpa.23061.

Stephens, N. B., Kivell, T. L., Pahr, D. H., Hublin, J. J., & Skinner, M. M. (2018). Trabecular bone patterning across the human hand. *Journal of Human Evolution*, 123, 1–23. https://doi.org/10.1016/j.jhevol.2018.05.004.

Stewart, T. J., Louys, J., & Miszkiewicz, J. J. (2021). Intra-skeletal vascular density in a bipedal hopping macropod with implications for analyses of rib histology. *Anatomical Science International*, 96(3), 386–399. https://doi. org/10.1007/s12565-020-00601-0.

Stieglitz, J., Madimenos, F., Kaplan, H., & Gurven, M. (2016). Calcaneal quantitative ultrasound indicates reduced bone status among physically active adult forager-horticulturalists. *Journal of Bone and Mineral Research*, 31(3), 663–671. https://doi.org/10.1002/jbmr.2730.

Stock, J. T., & Shaw, C. N. (2007). Which measures of diaphyseal robusticity are robust? A comparison of external methods of quantifying the strength of long bone diaphyses to cross-sectional geometric properties. *American Journal of Physical Anthropology*, 134(3), 412–423. https://doi.org/10.1002/ajpa.20686.

Stout, S. D. (1978). Histological structure and its preservation in ancient bone. *Current Anthropology*, 19(3), 601–604. https://doi.org/10.1086/202141.

Stout, S. D., Cole, M. E., & Agnew, A. M. (2019). Histomorphology: Deciphering the metabolic record. In *Ortner's identification of pathological conditions in human skeletal remains* (pp. 91–167). Academic Press.

Stout, S. D., & Crowder, C. (2011). Bone remodeling, histomorphology, and histomorphometry. In *Bone histology: An anthropological perspective* (pp. 1–21). CRC Press.

Stout, S., & Lueck, R. (1995). Bone remodeling rates and skeletal maturation in three archaeological skeletal populations. *American Journal of Physical Anthropology*, 98(2), 161–171. https://doi.org/10.1002/ajpa.1330980206.

Stout, S. D., & Paine, R. R. (1994). Bone remodeling rates: A test of an algorithm for estimating missing osteons. *American Jurnal of Physical Anthropology*, 93(1), 123–129. https://doi.org/10.1002/ajpa.1330930109.

Su, A., & Carlson, K. J. (2017). Comparative analysis of trabecular bone structure and orientation in South African hominin tali. *Journal of Human Evolution*, 106, 1–18. https://doi.org/10.1016/j.jhevol.2016.12.006.

Su, X., Sun, K., Cui, F., & Landis, W. (2003). Organization of apatite crystals in human woven bone. *Bone*, 32(2), 150–162. https://doi.org/10.1016/S8756-3282(02)00945-6.

Sukhdeo, S., Parsons, J., Niu, X. M., & Ryan, T. M. (2018). Trabecular bone structure in the distal femur of humans, apes, and baboons. *Anatomical Record*, 149, 129–149. https://doi.org/10.1002/ar.24050.

Sylvester, A. D., & Terhune, C. E. (2017). Trabecular mapping: Leveraging geometric morphometrics for analyses of trabecular structure. *American Journal of Physical Anthropology*, 163(3), 553–569. https://doi.org/10.1002/ajpa.23231.

Tella, S. H., & Gallagher, J. C. (2014). Prevention and treatment of postmenopausal osteoporosis. *The Journal of Steroid Biochemistry and Molecular Biology, 142*, 155–170. https://doi.org/10.1016/j.jsbmb.2013.09.008.

Thompson, D. D., & Gunness-Hey, M. (1981). Bone mineral-osteon analysis of Yupik-inupiaq skeletons. *American Journal of Physical Anthropology, 55*(1), 1–7. https://doi.org/10.1002/ajpa.1330550102.

Tsegai, Z. J., Kivell, T. L., Gross, T., Huynh Nguyen, N., Pahr, D. H., Smaers, J. B., et al. (2013). Trabecular bone structure correlates with hand posture and use in hominoids. *PLoS One, 8*(11). https://doi.org/10.1371/journal.pone.0078781, e78781.

Tsegai, Z. J., Skinner, M. M., Gee, A. H., Pahr, D. H., Treece, G. M., Hublin, J. J., et al. (2017). Trabecular and cortical bone structure of the talus and distal tibia in Pan and Homo. *American Journal of Physical Anthropology, 163*(4), 784–805. https://doi.org/10.1002/ajpa.23249.

Turner, C. (1992). On Wolff's law of trabecular architecture. *Journal of Biomechanics, 25*(1), 1–9. https://doi.org/10.1016/0021-9290(92)90240-2.

Turner, C. (1998). Three rules for bone adaptation to mechanical stimuli. *Bone, 23*(5), 399–407. https://doi.org/10.1016/S8756-3282(98)00118-5.

Umemura, Y., Ishiko, T., Yamauchi, T., Kurono, M., & Mashiko, S. (1997). Five jumps per day increase bone mass and breaking force in rats. *Journal of Bone and Mineral Research, 12*(9), 1480–1485. https://doi.org/10.1359/jbmr.1997.12.9.1480.

van Oers, R. F. M., Ruimerman, R., van Rietbergen, B., Hilbers, P. A. J., & Huiskes, R. (2008). Relating osteon diameter to strain. *Bone, 43*(3), 476–482. https://doi.org/10.1016/j.bone.2008.05.015.

Villotte, S., & Knüsel, C. J. (2013). Understanding entheseal changes: Definition and life course changes. *International Journal of Osteoarchaeology, 23*(2), 135–146. https://doi.org/10.1002/oa.2289.

von Meyer, G. H. (1867). Die architecktur der spongiosa. *Archiv Fur Anatomie, Physiologie Und Weissenschaftliche Medicin, 34*, 615–628.

von Scheven, E. (2007). Pediatric bone density and fracture. *Current Osteoporosis Reports, 5*(3), 128–134. https://doi.org/10.1007/s11914-007-0028-7.

Walker, M. M., Oxenham, M. F., Huong Nguyen, T. M., Trinh, H. H., Minh, T. T., Cuong Nguyen, L., et al. (2022). Primary bone retention in a young adult male with limb disuse: A bioarchaeological case study. *Historical Biology*. https://doi.org/10.1080/08912963.2022.2032027. In press.

Wallace, I. J., Judex, S., & Demes, B. (2015). Effects of load-bearing exercise on skeletal structure and mechanics differ between outbred populations of mice. *Bone, 72*, 1–8. https://doi.org/10.1016/j.bone.2014.11.013.

Walsh, J. S. (2015). Normal bone physiology, remodelling and its hormonal regulation. *Surgery, 33*(1), 1–6. https://doi.org/10.1016/j.mpsur.2014.10.010.

Wang, Q., & Seeman, E. (2008). Skeletal growth and peak bone strength. *Best Practice & Research Clinical Endocrinology & Metabolism, 22*(5), 687–700. https://doi.org/10.1016/j.beem.2008.07.008.

Ward, F. O. (1838). *Outlines of human osteology*. Henry Renshaw.

Wein, M. N. (2017). Bone lining cells: Normal physiology and role in response to anabolic osteoporosis treatments. *Current Molecular Biology Reports, 3*(2), 79–84. https://doi.org/10.1007/s40610-017-0062-x.

Weiner, S., Traub, W., & Wagner, H. D. (1999). Lamellar bone: Structure-function relations. *Journal of Structural Biology, 126*(3), 241–255.

Williamson, L., Hayes, A., Hanson, E. D., Pivonka, P., Sims, N. A., & Gooi, J. H. (2017). High dose dietary vitamin D3 increases bone mass and strength in mice. *Bone Reports, 6*, 44–50. https://doi.org/10.1016/j.bonr.2017.02.001.

Wolff, J. (1892). *Das Gesetz der Transformation der Knochen*. A. Hirchwild.

Wolff, J. (1986). *The Law of Bone Remodeling (English Translation)*. Springer-Verlag.

Wozney, J. M., Rosen, V., Byrne, M., Celeste, A. J., Moutsatsos, I., & Wang, E. A. (1990). Growth factors influencing bone development. *Journal of Cell Science, 13*, 149–156. https://doi.org/10.1242/jcs.1990.Supplement_13.14.

Yang, J., Zhou, S., Wei, M., Fang, Y., & Shang, P. (2021). Moderate static magnetic fields prevent bone architectural deterioration and strength reduction in ovariectomized mice. *IEEE Transactions on Magnetics, 57*(7), 1–9. https://doi.org/10.1109/TMAG.2021.3072148.

Yazdani, A., Stephens, N. B., Cherukuri, V., Ryan, T., & Monga, V. (2019). Domain-enriched deep network for micro-CT image segmentation. In *53rd Asilomar conference on signals, systems, and computers* (pp. 1867–1871).

Biosocial complexity and the skull

Suzy White[a,b] and Lumila Paula Menéndez[c,d]

[a]School of Biological Sciences, University of Reading, Reading, United Kingdom, [b]Department of Anthropology, University College London, London, United Kingdom, [c]Konrad Lorenz Institute for Evolution and Cognition Research, Klosterneuburg, Austria, [d]Department Anthropology of the Americas, University of Bonn, Bonn, Germany

3.1 Introduction

The skull is a highly complex structure. It is composed of many bones that have different embryological origins, perform various functions, and house numerous sensory organs, as well as the brain. The face has a relevant role in verbal and nonverbal communication, and head shaping practices can be considered as a marker of identity (Lieberman, 2011; Torres-Rouff, 2002), thus our face and head offer key social and behavioural signals to communicate with our contemporaries. Given this complexity, it is unsurprising that the skull is shaped by both biological and sociocultural factors. In addition to morphological differences due to sexual dimorphism (Garvin and Ruff, 2012; Gonzalez et al., 2011; Rosas and Bastir, 2002; Toledo Avelar et al., 2017), the strong effect of population and evolutionary history on differentiating individuals from different geographical groups (Harvati and Weaver, 2006a; Roseman, 2016; von Cramon-Taubadel, 2014), phenotypic plasticity (Collard and Wood, 2007; Menéndez et al., 2014; von Cramon-Taubadel, 2009b), and the influence of environmental factors, both ecological and cultural (Freidline et al., 2015; Vioarsdóttir et al., 2002; Williams and Cofran, 2016), play relevant roles in shaping the skull. All of these aspects have the potential to inform us about how biological processes and human behaviour, past and present, have interacted to influence our cranial morphology.

Human behaviour encompasses many different scenarios, some of which can be accessed through the study of skull morphology, as we will illustrate in this chapter. For instance, the study of some cranial structures provides a window into the history of our species' demographic expansions and migrations. Evidence of hybridisation preserved in the cranium tells us about the social interactions and mating relationships of the earlier members of our species with individuals from other species (e.g. Neanderthals, Denisovans). When comparing different populations inhabiting different environments, we can find some morphological changes that are associated with different

Behaviour in our Bones
https://doi.org/10.1016/B978-0-12-821383-4.00008-5

climates, altitudes, and diets. And phenomena such as intentional modification and self-domestication show us how culture and communication interact with biology to change craniofacial morphology across different timescales.

In this chapter, we start by reviewing some foundational notions of embryological development, modularity, and integration, to gain a better understanding of the major components of the skull. Then, we explore the main evolutionary processes associated with the origin and dispersals of humans (e.g. population history, gene flow, hybridisation), which should be taken into consideration before exploring changes that may result from ecological or behavioural effects. Next, we discuss the ecological factors that have the strongest impact on shaping the evolution of the skull (e.g. diet shifts, climate, and altitudinal differences). Lastly, we introduce examples of behavioural changes such as cultural modifications of the skull, as well as models proposed to explain the trend of craniofacial gracilisation present in our species, and the impact of nonverbal and verbal communication

3.2 The skull

3.2.1 Anatomy of the skull

The skull represents the most conspicuous derived trait that defines vertebrates as a taxon (Kuratani, 2019). It derives from the neural crest, a key innovation that evolved about 500 million years ago (Hall, 1999). This complex structure is the bony foundation for the senses of sight, smell, taste, and hearing, in addition to housing and protecting the brain, and forming the framework for the masticatory apparatus (White et al., 2011). The skull is comprised of 29 independent bones, including the hyoid and the three pairs of ear ossicles, which are often considered as part of the cranium due to their embryological and developmental origin (Meikle, 2002). Additionally, some individuals may also have intrasutural bones (i.e. irregular ossicles occurring along the sutures), which are especially common in skulls that are shaped by cultural modifications (see Section 3.6.1).

Although many terms can refer to specific parts of the skull, based on embryological and developmental origin, anatomists generally distinguish three major regions: the cranial base (basicranium), cranial vault (neurocranium), and the facial skeleton (viscerocranium or splanchnocranium) (Atchley and Hall, 1991). The cranial base includes parts of the occipital, temporals, sphenoid, and ethmoid, and ossifies endochondrally; the cranial vault includes the temporals, parietals, sphenoid, frontal, and the superior part of the occipital, and grows via intramembranous ossification; the facial skeleton incorporates the maxillae, palatines, vomer, inferior nasal conchae, lacrimals, nasals, zygomatics, ethmoid, mandible, and parts of the sphenoid, and derives both endochondrally and from intramembranous bone and cartilage (Lieberman, 2011). Alternative and widely applied terminology, mostly used to illustrate preservation and/or completeness, include the following: cranium, referring to the skull without the mandible; calvaria, which depicts the cranium without the face; and calotte, for describing the calvaria without the base (White et al., 2011).

The embryological development of the skull starts when the single diploid cell (zygote) divides repeatedly, creating the three layers of cells that give rise to all the tissues of the body: the ectoderm, mesoderm, and endoderm (Gilbert, 2006; Lieberman, 2011). The ectoderm cells differentiate into a central neural plate, whose edges thicken and move upward to form the neural folds (Gilbert, 2006). The neural folds migrate towards the midline of the embryo and fuse to form the neural tube. The dorsal part differentiates into neural crest cells, while the more externally positioned part becomes the epidermis. The neural crest is considered such an important structure that it has sometimes been identified as the fourth germ layer (Hall, 1999). The cranial neural crest cells migrate between the epidermis and the deeper mesoderm, differentiating into mesenchymal cells (Gilbert, 2006). The more rostral ones give rise to the frontonasal prominences that contribute to the middle and upper face, while the most caudal ones form the branchial arches that will generate: the lower face; bones of the middle ear; hyoid cartilages; and the thymus, parathyroid, and thyroid glands. The paraxial mesodermal cells split into 42 blocks called somites, from which the first four form the occipital bone, tooth roots, and related tissues (Gilbert, 2006; Lieberman, 2011).

3.2.2 Modularity and integration

Skull morphology is affected by modularity and integration, two closely related concepts within biology. A module is a unit or subset of elements that are closely related by interactions within the subset, yet are separated from other distinct modules by few or weak interactions between them (Klingenberg, 2008). In contrast, integration refers to the cohesion and covariation among traits resulting from interactions of common biological processes affecting them (e.g. pleiotropy, ontogeny, allometry) (Cheverud, 1996; Klingenberg, 2008; Olson and Miller, 1958). The cranial base, vault, and facial skeleton are examples of modules at a broad scale: they develop from different parts of the embryo, have different functionalities, and, although they become integrated during development, remain structurally independent (Cheverud, 1996; Hallgrímsson et al., 2007). Morphological integration is inevitable within the skull due to the number of bones incorporated within it and the multiple functions shared between them (Klingenberg, 2013). For instance, the bony orbit is formed from various aspects of seven different bones and is the area of connection between the neurocranial vault and the face, with the upper border also acting as the inferior aspect of the anterior cranial fossa (Lieberman, 2011). As a result, selective pressures acting on one element (e.g. the eye) are going to impact numerous others even if they are under separate functional, developmental, or mechanical constraints, leading to a cascading effect of changes on skull morphology. Modularity and integration are suggested to increase the evolvability of the skull, among other complex structures such as the hands and feet (Cheverud, 1996; Klingenberg, 2005; Rolian, 2009; Wagner and Altenberg, 1996), adding to the complexity and mosaicism of morphological variation in this region (Cheverud, 1982). Modularity and integration interact with other biosocial factors discussed below, such as hybridisation (Parr et al., 2016), dietary habits (Lieberman, 2011), climatic adaptation (Bastir and Rosas, 2013), cultural cranial modifications (Cheverud, 1996; Püschel et al., 2020), and the evolution of recent modern human skull morphology (Profico et al., 2017). As a consequence, different patterns of intra- and inter-population variation are produced, which contribute to the diversification between populations and taxa (Goswami and Polly, 2010; Goswami et al., 2014; Gunz and Harvati, 2007).

3.3 Origins and dispersals

3.3.1 Evolution of the human skull

The modern human skull is particularly distinctive among both hominins and extant primates. We have an expanded, globular braincase; a high position of maximum cranial breadth (on the parietals), frequently associated with parietal bossing; a noncontinuous ('bipartite') browridge, divided into lateral and medial portions by an oblique sulcus; a reduced dental arcade with small teeth; increased overall gracility, including a thin cranial vault; and relatively small, flat faces (Bookstein et al., 1999; Day and Stringer, 1982; Lieberman, 1996; Lieberman et al., 2002; Schwartz, 2016; Tattersall and Schwartz, 2008). Many, and possibly all, of these traits may be explained at least partially by two key developments in our evolution: neurocranial globularity and retraction of the face underneath the anterior cranial fossa (Lieberman et al., 2002). The first of these appears to occur through an early phase of 'globularisation' during development (Gunz et al., 2012); the second may be a consequence of the first, which requires structural changes in the cranial base (Bruner et al., 2011; Lieberman et al., 2004a; McCarthy and Lieberman, 2001; Neaux et al., 2015); and both may be linked to our expanded brains (Ruff et al., 1997).

Despite these unifying cranial features, our fossil record encompasses a fair amount of variability. For instance, the fossils from Jebel Irhoud, Morocco (315 thousand years ago; kya) (Richter et al., 2017), suggested to be possible early members of *Homo sapiens*, do not show all of the traits listed above; while their faces are relatively modern, their neurocrania are not as globular as later members of our species (Hublin et al., 2017). The earliest east African fossils from Omo Kibish, Ethiopia (230 kya) (Vidal et al., 2022) indicate that early *H. sapiens* may have been highly polymorphic, with some individuals retaining 'archaic' skull morphology (e.g. long, low neurocrania, and continuous browridges; see Fig. 3.1) alongside their more modern contemporaries. Anatomical modernity was not fixed in our species until quite late, with fossils from Skhūl and Qafzeh (Israel) showing a mosaic of derived and archaic features up until 90 kya (Schwartz, 2016; Tattersall and Schwartz, 2008). Such differences have led researchers to separate 'early' and 'anatomically modern' members of our species (Pearson, 2008; Schwartz, 2016; Trinkaus, 1982a), with the latter being those which show the full suite of our derived craniofacial characteristics (but see Stringer and Buck, 2014, for a discussion of these concepts).

It is clear that, as with all aspects of biology, our cranial evolution has not been a simple linear progression from a more archaic to a more modern form. For instance, late archaic (i.e. not anatomically modern) humans appear to have persisted at Ishango (Democratic Republic of the Congo) around 20 kya (Crevecoeur et al., 2016), Maludong (China) at 14 kya (Curnoe et al., 2012, 2015), and Iwo Eleru (Nigeria) at 12 kya (Harvati et al., 2011; Stojanowski, 2014), indicating deep population substructure in our recent past. The picture is equally complex in Australia, with the earliest fossils showing a coexistence of individuals with more archaic and more derived traits between 40 and 15 kya (Thorne and Wilson, 1977), some of which do not fulfil one of the only proffered morphological definitions of our species (Day and Stringer, 1982), despite clearly being *H. sapiens* (Wolpoff, 1986). This pattern of morphological variability also continues to a lesser extent into the current day, with marked variability in expression of our species-specific cranial traits (Lahr, 1996). This highlights the complexity of the human skull, and the difficulties in creating

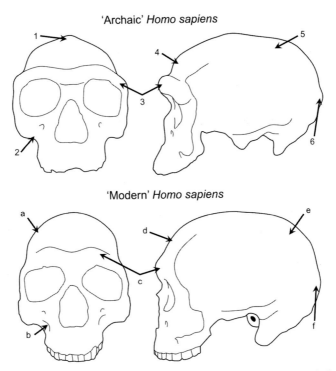

FIG. 3.1   Comparison of 'archaic' (top) and 'modern' (bottom) *Homo sapiens* skull morphology from front and left lateral view. In contrast to modern *H. sapiens*, archaic *H. sapiens* have (1) variable presence of sagittal keeling, vs (a) domed neurocrania with more pronounced parietal bossing; (2) absent or slight development of the canine fossa, vs (b) frequent presence of deep canine fossae; (3) strong supraorbital tori which are continuous across the brow, vs (c) browridges which are mainly limited to development of the supraciliary arches, and separated from the lateral trigone by a sulcus (i.e. bipartite or noncontinuous); (4) more receding frontal squamae, vs (d) more vertical frontal squamae; (5) longer, lower neurocrania, vs (e) globular neurocrania; (6) more angled occipital bones with variable presence of hemibunning and false suprainiac fossae, vs (f) rounded occipital squamae.

a morphological definition of our species that is universally applicable to recent modern humans, let alone earlier members of our species.

3.3.2 Population history

Natural selection, gene flow, developmental plasticity, and random evolutionary processes such as mutations and genetic drift have impacted our cranial evolution (Roseman, 2016; von Cramon-Taubadel, 2014). All of these factors have driven taxonomic and within-species diversification in primates and hominins (Ackermann and Cheverud, 2004; Schroeder and Ackermann, 2017; Schroeder and von Cramon-Taubadel, 2017; Zichello et al., 2018), although they have not always acted in conjunction or with the same magnitude. For instance, based on modelling of accumulation of differences in craniofacial morphology, the divergence between Neanderthal and modern human craniofacial morphology appears to have been mostly the result of genetic drift (Pearson, 2013; Weaver et al., 2007, 2008; Weaver and Stringer, 2015) (but see Box 3.1).

BOX 3.1

Is the Neanderthal face adaptive to cold environments, heavy biting, or the result of random factors?

The Neanderthal face presents distinctive features when compared to contemporary humans (Fig. 3.2). It is characterised by prognathism (i.e. forward protrusion of the lower face), inflation of the midface, a wide and tall nasal aperture, a depressed nasal floor, a wide, projecting nasal bridge, a retromolar gap, a continuous and double-arched browridge, 'swept back' zygomatic arches, and large anterior teeth (Franciscus, 1999; Harvati, 2010; Trinkaus, 1987; Wroe et al., 2018). There is a long-standing discussion among palaeoanthropologists on whether such an autapomorphic (i.e. derived traits that are unique to a given taxon) face is the result of genetic drift or if it represents adaptations relating to cold temperatures and/or heavy biting demands.

The anterior dental loading hypothesis states that the Neanderthal face could be interpreted as a biomechanical consequence of intense paramasticatory behaviour (e.g. using teeth as tools) (e.g. Smith, 1983), as evidenced by their unusual anterior dental wear pattern (Trinkaus, 1987) equivalent to modern human populations from the arctic who eat hard foods (Hylander, 1977). Some scholars argue that their midfacial prognathism represents a trade-off between demands for high bite force at the anterior teeth and increasing the functional surface area of the molars for mastication of resistant foods (Spencer and Demes, 1993). Quantitative studies indicate that Neanderthal skull and muscular morphology was unable to produce, and inefficient at resisting, high

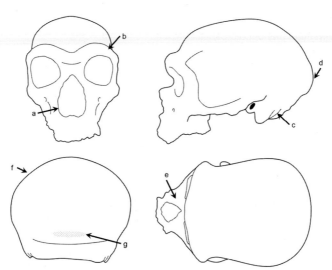

FIG. 3.2 Diagram of *Homo neanderthalensis* skull from (clockwise from top left) frontal, left lateral, superior, and posterior view. When compared to *Homo sapiens*, Neanderthal traits include (a) enlarged nose; (b) rounded, double-arched brow ridges; (c) large juxtamastoid processes; (d) occipital bunning; (e) projecting/inflated midface; (f) rounded 'en bombe' cranial profile in rear view; (g) suprainiac fossa.

(Continued)

BOX 3.1 (*cont'd*)

bite forces (Antón, 1994; Trinkaus, 1987). This, in combination with the finding that the wear on their anterior teeth was comparable to Late Pleistocene modern humans (Clement et al., 2012), weakens the explanatory power of this hypothesis.

The large nasal opening and voluminous nasal cavity in *Homo neanderthalensis* has long been proposed to relate to cold climate adaptations, since it might function in warming and humidifying inspired air, as well as dissipating heat (Coon, 1962; Hubbe et al., 2009). In fact, their midfacial prognathism has been interpreted as the result of enlarged paranasal sinuses that could warm and humidify inspired air (de Azevedo et al., 2017; Wroe et al., 2018). However, it has been argued that a narrower nasal cavity, such as that present

in recent human populations from cold climates, is more effective in warming and humidifying air, contradicting the traditional model of cold adaptation of the Neanderthal nose (Evteev et al., n.d.; Noback et al., 2016; Rae et al., 2011).

Finally, other researchers argue that, instead of selective pressures, random changes over generations (i.e. genetic drift; see Section 3.3.2) may have influenced the Neanderthal facial morphology (Hublin, 1998; Weaver et al., 2007). Despite several decades of research, the factors behind the Neanderthal face shape remain unclear, although the idea that the Neanderthal face may have arisen through a mosaic of evolutionary processes and specific behaviours is gaining recognition.

Numerous studies have shown that recent human cranial morphology matches neutral molecular variation and thus reflects stochastic evolutionary processes (Harvati and Weaver, 2006a,b; Relethford, 2002; Roseman, 2004; Roseman and Weaver, 2004, 2007). This means that that we can use cranial morphology to reconstruct previous population movement and dispersals. Particularly, both genetic (Cann et al., 1987; Nielsen et al., 2017) and cranial variation (Stringer and Andrews, 1988) support a single African origin model, with a possibility of multiple waves of dispersal to the rest of the world (Manica et al., 2007; Reyes-Centeno et al., 2014). Within-group variation in recent human skulls decreases with increasing geographical distance from Africa, which supports a model of isolation by distance from an African 'source' (i.e. successive geographically limited dispersals, leading to the increasing biological distance between the source and locations of later dispersals) (Betti et al., 2009; Manica et al., 2005, 2007; von Cramon-Taubadel and Lycett, 2008). According to this model, genetic drift explains most of the variation between human populations across the world. Our species went through a series of founder and bottleneck events during our dispersal from Africa, followed by rapid population expansion after 100 kya (Harpending and Rogers, 2000; Ramachandran et al., 2005; von Cramon-Taubadel and Lycett, 2008), making our population history arguably unique among extant apes (Zichello et al., 2018).

Neutral evolutionary processes only explain part of our history, however, and account for less variance in cranial morphology than in neutral molecular data (Manica et al., 2007; Smith, 2009). This is partly due to the lower heritability (h^2) of cranial traits, which is estimated to be around 0.55 (i.e. 55% of cranial variability can be solely attributed to genetic differences among individuals) (Carson, 2006; Martínez-Abadías et al., 2009). The environment also plays a relevant role, with differing selective pressures detected between groups (see Section 3.5) (Roseman, 2004). However, as the influence of environmental factors is also geographically structured, it could be difficult to separate out neutral evolution from natural selection (Smith, 2009). There is also differential preservation of population history signals across the skull. For instance, studies suggest that the facial skeleton is more influenced by selective pressures, and is thus a poorer indicator of population history, than the neurocranial vault or temporal bone (Harvati and Weaver, 2006a; Ponce de León et al., 2018; Roseman and Weaver, 2004). Differential patterning of evolutionary information across the skull is, unsurprisingly, intricate with separate traits in the same regions preserving their own signatures of population history (Betti et al., 2009; Harvati and Weaver, 2006a; Smith, 2009; von Cramon-Taubadel, 2009a). Due to the complexity of the skull, factors such as convergence and parallelism have to be considered (Collard and Lycett, 2008; Collard and Wood, 2000), along with natural selection, plasticity, and neutral evolutionary processes, when attempting to glean biological information from the skull.

3.3.3 Gene flow and hybridisation

As a result of frequent human migrations across the world, gene flow between individuals from different populations and hybridisation between individuals from different hominin species during the Pleistocene have increased the genotypic and phenotypic variation within our species. This gives us important insight into the social and mating interactions between our ancestors and our close relatives when we were not the only hominin species on the planet. Following the identification of Neanderthal DNA in our genome in 2010 (Green et al., 2010), numerous ancient DNA studies have revealed a complex history of interbreeding between early modern humans, Neanderthals, Denisovans, and some as-of-yet unidentified 'ghost' hominin lineages (Fu et al., 2015; Hammer et al., 2011; Meyer et al., 2012; Racimo et al., 2015; Slon et al., 2018). Current estimates indicate that modern humans have between 1.6% and 2.1% of DNA deriving from Neanderthal lineages (Prüfer et al., 2014) with up to 70% of the Neanderthal genome being preserved across *H. sapiens* (Vernot and Akey, 2014). In fact, some researchers have suggested that interbreeding with other hominins may have led to the development and success of our own species, both in terms of culture and physiology (Ackermann et al., 2016; Enard and Petrov, 2018; Gittelman et al., 2016; Huerta-Sánchez et al., 2014) (see Box 3.2).

At present, interdisciplinary work combining morphological and genetic analysis is the best way to identify signals of past hybridisation in fossil humans. The morphological changes caused by hybridisation can serve as a guide to identifying admixture in the fossil record using phenotypes alone, a necessary preliminary step to understanding the dynamics of past gene exchange (Ackermann et al., 2019). Fortunately, the skull has been shown to have reliable morphological indicators of recent admixture in macaques, baboons, and gorillas (Ackermann and Bishop, 2010; Ackermann et al., 2006, 2014; Boel et al., 2019; Eichel and Ackermann, 2016), as interbreeding between distinct lineages leads to disruption in

BOX 3.2

Denisovans and skeletal adaptations to high altitude

During the last decade, genetic studies have shown that hybridisation between humans and other hominin species occurred fairly frequently in Eurasia (see Section 3.3.3) (Jacobs et al., 2019; Villanea and Schraiber, 2019). As a result of such admixture, humans acquired genetic variants that contributed to our current phenotypic variation (Ackermann et al., 2016; Enard and Petrov, 2018; Gittelman et al., 2016). If alleles are maintained through selection, they have great potential to mediate rapid genetic adaptations when the population encounters new environments (Rees et al., 2020). In fact, some scholars argue that most of the introgressed variants in recent humans are functionally involved in immunity and metabolism-related traits (Gokcumen, 2020). For instance, some of the high-altitude adaptation features in Tibetans are linked to a hypoxia pathway gene, EPAS1, which is associated with differences in haemoglobin concentration at high altitude. Huerta-Sánchez et al. (2014) showed that the EPAS1 gene is present in high frequencies in populations from the Tibetan plateau, very low in Han Chinese, and absent in all other populations. Moreover, since this genetic region has an unusual haplotype structure, and the divergence between the Tibetan and the Han haplotypes is greater than expected for comparisons among recent humans, the authors conclude that its presence can only be interpreted as result of DNA introgression from hybridisation between Denisovans and early *H. sapiens*.

Despite Melanesians having the largest proportions of Denisovan components in their genome (Meyer et al., 2012; Reich et al., 2010), the EPAS1 haplotype is not observed among contemporary Melanesians, nor in the high-coverage Neanderthal genome sequence from the Altai mountains (Prüfer et al., 2014), where Denisova cave is located. This confirms that the persistence of this beneficial high-altitude variant among Tibetans can be explained as a result of recent positive selection for the traits. Unfortunately, Denisovan fossils are scarce, currently represented by a hemimandible (Chen et al., 2019), a partial parietal (Viola et al., 2019), and various teeth and postcranial fragments. Future studies on hybrid or potentially Denisovan individuals (Martinón-Torres et al., 2019), as well as more complete fossils, will allow further exploration of craniofacial and/ or skeletal adaptations to high altitude. For instance, current Tibetan populations living in high altitude present relatively larger internal nasal breadths and heights for their facial size (Butaric and Klocke, 2018), as well as largest chest circumferences (i.e. wider and/or deeper thorax) (Weitz et al., 2004) when compared to nearby low-land populations. Systematic excavations in Asia would allow recovering more Denisovan fossils to test hypothesis on how these morphological features originated and developed throughout human history.

developmental processes. This can manifest itself in quantifiable ways, such as an increase in variability in both morphology and size (heterosis), and therefore robusticity, and the presence of discrete nonmetric traits, including additional sutures, ossicles, or supernumerary teeth (Ackermann et al., 2019). Assessment of recent *H. sapiens* fossils shows a relatively high incidence of developmental anomalies, suggested to be linked to both inbreeding and interbreeding (Ackermann et al., 2016; Trinkaus, 2018). While genetic analyses are limited by the preservation of ancient DNA, which decays with increasing time depth and in relation to ecological factors such as warmth and humidity, they provide direct evidence of gene introgression. Over the next decades, promising studies will be based on interdisciplinary research investigating morphological signatures of hybridisation further, offering a window into the complex history of interactions between our species and our close relatives when DNA evidence is not available.

3.4 Transition to agriculture

3.4.1 Craniofacial gracilisation and globularisation

The transition to agriculture was a complex process that occurred independently in different regions of the world between 10,000 and 4000 years BP. The development of agriculture resulted in a dietary shift that initiated several changes to our skulls, especially in craniofacial and mandibular features directly involved in mastication (Fig. 3.3). Overall, compared to preagricultural people, farmers present shorter, more posteriorly and inferiorly placed temporalis muscles, and smaller mastoid processes and nuchal planes (Cheronet et al., 2016; Noback and Harvati, 2015; Paschetta et al., 2010; Sardi et al., 2004, 2006). Also, they present a reduction in the maxillomandibular complex, with narrower alveolar processes, modestly taller palates, shorter dental arcades, and more posteriorly displaced dentition due to the reduced length of the maxilla (Katz et al., 2017; Perez et al., 2011). The neurocranial vault becomes taller and more globular (see Section 3.3.1) relative to the lower, narrower, and smaller faces (Carlson and Van Gerven, 1977; González-José et al., 2005; Larsen, 1995; Menéndez et al., 2014; von Cramon-Taubadel, 2011). Complementarily, the mandible presents a reduction in size and a trend towards a taller ramus (Pokhojaev et al., 2019; Sella-Tunis et al., 2018). In addition, a gracilisation trend has been described in populations that have shifted to an agricultural diet (Fukase and Suwa, 2008; Galland et al., 2016; Kaifu, 1997; Katz et al., 2017; Pokhojaev et al., 2019) (see Section 3.5.1).

The morphological changes associated with the transition to agriculture have been explained by two main mechanisms: biomechanical changes due to alterations in loading, and/or systematic changes due to adjustments in diet or physical activity (Menéndez and Buck, 2022). Biomechanical explanations usually invoke a decrease in masticatory stress due to changes in food preparation techniques towards producing softer food (cooking, grinding, etc.) and/or the reduction or complete replacement of meat and fibre, resulting in a diet with a large proportion of softer, grain-based foods (food consistency) (e.g. the Masticatory-Functional Hypothesis, Carlson and Van Gerven, 1977). Systematic explanations propose that the morphological changes in populations transitioning to an agricultural diet may be related to a lack of protein and other nutrients experienced during development, at the expense of an increase in the consumption of carbohydrates, as well as changes in activity patterns (Sardi et al., 2004; Smith et al., 1984; Stini,

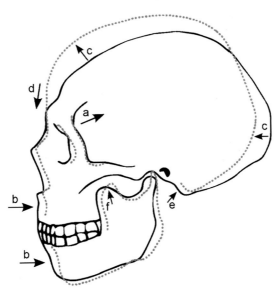

FIG. 3.3 Morphological changes in the skull and mandible observed in hunter-gatherers (black solid line) when compared to agriculturalists (dashed grey line). The latter present (a) a reduction in the size and a relatively more posterior site of origin of the muscles of mastication; (b) a reduced anteroposterior growth of the maxillomandibular complex; (c) a reduction in the relative length and increase in the relative height of the cranial vault, which becomes more globular, and reduction of the nuchal crest; (d) a reduction in the size of the face which becomes more infer-oposteriorly located; (e) an overall reduction, expressed in some parts such as the mastoid; and (f) an increase in mandibular ramus height. Reduction in teeth size is not shown here. *From Menéndez, L., & Buck, L. (2022). Evaluating potential proximate and ultimate causes of phenotypic change in the human skeleton over the agricultural transition. In T. Schultz, P. Peregrine, & R. Gawne (Eds.), The convergent evolution of agriculture in humans and insects (pp. 225–256). Cambridge, MA: MIT Press.*

1975). According to the former, morphological changes are interpreted to be a result of differences in the masticatory apparatus overloading, while the latter emphasises the nutritional composition and its effect on bone growth (Menéndez and Buck, 2022; Menéndez et al., 2014).

3.4.2 Dental size reduction

One of the most widespread changes after the adoption of agriculture is a consistent re-duction in tooth size among agriculturalist groups. In some cases, this pattern is limited to buccolingual dimensions (Pinhasi et al., 2008), while in others there is also a decrease in the overall size of the teeth (Calcagno, 1986). Different interpretations have been proposed for explaining the teeth size reduction, such as: a secondary consequence of facial reduction (Carlson and Van Gerven, 1977; Lieberman, 2011; Smith et al., 1984); the result of directional selective pressures (Calcagno and Gibson, 1988; Martin et al., 1984; Pinhasi et al., 2015); and even phenotypic plasticity (Hillson and Trinkaus, 2002; Lieberman, 2011).

The most accepted explanation is that consuming softer food materials meant that we no longer needed large robust jaws suited to the mastication of tough forager diets. This also led to individuals having smaller and less complex teeth and jaws (Greene et al., 1967; Y'edynak and Fleisch, 1983). According to this explanation, dental size reduction could then be a consequence of directional selection towards teeth with smaller surfaces for potential dental decay. Carlson and Van Gerven (1977) proposed a dual-mechanism explanatory model in which, in addition to selective pressures acting to reduce the overall size and morphological complexity of the dentition, the reduction in the functional demands on the masticatory complex would have also led to alterations in growth in such a way that the face became smaller, less robust, and oriented more inferoposteriorly. Such decreases in maxillary and jaw size (Katz et al., 2017; Larsen, 2006; Lieberman, 2011; von Cramon-Taubadel, 2011) may proceed at a faster rate than changes in tooth size and increase the risk of malocclusions.

Despite the fact that teeth are less developmentally plastic in response to environmental stimuli than bones (von Cramon-Taubadel, 2017), the lack of a membrane separating the developing teeth from the alveolus might have contributed to dental reduction because teeth are influenced by depletion of mastication in a similar way to bone (Lieberman, 2011). This is supported by the fact that tooth crown reduction has been reported to be much greater among permanent teeth than deciduous ones (Hillson and Trinkaus, 2002). Therefore chewing highly processed food may have contributed to the size decline observed in the permanent teeth during human evolution. Overall, it seems that dental size reduction among agriculturalists resulted from a complex process in which different evolutionary, health, and biomechanical factors were involved.

3.4.3 Plasticity, selection, and lifestyle changes

As discussed before (Section 3.3.1), selection and developmental plasticity have been proposed to explain most of the craniofacial changes interpreted as resulting from the transition to agriculture (Menéndez and Buck, 2022). On the one side, studies comparing populations worldwide suggest that the morphological changes observed in populations with different diets are not genetically fixed; in which case, plasticity becomes the most plausible explanation (González-José et al., 2005; Holmes and Ruff, 2011; Larsen, 1984; Perez et al., 2011). The role of plasticity in explaining morphological variation between populations with different subsistence practices is further supported by palaeopathological and experimental studies. The greater malocclusion and dental crowding among farmers (Katz et al., 2017; Larsen, 2006; Lieberman, 2011) might be a consequence of an inadequate coordination between facial and dental growth, since bone responds directly to biomechanical forces, while teeth do not (von Cramon-Taubadel, 2017). Experimental studies conducted in different mammal species show rapid changes in the size and shape of the maxilla and orbital plane depending on whether animals were fed with hard/soft food or exposed to high/low masticatory stress (Ciochon et al., 1997; Corruccini and Beecher, 1982, 1984; Lieberman et al., 2004b; Scott et al., 2014). Conversely, the morphological changes occurring on specific mandibular traits (e.g. mandibular condyle, angle, ramus) have been interpreted as resulting from selection (Hinton and Carlson, 1979; Kanazawa and Kasai, 1998; May et al., 2018; Pinhasi et al., 2008), in addition to the reduction in teeth size, as developed before (see Section 3.4.2). The explanation that selection shapes mandibular changes in association with diet shifts is supported by studies showing that some morphological differences consistent with dietary variation are manifested very early in development (Fukase and Suwa, 2008; Gonzalez et al., 2010; Katz et al., 2017).

3.5 Adaptations to environmental changes

3.5.1 The impact of climate on the skull

Cranial variation associated with climate has been thoroughly studied in humans for almost a century (Beals et al., 1984; Evteev et al., 2014, 2017; Franciscus, 1995; Harvati and Weaver, 2006a; Hubbe et al., 2009; Newman, 1953; Thomson and Buxton, 1923). As a result of humans expanding to multiple environments, different populations have developed biological and behavioural adaptations to cope with extreme environmental conditions. For instance, individuals living in cold environments have facial morphology that enable them to condition the inspired air by warming and humidifying it before it enters the lungs (Evteev and Grosheva, 2019; Frisancho, 1993). In addition, individuals living in high altitudes are also exposed to hypoxia and have developed physiological mechanisms for improving oxygen delivery and utilisation (Beall, 2001; Frisancho, 1993). As a result, climate has been described as having had a strong impact on human evolution, so it should be taken into account when conducting studies into inter-population relationships, but also for inferring functional adaptations in hominins (see Boxes 3.1 and 3.2).

Multiple factors have been proposed to account for most of the morphological changes associated with climate, such as variation in the degree of humidity, precipitation, solar radiation, temperature, and altitude. In particular, temperature has been one of the main variables analysed, often as an attempt to test the ecological interpretations provided by Bergmann (1847) (i.e. the existence of an inverse relationship between body size and temperature in populations of the same species) and Allen (1877) (i.e. a pattern of shorter limbs in individuals from colder climates), which generalise morphological trends in the postcranial skeleton of mammals and birds. Additionally, cranial structures such as the vault (Katz et al., 2016), facial skeleton (Evteev et al., 2014; Harvati and Weaver, 2006a; Hubbe et al., 2009), orbital area (Tomaszewska et al., 2015), maxillary and zygomatic bones (Maddux and Butaric, 2017), and nasal region (Fabra and Demarchi, 2011; Noback et al., 2011) have also shown a significant association with temperature diversity, and the latter with altitudinal variation as well. Due to the strong association between temperature and altitude (i.e. high-altitude environments present cold-temperate conditions) it becomes difficult to disentangle the effect of these factors, thus, some interpretations of these associations could be biased. This could be overcome statistically by modelling multiple variables and analysing their differential contributions in a mixed model.

3.5.2 Extreme cold environments

Overall, most climate-related morphological changes have been described in the shape and form of the cranial vault and nasal aperture. However, some scholars have interpreted the strong association between the overall cranial morphology and climate as being mostly driven by populations from extremely cold climatic zones (i.e. Inuit in the Northern Hemisphere, Fuegians in the Southern Hemisphere), when they are compared to groups from the tropics (Harvati and Weaver, 2006a; Hubbe et al., 2009; Roseman, 2004). A recent comprehensive study by Evteev and collaborators (in press), which analysed populations from temperate and cold environments from North America, South America, and Asia, found that multiple populations living in cold environments present a convergent morphological pattern in the form of the facial skeleton and, to a lesser extent, the cranial vault, when compared to populations from temperate environments. Populations inhabiting cold environments present an increase in nasal height; orbital width; facial and orbital heights; and larger, longer, and lower cranial vaults. Those similar craniofacial changes are interpreted as convergent because if those populations descend from a common ancestor, they split from each other at least 40 thousand years ago (Fu et al., 2013). On the contrary, the distinct craniofacial changes could be the result of the different severity of the climate and the diverse cultural responses to coping with cold conditions.

The most recurrent morphological trend is a change in the nasal aperture, which is taller and narrower in individuals living in colder climates (i.e. the "Thompson-Buxton rule") (Evteev et al., 2014; Fukase et al., 2016; Noback et al., 2011). Additionally, an increase in nasal protrusion has been described in individuals inhabiting cold-dry climate conditions (Franciscus, 1995). These changes in the nasal aperture (and the nasal cavity in general) enhance the contact with the mucosa of the cavity, improving the cavity's air-conditioning capacities (Butaric et al., 2010; Evteev and Grosheva, 2019; Noback et al., 2011; Yokley, 2009). Changes are also noted in the cranial vault: populations from cold climates display a larger, more robust skull, accompanied by anteroposterior shortening and rounding (Beals et al., 1984; Hubbe et al., 2009; Katzmarzyk and Leonard, 1998; Roseman, 2004), despite some populations such as the Inuit or Fuegians presenting a very long, tall, and narrow skull. Such increments in size provide support to Bergmann's rule (i.e. an inverse relationship between cranial vault size and temperature) and could be interpreted as an adaptation towards the reduction in heat loss due to a decrease in the body surface area/volume ratio (Ruff et al., 1997).

3.5.3 High-altitude adaptations

Earth's long-lasting highest-altitude inhabited lands are located in the Tibetan, Andean, and Ethiopian plateaus. Archaeological evidence suggests that humans may have permanently settled in these high-altitude environments at the end of the Pleistocene (Aldenderfer, 2006; Alkorta-Aranburu et al., 2012; Jeong et al., 2016; Rademaker et al., 2014). The main environmental change tied to the increase in altitude is the low oxygen availability that results from the decrease in barometric pressure. Consequently, many physiological changes (i.e. longer skeletal growth, larger lung volumes) occur to enhance oxygen transport to other tissues from the body. Individuals might acclimatise (i.e. the organism adjusts to an environmental change while maintaining its performance) to these conditions either permanently (within their lifetime) or for short periods of time (Frisancho, 1993). As a result, some of the morphological changes that have been described in individuals born (or descending from parents born) in high altitude (i.e. large thorax, large nasal cavities) could be interpreted as a result of adaptations (e.g. genetically fixed), while others might result from plasticity, acquired by individuals who were born and raised at low altitudes when moving to high-altitude environments (Butaric and Klocke, 2018; Weinstein, 2017).

Despite the fact that the most conspicuous skeletal changes associated with high-altitude adaptations are concentrated in the postcranial skeleton (e.g. enlarged thorax in relation to stature), some anatomical changes have been described in the skull, such as the length of the nose, internal nasal breadth, and nasal height (Butaric and Klocke, 2018). These changes are associated with increasing the performance of oxygen uptake and conditioning the inhaled air. Experimental and comparative studies suggested that hypoxia alters the relative growth of different tissues, with the lungs and brain being most susceptible to change

(Hammond et al., 2001; Petajan, 1973). For that reason, and as a result of the low availability of oxygen during brain development, changes in other parts of the skull cannot be disregarded and more studies exploring this hypothesis are needed.

3.6 Culture and communication

3.6.1 Cultural modifications of the skull

Cultural modifications of the body are a widespread practice among human groups from every continent and include permanent (i.e. not reversible) changes in the teeth and skull, as well as other parts of the body (Dembo and Imbelloni, 1938). Particularly, the cultural modifications of the skull (also known as artificial cranial deformations) have been a common practice among humans across the world. They illustrate the intentional impact of behaviour on shaping the skull. They have been more recurrent in the late Holocene, and particularly in groups from Oceania and America (Broca, 1879; Dembo and Imbelloni, 1938; Dingwall, 1931). However, this has also been described in Late Pleistocene groups from Australia (Antón and Weinstein, 1999; Durband, 2011), contemporary Shipibo from the Peruvian Amazonia (Tommaseo and Drusini, 1984), and in Neanderthals (Trinkaus, 1982b).

Cultural modifications arise through the application of external compressive forces on the skull during the first years of life (Brothwell, 1981; Ubelaker, 1984). This leads to permanent changes in morphology: the sagittal plane is altered, the axis of symmetry changes, and new angles are created, altering the magnitude and direction of the vectors that guide the general shape of the skull during development (Manríquez et al., 2006; Munizaga, 1987). Most of the mechanical pressures are applied to the frontal and occipital bones, which acquire different shapes according to the tools used—either cords bandaging or binding between two wood boards (Fig. 3.4). In general, according to the resultant skull shape, scholars can infer the instruments that have been used for performing the skull modification. For instance, skulls described as presenting an "annular" modification (Fig. 3.4) are the result of cord bandaging (O'Brien and Stanley, 2011).

These modifications are frequently the result of intrinsic motivations, but they can also occur secondarily as a consequence of another activity (i.e. use of cradles for transportation). They are a recognisable part of a person's appearance from an early age, which allows expression of social meanings as a symbol of a powerful and visually remarkable identity (Blom, 1999, 2005; Munizaga, 1992; Tiesler, 2014; Torres-Rouff, 2002, 2003). Diverse interpretations have been proposed to explain the original motivations, in each culture, including lineage, gender, or ethnic ascription (Blom, 2005; Tiesler, 2014; Torres-Rouff and Yablonsky, 2005; Weiss, 1961); and indicators of socioeconomic status, aesthetics, health, and cosmological differences among the individuals of a group (Blom, 1999; Dembo and Imbelloni, 1938; Dingwall, 1931; Tiesler, 2014; Torres-Rouff, 2002).

Early studies on cultural cranial modifications were based on qualitative descriptions by which skulls were classified into types according to the main bone on which the mechanical pressure was applied (Dingwall, 1931; Hrdlicka, 1912; Imbelloni, 1925; Neumann, 1942; Topinard, 1879). More recently, various approaches have been utilised to describe the morphological variation in a quantitative way using discriminant functions (Clark et al., 2007; O'Brien and Stanley, 2011) or combining geometric morphometrics and multivariate statistics (Kuzminsky et al., 2016; Manríquez et al., 2006; Menéndez and Lotto, 2016; Perez, 2007; Serna et al., 2019). While the main morphological changes are visible

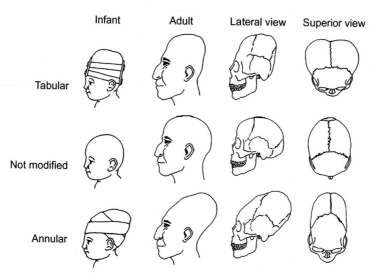

FIG. 3.4 Schematic representation of cultural modifications of the skull according to the instruments used. The rows indicate from the first to the third, tabular modification (as a result of binding between two wood boards), no modification, and annular modification (as a result of cords bandaging). The columns indicate morphological changes, the first one, the infant head together with the modification device; the second one, the resulting changes in the adult head; the third one, changes on the lateral view from the adult skull; and the last column, the morphological changes in the superior view of the adult skull. *Modified from O'Brien, T., & Stanley, A. (2011). Boards and Cords: Discriminating Types of Artificial Cranial Deformation in Prehispanic South Central Andean Populations. International Journal of Osteoarchaeology, 23(4), 459–470. https://doi.org/10.1002/oa.1269.*

in the cranial vault, researchers have suggested that the shape of the face (Antón, 1989; Cheverud and Midkiff, 1992; Hrdlicka, 1912; Manríquez et al., 2006; Pomeroy et al., 2010), cranial base (Moss, 1958), and even the mandible (Cheverud and Midkiff, 1992) could also be altered, and recommend taking cranial modifications into account when conducting inter-population studies.

3.6.2 Self-domestication and 'modernity'

Historic societal selection for certain behaviours may have been the main driving factor shaping our current ('modern') cranial morphology, through self-domestication. Domestication can be defined as "the process by which humans transformed wild animals and plants into more useful products through the control of their breeding" (Leach, 2003, p. 349). The idea of self-domestication, i.e. the process in which we domesticated ourselves as a species, has long been linked to human evolution (Blumenbach, 1799; Boas, 1938; Darwin, 1868), although sometimes in problematic ways linked to eugenics and degeneration theories around the origin of biological 'races' (see Brüne, 2007 for more detail; Fischer, 1914). Broadly speaking, domestication is a two-stage process, with an unintentional first stage (human habituation or anthropophily) and a second stage of intentional selective breeding (Hare et al., 2012; Sánchez-Villagra and van Schaik, 2019). It is understood to be linked to a range of physiological, behavioural, cognitive, and anatomical changes, which are referred to as 'Domestication

Syndrome' and are noted across many domesticated mammalian species (Geiger et al., 2017; Hare et al., 2012; Trut, 1999; Wrangham and Pilbeam, 2002). The anatomical aspects of domestication parallel changes seen in the evolution of the human skull: decreased robusticity, size reduction, decreased absolute cranial capacity, and decreased sexual dimorphism, as well as shortening and gracilisation of the facial region, sometimes linked to dental crowding and size reduction (Clark and Henneberg, 2015; Gariépy et al., 2001; Sánchez-Villagra and van Schaik, 2019; Trut et al., 2004, 2009).

The proximate driver of domestication seems to be selection for reduced reactive aggression and increased tameness (e.g. Belyaev, 1969; Trut, 1999; Wrangham, 2018). There are many suggestions as to the ultimate mechanism that produces the changes we identify as the result of domestication, with a number focusing on heterochrony (i.e. changes in timing of development), as many of the craniofacial traits are linked to paedomorphism and neoteny (i.e. retention of juvenile characteristics). The most widely accepted hypothesis at present is mild deficits of neural crest cells and derived tissues (Wilkins et al., 2014). Neural crest cells are multipotent cells which are unique to vertebrates and central to the development of many cell lineages, including neurons and glial cells of the peripheral sensory and automatic ganglia, Schwann cells, melanocytes, endocrine cells, and smooth muscle-skeletal and connective tissue cells of the craniofacial complex (Motohashi and Kunisada, 2015). As such, it is argued that these may have a direct role in the phenotypic signatures of domestication through hypoplasia of the neural crest cells or dampening of the migration of these cells, and possibly an indirect impact through alterations to the development of the pituitary gland (Gleeson, 2020; Wilkins et al., 2014). The latter would allow this hypothesis to be linked with the reported heterochronic, and specifically neotenic, nature of many domestication syndrome traits. Thus, while selection may have been focused on aggression, the many traits of 'domestication syndrome' likely arose in domesticates as a secondary by-product (Gould and Lewontin's 'spandrels', 1979; Sánchez-Villagra and van Schaik, 2019).

The parallels between domestication syndrome and the neotenic traits uniquely found in anatomically modern humans, such as increased gracility and shortening of the face, have led to the formulation of the Human Self-Domestication hypothesis (HSD) (Cieri et al., 2014; Hare, 2017; Leach, 2003). Leach (2003) argued that these traits appear in Late Pleistocene humans, and increase in frequency and degree in the Holocene. Cieri et al. (2014) identified a temporal trend towards increasing 'gracilisation' of the human skull in the last 200 kya, as recorded through morphological changes discussed in Sections 3.3.1 and 3.4.1, and a trend of decreasing endocranial volume (Beals et al., 1984; Henneberg, 1988; Ruff et al., 1997). They link this to selection for decreased aggression as a consequence of increasing population density and expanded social networks after 80 kya, and sedentism following the transition to agriculture. This decrease in aggression is hypothesised to have resulted in higher fitness due to social benefits from more cooperative group members, intersexual selection through female choice for less aggressive mates (Cieri et al., 2014; Gleeson and Kushnick, 2018), group advantages through ostracisation of excessively aggressive individuals (Wrangham, 2018, 2019), or a combination of these factors (see also Gleeson, 2020). Sánchez-Villagra and van Schaik (2019) make a similar connection with demographic changes, extending the HSD to argue that selection against reactive aggression would have enabled social tolerance in increasingly frequent between-group interactions, with the associated craniofacial traits possibly being secondarily selected for as 'honest signals' of friendliness.

As such, the HSD has been argued to link the evolution of our anatomically modern cranial features to the increasing intensity of signatures of behavioural 'modernity' (Mellars and Stringer, 1989) and cumulative technological evolution since the Middle Stone Age (Cieri et al., 2014). This is based on the argument that the observed cultural ratcheting from this period was related to the same demographic expansion that led to human self-domestication (Hare, 2017; Henrich, 2015). Nevertheless, behavioural 'modernity' is a highly divisive concept and behaviours linked to this potentially have much deeper routes (McBrearty and Brooks, 2000; Shea, 2011), calling into question a link between cultural innovation and the appearance of anatomical modernity. It should also be noted that, from the early Holocene, there is an overlap between the morphological changes interpreted as a result of HSD with those that resulted from the impact of agriculture. They both produced similar morphological changes in the skull (i.e. gracilisation, see Section 3.4.1), making it difficult to separate the possible impact of these two behavioural changes on the morphology of the modern human skull.

3.6.3 Communication and the face

There have been advances both in terms of inferring verbal and nonverbal communication in the past. For example, several studies have used biomechanical simulations that allow reconstruction of Neanderthal soft tissue and vowel production (Barney et al., 2012; Boë et al., 2002; Dediu and Levinson, 2013, 2018) as well as prehistoric languages of more recent human populations (Dediu et al., 2021). These studies use detailed anatomical information concerning some components of the vocal tract such as the lower jaw, the hard palate, and the relative position of the hyoid, as well as other features in the mandible and cranial base that can be used to predict the laryngeal and tongue root position. Overall, they indicate that Neanderthals had the vocal tract, breathing control, and acoustic sensitivity that are required and involved in modern human speech. However, there are several criticisms to these approaches in terms of unfounded assumptions and lack of empirical tests to verify them, as well as other simulation studies showing that language is a more recent phenomenon (de Boer and Tecumseh Fitch, 2010; Hauser et al., 2014; Lieberman and Crelin, 1971; Lieberman et al., 1972). Further studies are needed to address the language capacity in fossil *Homo* species.

Complementarily, humans have a number of derived facial features that have been linked to nonverbal communication, such as our depigmented (white to light yellow) sclera and our conspicuous eyebrows (Lieberman, 2011). Recently, the latter feature has been tentatively associated with the evolution of our reduced browridges and vertical frontal squamae; it has been argued that these features would have allowed us to increase the range and subtlety of communication available through more visible eyebrow movements (Godinho et al., 2018). Many social and adaptive functions have been put forward to explain the expanded browridge documented within *Homo*, including dominance and intimidation displays, sexual selection, protection from the sun, rain or blows to the head, as well as hair maintenance (Davies, 1972; Guthrie, 1974; Krantz, 1973; Kurten, 1979; Tappen, 1973). In contrast, the unique reduction of this cranial superstructure recorded in our species has received relatively little attention, outside of structural, biomechanical, and integrative hypotheses (Lieberman et al., 2002) (see Sections 3.3.1, 3.4.1, and following).

While speculative at present, social hypotheses for humans' distinctive foreheads are tantalising. Baboons, macaques, and humans are known to use eyebrow 'flashes' and raises in both aggressive and nonaggressive communication, sometimes to emphasise other facial expressions (Dube, 2013; Grammer et al., 1988; Hesler and Fischer, 2007; Maestripieri, 1996). Browridge development is linked to androgen production, and it is possible to relate reduction of this feature to selection against aggression, and therefore permanent signals of dominance, via the HSD, as the derived human browridge becomes fixed in the palaeontological record after 90 kya (Cieri et al., 2014) (see earlier). Nevertheless, both the expansion and reduction of the hominin browridge are currently more satisfactorily explained via biomechanical or spatial hypotheses relating to para/masticatory strain or structural responses to changing relationships between the brain and the face, respectively (Bruner et al., 2013; Godinho and O'Higgins, 2018; Russell, 1985). Any advantage for nonverbal communication conferred by our globular cranium and flat faces is likely to be exaptive (sensu Gould and Vrba, 1982), although we could hypothesise that this would not preclude secondary selection for more vertical foreheads and visible eyebrows once these had started to appear.

3.7 Summary and conclusion

The high degree of biosocial complexity of the human skull is due not only to the numerous elements that make up this region, but also to the different parts being shaped by both biological processes and ecological factors during development, as well as the individuals' agency, and their social and cultural environment. Different modules of the skull have specific developmental and embryological origins by which they acquire their final adult shape. As a result, those elements that finish their growth later in life will be more subjected to environmental influences and behaviour than those that acquire the adult form at an earlier developmental stage, with the latter preserving a stronger signal of population and evolutionary history as a result. But even after development finishes, the skull is still influenced by environmental and cultural factors (i.e. behaviour) throughout an individual's life. Thus some practices that are triggered by either an individual's agency and/or their social environment during childhood and adulthood could produce changes in the skull. On top of that, some bones are associated with, and thus influenced by, multiple functions (e.g. the maxillae, which form the upper jaw, the floor of the eye sockets, and the lower sections and sides of the nasal cavity), some functions are shared by several bones (e.g. the maxillae, parietals, and mandible are linked to mastication) while others, such as hearing, are unique to particular bones (e.g. the ear ossicles, temporal bone). Overall, the skull is a mosaic that represents the complex interactions of biological processes and behavioural factors shaping our evolutionary history (see Table 3.1 for a summary of key studies exploring these factors), as well as the particular history of that individual. As a consequence, when studying the human skull, it is necessary to consider the multiple spectra of interacting factors, either when analysing differences between and among taxa, populations, ontogenetic series, or an assemblage of individuals from a shared burial or single palaeontological context.

Sections	Reference	Sample	Method	Key results
Section 3.2.2. Modularity and integration	Barbeito-Andrés et al. (2016)	CT data for 122 living *Homo sapiens*, aged 0 to 31 years	3D GMM, 30 landmarks and semilandmarks on orbital margin, zygomatic, and frontal bone	The bony orbit is more integrated with the neurocranium than the zygomatic; adults are more different to each other in the integration of the orbit with the frontal and zygomatic bones than younger individuals
Section 3.3.1. Evolution of the human skull	Lieberman et al. (2002)	Recent (*n* = 100), Late Pleistocene (*n* = 10) and archaic *H. sapiens* (*n* = 9), Middle Pleistocene hominins (*n* = 4), and ontogenetic series of *H. sapiens* (*n* = 12) and *Pan troglodytes* (*n* = 61)	2D GMM, 17 landmarks	Anatomically modern *H. sapiens* are distinguished by relatively smaller, retracted faces, globular neurocrania, and longer and more flexed basicrania
Section 3.3.2. Population history	Roseman (2004)	Global sample of 955 recent *H. sapiens* from 10 populations, and genetic data from 413 individuals from matched populations	57 linear measurements, 377 microsatellite loci, climatic variables	Cranial morphological distance correlates with neutral genetic distance, especially in measurements of length
	Betti et al. (2009)	Global sample of 3245 crania of recent humans	37 linear measurements, climatic variables	Cranial morphology correlates with distance from sub-Saharan Africa, not climate
Section 3.3.3. Gene flow and hybridisation	Ackermann et al. (2006)	169 crania of *Papio cynocephalus* (Yellow baboons), *Papio anubis* (Olive baboons), and their hybrids	3D GMM, 39 Euclidean distances from 36 landmarks; qualitative assessment	Hybrid crania show heterosis through increased size in a number of cranial traits, supernumerary teeth, large, rugose faces, and additional or unusual zygomaxillary sutures
Section 3.4.1. Craniofacial gracilisation and globularisation	Carlson and Van Gerven (1977)	437 skulls from a diachronic sequence of 10,000 years (Mesolithic to Neolithic) from Nubia, Africa.	16 linear measurements	A reduction in the functional demand on the masticatory complex led to an alteration in its growth such that the face became less robust and more inferoposteriorly located relative to the cranial vault, which became more "globular"
Section 3.4.2. Dental size reduction	y'Edynak and Fleisch (1983)	58 individuals from ex-Yugoslavia (Mesolithic = 42, Neolithic = 16)	Linear measurements in the mandible, metric and nonmetric traits in the teeth	The adoption of a Neolithic way of life which resulted in a softer diet producing a biomechanically reduced jaw that could no longer accommodate the large Mesolithic teeth. Selection explains the reduction of the size of the teeth
Section 3.4.3. Plasticity, selection, and lifestyle changes	Katz et al. (2017)	559 crania from 25 worldwide populations, 534 mandibles from 24 groups	3D landmarks (37 in the whole skull, 23 in the mandible)	Despite neutral evolutionary processes shape most of the human skull diversity, there is a small effect of diet when comparing foragers and farmers as a result of modest directional differences

Section	Study	Sample	Methods	Findings
Section 3.5.1. The impact of climate on the skull	Fabra and Demarchi (2011)	301 crania from 17 South American populations	10 linear measurements, 3 climatic variables	Craniometric variation is significantly influenced by geography, and there is a significant effect of altitude modelling nasal shape
Section 3.5.2. Cold environments	Hubbe et al. (2009)	7423 crania from 135 worldwide populations	33 linear measurements, 10 climatic variables	Neurocranial morphology is phylogenetically informative, while the face and cranium are subject to selection related to climatic factors. Also, selection to climate is largely restricted to groups living in extremely cold environments
	Evteev and Grosheva (2019)	173 skulls from 15 Asian populations	16 midfacial external linear measurements, 20 linear measurements from nasal cavity and maxillary sinuses, 7 climatic variables	Asian populations living at cold environments present a strong association between the form of nasal cavity and climatic variables. The external midfacial morphology presents a higher correlation with climate than the shape of the internal nasal cavity
Section 3.5.3. High-altitude adaptations	Butaric and Klocke (2018)	130 skulls from Tibetans and Peruvians living at low and high altitude	17 linear measurements describing the external midface and nasal region	Tibetans living at high altitude display relatively larger nasal cavities, allowing increased oxygen uptake. Specific nasal adaptations were not identified among Peruvians, interpreted as a result of their relatively recent migration history and population structure
Section 3.6.1. Cultural modifications of the skull	O'Brien and Stanley (2011)	469 crania from South America	33 linear measurements	A new method to quantitative study and classify culturally modified crania, based on discriminant functions
Section 3.6.2. Self-domestication and 'modernity'	Cieri et al. (2014)	Pre-80 ka ($n=13$), post-80 ka ($n=41$) Pleistocene human crania, and global sample of recent Holocene humans ($n=1367$)	Linear craniometrics (upper facial height, supraorbital projection, glabellar projection, endocranial volume)	Browridge projection decreases over time, as does facial length after 80 kya. Endocranial volume increases between the pre- and post-80 kya samples, then decreases
Section 3.6.3. Communication and the face	Godinho et al. (2018)	CT data for Kabwe/Broken Hill 1 (*Homo rhodesiensis/heidelbergensis*) fossil hominin	Virtual anthropology, finite element analysis	Kabwe 1's browridge is larger than necessary under both a spatial (neuro-orbital disjunction) hypothesis and a biomechanical (masticatory stress) model

Key studies investigating the effects of the various biosocial factors understood to affect craniofacial morphology.

References

Ackermann, R. R., Arnold, M. L., Baiz, M. D., Cahill, J. A., Cortés-Ortiz, L., Evans, B. J., et al. (2019). Hybridization in human evolution: Insights from other organisms. *Evolutionary Anthropology, 28*(4), 189–209. https://doi.org/10.1002/evan.21787.

Ackermann, R. R., & Bishop, J. M. (2010). Morphological and molecular evidence reveals recent hybridization between gorilla taxa. *Evolution, 64*(1), 271–290. https://doi.org/10.1111/j.1558-5646.2009.00858.x.

Ackermann, R. R., & Cheverud, J. M. (2004). Detecting genetic drift versus selection in human evolution. *Proceedings of the National Academy of Sciences of the United States of America, 101*(52), 17946–17951. https://doi.org/10.1073/pnas.0405919102.

Ackermann, R. R., Mackay, A., & Arnold, M. L. (2016). The hybrid origin of "modern" humans. *Evolutionary Biology, 43*(1), 1–11. https://doi.org/10.1007/s11692-015-9348-1.

Ackermann, R. R., Rogers, J., & Cheverud, J. M. (2006). Identifying the morphological signatures of hybridization in primate and human evolution. *Journal of Human Evolution, 51*(6), 632–645. https://doi.org/10.1016/j.jhevol.2006.07.009.

Ackermann, R. R., Schroeder, L., Rogers, J., & Cheverud, J. M. (2014). Further evidence for phenotypic signatures of hybridization in descendant baboon populations. *Journal of Human Evolution, 76*(C), 54–62. https://doi.org/10.1016/j.jhevol.2014.05.004.

Aldenderfer, M. (2006). Modelling plateau peoples: The early human use of the world's high plateaux. *World Archaeology, 38*(3), 357–370.

Alkorta-Aranburu, G., Beall, C. M., Witonsky, D. B., Gebremedhin, A., Pritchard, J. K., & Di Rienzo, A. (2012). The genetic architecture of adaptations to high altitude in Ethiopia. *PLoS Genetics, 8*(12). https://doi.org/10.1371/journal.pgen.1003110, e1003110.

Allen, J. A. (1877). The influence of physical conditions in the genesis of species. *Radical Review, 1*, 108–140.

Antón, S. C. (1989). Intentional cranial vault deformation and induced changes of the cranial base and face. *American Journal of Physical Anthropology, 79*(2), 253–267. https://doi.org/10.1002/ajpa.1330790213.

Antón, S. C. (1994). Biomechanical and other perspectives on the Neandertal face. In R. S. Corruccini, & R. L. Ciochon (Eds.), *Integrated pathways to the past: Paleoanthropological advances in honor of F. Clark Howell* (pp. 677–695). Englewood Cliffs, NJ: Prentice Hall.

Antón, S. C., & Weinstein, K. J. (1999). Artificial cranial deformation and fossil Australians revisited. *Journal of Human Evolution, 36*(2), 195–209. https://doi.org/10.1006/jhev.1998.0266.

Atchley, W. R., & Hall, B. K. (1991). A model for development and evolution of complex morphological structures. *Biological Reviews, 66*(2), 101–157. https://doi.org/10.1111/j.1469-185X.1991.tb01138.x.

Barbeito-Andrés, J., Anzelmo, M., Ventrice, F., Pucciarelli, H. M., & Sardi, M. L. (2016). Morphological integration of the orbital region in a human ontogenetic sample. *The Anatomical Record, 299*(1), 70–80. https://doi.org/10.1002/ar.23282.

Barney, A., Martelli, S., Serrurier, A., & Steele, J. (2012). Articulatory capacity of Neanderthals, a very recent and human-like fossil hominin. *Philosophical Transactions of The Royal Society B Biological Sciences, 367*(1585), 88–102. https://doi.org/10.1098/rstb.2011.0259.

Bastir, M., & Rosas, A. (2013). Cranial airways and the integration between the inner and outer facial skeleton in humans. *American Journal of Physical Anthropology, 152*(2), 287–293. https://doi.org/10.1002/ajpa.22359.

Beall, C. M. (2001). Adaptations to altitude: A current assessment. *Annual Review of Anthropology, 30*(1), 423–456. https://doi.org/10.1146/annurev.anthro.30.1.423.

Beals, K. L., Smith, C. L., & Dodd, S. M. (1984). Brain size, cranial morphology, climate and time machines. *Current Anthropology, 25*(3), 301–330. https://doi.org/10.1086/203138.

Belyaev, D. K. (1969). Domestication of animals. *Science Journal, 5*, 47–52.

Bergmann, C. (1847). Über die Verhältnisse der Wärmeökonomie der Thiere zu ihrer Größe. *Gottinger Studien, 3*, 595–708.

Betti, L., Balloux, F., Amos, W., Hanihara, T., & Manica, A. (2009). Distance from Africa, not climate, explains within-population phenotypic diversity in humans. *Proceedings of the Royal Society B: Biological Sciences, 276*(1658), 809–814. https://doi.org/10.1098/rspb.2008.1563.

Blom, D. (1999). *Tiwanaku regional interaction and social identity: A bioarchaeological approach* (PhD thesis). University of Chicago.

Blom, D. (2005). A bioarchaeological approach to Tiwanaku group dynamics. In R. Reycraft (Ed.), *Us and them: Archaeology and ethnicity in the Andes* (pp. 153–182). Los Angeles, CA: Cotsen Institute of Archaeology Press.

Blumenbach, J. F. (1799). Comparison between the human race and that of swine. *Philosophical Magazine*, 3, 284–290.

Boas, F. (1938). *The mind of primitive man*. New York: Macmillan.

Boë, L. J., Heim, J. L., Honda, K., & Maeda, S. (2002). The potential Neandertal vowel space was as large as that of modern humans. *Journal of Phonetics*, 30(3), 465–484. https://doi.org/10.1006/jpho.2002.0170.

Boel, C., Curnoe, D., & Hamada, Y. (2019). Craniofacial shape and nonmetric trait variation in hybrids of the Japanese macaque (*Macaca fuscata*) and the Taiwanese macaque (*Macaca cyclopis*). *International Journal of Primatology*. https://doi.org/10.1007/s10764-019-00081-2.

Bookstein, F., Schäfer, K., Prossinger, H., Seidler, H., Fieder, M., Stringer, C., et al. (1999). Comparing frontal cranial profiles in archaic and modern *Homo* by morphometric analysis. *The Anatomical Record*, 257(6), 217–224.

Broca, P. (1879). Sur un mode peu connu de déformation toulousaine. *Bulletins de La Société d'Anthropologie de Paris*, 2(1), 699–701.

Brothwell, D. (1981). *Digging up bones: The excavation, treatment, and study of human skeletal remains*. Ithaca, NY: Cornell University Press.

Brüne, M. (2007). On human self-domestication, psychiatry, and eugenics. *Philosophy, Ethics, and Humanities in Medicine*, 2, 21. https://doi.org/10.1186/1747-5341-2-21.

Bruner, E., Athreya, S., De La Cuétara, J. M., & Marks, T. (2013). Geometric variation of the frontal squama in the genus *Homo*: Frontal bulging and the origin of modern human morphology. *American Journal of Physical Anthropology*, 150(2), 313–323. https://doi.org/10.1002/ajpa.22202.

Bruner, E., de la Cuétara, J. M., & Holloway, R. (2011). A bivariate approach to the variation of the parietal curvature in the genus *Homo*. *Anatomical Record*, 294(9), 1548–1556. https://doi.org/10.1002/ar.21450.

Butaric, L. N., & Klocke, R. (2018). Nasal variation in relation to high-altitude adaptations among Tibetans and Andeans. *American Journal of Human Biology*, 30(3), e23104.

Butaric, L. N., McCarthy, R. C., & Broadfield, D. C. (2010). A preliminary 3D computed tomography study of the human maxillary sinus and nasal cavity. *American Journal of Physical Anthropology*, 143(3), 426–436. https://doi.org/10.1002/ajpa.21331.

Calcagno, J. M. (1986). Dental reduction in post-Pleistocene Nubia. *American Journal of Physical Anthropology*, 70(3), 349–363. https://doi.org/10.1002/ajpa.1330700310.

Calcagno, J. M., & Gibson, K. R. (1988). Human dental reduction: Natural selection or the probable mutation effect. *American Journal of Physical Anthropology*, 77(4), 505–517. https://doi.org/10.1002/ajpa.1330770411.

Cann, R. L., Stoneking, M., & Wilson, A. C. (1987). Mitochondrial DNA and human evolution. *Nature*, 325(6099), 31–36. https://doi.org/10.1038/325031a0.

Carlson, D. S., & Van Gerven, D. P. (1977). Masticatory function and post-Pleistocene evolution in Nubia. *American Journal of Physical Anthropology*, 46(3), 495–506. https://doi.org/10.1002/ajpa.1330460316.

Carson, E. A. (2006). Maximum likelihood estimation of human craniometric heritabilities. *American Journal of Physical Anthropology*, 131(2), 169–180. https://doi.org/10.1002/ajpa.20424.

Chen, F., Welker, F., Shen, C.-C., Bailey, S. E., Bergmann, I., Davis, S., et al. (2019). A late Middle Pleistocene Denisovan mandible from the Tibetan Plateau. *Nature*, 569(7756), 409–412. https://doi.org/10.1038/s41586-019-1139-x.

Cheronet, O., Finarelli, J. A., & Pinhasi, R. (2016). Morphological change in cranial shape following the transition to agriculture across western Eurasia. *Scientific Reports*, 6, 33316. https://doi.org/10.1038/srep33316.

Cheverud, J. M. (1982). Phenotypic, genetic, and environmental morphological integration in the cranium. *Evolution*, 36(3), 499–516. https://doi.org/10.1111/j.1558-5646.1982.tb05070.x.

Cheverud, J. M. (1996). Developmental integration and the evolution of pleiotropy. *American Zoologist*, 36(1), 44–50. https://doi.org/10.1093/icb/36.1.44.

Cheverud, J. M., & Midkiff, J. E. (1992). Effects of fronto-occipital cranial reshaping on mandibular form. *American Journal of Physical Anthropology*, 87(2), 167–171. https://doi.org/10.1002/ajpa.1330870205.

Cieri, R. L., Churchill, S. E., Franciscus, R. G., Tan, J., & Hare, B. (2014). Craniofacial feminization, social tolerance, and the origins of behavioral modernity. *Current Anthropology*, 55(4), 419–443. https://doi.org/10.1086/677209.

Ciochon, R. L., Nisbett, R. A., & Corruccini, R. S. (1997). Dietary consistency and craniofacial development related to masticatory function in minipigs. *Journal of Craniofacial Genetics and Developmental Biology*, 17(2), 96–102.

Clark, J. L., Dobson, S. D., Antón, S. C., Hawks, J., Hunley, K. L., & Wolpoff, M. H. (2007). Identifying artificially deformed crania. *International Journal of Osteoarchaeology*, 17(6), 596–607. https://doi.org/10.1002/oa.910.

Clark, G., & Henneberg, M. (2015). The life history of *Ardipithecus ramidus*: A heterochronic model of sexual and social maturation. *Anthropological Review*, 78(2), 109–132. https://doi.org/10.1515/anre-2015-0009.

Clement, A. F., Hillson, S. W., & Aiello, L. C. (2012). Tooth wear, Neanderthal facial morphology and the anterior dental loading hypothesis. *Journal of Human Evolution, 62*(3), 367–376. https://doi.org/10.1016/j.jhevol.2011.11.014.

Collard, M., & Lycett, S. J. (2008). Does phenotypic plasticity confound attempts to identify hominin fossil species? An assessment using extant Old World monkey craniodental data. *Folia Primatologica, 79*(3), 111–122. https://doi.org/10.1159/000110680.

Collard, M., & Wood, B. (2000). How reliable are human phylogenetic hypotheses? *Proceedings of the National Academy of Sciences of the United States of America, 97*(9), 5003–5006. https://doi.org/10.1073/pnas.97.9.5003.

Collard, M., & Wood, B. (2007). Hominin homoiology: An assessment of the impact of phenotypic plasticity on phylogenetic analyses of humans and their fossil relatives. *Journal of Human Evolution, 52*(5), 573–584. https://doi.org/10.1016/j.jhevol.2006.11.018.

Coon, C. S. (1962). *The origin of races.* London: Jonathan Cape.

Corruccini, R. S., & Beecher, R. M. (1982). Occlusal variation related to soft diet in a nonhuman primate. *Science, 218*(4567), 74–76. https://doi.org/10.1126/science.7123221.

Corruccini, R. S., & Beecher, R. M. (1984). Occlusofacial morphological integration lowered in baboons raised on soft diet. *Journal of Craniofacial Genetics and Developmental Biology, 4*(2), 135–142.

Crevecoeur, I., Brooks, A., Ribot, I., Cornelissen, E., & Semal, P. (2016). Late Stone age human remains from Ishango (Democratic Republic of Congo): New insights on Late Pleistocene modern human diversity in Africa. *Journal of Human Evolution, 96*, 35–57. https://doi.org/10.1016/j.jhevol.2016.04.003.

Curnoe, D., Ji, X., Taçon, P. S. C., & Yaozheng, G. (2015). Possible signatures of hominin hybridization from the early Holocene of Southwest China. *Scientific Reports, 5*, 12408. https://doi.org/10.1038/srep12408.

Curnoe, D., Xueping, J., Herries, A. I. R., Kanning, B., Taçon, P. S. C., Zhende, B., et al. (2012). Human remains from the Pleistocene-Holocene transition of southwest China suggest a complex evolutionary history for East Asians. *PLoS One, 7*(3). https://doi.org/10.1371/journal.pone.0031918, e31918.

Darwin, C. (1868). *The variation of animals and plants under domestication.* London: John Murray.

Davies, D. M. (1972). *The influence of the teeth, diet, and habits on the human face.* London: William Heinemann Medical Books.

Day, M., & Stringer, C. B. (1982). A reconsideration of the Omo-Kibish remains and the erectus-sapiens transition. In *Congrès international de paléontologie humaine Human Palaeontology* (pp. 814–846).

de Azevedo, S., González, M. F., Cintas, C., Ramallo, V., Quinto-Sánchez, M., Márquez, F., et al. (2017). Nasal airflow simulations suggest convergent adaptation in Neanderthals and modern humans. *Proceedings of the National Academy of Sciences of the United States of America, 114*(47), 12442–12447. https://doi.org/10.1073/pnas.1703790114.

de Boer, B., & Tecumseh Fitch, W. (2010). Computer models of vocal tract evolution: An overview and critique. *Adaptive Behavior, 18*(1), 36–47. https://doi.org/10.1177/1059712309350972.

Dediu, D., & Levinson, S. C. (2013). On the antiquity of language: The reinterpretation of Neandertal linguistic capacities and its consequences. *Frontiers in Psychology, 4*, 397. https://doi.org/10.3389/fpsyg.2013.00397.

Dediu, D., & Levinson, S. C. (2018). Neanderthal language revisited: Not only us. *Current Opinion in Behavioral Sciences, 21*, 49–55. https://doi.org/10.1016/j.cobeha.2018.01.001.

Dediu, D., Moisik, S. R., Baetsen, W. A., Bosman, A. M., & Waters-Rist, A. L. (2021). The vocal tract as a time machine: Inferences about past speech and language from the anatomy of the speech organs. *Philosophical Transactions of the Royal Society B, 376*(1824), 20200192. https://doi.org/10.1098/rstb.2020.0192.

Dembo, A., & Imbelloni, J. (1938). *Deformaciones Intencionales del Cuerpo Humano de Carácter Étnico.* Humanior, Biblioteca del Americanista Moderno.

Dingwall, E. (1931). *Artificial cranial deformation.* London: Bale, Sons & Danielsson.

Dube, F. A. (2013). *Visual and tactile communication of a captive hamadryas baboon group (Papio hamadryas hamadryas) with special regard to their intentionality* (PhD Thesis). Martin-Luther-Universität Halle-Wittenberg.

Durband, A. C. (2011). Is there evidence for artificial cranial deformation at the Willandra Lakes? *Australian Archaeology, 73*, 62–64.

Eichel, K. A., & Ackermann, R. R. (2016). Variation in the nasal cavity of baboon hybrids with implications for late Pleistocene hominins. *Journal of Human Evolution, 94*, 134–145. https://doi.org/10.1016/j.jhevol.2016.02.007.

Enard, D., & Petrov, D. A. (2018). Evidence that RNA viruses drove adaptive introgression between Neanderthals and modern humans. *Cell, 175*(2), 360–371.e13. https://doi.org/10.1016/j.cell.2018.08.034.

Evteev, A. A., Cardini, A. L., Morozova, I., & O'Higgins, P. (2014). Extreme climate, rather than population history, explains mid-facial morphology of Northern Asians. *American Journal of Physical Anthropology, 153*(3), 449–462. https://doi.org/10.1002/ajpa.22444.

Evteev, A. A., & Grosheva, A. N. (2019). Nasal cavity and maxillary sinuses form variation among modern humans of Asian descent. *American Journal of Physical Anthropology*, 169(3), 513–525. https://doi.org/10.1002/ajpa.23841.

Evteev, A.A., Grosheva, A., Syutkina, T., Santos, P., Ghirotto, S., Hanihara, T.H., Hubbe, M., & Menéndez, L.P. (n.d.). Convergent ecogeographic cranial changes in human populations from Asia and America living at cold to temperate climates.

Evteev, A. A., Movsesian, A. A., & Grosheva, A. N. (2017). The association between mid-facial morphology and climate in northeast Europe differs from that in north Asia: Implications for understanding the morphology of Late Pleistocene *Homo sapiens*. *Journal of Human Evolution*, 107, 36–48. https://doi.org/10.1016/j.jhevol.2017.02.008.

Fabra, M., & Demarchi, D. A. (2011). Geographic patterns of craniofacial variation in pre-hispanic populations from the southern cone of South America. *Human Biology*, 83(4), 491–507. https://doi.org/10.3378/027.083.0404.

Fischer, E. (1914). Die rassenmerkmale des menschen als domesticationserscheinungen. *Zeitschrift für Morphologie und Anthropologie*, 18, 479–524.

Franciscus, R. G. (1995). *Later Pleistocene nasofacial variation in western Eurasia and Africa and modern human origins* (PhD thesis). University of New Mexico.

Franciscus, R. G. (1999). Neandertal nasal structures and upper respiratory tract "specialization". *Proceedings of the National Academy of Sciences of the United States of America*, 96(4), 1805–1809. https://doi.org/10.1073/pnas.96.4.1805.

Freidline, S. E., Gunz, P., & Hublin, J.-J. (2015). Ontogenetic and static allometry in the human face: Contrasting Khoisan and Inuit. *American Journal of Physical Anthropology*, 158(1), 116–131. https://doi.org/10.1002/ajpa.22759.

Frisancho, A. R. (1993). *Human adaptation and accommodation*. University of Michigan Press.

Fu, Q., Hajdinjak, M., Moldovan, O. T., Constantin, S., Mallick, S., Skoglund, P., et al. (2015). An early modern human from Romania with a recent Neanderthal ancestor. *Nature*, 524(7564), 216–219. https://doi.org/10.1038/nature14558.

Fu, Q., Meyer, M., Gao, X., Stenzel, U., Burbano, H. A., Kelso, J., et al. (2013). DNA analysis of an early modern human from Tianyuan cave, China. *Proceedings. National Academy of Sciences. United States of America*, 110(6), 2223–2227. https://doi.org/10.1073/pnas.1221359110.

Fukase, H., Ito, T., & Ishida, H. (2016). Geographic variation in nasal cavity form among three human groups from the Japanese Archipelago: Ecogeographic and functional implications. *American Journal of Human Biology*, 28(3), 343–351. https://doi.org/10.1002/ajhb.22786.

Fukase, H., & Suwa, G. (2008). Growth-related changes in prehistoric Jomon and modern Japanese mandibles with emphasis on cortical bone distribution. *American Journal of Physical Anthropology*, 136(4), 441–454. https://doi.org/10.1002/ajpa.20828.

Galland, M., Van Gerven, D. P., von Cramon-Taubadel, N., & Pinhasi, R. (2016). 11,000 years of craniofacial and mandibular variation in Lower Nubia. *Scientific Reports*, 6, 31040. https://doi.org/10.1038/srep31040.

Gariépy, J. L., Bauer, D. J., & Cairns, R. B. (2001). Selective breeding for differential aggression in mice provides evidence for heterochrony in social behaviours. *Animal Behaviour*, 61(5), 933–947. https://doi.org/10.1006/anbe.2000.1700.

Garvin, H. M., & Ruff, C. B. (2012). Sexual dimorphism in skeletal browridge and chin morphologies determined using a new quantitative method. *American Journal of Physical Anthropology*, 147(4), 661–670. https://doi.org/10.1002/ajpa.22036.

Geiger, M., Evin, A., Sánchez-Villagra, M. R., Gascho, D., Mainini, C., & Zollikofer, C. P. E. (2017). Neomorphosis and heterochrony of skull shape in dog domestication. *Scientific Reports*, 7(1), 13443. https://doi.org/10.1038/s41598-017-12582-2.

Gilbert, S. F. (2006). *Developmental biology* (8th ed.). Sunderland, MA: Sinauer Associates.

Gittelman, R. M., Schraiber, J. G., Vernot, B., Mikacenic, C., Wurfel, M. M., & Akey, J. M. (2016). Archaic hominin admixture facilitated adaptation to Out-of-Africa environments. *Current Biology*, 26(24), 3375–3382. https://doi.org/10.1016/j.cub.2016.10.041.

Gleeson, B. T., & Kushnick, G. (2018). Female status, food security, and stature sexual dimorphism: Testing mate choice as a mechanism in human self-domestication. *American Journal of Physical Anthropology*, 167(3), 458–469. https://doi.org/10.1002/ajpa.23642.

Gleeson, B. T. (2020). Masculinity and the mechanisms of human self-domestication. *Adaptive Human Behavior and Physiology*, 6(1), 1–29. https://doi.org/10.1007/s40750-019-00126-z.

Godinho, R. M., & O'Higgins, P. (2018). The biomechanical significance of the frontal sinus in Kabwe 1 (*Homo heidelbergensis*). *Journal of Human Evolution*, 114, 141–153. https://doi.org/10.1016/j.jhevol.2017.10.007.

Godinho, R. M., Spikins, P., & O'Higgins, P. (2018). Supraorbital morphology and social dynamics in human evolution. *Nature Ecology and Evolution, 2*(6), 956–961. https://doi.org/10.1038/s41559-018-0528-0.

Gokcumen, O. (2020). Archaic hominin introgression into modern human genomes. *American Journal of Physical Anthropology, 171*(S70), 60–73. https://doi.org/10.1002/ajpa.23951.

Gonzalez, P. N., Bernal, V., & Perez, S. I. (2011). Analysis of sexual dimorphism of craniofacial traits using geometric morphometric techniques. *International Journal of Osteoarchaeology, 21*(1), 82–91. https://doi.org/10.1002/oa.1109.

Gonzalez, P. N., Perez, S. I., & Bernal, V. (2010). Ontogeny of robusticity of craniofacial traits in modern humans: A study of South American populations. *American Journal of Physical Anthropology, 142*(3), 367–379. https://doi.org/10.1002/ajpa.21231.

González-José, R., Ramírez-Rozzi, F., Sardi, M., Martínez-Abadías, N., Hernández, M., & Pucciarelli, H. M. (2005). Functional-cranial approach to the influence of economic strategy on skull morphology. *American Journal of Physical Anthropology, 128*(4), 757–771. https://doi.org/10.1002/ajpa.20161.

Goswami, A., & Polly, P. D. (2010). The influence of modularity on cranial morphological disparity in Carnivora and Primates (Mammalia). *PLoS One, 5*(3), e 9517. https://doi.org/10.1371/journal.pone.0009517.

Goswami, A., Smaers, J. B., Soligo, C., & Polly, P. D. (2014). The macroevolutionary consequences of phenotypic integration: From development to deep time. *Philosophical Transactions of the Royal Society of London. Series B, Biological Sciences, 369*(1649), 20130254. https://doi.org/10.1098/rstb.2013.0254.

Gould, S. J., & Lewontin, R. C. (1979). The spandrels of San Marco and the Panglossian paradigm: A critique of the adaptationist programme. *Proceedings of the Royal Society of London B: Biological Sciences, 205*(1161), 581–598.

Gould, S. J., & Vrba, E. S. (1982). Exaptation—A missing term in the science of form. *Paleobiology, 1*, 4–15. https://doi.org/10.1017/S0094837300004310.

Grammer, K., Schiefenhövel, W., Schleidt, M., Lorenz, B., & Eibl-Eibesfeldt, I. (1988). Patterns on the face: The eyebrow flash in crosscultural comparison. *Ethology, 77*(4), 279–299. https://doi.org/10.1111/j.1439-0310.1988.tb00211.x.

Green, R. E., Krause, J., Briggs, A. W., Maricic, T., Stenzel, U., Kircher, M., et al. (2010). A draft sequence of the Neandertal genome. *Science, 328*(5979), 710–722. https://doi.org/10.1126/science.1188021.

Greene, D. L., Ewing, G. H., & Armelagos, G. J. (1967). Dentition of a Mesolithic population from Wadi Halfa, Sudan. *American Journal of Physical Anthropology, 27*(1), 41–55. https://doi.org/10.1002/ajpa.1330270107.

Gunz, P., & Harvati, K. (2007). The Neanderthal "chignon": Variation, integration, and homology. *Journal of Human Evolution, 52*(3), 262–274. https://doi.org/10.1016/j.jhevol.2006.08.010.

Gunz, P., Neubauer, S., Golovanova, L., Doronichev, V., Maureille, B., & Hublin, J.-J. (2012). A uniquely modern human pattern of endocranial development. Insights from a new cranial reconstruction of the Neandertal newborn from Mezmaiskaya. *Journal of Human Evolution, 62*(2), 300–313. https://doi.org/10.1016/j.jhevol.2011.11.013.

Guthrie, R. D. (1974). Evolution of human threat display organs. *Evolutionary Biology, 4*, 257–301.

Hall, B. K. (1999). *The neural crest in development and evolution*. New York: Springer-Verlag.

Hallgrímsson, B., Lieberman, D. E., Liu, W., Ford-Hutchinson, A. F., & Jirik, F. R. (2007). Epigenetic interactions and the structure of phenotypic variation in the cranium. *Evolution & Development, 9*(1), 76–91. https://doi.org/10.1111/j.1525-142X.2006.00139.x.

Hammer, M. F., Woerner, A. E., Mendez, F. L., Watkins, J. C., & Wall, J. D. (2011). Genetic evidence for archaic admixture in Africa. *Proceedings of the National Academy of Sciences of the United States of America, 108*(37), 15123–15128. https://doi.org/10.1073/pnas.1109300108.

Hammond, K. A., Szewczak, J., & Król, E. (2001). Effects of altitude and temperature on organ phenotypic plasticity along an altitudinal gradient. *The Journal of Experimental Biology, 204*(Pt 11), 1991–2000.

Hare, B. (2017). Survival of the friendliest: *Homo sapiens* evolved via selection for prosociality. *Annual Review of Psychology, 68*, 155–186. https://doi.org/10.1146/annurev-psych-010416-044201.

Hare, B., Wobber, V., & Wrangham, R. (2012). The self-domestication hypothesis: Evolution of bonobo psychology is due to selection against aggression. *Animal Behaviour, 83*(3), 573–585. https://doi.org/10.1016/j.anbehav.2011.12.007.

Harpending, H., & Rogers, A. (2000). Genetic perspectives on human origins and differentiation. *Annual Review of Genomics and Human Genetics, 1*(2000), 361–385. https://doi.org/10.1146/annurev.genom.1.1.361.

Harvati, K. (2010). Neanderthals. *Evolution: Education and Outreach, 3*, 367–376. https://doi.org/10.1007/s12052-010-0250-0.

Harvati, K., Stringer, C., Grün, R., Aubert, M., Allsworth-Jones, P., & Folorunso, C. A. (2011). The later stone age calvaria from Iwo Eleru, Nigeria: Morphology and chronology. *PLoS One, 6*(9). https://doi.org/10.1371/journal.pone.0024024.

Harvati, K., & Weaver, T. D. (2006a). Human cranial anatomy and the differential preservation of population history and climate signatures. *Anatomical Record, 288*(12), 1225–1233. https://doi.org/10.1002/ar.a.20395.

Harvati, K., & Weaver, T. D. (2006b). Reliability of cranial morphology in reconstructing Neanderthal phylogeny. In J.-J. Hublin, K. Harvati, & T. Harrison (Eds.), *Neanderthals revisited: New approaches and perspectives* Springer Netherlands. https://doi.org/10.1007/978-1-4020-5121-0_13.

Hauser, M. D., Yang, C., Berwick, R. C., Tattersall, I., Ryan, M. J., Watumull, J., et al. (2014). The mystery of language evolution. *Frontiers in Psychology, 5*, 401. https://doi.org/10.3389/fpsyg.2014.00401.

Henneberg, M. (1988). Decrease of human skull size in the Holocene. *Human Biology, 60*(3), 395–405.

Henrich, J. (2015). *The secret of our success: How culture is driving human evolution, domesticating our species, and making us smarter.* Princeton, NJ: Princeton University Press.

Hesler, N., & Fischer, J. (2007). Gestural communication in Barbary macaques (*Macaca sylvanus*): An overview. In J. Call, & M. Tomasello (Eds.), *The gestural communication of apes and monkeys* (pp. 159–195). Lawrence Erlbaum.

Hillson, S., & Trinkaus, E. (2002). Comparative dental crown metrics. In J. Zilhao, & E. Trinkaus (Eds.), *Portrait of the artist as a child. The Gravettian Human Skeleton from the Abrigo do Lagar Velho and its Archeological Context* (pp. 355–364). Instituto Português de Arqueologia.

Hinton, R. J., & Carlson, D. S. (1979). Temporal changes in human temporomandibular joint size and shape. *American Journal of Physical Anthropology, 50*(3), 325–333. https://doi.org/10.1002/ajpa.1330500305.

Holmes, M. A., & Ruff, C. B. (2011). Dietary effects on development of the human mandibular corpus. *American Journal of Physical Anthropology, 145*(4), 615–628. https://doi.org/10.1002/ajpa.21554.

Hrdlicka, A. (1912). Artificial deformations of the human skull with special reference to America. In *Actas del 17mo Congreso Internacional de Americanistas* (pp. 147–149).

Hubbe, M., Hanihara, T., & Harvati, K. (2009). Climate signatures in the morphological differentiation of worldwide modern human populations. *Anatomical Record, 292*(11), 1720–1733. https://doi.org/10.1002/ar.20976.

Hublin, J.-J. (1998). Neanderthal acculturation in Western Europe. *Current Anthropology, 39*, S24–S25.

Hublin, J.-J., Ben-Ncer, A., Bailey, S. E., Freidline, S. E., Neubauer, S., Skinner, M. M., et al. (2017). New fossils from Jebel Irhoud, Morocco and the pan-African origin of *Homo sapiens. Nature, 546*(7657), 289–292. https://doi.org/10.1038/nature22336.

Huerta-Sánchez, E., Jin, X., Asan, Bianba, Z., Peter, B. M., Vinckenbosch, N., et al. (2014). Altitude adaptation in Tibetans caused by introgression of Denisovan-like DNA. *Nature, 512*(7513), 194–197. https://doi.org/10.1038/nature13408.

Hylander, W. L. (1977). Morphological changes in human teeth and jaws in a high-attrition environment. In A. A. Dahlberg, & T. M. Graber (Eds.), *Orofacial growth and development* (pp. 301–330). Mouton Publishers.

Imbelloni, J. (1925). Deformaciones intencionales del cráneo en Sudamérica: Polígonos craneanos aberrantes. *Revista Del Museo de La Plata, 38*, 329–407.

Jacobs, Z., Li, B., Shunkov, M. V., Kozlikin, M. B., Bolikhovskaya, N. S., Agadjanian, A. K., et al. (2019). Timing of archaic hominin occupation of Denisova Cave in southern Siberia. *Nature, 565*(7741), 594–599. https://doi.org/10.1038/s41586-018-0843-2.

Jeong, C., Ozga, A. T., Witonsky, D. B., Malmström, H., Edlund, H., Hofman, C. A., et al. (2016). Long-term genetic stability and a high-altitude East Asian origin for the peoples of the high valleys of the Himalayan arc. *Proceedings of the National Academy of Sciences of the United States of America, 113*(27), 7485–7490. https://doi.org/10.1073/pnas.1520844113.

Kaifu, Y. (1997). Changes in mandibular morphology from the Jomon to modern periods in eastern Japan. *American Journal of Physical Anthropology, 104*(2), 227–243. https://doi.org/10.1002/(SICI)1096-8644(199710)104:2<227::AID-AJPA9>3.0.CO;2-V.

Kanazawa, E., & Kasai, K. (1998). Comparative study of vertical sections of the Jomon and modern Japanese mandibles. *Anthropological Science, 106*, 107–118. https://doi.org/10.1537/ase.106.supplement_107.

Katz, D. C., Grote, M. N., & Weaver, T. D. (2016). A mixed model for the relationship between climate and human cranial form. *American Journal of Physical Anthropology, 160*(4), 593–603. https://doi.org/10.1002/ajpa.22896.

Katz, D. C., Grote, M. N., & Weaver, T. D. (2017). Changes in human skull morphology across the agricultural transition are consistent with softer diets in preindustrial farming groups. *Proceedings of the National Academy of Sciences of the United States of America, 114*(34), 9050–9055. https://doi.org/10.1073/pnas.1702586114.

Katzmarzyk, P. T., & Leonard, W. R. (1998). Climatic influences on human body size and proportions: Ecological adaptations and secular trends. *American Journal of Physical Anthropology, 106*(4), 483–503. https://doi.org/10.1002/(SICI)1096-8644(199808)106:4<483::AID-AJPA4>3.0.CO;2-K.

Klingenberg, C. P. (2005). Developmental constraints, modules, and evolvability. In B. Hallgrímsson, & B. K. Hall (Eds.), *Variation* (pp. 219–247). Academic Press. https://doi.org/10.1016/B978-012088777-4/50013-2.

Klingenberg, C. P. (2008). Morphological integration and developmental modularity. *Annual Review of Ecology, Evolution, and Systematics, 39*(1), 115–132. https://doi.org/10.1146/annurev.ecolsys.37.091305.110054.

Klingenberg, C. P. (2013). Cranial integration and modularity: Insights into evolution and development from morphometric data. *Hystrix, the Italian Journal of Mammalogy, 24*(1), 43–58. https://doi.org/10.4404/hystrix-24.1-6367.

Krantz, G. S. (1973). Cranial hair and brow ridges. *Man, 9*(2), 109–111. https://doi.org/10.1111/j.1835-9310.1973.tb01381.x.

Kuratani, S. (2019). Evolution and development of the vertebrate cranium. In L. Nuño de la Rosa, & G. Müller (Eds.), *Evolutionary developmental biology* (pp. 1–15). Springer Nature Switzerland.

Kurten, B. (1979). The shadow of the brow. *Current Anthropology, 20*(1), 229–230. https://doi.org/10.1086/202246.

Kuzminsky, S. C., Tung, T. A., Hubbe, M., & Villaseñor-Marchal, A. (2016). The application of 3D geometric morphometrics and laser surface scanning to investigate the standardization of cranial vault modification in the Andes. *Journal of Archaeological Science, 10*, 507–513. https://doi.org/10.1016/j.jasrep.2016.11.007.

Lahr, M. M. (1996). *The evolution of modern human diversity: A study of cranial variation*. Cambridge University Press.

Larsen, C. S. (1984). Health and disease in prehistoric Georgia: The transition to agriculture. In M. N. Cohen, & G. J. Armelagos (Eds.), *Paleopathology at the origins of agriculture* (pp. 367–392). Academic Press.

Larsen, C. S. (1995). Biological changes in human populations with agriculture. *Annual Review of Anthropology, 24*, 185–213. https://doi.org/10.1146/annurev.an.24.100195.001153.

Larsen, C. S. (2006). The agricultural revolution as environmental catastrophe: Implications for health and lifestyle in the Holocene. *Quaternary International, 150*(1), 12–20. https://doi.org/10.1016/j.quaint.2006.01.004.

Leach, H. M. (2003). Human domestication reconsidered. *Current Anthropology, 44*(3), 349–368. https://doi.org/10.1086/368119.

Lieberman, D. E. (1996). How and why humans grow thin skulls: Experimental evidence for systemic cortical robusticity. *American Journal of Physical Anthropology, 101*(2), 217–236. https://doi.org/10.1002/(SICI)1096-8644(199610)101:2<217::AID-AJPA7>3.0.CO;2-Z.

Lieberman, D. E. (2011). *The evolution of the human head*. The Belknap Press of Harvard University Press.

Lieberman, P., & Crelin, E. S. (1971). On the speech of Neanderthal man. *Linguistic Inquiry, 2*(2), 203–222.

Lieberman, P., Crelin, E. S., & Klatt, D. H. (1972). Phonetic ability and related anatomy of the newborn and adult human, Neanderthal man, and the chimpanzee. *American Anthropologist, 74*(3), 287–307.

Lieberman, D. E., Krovitz, G. E., & McBratney-Owen, B. (2004a). Testing hypotheses about tinkering in the fossil record: The case of the human skull. *Journal of Experimental Zoology Part B: Molecular and Developmental Evolution, 302*(3), 284–301. https://doi.org/10.1002/jez.b.21004.

Lieberman, D. E., Krovitz, G. E., Yates, F. W., Devlin, M., & St. Claire, M. (2004b). Effects of food processing on masticatory strain and craniofacial growth in a retrognathic face. *Journal of Human Evolution, 46*(6), 655–677. https://doi.org/10.1016/j.jhevol.2004.03.005.

Lieberman, D. E., McBratney, B. M., & Krovitz, G. (2002). The evolution and development of cranial form in *Homo sapiens*. *Proceedings of the National Academy of Sciences of the United States of America, 99*(3), 1134–1139.

Maddux, S. D., & Butaric, L. N. (2017). Zygomaticomaxillary morphology and maxillary sinus form and function: How spatial constraints influence pneumatization patterns among modern humans. *Anatomical Record, 300*(1), 209–225. https://doi.org/10.1002/ar.23447.

Maestripieri, D. (1996). Gestural communication and its cognitive implications in pigtail macaques (*Macaca nemestrina*). *Behaviour, 133*(13–14), 997–1022. https://doi.org/10.1163/156853996X00576.

Manica, A., Amos, W., Balloux, F., & Hanihara, T. (2007). The effect of ancient population bottlenecks on human phenotypic variation. *Nature, 448*(7151), 346–348. https://doi.org/10.1038/nature05951.

Manica, A., Prugnolle, F., & Balloux, F. (2005). Geography is a better determinant of human genetic differentiation than ethnicity. *Human Genetics, 118*(3–4), 366–371. https://doi.org/10.1007/s00439-005-0039-3.

Manríquez, G., González-Bergás, F. E., Salinas, J. C., & Espoueys, O. (2006). Deformación intencional del cráneo en poblaciones arqueológicas de Arica, Chile: Análisis preliminar de morfometría geométrica con uso de radiografías craneofaciales. *Chungará, 38*(1), 13–34. https://doi.org/10.4067/S0717 73562006000100004.

Martin, D. L., Goodman, A. H., & Gerven, D. P. V. (1984). The effects of socioeconomic change in prehistoric Africa: Sudanese Nubia as a case study. In M. N. Cohen, & G. J. Armelagos (Eds.), *Paleopathology at the origins of agriculture* (pp. 193–214). Academic Press.

Martínez-Abadías, N., Esparza, M., Sjøvold, T., González-José, R., Santos, M., & Hernández, M. (2009). Heritability of human cranial dimensions: Comparing the evolvability of different cranial regions. *Journal of Anatomy, 214*(1), 19–35. https://doi.org/10.1111/j.1469-7580.2008.01015.x.

Martinón-Torres, M., Castro, B., Xing, S., Wu, X., & Liu, W. (2019). What do Denisovans look like? Looking into the Middle and Late Pleistocene hominin fossil record from Asia. *American Journal of Physical Anthropology, 168*(S68), 156–157.

May, H., Sella-Tunis, T., Pokhojaev, A., Peled, N., & Sarig, R. (2018). Changes in mandible characteristics during the terminal Pleistocene to Holocene Levant and their association with dietary habits. *Journal of Archaeological Science: Reports, 22*, 413–419. https://doi.org/10.1016/j.jasrep.2018.03.020.

McBrearty, S., & Brooks, A. S. (2000). The revolution that wasn't: A new interpretation of the origin of modern human behavior. *Journal of Human Evolution, 39*(5), 453–563. https://doi.org/10.1006/jhev.2000.0435.

McCarthy, R. C., & Lieberman, D. E. (2001). Posterior maxillary (PM) plane and anterior cranial architecture in primates. *Anatomical Record, 264*(3), 247–260. https://doi.org/10.1002/ar.1167.

Meikle, M. C. (2002). *Craniofacial development, growth and evolution*. Bressingham: Bateson.

Mellars, P., & Stringer, C. (1989). *The human revolution: Behavioural and biological perspectives in the origins of modern humans*. Edinburgh University Press.

Menéndez, L., Bernal, V., Novellino, P., & Perez, S. I. (2014). Effect of bite force and diet composition on craniofacial diversification of Southern South American human populations. *American Journal of Physical Anthropology, 155*(1), 114–127. https://doi.org/10.1002/ajpa.22560.

Menéndez, L., & Buck, L. T. (2022). Evaluating potential proximate and ultimate causes of phenotypic change in the human skeleton over the agricultural transition. In T. R. Schultz, P. Peregrine, & R. Gawne (Eds.), *The convergent evolution of agriculture in humans and insects* (pp. 225–256). Cambridge, MA: MIT Press.

Menéndez, L., & Lotto, F. (2016). Evaluating potential proximate and ultimate causes of phenotypic change in the human skeleton over the agricultural transition. *Comechingonia, 20*(1), 143–174.

Meyer, M., Kircher, M., Gansauge, M.-T., Li, H., Racimo, F., Mallick, S., et al. (2012). A high-coverage genome sequence from an archaic Denisovan individual. *Science, 338*(6104), 222–226. https://doi.org/10.1126/science.1224344.

Moss, M. L. (1958). The pathogenesis of artificial cranial deformation. *American Journal of Physical Anthropology, 16*(3), 269–286. https://doi.org/10.1002/ajpa.1330160302.

Motohashi, T., & Kunisada, T. (2015). Extended multipotency of neural crest cells and neural crest derived cells. *Current Topics in Developmental Biology, 111*, 69–95. https://doi.org/10.1016/bs.ctdb.2014.11.003.

Munizaga, J. (1987). Deformación craneana intencional en América. *Revista Chilena de Antropología, 6*, 113–147.

Munizaga, J. (1992). Antropología física de los Andes del sur. In B. Meggers (Ed.), *Prehistoria Sudamericana. Nuevas Perspectivas* (pp. 65–75). (Taraxacum).

Neaux, D., Gilissen, E., Coudyzer, W., & Guy, F. (2015). Implications of the relationship between basicranial flexion and facial orientation for the evolution of hominid craniofacial structures. *International Journal of Primatology, 36*(6), 1120–1131. https://doi.org/10.1007/s10764-015-9886-5.

Neumann, G. K. (1942). Types of artificial cranial deformation in the eastern United States. *American Antiquity, 7*(3), 306–310. https://doi.org/10.2307/275486.

Newman, M. T. (1953). The application of ecological rules to the racial anthropology of the aboriginal new world. *American Anthropologist, 55*(3), 311–327. https://doi.org/10.1525/aa.1953.55.3.02a00020.

Nielsen, R., Akey, J. M., Jakobsson, M., Pritchard, J. K., Tishkoff, S., & Willerslev, E. (2017). Tracing the peopling of the world through genomics. *Nature, 541*(7637), 302–310. https://doi.org/10.1038/nature21347.

Noback, M. L., & Harvati, K. (2015). The contribution of subsistence to global human cranial variation. *Journal of Human Evolution, 80*, 34–50. https://doi.org/10.1016/j.jhevol.2014.11.005.

Noback, M. L., Harvati, K., & Spoor, F. (2011). Climate-related variation of the human nasal cavity. *American Journal of Physical Anthropology, 145*(4), 599–614. https://doi.org/10.1002/ajpa.21523.

Noback, M. L., Samo, E., van Leeuwen, C. H. A., Lynnerup, N., & Harvati, K. (2016). Paranasal sinuses: A problematic proxy for climate adaptation in Neanderthals. *Journal of Human Evolution, 97*, 176–179. https://doi.org/10.1016/j.jhevol.2016.06.003.

O'Brien, T., & Stanley, A. (2011). Boards and cords: Discriminating types of artificial cranial deformation in prehispanic South Central Andean populations. *International Journal of Osteoarchaeology, 34*(4), 459–470. https://doi.org/10.1002/oa.1269.

Olson, E. C., & Miller, R. L. (1958). *Morphological integration*. University of Chicago Press.

Parr, W. C. H., Wilson, L. A. B., Wroe, S., Colman, N. J., Crowther, M. S., & Letnic, M. (2016). Cranial shape and the modularity of hybridization in dingoes and dogs; hybridization does not spell the end for native morphology. *Evolutionary Biology, 43*(2), 171–187. https://doi.org/10.1007/s11692-016-9371-x.

Paschetta, C., De Azevedo, S., Castillo, L., Martínez-Abadías, N., Hernández, M., Lieberman, D. E., et al. (2010). The influence of masticatory loading on craniofacial morphology: A test case across technological transitions in the Ohio valley. *American Journal of Physical Anthropology, 141*(2), 297–314. https://doi.org/10.1002/ajpa.21151.

Pearson, O. M. (2008). Statistical and biological definitions of "anatomically modern" humans: Suggestions for a unified approach to modern morphology. *Evolutionary Anthropology: Issues, News, and Reviews*, 38–48. https://doi.org/10.1002/evan.20155.

Pearson, O. M. (2013). Hominin evolution in the Middle-late Pleistocene: Fossils, adaptive scenarios, and alternatives. *Current Anthropology, 54*, S221–S233. https://doi.org/10.1086/673503.

Perez, S. I. (2007). Artificial cranial deformation in South America: A geometric morphometrics approximation. *Journal of Archaeological Science, 34*(10), 1649–1658. https://doi.org/10.1016/j.jas.2006.12.003.

Perez, S. I., Lema, V., Diniz-Filho, J. A. F., Bernal, V., Gonzalez, P. N., Gobbo, D., et al. (2011). The role of diet and temperature in shaping cranial diversification of South American human populations: An approach based on spatial regression and divergence rate tests. *Journal of Biogeography, 38*(1), 148–163. https://doi.org/10.1111/j.1365-2699.2010.02392.x.

Petajan, J. H. (1973). Neuropsychological acclimatization to high altitude. *Journal of Human Evolution, 2*(2), 105–115. https://doi.org/10.1016/0047-2484(73)90058-4.

Pinhasi, R., Eshed, V., & Shaw, P. (2008). Evolutionary changes in the masticatory complex following the transition to farming in the southern Levant. *American Journal of Physical Anthropology, 135*(2), 136–148. https://doi.org/10.1002/ajpa.20715.

Pinhasi, R., Eshed, V., & von Cramon-Taubadel, N. (2015). Incongruity between affinity patterns based on mandibular and lower dental dimensions following the transition to agriculture in the Near East, Anatolia and Europe. *PLoS One, 10*(2). https://doi.org/10.1371/journal.pone.0117301, e0117301.

Pokhojaev, A., Avni, H., Sella-Tunis, T., Sarig, R., & May, H. (2019). Changes in human mandibular shape during the Terminal Pleistocene-Holocene Levant. *Scientific Reports, 9*(1), 8799. https://doi.org/10.1038/s41598-019-45279-9.

Pomeroy, E., Stock, J. T., Zakrzewski, S. R., & Lahr, M. M. (2010). A metric study of three types of artificial cranial modification from north-central Peru. *International Journal of Osteoarchaeology, 20*(3), 317–334. https://doi.org/10.1002/oa.1044.

Ponce de León, M. S., Koesbardiati, T., Weissmann, J. D., Milella, M., Reyna-Blanco, C. S., Suwa, G., et al. (2018). Human bony labyrinth is an indicator of population history and dispersal from Africa. *Proceedings of the National Academy of Sciences of the United States of America, 115*(16), 4128–4133. https://doi.org/10.1073/pnas.1717873115.

Profico, A., Piras, P., Buzi, C., Di Vincenzo, F., Lattarini, F., Melchionna, M., et al. (2017). The evolution of cranial base and face in Cercopithecoidea and Hominoidea: Modularity and morphological integration. *American Journal of Primatology, 79*(12). https://doi.org/10.1002/ajp.22721, e22721.

Prüfer, K., Racimo, F., Patterson, N., Jay, F., Sankararaman, S., Sawyer, S., et al. (2014). The complete genome sequence of a Neanderthal from the Altai Mountains. *Nature, 505*(7481), 43–49. https://doi.org/10.1038/nature12886.

Püschel, T. A., Friess, M., & Manríquez, G. (2020). Morphological consequences of artificial cranial deformation: Modularity and integration. *PLoS One, 15*(1). https://doi.org/10.1371/journal.pone.0227362, e0227362.

Racimo, F., Sankararaman, S., Nielsen, R., & Huerta-Sánchez, E. (2015). Evidence for archaic adaptive introgression in humans. *Nature Reviews Genetics, 16*(6), 359–371. https://doi.org/10.1038/nrg3936.

Rademaker, K., Hodgins, G., Moore, K., Zarrillo, S., Miller, C., Bromley, G. R. M., et al. (2014). Paleoindian settlement of the high-altitude Peruvian Andes. *Science, 346*(6208), 466–469. https://doi.org/10.1126/science.1258260.

Rae, T. C., Koppe, T., & Stringer, C. B. (2011). The Neanderthal face is not cold adapted. *Journal of Human Evolution, 60*(2), 234–239. https://doi.org/10.1016/j.jhevol.2010.10.003.

Ramachandran, S., Deshpande, O., Roseman, C. C., Rosenberg, N. A., Feldman, M. W., & Cavalli-Sforza, L. L. (2005). Support from the relationship of genetic and geographic in human populations for a serial founder effect originating in Africa. *Proceedings of the National Academy of Sciences of the United States of America, 102*(44), 15942–15947. https://doi.org/10.1073/pnas.0507611102.

Rees, J. S., Castellano, S., & Andrés, A. M. (2020). The genomics of human local adaptation. *Trends in Genetics, 36*(6), 415–428. https://doi.org/10.1016/j.tig.2020.03.006.

Reich, D., Green, R. E., Kircher, M., Krause, J., Patterson, N., Durand, E. Y., et al. (2010). Genetic history of an archaic hominin group from Denisova Cave in Siberia. *Nature, 468*(7327), 1053–1060. https://doi.org/10.1038/nature09710.

Relethford, J. H. (2002). Apportionment of global human genetic diversity based on craniometrics and skin color. *American Journal of Physical Anthropology, 118*(4), 393–398. https://doi.org/10.1002/ajpa.10079.

Reyes-Centeno, H., Ghirotto, S., Détroit, F., Grimaud-Hervé, D., Barbujani, G., & Harvati, K. (2014). Genomic and cranial phenotype data support multiple modern human dispersals from Africa and a southern route into Asia. *Proceedings of the National Academy of Sciences of the United States of America*, 111(20), 7248–7253. https://doi.org/10.1073/pnas.1323666111.

Richter, D., Grün, R., Joannes-Boyau, R., Steele, T. E., Amani, F., Rué, M., et al. (2017). The age of the hominin fossils from Jebel Irhoud, Morocco, and the origins of the Middle Stone Age. *Nature*, 546(7657), 293–296. https://doi.org/10.1038/nature22335.

Rolian, C. (2009). Integration and evolvability in primate hands and feet. *Evolutionary Biology*, 36, 100–117. https://doi.org/10.1007/s11692-009-9049-8.

Rosas, A., & Bastir, M. (2002). Thin-plate spline analysis of allometry and sexual dimorphism in the human craniofacial complex. *American Journal of Physical Anthropology*, 117(3), 236–245. https://doi.org/10.1002/ajpa.10023.

Roseman, C. C. (2004). Detecting interregionally diversifying natural selection on modern human cranial form by using matched molecular and morphometric data. *Proceedings of the National Academy of Sciences of the United States of America*, 101(35), 12824–12829. https://doi.org/10.1073/pnas.0402637101.

Roseman, C. C. (2016). Random genetic drift, natural selection, and noise in human cranial evolution. *American Journal of Physical Anthropology*, 160(4), 582–592. https://doi.org/10.1002/ajpa.22918.

Roseman, C. C., & Weaver, T. D. (2004). Multivariate apportionment of global human craniometric diversity. *American Journal of Physical Anthropology*, 125(3), 257–263. https://doi.org/10.1002/ajpa.10424.

Roseman, C. C., & Weaver, T. D. (2007). Molecules versus morphology? Not for the human cranium. *Bio Essays*, 29(12), 1185–1188. https://doi.org/10.1002/bies.20678.

Ruff, C. B., Trinkaus, E., & Holliday, T. W. (1997). Body mass and encephalization in Pleistocene *Homo*. *Nature*, 387(6629), 173–176. https://doi.org/10.1038/387173a0.

Russell, M. D. (1985). The supraorbital torus: "A most remarkable peculiarity.". *Current Anthropology*, 26(3), 337–360. https://doi.org/10.1086/203279.

Sánchez-Villagra, M. R., & van Schaik, C. P. (2019). Evaluating the self-domestication hypothesis of human evolution. *Evolutionary Anthropology*, 28(3), 133–143. https://doi.org/10.1002/evan.21777.

Sardi, M. L., Novellino, P. S., & Pucciarelli, H. M. (2006). Craniofacial morphology in the Argentine Center-West: Consequences of the transition to food production. *American Journal of Physical Anthropology*, 130(3), 333–343. https://doi.org/10.1002/ajpa.20379.

Sardi, M. L., Ramírez Rozzi, F., & Pucciarelli, H. M. (2004). The Neolithic transition in Europe and North Africa. The functional craneology contribution. *Anthropologischer Anzeiger; Bericht Über Die Biologisch-Anthropologische Literatur*, 62(2), 129–145. https://doi.org/10.1127/anthranz/62/2004/129.

Schroeder, L., & Ackermann, R. R. (2017). Evolutionary processes shaping diversity across the *Homo* lineage. *Journal of Human Evolution*, 111, 1–17. https://doi.org/10.1016/j.jhevol.2017.06.004.

Schroeder, L., & von Cramon-Taubadel, N. (2017). The evolution of hominoid cranial diversity: A quantitative genetic approach. *Evolution*, 71(11), 2634–2649. https://doi.org/10.1111/evo.13361.

Schwartz, J. H. (2016). What constitutes *Homo sapiens*? Morphology versus received wisdom. *Journal of Anthropological Sciences*, 94, 65–80. https://doi.org/10.4436/jass.94028.

Scott, J. E., McAbee, K. R., Eastman, M. M., & Ravosa, M. J. (2014). Teaching an old jaw new tricks: Diet-induced plasticity in a model organism from weaning to adulthood. *Journal of Experimental Biology*, 217(22), 4099–4107. https://doi.org/10.1242/jeb.111708.

Sella-Tunis, T., Pokhojaev, A., Sarig, R., O'Higgins, P., & May, H. (2018). Human mandibular shape is associated with masticatory muscle force. *Scientific Reports*, 8(1), 6042. https://doi.org/10.1038/s41598-018-24293-3.

Serna, A., Prates, L., Flensborg, G., Martínez, G., Favier Dubois, C., & Perez Sergio, I. (2019). Does the shape make a difference? Evaluating the ethnic role of cranial modification in the Pampa-Patagonia region (Argentina) during the late Holocene. *Archaeological and Anthropological Sciences*, 11(6), 2597–2610. https://doi.org/10.1007/s12520-018-0687-6.

Shea, J. J. (2011). *Homo sapiens* is as *Homo sapiens* was: Behavioral variability versus "behavioral modernity" in Paleolithic archaeology. *Current Anthropology*, 52(1), 1–35. https://doi.org/10.1086/658067.

Slon, V., Mafessoni, F., Vernot, B., de Filippo, C., Grote, S., Viola, B., et al. (2018). The genome of the offspring of a Neanderthal mother and a Denisovan father. *Nature*, 561(7721), 113–116. https://doi.org/10.1038/s41586-018-0455-x.

Smith, F. H. (1983). Behavioural interpretations of changes in craniofacial morphology across the archaic/modern *Homo sapiens* transition. In E. Trinkaus (Ed.), *British archaeological reports, international series 164. Oxford. The mousterian legacy: Human biocultural change in the upper pleistocene* (pp. 141–163).

Smith, H. F. (2009). Which cranial regions reflect molecular distances reliably in humans? Evidence from three-dimensional morphology. *American Journal of Human Biology, 21*(1), 36–47. https://doi.org/10.1002/ajhb.20805.

Smith, P., Bar-Yosef, O., & Sillen, A. (1984). Archaeological and skeletal evidence for dietary change during the late Pleistocene/early Holocene in the Levant. In M. N. Cohen, & G. J. Armelagos (Eds.), *Paleopathology at the origins of agriculture* (pp. 101–136). Academic Press.

Spencer, M. A., & Demes, B. (1993). Biomechanical analysis of masticatory system configuration in Neandertals and Inuits. *American Journal of Physical Anthropology, 91*(1), 1–20. https://doi.org/10.1002/ajpa.1330910102.

Stini, W. A. (1975). Adaptive strategies of human populations under nutritional stress. In E. S. Watts, F. E. Johnston, & G. W. Lasker (Eds.), *Biosocial interrelations in population adaptation* (pp. 19–42). Mouton & Co.

Stojanowski, C. M. (2014). Iwo Eleru's place among Late Pleistocene and Early Holocene populations of North and East Africa. *Journal of Human Evolution, 75*, 80–89. https://doi.org/10.1016/j.jhevol.2014.02.018.

Stringer, C. B., & Andrews, P. (1988). Genetic and fossil evidence for the origin of modern humans. *Science, 239*(4846), 1263–1268. https://doi.org/10.1126/science.3125610.

Stringer, C. B., & Buck, L. T. (2014). Diagnosing *Homo sapiens* in the fossil record. *Annals of Human Biology, 41*(4), 312–322. https://doi.org/10.3109/03014460.2014.922616.

Tappen, N. C. (1973). Structure of bone in the skulls of Neanderthal fossils. *American Journal of Physical Anthropology, 38*(1), 93–97. https://doi.org/10.1002/ajpa.1330380123.

Tattersall, I., & Schwartz, J. H. (2008). The morphological distinctiveness of *Homo sapiens* and its recognition in the fossil record: Clarifying the problem. *Evolutionary Anthropology, 17*(1), 49–54. https://doi.org/10.1002/evan.20153.

Thomson, A., & Buxton, L. H. D. (1923). Man's nasal index in relation to certain climatic conditions. *The Journal of the Royal Anthropological Institute of Great Britain and Ireland, 53*, 92–122. https://doi.org/10.2307/2843753.

Thorne, A. G., & Wilson, S. R. (1977). Pleistocene and recent Australians: A multivariate comparison. *Journal of Human Evolution, 6*(4), 393–402. https://doi.org/10.1016/S0047-2484(77)80007-9.

Tiesler, V. (2014). Head shapes in classic period Mesoamerica. In V. Tielser (Ed.), *The bioarchaeology of artificial cranial modifications* (pp. 185–207). Springer.

Toledo Avelar, L. E., Cardoso, M. A., Santos Bordoni, L., de Miranda Avelar, L., & de Miranda Avelar, J. V. (2017). Aging and sexual differences of the human skull. *Plastic and Reconstructive Surgery, 5*(4). https://doi.org/10.1097/GOX.0000000000001297, e1297.

Tomaszewska, A., Kwiatkowska, B., & Jankauskas, R. (2015). Is the area of the orbital opening in humans related to climate? *American Journal of Human Biology, 27*(6), 845–850. https://doi.org/10.1002/ajhb.22735.

Tommaseo, M., & Drusini, A. (1984). Physical anthropology of two tribal groups of Amazonic Peru (with reference to artificial cranial deformation). *Zeitschrift Fur Morphologie Und Anthropologie, 74*(3), 315–333.

Topinard, P. (1879). Des déformations ethniques du crane. *Revue d'Anthropologie, 2*, 496–506.

Torres-Rouff, C. (2002). Cranial vault modification and ethnicity in middle horizon San Pedro de Atacama, Chile. *Current Anthropology, 43*(1), 163–171. https://doi.org/10.1086/338290.

Torres-Rouff, C. (2003). *Shaping identity: Cranial vault modification in the pre-Columbian Andes* (PhD thesis). University of California.

Torres-Rouff, C., & Yablonsky, L. T. (2005). Cranial vault modification as a cultural artifact: A comparison of the Eurasian steppes and the Andes. *HOMO- Journal of Comparative Human Biology, 56*(1), 1–16. https://doi.org/10.1016/j.jchb.2004.09.001.

Trinkaus, E. (1982a). A history of *Homo erectus* and *Homo sapiens* paleontology in America. In F. Spencer (Ed.), *A history of American Physical Anthropology 1930–1980* (pp. 261–280). Academic Press.

Trinkaus, E. (1982b). Artificial cranial deformation in the Shanidar 1 and 5 Neandertals. *Current Anthropology, 23*(2), 198–199. https://doi.org/10.1086/202808.

Trinkaus, E. (1987). The Neandertal face: Evolutionary and functional perspectives on a recent hominid face. *Journal of Human Evolution, 16*(5), 429–443. https://doi.org/10.1016/0047-2484(87)90071-6.

Trinkaus, E. (2018). An abundance of developmental anomalies and abnormalities in Pleistocene people. *Proceedings of the National Academy of Sciences of the United States of America, 115*(47), 11941–11946. https://doi.org/10.1073/pnas.1814989115.

Trut, L. N. (1999). Early canid domestication: The farm-fox experiment: Foxes bred for tamability in a 40-year experiment exhibit remarkable transformations that suggest an interplay between behavioral genetics and development. *American Scientist, 87*(2), 160–169. https://doi.org/10.1511/1999.2.160.

Trut, L. N., Oskina, I., & Kharlamova, A. (2009). Animal evolution during domestication: The domesticated fox as a model. *BioEssays, 31*(3), 349–360. https://doi.org/10.1002/bies.200800070.

Trut, L. N., Plyusnina, I. Z., & Oskina, I. N. (2004). An experiment on fox domestication and debatable issues of evolution of the dog. *Russian Journal of Genetics*, 40(6), 644–655. https://doi.org/10.1023/B:RUGE.0000033312.92773.c1.

Ubelaker, D. (1984). *Human skeletal remains: Excavation, analysis, interpretation*. Washington, DC: Taraxacum.

Vernot, B., & Akey, J. M. (2014). Resurrecting surviving Neandertal lineages from modern human genomes. *Science*, 343(6174), 1017–1021. https://doi.org/10.1126/science.1245938.

Vidal, C. M., Lane, C. S., Asrat, A., Barfod, D. N., Mark, D. F., Tomlinson, E. L., et al. (2022). Age of the oldest known *Homo sapiens* from eastern Africa. *Nature*, 601(7894), 579–583. https://doi.org/10.1038/s41586-021-04275-8.

Villanea, F. A., & Schraiber, J. G. (2019). Multiple episodes of interbreeding between Neanderthal and modern humans. *Nature Ecology and Evolution*, 3(1), 39–44. https://doi.org/10.1038/s41559-018-0735-8.

Vioarsdóttir, U. S., O'Higgins, P., & Stringer, C. (2002). A geometric morphometric study of regional differences in the ontogeny of the modern human facial skeleton. *Journal of Anatomy*, 201(3), 211–229. https://doi.org/10.1046/j.1469-7580.2002.00092.x.

Viola, B. T., Gunz, P., Neubauer, S., Slon, V., Kozlikin, M. B., Shunkov, M. V., et al. (2019). A parietal fragment from Denisova cave. *American Journal of Physical Anthropology*, 168(S68), 258.

von Cramon-Taubadel, N. (2009a). Congruence of individual cranial bone morphology and neutral molecular affinity patterns in modern humans. *American Journal of Physical Anthropology*, 140(2), 205–215. https://doi.org/10.1002/ajpa.21041.

von Cramon-Taubadel, N. (2009b). Revisiting the homoiology hypothesis: The impact of phenotypic plasticity on the reconstruction of human population history from craniometric data. *Journal of Human Evolution*, 57(2), 179–190. https://doi.org/10.1016/j.jhevol.2009.05.009.

von Cramon-Taubadel, N. (2011). Global human mandibular variation reflects differences in agricultural and hunter-gatherer subsistence strategies. *Proceedings of the National Academy of Sciences of the United States of America*, 108(49), 19546–19551. https://doi.org/10.1073/pnas.1113050108.

von Cramon-Taubadel, N. (2014). Evolutionary insights into global patterns of human cranial diversity: Population history, climatic and dietary effects. *Journal of Anthropological Sciences*, 92, 43–77. https://doi.org/10.4436/jass.91010.

von Cramon-Taubadel, N. (2017). Measuring the effects of farming on human skull morphology. *Proceedings of the National Academy of Sciences of the United States of America*, 114(34), 8917–8919. https://doi.org/10.1073/pnas.1711475111.

von Cramon-Taubadel, N., & Lycett, S. J. (2008). Brief communication: Human cranial variation fits iterative founder effect model with African origin. *American Journal of Physical Anthropology*, 136(1), 108–113. https://doi.org/10.1002/ajpa.20775.

Wagner, G. P., & Altenberg, L. (1996). Complex adaptations and the evolution of evolvability. *Evolution*, 50(3), 967–976. https://doi.org/10.1111/j.1558-5646.1996.tb02339.x.

Weaver, T. D., Roseman, C. C., & Stringer, C. B. (2007). Were Neandertal and modern human cranial differences produced by natural selection or genetic drift? *Journal of Human Evolution*, 53(2), 135–145. https://doi.org/10.1016/j.jhevol.2007.03.001.

Weaver, T. D., Roseman, C. C., & Stringer, C. B. (2008). Close correspondence between quantitative- and molecular-genetic divergence times for Neandertals and modern humans. *Proceedings of the National Academy of Sciences of the United States of America*, 105(12), 4645–4649. https://doi.org/10.1073/pnas.0709079105.

Weaver, T. D., & Stringer, C. B. (2015). Unconstrained cranial evolution in Neandertals and modern humans compared to common chimpanzees. *Proceedings of the Royal Society B: Biological Sciences*, 282(1817), 20151519. https://doi.org/10.1098/rspb.2015.1519.

Weinstein, K. J. (2017). Morphological signatures of high-altitude adaptations in the Andean archaeological record: Distinguishing developmental plasticity and natural selection. *Quaternary International*, 461, 14–24. https://doi.org/10.1016/j.quaint.2017.06.004.

Weiss, P. (1961). *Osteología Cultural, Prácticas Cefálicas: 2da Parte, Tipología de las Deformaciones Cefálicas—Estudio Cultural de los Tipos Cefálicos y de Algunas Enfermedades Oseas*. Universidad Nacional Mayor de San Marcos.

Weitz, C. A., Garruto, R. M., Chin, C. T., & Liu, J. C. (2004). Morphological growth and thorax dimensions among Tibetan compared to Han children, adolescents and young adults born and raised at high altitude. *Annals of Human Biology*, 31(3), 292–310. https://doi.org/10.1080/0301446042000196316.

White, T. D., Black, M. T., & Folkens, P. A. (2011). *Human osteology* (3rd ed.). Academic Press.

Wilkins, A. S., Wrangham, R. W., & Tecumseh Fitch, W. (2014). The "Domestication Syndrome" in mammals: A unified explanation based on neural crest cell behavior and genetics. *Genetics*, 197(3), 795–808. https://doi.org/10.1534/genetics.114.165423.

Williams, F. L., & Cofran, Z. (2016). Postnatal craniofacial ontogeny in Neandertals and modern humans. *American Journal of Physical Anthropology, 159*(3), 394–409. https://doi.org/10.1002/ajpa.22895.

Wolpoff, M. H. (1986). Describing anatomically modern *Homo sapiens*: A distinction without a definable difference. In V. V. Novothy, & A. Mizerova (Eds.), *Vol. 23. Fossil man- new facts, new ideas* (pp. 41–53). Anthropos Institute-Moravian Museum.

Wrangham, R. W. (2018). Two types of aggression in human evolution. *Proceedings of the National Academy of Sciences of the United States of America, 115*(2), 245–253. https://doi.org/10.1073/pnas.1713611115.

Wrangham, R. W. (2019). Hypotheses for the evolution of reduced reactive aggression in the context of human self-domestication. *Frontiers in Psychology, 10*, 1914. https://doi.org/10.3389/fpsyg.2019.01914.

Wrangham, R. W., & Pilbeam, D. (2002). African apes as time machines. In B. M. F. Galdikas, N. E. Briggs, L. K. Sheeran, G. L. Shapiro, & J. Goodall (Eds.), *All apes great and small* (pp. 5–17). Springer.

Wroe, S., Parr, W. C. H., Ledogar, J. A., Bourke, J., Evans, S. P., Fiorenza, L., et al. (2018). Computer simulations show that Neanderthal facial morphology represents adaptation to cold and high energy demands, but not heavy biting. *Proceedings of the Royal Society B: Biological Sciences, 285*(1876), 20180085. https://doi.org/10.1098/rspb.2018.0085.

Y'edynak, G., & Fleisch, S. (1983). Microevolution and biological adaptability in the transition from food-collecting to food-producing in the Iron Gates of Yugoslavia. *Journal of Human Evolution, 12*(3), 279–296. https://doi.org/10.1016/S0047-2484(83)80150-X.

Yokley, T. R. (2009). Ecogeographic variation in human nasal passages. *American Journal of Physical Anthropology, 138*(1), 11–22. https://doi.org/10.1002/ajpa.20893.

Zichello, J. M., Baab, K. L., McNulty, K. P., Raxworthy, C. J., & Steiper, M. E. (2018). Hominoid intraspecific cranial variation mirrors neutral genetic diversity. *Proceedings of the National Academy of Sciences of the United States of America, 115*(45), 11501–11506. https://doi.org/10.1073/pnas.1802651115.

Activity and the shoulder: From soft tissues to bare bones

Francisca Alves Cardoso[a,b] *and Aaron Gasparik*[c]

[a]LABOH—Laboratory of Biological Anthropology and Human Osteology, CRIA—Center for Research in Anthropology, NOVA University of Lisbon—School of Social Sciences and Humanities (NOVA FCSH), Lisbon, Portugal, [b]Cranfield Forensic Institute, Cranfield University, Defence Academy of the United Kingdom, Shrivenham, United Kingdom, [c]Institute of Archaeology, University College London, London, United Kingdom

4.1 Introduction

Humans perform a variety of activities using the upper limb, and the shoulder, with its many degrees of freedom, facilitates these arm movements with an almost unlimited range of motion (McGeough, 2013; Tytherleigh-Strong et al., 2001). Analyses of the glenohumeral joint—including any pathological or morphological changes that affect it—can potentially provide insight into a persons' changing activity patterns and, at a population scale, human activity patterns, habitual behaviours, the morphological evolution of the human arm, and/ or the body's adaptive responses to musculoskeletal stressors throughout history. As such, the shoulder is a key joint to consider in the analysis and interpretation of activity-related bone changes in human osteological remains. In this regard, the current chapter focuses on how behaviour in the past has been inferred from morphological changes to the shoulder complex.

In bioarchaeology and biological anthropology most of the methodological and interpretative approaches to inferring behaviour from skeletal changes in the shoulder and upper limb have focused on what are frequently referred to as markers of occupational stress (i.e., Havelková et al., 2011; Larsen, 2015; Molnar, 2006; Villotte et al., 2010). The concept of interpreting behaviour from musculoskeletal stress markers has been highlighted at length in earlier works that set out to improve theoretical and methodological approaches to the study of activity-related osseous changes (Henderson and Alves-Cardoso, 2013; Kennedy, 1998; Peterson and Hawkey, 1998). In 2012 Henderson and Alves-Cardoso addressed the technical and theoretical advances for the analysis of entheses at a symposium held at the 81st Annual

Meeting of the American Association of Physical Anthropology. The papers presented at the symposium resulted in a special issue published by the *International Journal of Osteoarchaeology* entitled "Entheseal Changes and Occupation: Technical and Theoretical Advances and Their Applications" (Henderson and Alves-Cardoso, 2013). This work is essential reading in the study of entheseal changes in the skeleton, as it sets the pace for studies of activity-related osseous change, discussing how entheseal changes have been used to infer behaviour and exploring how to better interpret the effects of activity on the skeleton.

Building on the existing literature discussing the reconstruction of behaviour and activities from the analysis of osseous changes, particularly the research published in the aforementioned special issue, the current work focuses specifically on the relationship between activity and changes in the shoulder joint. The general features of entheses will first be highlighted, as these features may show signs of forces exerted on the bone during life and are one of the key elements examined in archaeological studies of activity in the past. This chapter will then discuss the soft tissue structures that move and stabilise the glenohumeral joint to better understand the biomechanics of these soft tissue stabilisers and, in turn, how these components break down. Following these discussions, the chapter will provide a summary of current archaeological studies that have used osseous changes in the shoulder to interpret and discuss activity patterns, as well as present the ways in which this topic is approached in the clinical literature. The goal of this section is to illustrate the ways in which these two fields, bioarchaeology and clinical research, differ in their approaches to studies of activity and highlight how bioarchaeological analysis of the relationship between activity and osseous changes can continue to improve going forward. Furthermore, though changes in the shoulder joint are the primary focus of this chapter, it is important to emphasise that the methods and approaches highlighted here (as well as the shortcomings associated with these approaches) are applicable when analysing other synovial joints in the skeleton.

4.2 Activity versus occupation

In bioarchaeology, the terms 'occupation' and 'activity' are frequently used interchangeably. However, understanding and clarifying the meaning of these terms in the context of bioarchaeological research is an ongoing challenge. The term 'occupation' tends to be synonymous with a formal profession or paid employment—a carpenter, a factory worker, or a maid, for instance—rather than encompassing the broad scope of movements and physical tasks one can perform. An occupation is also always embedded in the social and cultural fabric of its context of origin, and although investigating past occupations is a key part of exploring cultural, social, and economic inequalities between populations and individuals in different temporal and geographic contexts, this contextual specificity must be kept in mind. What may be considered an occupation in one time period or place may not be understood as such in another setting, while the same occupation in different contexts may incorporate a different range of tasks. Furthermore, many activities undertaken during an individual's life may not relate to an occupation at all and could include hobbies, recreational sports, and/or the activities of daily life, such as personal grooming, childcare, feeding oneself, or locomotion. The categorisation of various 'occupations' directly relates to the methodologies used in a study to assess the remains, as well as the general aims of that specific study (Perréard Lopreno et al., 2013). Unfortunately, definitions often vary between researchers, making comparison between studies less straightforward. Different biological, social, and cultural ideas of 'occupation' impact the analysis and interpretation of findings across numerous studies and may bias results—ultimately demonstrating a need for the standardisation of the concept of 'occupation' in bioarchaeological studies (Alves-Cardoso and Henderson, 2013).

Conversely, the use of the term 'activity' takes into account the considerable range of motion of the shoulder joint and the array of physical activities that this mobility allows one to perform with the upper limb. It is possible that activity could be inferred from joint changes if the focus is placed on the frequency of the movements undertaken, the intensity of the movements (i.e., the amount of effort required to perform the activity), and their duration (i.e., how long the movements are performed for). Though an association between particular movements and activities is sometimes discussed in studies of specific diseases—such as with cases of lateral epicondylitis or 'tennis elbow' (Cutts et al., 2020)—clinical literature also shows that numerous musculoskeletal conditions have a multifactorial aetiology (Chourasia et al., 2013). That is not to say, however, that the relationship between 'activity' and joint changes is simple and straightforward. Even if morphological changes in the skeleton *are* related in some way to the activities that an individual performed during life, a range of other factors—age, biological sex, or pre-existing diseases, for instance—might affect the types of changes visible in the shoulder joint and must also be considered.

Given the confounding factors discussed here, the use of 'occupation' may limit osteological interpretations of the myriad factors that affect shoulder morphology and joint breakdown by excluding the effects of physical activities unrelated to employment. Avoiding this term can therefore help to avoid the potential interpretative biases associated with it, and for the purposes of this chapter, the authors will consider the concept of 'activity' rather than 'occupation'.

4.3 Entheses

Tendons connect to the skeleton at the osteotendinous junction, or *enthesis*, which is a specialised and complex anatomical feature that provides an interface between a highly elastic tendon and relatively inelastic bone surface (Benjamin et al., 2002, 2006; Thorpe et al., 2015). To balance the different mechanical properties of these materials and facilitate a stable tendon attachment into the bone, the fibres of most tendons and ligaments flare out at the insertion sites (Benjamin et al., 2006). This arrangement also creates a broader insertion base, allowing tensile forces to be dissipated over a wider surface area and mitigating the effects of movement-induced changes to the insertional angle (Benjamin et al., 2006; Schlecht, 2012). There are two types of entheses, *fibrous* and *fibrocartilaginous*, which are defined by the varieties of tissue present at the insertion (Fig. 4.1). The type of enthesis present determines the way in which the tendon connects to the bone (Apostolakos et al., 2014; Benjamin et al., 2006; Schlecht, 2012).

Fibrous entheses either attach muscles directly to the bone or blend with the periosteum where collagen fibres from the periosteum and tendon anchor the muscle to the bone

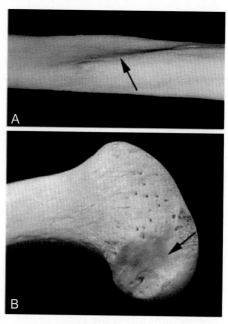

FIG. 4.1   Examples of fibrous (A) and fibrocartilaginous (B) entheses in the human body. Image A shows the midshaft of the humerus, where there is a roughened ridge created by the bony fibrous attachment tendon of the deltoid muscle. Image B shows the lateral condyle of the distal femur, where there is a smooth, well-circumscribed marking left by the tendon of the popliteus muscle. This marking is typical of those left on dried bones by fibrocartilaginous entheses. *Adapted from Benjamin, M., Kumai, T., Milz, S., Boszczyk, B. M., Boszczyk, A. A., & Ralphs, J. R. (2002). The skeletal attachment of tendons—Tendon "entheses". Comparative Biochemistry and Physiology. Part A, Molecular & Integrative Physiology, 133(4), 931–945. https://doi.org/10.1016/s1095-6433(02)00138-1, 939, used with permission.*

(Benjamin et al., 2002; Schlecht, 2012). These types of insertions are generally poorly demarcated, larger than their fibrocartilaginous counterparts, and more commonly found on the diaphyses of long bones and on the vertebral column (Apostolakos et al., 2014; Villotte and Knüsel, 2013). Fibrous entheses typically manifest as ridges or roughened patches on the bone, as observed at the deltoid tuberosity on the humerus, the *pronator teres* muscle attachment on the radius, or the interosseous crests on the radius and ulna. However, they may also leave smooth markings on the bone, though the boundaries of these impressions typically cover extensive areas and are less well circumscribed than fibrocartilaginous insertions (Benjamin et al., 2002).

Fibrocartilaginous entheses are typically located nearer to the joints. They insert at long bone epiphyses and apophyses (protuberances), in the hands and feet, and at the interspinous ligaments and *ligamentum flava* in the vertebrae (Apostolakos et al., 2014; Thorpe et al., 2015; Villotte and Knüsel, 2013). Examples include the rotator cuff tendon insertions in the proximal humerus, the bicipital tuberosity on the proximal radius, or the *Achilles* tendon insertion in the calcaneus. Fibrocartilaginous entheses have smoother surfaces and more well-defined borders compared to their fibrous equivalents (Villotte et al., 2010). The bone–tendon interface in a fibrocartilaginous enthesis is mediated by a specialised region of tissue that gradually transitions from tendon to fibrocartilage, then to calcified cartilage, and finally to bone (Apostolakos et al., 2014; Benjamin et al., 1986; Thorpe et al., 2015). The boundary between avascular fibrocartilage and calcified cartilage is known as the tidemark—a smooth and relatively straight border region that gives healthy fibrocartilaginous entheses their characteristically smooth, well-defined appearance and is the location where the soft tissue separates from macerated bone (Benjamin et al., 1986; Schlecht, 2012; Villotte et al., 2010). Functionally, the four-part tissue gradient distributes strains on the tendon–bone interface, helping to maintain integrity of the insertion while force is exerted on the joint during movement (Benjamin et al., 1986; Schlecht, 2012). It must also be noted that the collagen transition layers are thinner—or entirely absent—at the periphery of the enthesis; here, fibres fuse directly with the surrounding periosteal layer and there may be more vascular foramina present (Benjamin et al., 2002; Villotte et al., 2010). Because of these tissue differences, the centre and margins of fibrocartilaginous insertions can react differently to mechanical forces and be subject to distinct patterns of entheseal changes.

4.4 Shoulder joint anatomy

The shoulder joint complex possesses a large range of motion and allows humans to perform a diverse range of activities with the upper limb (Fig. 4.2); it is made up of three interrelated synovial joints—the sternoclavicular, acromioclavicular, and glenohumeral joints, as well as two physiological articulations—the scapulothoracic and subacromial 'joints' (Felstead and Ricketts, 2017). The mobility of the shoulder is largely facilitated by a shallow ball-and-socket articulation between the small glenoid fossa and the comparatively large humeral head at the glenohumeral joint (GHJ) (Beltran et al., 2002; Felstead and Ricketts, 2017; McGeough, 2013). However, the trade-off for the wide range of motion of the GHJ is that it is inherently unstable, with only about 25%–30% of the humeral head in contact with the glenoid during normal movement.

As a result of this bony incongruity, the GHJ requires numerous static and dynamic soft tissue supports to maintain humeral stability throughout the movement arc (Fig. 4.3) (Prescher, 2000; Tytherleigh-Strong et al., 2001). The static stabilisers provide a passive physical barrier to joint translation or instability and include the glenoid labrum and the glenohumeral ligaments. Dynamic stabilisers are the structures that exert a specific muscle action to limit joint instability, including the rotator cuff muscles and the long head of the biceps tendon (LHB) (Lugo et al., 2008; Wilk et al., 1997).

Many additional muscles also affect shoulder joint stability and mobility. They can be classified as extrinsic muscles (e.g., *trapezius*, *rhomboids*, *latissimus dorsi*, and *levator scapulae*) that originate from the torso and attach to the shoulder bones, intrinsic muscles (e.g., *deltoideus* and *teres major*) originating from the clavicle and/or scapula that attach to the humerus, and muscles (e.g., *pectoralis major* and *pectoralis minor*) originating from the pectoral girdle (Felstead and Ricketts, 2017). The relatively large size and distance of these muscles from the joint rotational centre at the glenoid means they can generate large amounts of torque, which can cause instability in the GHJ if the shoulder stabilisers cannot sufficiently counteract their forces (Lugo et al., 2008).

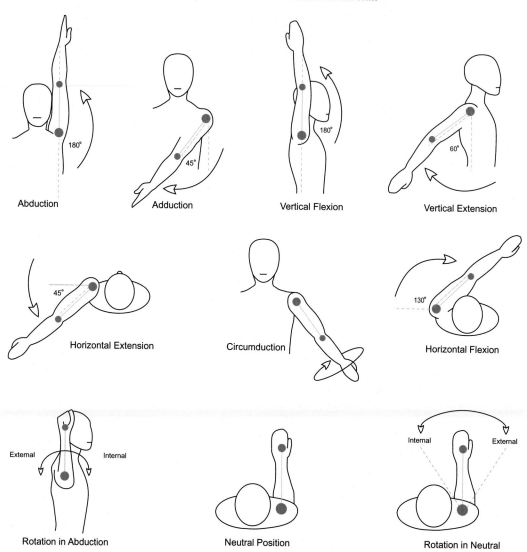

FIG. 4.2   Motions of the glenohumeral joint. *Images courtesy of Aaron Gasparik.*

4.4.1 The static stabilisers

The glenoid labrum is a triangularly shaped fibrocartilaginous structure affixed to the glenoid rim via fibrocartilage and fibrous bone (Lugo et al., 2008). The base of the labrum attaches to the circumference of the glenoid fossa, the superior aspect blends with the LHB at the supraglenoid tubercle, and the free edge projects outwards into the joint cavity (Lugo et al., 2008). The labrum nearly doubles the glenoid fossa depth, reducing chances of humeral subluxation and/or abnormal translation, while also creating a suction effect around the humeral head to keep it seated in the fossa. Additionally, it buffers against shear forces on the

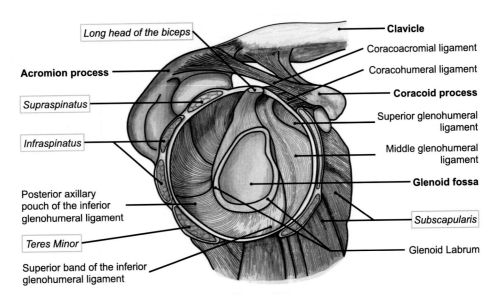

FIG. 4.3 The soft tissue stabilisers and osseous structures of the shoulder joint (humerus not pictured). The skeletal elements are in bold, the static stabilisers are in plain text, and the dynamic stabilisers are outlined in yellow. *Image courtesy of Aaron Gasparik.*

humeral head—especially during mid-ranges of motion—and improves glenohumeral congruity by increasing the glenoid surface area (Wilk et al., 1997; Yamamoto and Itoi, 2015). The labrum also serves as an attachment site for the glenohumeral ligaments—superior, middle, and inferior—that make up the joint capsule (Beltran et al., 2002; Lugo et al., 2008).

The ligamentous articular capsule is a large and relatively lax structure encircling the GHJ, which allows the humerus to move freely while still limiting how far it can separate from the glenoid (Wilk et al., 1997). The superior glenohumeral ligament originates from the supraglenoid tubercle, inserts into the proximal lesser tuberosity, and forms a U-shaped sling on the floor of the bicipital groove which supports the LHB (Krupp et al., 2009). The middle glenohumeral ligament originates from the supraglenoid tubercle and anterosuperior labrum and crosses the anterior capsular surface to insert on the lesser tuberosity, where it blends with the *subscapularis* tendon (Beltran et al., 2002; Lugo et al., 2008). The inferior glenohumeral ligament is split into three components: an anterior band, an axillary pouch fused with the labrum, and a variably present posterior band (Lugo et al., 2008; Passanante et al., 2017). This ligament is anchored on the anterior and posterior glenoid rim and an attachment on the inferomedial junction of the humeral neck/head (Passanante et al., 2017). It provides stability against inferior translation when the shoulder is abducted and is the most frequently injured part of the capsule (Beltran et al., 2002; Felstead and Ricketts, 2017). The coracohumeral ligament originates from a thin enthesis on the lateral coracoid and is separated into two strands: one that inserts into the superior edge of the lesser tuberosity and one that inserts into the anterior margin of the greater tuberosity (Arai et al., 2014). The coracohumeral ligament thus forms a roof over the intertubercular groove and tears here can potentially lead to biceps tendon subluxation (Marmery, 2013; Walch et al., 1994). The

TABLE 4.1   Capsular ligaments and their functions.

Ligament	Function
Superior glenohumeral ligament (SGHL)	- Resists inferior translation when the arm is adducted and neutrally rotated - Limits external rotation (in conjunction with the CHL) - Shields the LHB from anterior shear forces proximal to the bicipital groove
Middle glenohumeral ligament (MGHL)	- Anterior stabiliser for adducted arm - Stabilises abducted shoulder in mid-ranges of motion (30–45 degrees)
Inferior glenohumeral ligament (IGHL) Anterior band Axillary pouch Posterior band	- Resists anteroinferior translation, especially during external rotation, abduction, and extension - Anterior band tightens during abduction and external rotation - Anterior band is the main static stabiliser when the arm is at neutral position (0 degree abduction/30 degrees horizontal extension) - Posterior band is the main static stabiliser during flexion and internal rotation, resisting posterior subluxation at 90 degrees of flexion and internal rotation
Coracohumeral ligament (CHL)	- Resists posterior and inferior translation while the shoulder is suspended - Acts as an inferior stabiliser when the arm is adducted - Tightens with external rotation - Supports the *subscapularis* and *supraspinatus* at their insertions - Serves as a secondary stabiliser preventing anterosuperior translation when the deteriorated cuff can no longer contain the humeral head

Adapted from Lugo, R., Kung, P., & Ma, C. B. (2008). Shoulder biomechanics. European Journal of Radiology, 68(1), 16–24. https://doi. org/10.1016/j.ejrad.2008.02.051 with additional information from Arai, R., Nimura, A., Yamaguchi, K., Yoshimura, H., Sugaya, H., Saji, T., Matsuda, S., & Akita, K. (2014). The anatomy of the coracohumeral ligament and its relation to the subscapularis muscle. Journal of Shoulder and Elbow Surgery, 23(10), 1575–1581. https://doi.org/10.1016/j.jse.2014.02.009; Felstead, A. J., & Ricketts, D. (2017). Biomechanics of the shoulder and elbow. Orthopaedics and Traumatology, 31(5), 300–305. https://doi.org/10.1016/j.mporth.2017.07.004; Krupp, R. J., Kevern, M. A., Gaines, M. D., Kotara, S., & Singleton, S. B. (2009). Long head of the biceps tendon pain: Differential diagnosis and treatment. The Journal of Orthopaedic and Sports Physical Therapy, 39(2), 55–70. https://doi.org/10.2519/jospt.2009.2802; Passanante, G. J., Skalski, M. R., Patel, D. B., White, E. A., Schein, A. J., Gottsegen, C. J., & Matcuk, G. R. (2017). Inferior glenohumeral ligament (IGHL) complex: Anatomy, injuries, imaging features, and treatment options. Emergency Radiology, 24(1), 65–71. https://doi.org/10.1007/s10140-016-1431-0; Wilk, K. E., Arrigo, C. A., & Andrews, J. R. (1997). Current concepts: The stabilizing structures of the glenohumeral joint. Journal of Orthopaedic & Sports Physical Therapy, 25(6), 364–379. https://doi.org/10.2519/jospt.1997.25.6.364.

specific actions of these ligaments depend on arm position, but generally, they become taut at the limits of the humeral movement arc, inhibiting any potential abnormal motion (Lugo et al., 2008; Matsuhashi et al., 2013). The ligaments also form a pulley system between the tuberosities, stabilising the LHB and centring it within the bicipital groove (Bencardino et al., 2003; Edwards et al., 2005; Farid et al., 2008). Table 4.1 provides a summary of the capsular ligament functions.

4.4.2 The dynamic stabilisers

The rotator cuff is comprised of tendons from four muscles—the *supraspinatus, infraspinatus, teres minor,* and *subscapularis* (Fig. 4.3)—which insert via fibrocartilaginous enthesis into the humeral tuberosities. The *supraspinatus, infraspinatus,* and *teres minor* tendons insert into the greater tuberosity from anterosuperior to posteroinferior directions, respectively, while the *subscapularis* tendon inserts mostly into the lesser tuberosity, with some fibres extending over the bicipital groove and attaching to the greater tuberosity (Bencardino et al., 2003;

Day et al., 2012) (Fig. 4.4). The cuff tendons interdigitate with the capsular ligaments at the humeral entheses, which increases their insertional surface area, distributes mechanical forces across multiple musculotendinous units, and increases the cuff's resistance to tearing (Clark and Harryman, 1992). The long head of the biceps should also be considered when analysing the rotator cuff complex due to its close relationship with the cuff tendons and capsular ligaments, as well as its contributions to overall glenohumeral stability (Gellhorn et al., 2015). The rotator interval is a perforation in the joint capsule located between the *supraspinatus* and *subscapularis* tendons that allows the LHB to pass into the joint space (Day et al., 2012; Farid et al., 2008). Though these anatomical elements blend into a continuous structure, the cuff can be divided into three segments: the posterior cuff, including the *infraspinatus* and *teres minor*; the superior cuff, comprised of the *supraspinatus*; and the anterior cuff, made up of the *subscapularis*, the rotator interval, and the intraarticular part of the LHB (Edwards et al., 2005). These regions are important to note, as their varying anatomy, stabilising roles, and ability to manage strains may affect both the initial development of instability as well as the spread of degenerative changes across the shoulder (Edwards et al., 2005; Clark and Harryman, 1992).

The rotator cuff muscles help facilitate shoulder movement in all directions and act as a stable fulcrum from which the extrinsic, intrinsic, and pectoral girdle muscles can move the joint. Additionally, the cuff tendons stabilise the shoulder throughout its entire range of motion by resisting against proximal, anterior, or posterior subluxations and dislocations, keeping the humeral head pressed into the glenoid, and preventing the ligaments from exceeding their functional limits at the apex of motion (Gellhorn et al., 2015; Lugo et al., 2008). The fusion between the joint capsule and the cuff tendons creates a dually passive and active barrier against humeral head translation. The capsule itself statically opposes translation, while the cuff muscles actively contract to produce tension on the capsular ligaments, which

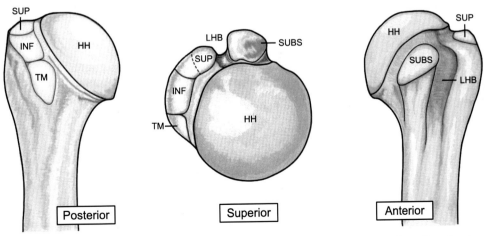

FIG. 4.4 Posterior, superior, and anterior views of the humeral head (left) illustrating the cuff tendon insertion points and the course of the long head of the bicep tendon. The anatomical features are defined as follows: *SUP*, supraspinatus; *INF*, infraspinatus; *TM*, teres minor; *SUBS*, subscapularis; *LHB*, long head of the biceps; *HH*, humeral head. The infraspinatus overlaps the posterior portion of the supraspinatus (area within dashed line). *Images courtesy of Aaron Gasparik.*

tightens the capsule and re-centres the humerus on the fossa (Wilk et al., 1997). However, the stability provided by the cuff muscles varies with arm position and they are less effective during prolonged overhead activity (Wilk et al., 1997). The muscles of the rotator cuff also work in tandem with the other shoulder muscles to maintain GHJ stability by exerting balanced forces in opposite directions in what are known as force couples (Lugo et al., 2008; Pandey and Jaap Willems, 2015). Disruption of these coupled relationships can alter shoulder biomechanics and lead to joint instability (Lugo et al., 2008). The anatomy and function of the dynamic stabilisers are summarised in Table 4.2.

TABLE 4.2   The dynamic shoulder stabilisers—anatomy and functions.

Structure	Anatomy	Function
Supraspinatus	- Originating at the supraspinous fossa, travelling laterally across the shoulder to insert on the anterosuperior greater tuberosity (Edwards et al., 2005) - Blends anteriorly with CHL and the *subscapularis* tendon and posteriorly with the *infraspinatus* tendon (Bencardino et al., 2003) - Anterior aspect thicker and band-like, while the posterior is thinner and membranous (Itoi et al., 1995)	- Contributes to abduction during first 30 degrees of arm elevation (Higgins and Warner, 2005) - Counteracts upward shear forces produced by *deltoideus* muscle, preventing posterosuperior humeral head translation (Mura et al., 2003) - Helps decelerate the arm during overhead activities (Oyama et al., 2010) - Forms a functional unit with *infraspinatus* and *subscapularis* to improve superior shoulder stability (Itoi et al., 1995)
Infraspinatus	- Triangular muscle originating in the infraspinous fossa (Bencardino et al., 2003) - Tendon passes over posterior aspect of joint capsule and inserts on middle facet of greater tuberosity where it blends inferiorly with the capsular ligaments and anteriorly with the *supraspinatus* tendon (Curtis et al., 2006)	- Works with the *teres minor* in external shoulder rotation and in drawing the arm backwards (Bitter et al., 2007) - Forms transverse force couple with *subscapularis* and *teres minor* to maintain anteroposterior stability, strength, and range of motion (Bitter et al., 2007; Mura et al., 2003) - Helps slow the arm during overhead activities and protect against internal hyper-rotation and horizontal adduction of the humerus during deceleration (Oyama et al., 2011)
Teres Minor	- Relatively narrow and elongated muscle originating from middle and upper thirds of lateral scapular body; fibres run posterolaterally to insert on posteroinferior aspect of greater tuberosity and humeral shaft (Curtis et al., 2006) - Blends with posterior joint capsule to provide strength/stability to the structure (Bencardino et al., 2003)	- Works with the *infraspinatus* to externally rotate the shoulder and to draw the arm backwards; *teres minor* provides between 20% and 45% of power during external rotation activities (Bitter et al., 2007; Melis et al., 2011) - Depresses the humeral head, limiting superior translation on the glenoid (Kikukawa et al., 2014) - Provides posterior joint stability by blending with capsuloligamentous tissue (Kikukawa et al., 2014) - Helps maintain shoulder stability and strength when there are large RC tears by exerting additional force (Melis et al., 2011)

TABLE 4.2   The dynamic shoulder stabilisers—anatomy and functions.—cont'd

Structure	Anatomy	Function
Subscapularis	- Multiple tendinous bands originate at medial two-thirds of subscapular fossa and lower axillary border of the scapula - Superior two-thirds of tendon insert into the lesser tuberosity, with some fibres attaching in bicipital groove and greater tuberosity, interdigitating with *supraspinatus* (Morag et al., 2011) - Inferior third of tendon attaches to bottom portion of lesser tuberosity and displays lower maximum load strength than upper part (Morag et al., 2011) - Blends with MGHL to stabilise anterior joint capsule and with CHL/SGHL to stabilise superior LHB (Bencardino et al., 2003; Longo et al., 2012)	- Primary role during internal rotation, but assists with flexion, extension, abduction, and adduction (Longo et al., 2012; Rathi et al., 2017) - The superior fibres contribute more to abduction, while the inferior fibres are more involved with adduction (Morag et al., 2011) - Contributes to anterior shoulder stability by providing a passive buttressing effect (preventing humeral head displacement) and also actively balances the muscle forces of *infraspinatus* and *teres minor* in the transverse force couple (Longo et al., 2012; Morag et al., 2011) - Inferior fibres provide a downward force, centring the humeral head on the glenoid (Rathi et al., 2017)
Long head of the biceps (LHB)	- Origin at supraglenoid tubercle, where it inserts into bone and merges with the glenoid labrum (Wilk et al., 1997) - Runs obliquely over humeral head before exiting joint capsule and running inferiorly through bicipital groove (Krupp et al., 2009) - -Stabilised within the groove by sling made up of fibres from CHL, SGHL, *supraspinatus*, and *subscapularis* (Krupp et al., 2009)	- Assists *supraspinatus* and *infraspinatus* in shoulder rotation (Krupp et al., 2009) - Coordinates actions between the shoulder and elbow joints; for instance, decelerating the arm and preventing elbow hyperextension in overhead throwing activities (Krupp et al., 2009) - Strengthens the superior articular capsule, depresses the humeral head, helps maintain proper tension in the glenohumeral ligaments, limits posterior humeral translation during external rotation as well as anterosuperior translation during internal rotation (Felstead and Ricketts, 2017; Lugo et al., 2008)
Rotator interval	- Roughly triangular area on anterior shoulder where the coracoid process perforates the anterosuperior cuff, allowing the LHB to pass into the joint capsule (Farid et al., 2008) - Comprised of thin, elastic membranous tissue and strengthened by the joint capsule, CHL and SGHL (Edwards et al., 2005; Lugo et al., 2008)	- Resists extreme glenohumeral flexion, extension, adduction, and external rotation (Felstead and Ricketts, 2017; Huri et al., 2019) - Limits internal rotation during adduction and prevents inferior humeral head translation (Felstead and Ricketts, 2017; Huri et al., 2019) - Provides stability against posterior dislocation (Felstead and Ricketts, 2017; Huri et al., 2019)

4.4.3 Anatomical summary

The shoulder joint complex incorporates numerous soft tissue and skeletal components in a complicated and interconnected way. The normal function and stability of the shoulder depends on the coordinated interaction between these structures to respond and adjust to the mechanical stimuli placed on the joint. Understanding this anatomy, the ways in which these structures may affect each other, and the myriad functions of the shoulder elements in a modern context are essential for analysing any activity-related or pathological changes in the shoulders of archaeological individuals.

4.5 Studies of shoulder activity in modern contexts

As illustrated in the preceding sections, shoulder joint anatomy is complex, comprising not only synovial joints (the sternoclavicular, acromioclavicular, and glenohumeral joints), but also a number of tendons and muscles to stabilise the bony structures. In bioarchaeological research, biomechanical effort is inferred via the analysis of bony changes identified in dried bone, specifically in joint articular surfaces and entheseal sites. Degenerative changes to the articular surfaces include marginal osteophytes (sometimes referred to as lipping), new bone on the joint surface, micro and macroporosity on the joint surface, or joint contour changes (Rogers and Waldron, 1995; Waldron, 2009). In some more marked cases, eburnation, which is pathognomonic of osteoarthritis, is also present on the joint surface. Eburnation appears as polished, shiny areas on the articular surface (rarely the entire surface) that develops when the articular cartilage of the joint wears down and the two joint surfaces have direct contact between them (Buikstra and Ubelaker, 1994; Rogers and Waldron, 1995). Entheseal changes are mainly described as including mineralised tissue formation, isolated bone protrusions (e.g., enthesophytes, raised margins, longitudinal or shapeless protrusions), surface discontinuities (e.g., fine and macro porosity, cortical defects, erosive lesions, and cavitations), and in extreme cases, the complete loss of enthesis original morphology can occur (Figs. 4.5 and 4.6; see also Villotte et al. (2016) for detailed discussion of entheseal change terminology). In living patients, the assessment of musculoskeletal injuries in clinical studies is done via conventional radiographic, arthrography and bursography imaging tests, ultrasound, and/or magnetic resonance imaging (MRI) (Mehta et al., 2003). These techniques allow for the observation of shoulder stabiliser anatomy, including any changes to their thickness, retractions, or atrophy. Bone changes, such as spur formations and other degenerative changes, are also observable via these techniques.

A word of caution is necessary when exploring bone changes ascribed to activity, as many are multifactorial. That is, they may result from a number of pathological conditions and events not necessarily related to activity. For example, acute traumatic injuries may cause secondary degenerative changes in the joint, while entheseal changes may be related to spondyloarthropathies (Jurmain, 1999; Jurmain et al., 2012; Maat et al., 1995; Rogers and Waldron, 1995). Senescence is a major contributor to the development of entheseal changes and degenerative joint changes, making it imperative to consider the age at death of the individuals being analysed (Cardoso and Henderson, 2010; Jurmain et al., 2012; Milella et al., 2012, 2015). Despite the interpretive limitations that result from

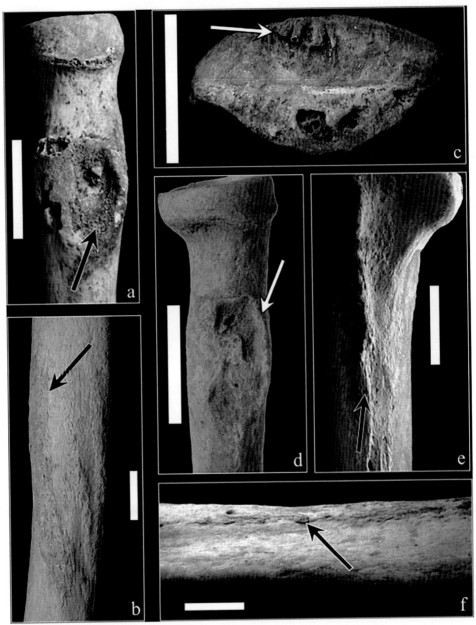

FIG. 4.5 Examples of entheseal changes noted by Villotte and colleagues: (A) Grained surface, (B) diffuse cortical irregularity, (C) enthesophyte, (D) raised margin, (E) longitudinal protrusion, (F) shapeless protrusion. Scale: 2 cm. *Photo credits: Villotte, S. (A, C, E, F); Assis, S. (B); Reichmann, W. and Pany-Kucera, D. (D). From Villotte, S., Assis, S., Cardoso, F. A., Henderson, C. Y., Mariotti, V., Milella, M., Pany-Kucera, D., Speith, N., Wilczak, C. A., & Jurmain, R. (2016). In search of consensus: Terminology for entheseal changes (EC). International Journal of Paleopathology, 13, 49–55. https://doi. org/10.1016/j.ijpp.2016.01.003, 50, used with permission.*

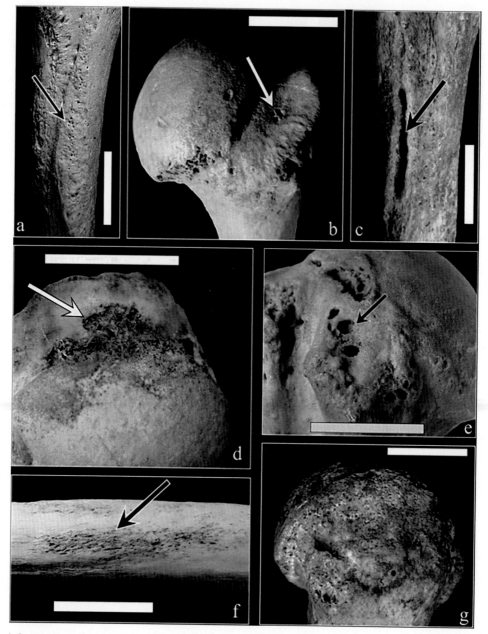

FIG. 4.6 Examples of entheseal changes noted by Villotte and colleagues: (A) Fine porosity, (B) macro porosity, (C) cortical defect, (D) erosive lesion, (E) cavitation, (F) furrowed surface, (G) complete loss of original morphology. Scale: 2cm. *Images courtesy of Mariotti, V. (A); Villotte, S. (B, C, F, G); Henderson, C. (D); Reichmann, W. and Pany-Kucera, D. (E). Image from Villotte, S., Assis, S., Cardoso, F. A., Henderson, C. Y., Mariotti, V., Milella, M., Pany-Kucera, D., Speith, N., Wilczak, C. A., & Jurmain, R. (2016). In search of consensus: Terminology for entheseal changes (EC). International Journal of Paleopathology, 13, 49–55. https://doi.org/10.1016/j.ijpp.2016.01.003, 52, used with permission.*

the multiaetiological nature of skeletal changes, when applied critically and with careful consideration of the context(s) in which the skeletal remains are situated, interpretation of past activities and lifestyles based on skeletal analysis may still be possible, at least at a broad level.

In biomechanical studies, shoulder activity is often studied with regard to how it can cause joint instability, pain, and dysfunction. Activity-related factors are suggested to play a role in the development of shoulder dysfunction. Some studies have proposed that specific occupational environments, or activities, may influence the appearance of pathological changes in this joint allowing for an association between specific pathological changes and activity (Mackinnon and Novak, 1997; van der Windt et al., 2000; Yamaguchi et al., 2006). Within a particular occupation, however, the ways in which the shoulder may be affected vary depending on the specific task being performed and its duration (Forde and Buchholz, 2004). Therefore, rather than discuss the particular occupations that might increase the risks of developing shoulder dysfunction, it is more pertinent to discern the *types of activities* that create a biomechanical environment where the strains on the shoulder joint complex are greater (Forde and Buchholz, 2004; Jones et al., 2007). In this regard, studies in the occupational health and safety literature describe working above shoulder height, carrying weights on one shoulder, awkward postural positions, exposure to vibration, repetitive tasks, and/or working with heavy loads as risk factors for developing shoulder dysfunction (Linaker and Walker-Bone, 2015; Shanahan and Sladek, 2011; van der Windt et al., 2000). The risks are increased especially when two or more of these movements are combined (Hanvold et al., 2015; Linaker and Walker-Bone, 2015). However, some types of strenuous activities are suggested to have a more detrimental effect on the shoulder than others, with degree of arm elevation, duration of activity, and repetitive motion cited as particularly important activity-related factors, while the level of force involved appears to have less direct impact on shoulder joint deterioration (Jones et al., 2007; Rhode and Rhode, 2015; Walker-Bone and Cooper, 2005). For example, working with the arm elevated above 60 degrees has consistently been shown to increase the risk of pathological degeneration in the shoulder stabilisers (Forde and Buchholz, 2004; Jones et al., 2007; Rhode and Rhode, 2015; Seitz et al., 2011). As the arm is elevated and the humerus translates under the acromion, there is an increase in the mechanical pressure on the *supraspinatus*, particularly between 60 degrees and 120 degrees of arm elevation (Herberts et al., 1984; Mehta et al., 2003; Rhode and Rhode, 2015). The increased intramuscular pressure may impair the blood flow to the cuff tendons and reduce their stabilising capabilities, potentially leading to complete tendon rupture (Herberts et al., 1984). The extent of the degeneration in the shoulder tissues and the onset of clinical symptoms are also linked to the length of time spent with arms in an overhead position (Rhode and Rhode, 2015). Even when the arm is only elevated above shoulder height for ~10% of the work period, the risk of developing a cuff tear is increased (Hanvold et al., 2015). Of course, the development of bone lesions will depend on the time in which a person is exposed to a stressor/injury, as well as the overall level of strain involved. Most musculoskeletal conditions require some degree of chronicity to manifest any subsequent bone changes.

While normal mechanical loading on the tendons and entheses can strengthen these tissues, repetitive strains applied without adequate recovery time (known as fatigue loading) can cause micro-tears and inflammation in the shoulder stabilisers that compound over time (Punnett and Herbert, 2000; Shanahan and Sladek, 2011; Wang, 2006). If excessive loading of the mechanically compromised tendon continues, a progressively worsening tendinopathy can develop, which renders the shoulder stabilisers less able to withstand further loading and can lead to eventual degeneration of the shoulder joint complex (Riggin et al., 2015; Shindle et al., 2011; Snedeker and Foolen, 2017; Wang, 2006). Even in healthy (undegenerated) tendons, repetitive strains approaching the ultimate failure limit of the tendon can accumulate and increase the risk of full rupture (Riggin et al., 2015). Joint overuse and fatigue loading are common in occupations requiring repetitive manual activities and can lead to the development of shoulder tendinopathy—a complex, multifaceted pathology affecting the tendon which is characterised by pain, decline in function, and reduced exercise tolerance (Millar et al., 2021; Rhode and Rhode, 2015). Repetitive upper limb activity can also cause mechanical wear in the articular cartilage, which may disrupt the normal shoulder biomechanics and lead to further dysfunction, rotator cuff disease, mechanical compression of the stabilisers within the joint space, and/or secondary osteoarthritic changes (Mehta et al., 2003). Although repetitive activity beyond the point of fatigue appears to negatively affect the shoulder joint structures, other detrimental factors such as overhead loading or elevated muscle force requirements often occur alongside repetitive movements in occupational settings. The overlap of these multiple factors may confound the relationship between repetitive activity and development of shoulder pathology (Jones et al., 2007).

Establishing a cause-and-effect relationship between certain movements and shoulder joint changes is especially difficult with older individuals. In younger age groups, mechanical stressors experienced during rigorous activities are suggested as one of the primary catalysts for the development of enthesopathies, while for older individuals, age-related tendon deterioration can destabilise the joint and lead to osseous changes regardless of the activities being performed (Pandey and Jaap Willems, 2015; Rhode and Rhode, 2015; Villotte and Knüsel, 2013). Thus age-related pathology in the shoulder joint may obscure activity-related differences between groups that might have been present earlier in life. However, increased enthesopathic changes in younger individuals or differences between young individuals from various assemblages could highlight possible differences in the types of activities those groups were performing (Hanvold et al., 2015; Rhode and Rhode, 2015). Individuals working in occupations requiring prolonged overhead activity showed both a higher frequency of shoulder instability and a younger age of onset, while cuff ruptures in elderly patients appeared more frequently in individuals who had been employed in heavy labour activities (Rhode and Rhode, 2015; Simons et al., 2019). Furthermore, degenerative stabiliser tears are relatively rare for individuals under 50 years old, with ruptures occurring more commonly following trauma or overuse (Rhode and Rhode, 2015). Despite the possibility that shoulder joint dysfunction is an inevitable part of ageing, these patterns imply that activity does factor, at least somewhat, into the development of shoulder tendinopathy.

4.6 Archaeological studies of activity and the shoulder

The assessment of behaviour in many bioarchaeological studies relies generally on the observation of alterations to skeletal structures. The analysis of both entheseal changes (e.g., surface or marginal new bone, pitting, cortical surface contour changes, and/or subchondral cysts at the enthesis (Villotte et al., 2016)) and degenerative changes to the joint surfaces (e.g., marginal lipping, porosity, osteophytes, and/or eburnation) plays a role in these assessments of activity patterns in the past (Buikstra and Ubelaker, 1994; Jurmain, 1999; Waldron, 2012). Bioarchaeological studies employing either of these two approaches have been numerous over the years (for some of the major bibliographic references, see Jurmain et al., 2012; Waldron, 2012; Weiss and Jurmain, 2007). More recent studies tend to employ the same methodological approaches, i.e., the analysis of entheseal changes and degenerative joint changes (Becker, 2019; Calce et al., 2018; Palmer and Waters-Rist, 2019), and investigations are no longer limited to the study of human remains and have expanded to consider activity-related osseous changes in zooarchaeological remains (Bindé et al., 2019; Salmi and Niinimäki, 2016).

Most bioarchaeological studies investigating activity do not focus on the shoulder alone. The aim is almost always to analyse the overall changes in the skeleton and compose an image of how the entire body may have been used to perform particular activities. To illustrate this point, a search of papers published between 1990 and 2020 was carried out on Scopus indexed journals (June 2020) using the keywords entheseal changes, degenerative joint changes, and/or activity-related osseous changes. From this search, a total of 56 publications were identified in journals that make up the core literature base for exploring activity-related patterns and past population analysis. Sources included the *International Journal of Osteoarchaeology* (n = 26), the *American Journal of Physical Anthropology* (n = 14), the *International Journal of Paleopathology* (n = 9), *Journal of Comparative Human Biology* (n = 4), and the *Archaeological and Anthropological Sciences* (n = 3). These 56 publications were analysed further to identify the most commonly used keywords used to describe articles on the subject of activity-related changes involving the shoulder. These additional analyses highlighted "Age", "Entheseal", "Musculoskeletal", "Male", "Stress", and "Markers" as the most commonly used terms (Fig. 4.6). Other frequently appearing words include those related to biographical data (e.g., sex and age category, socioeconomic status), disease association (e.g., rheumatic), and anatomy (e.g., radius, humerus). Fig. 4.7 summarises the major themes addressed in these articles. It is also important to note that the "top ten" most cited articles do not address any specific joint, which clarify the difficulty that bioarchaeologists may have in discussing how changes to a single joint—in this case, the shoulder—have been analysed (Table 4.3).

Overall, and as presented in Table 4.3, the Scopus search highlighted studies that covered a wide range of activity-related bone changes none pertaining to changes solely in the shoulder. Some of these articles discussed the sexual division of labour and changes in subsistence strategies (Ibáñez-Gimeno et al., 2015; Villotte and Knüsel, 2014; Wanner et al., 2007), while others explored the correlation between known occupations and osseous changes (Alves-Cardoso and Henderson, 2013; Cardoso and Henderson, 2010; Cunha and Umbelino, 1995; Milella et al., 2015; Perréard Lopreno et al., 2013). These latter studies were possible because they relied on Human Identified Skeletal Collections (Henderson and Cardoso, 2018), which are composed of human remains from identified individuals for whom biographical data are available. This biographical data includes sex, age at death, cause of death, and occupation at death, making it possible to relate osseous changes at the joints and entheses to occupation during life while controlling for sex and age at death (Alves-Cardoso and Henderson, 2013; Cardoso and Henderson, 2010; Cunha and Umbelino, 1995; Milella et al., 2015; Perréard Lopreno et al., 2013; Villotte and Knüsel, 2014).

Accident (2) **Activity** (13) Activity-related (3) Adaptation (2) Adolescent (3)
Adult (6) **Age** (38) Anatomy (2) Animals (2) Anthropometry (3) Arm (2) Arthritis
(2) Biomechanics (6) Controlled (5) Cortical (2) **Diseases** (7) Early
(2)   Elderly   (2) **Entheseal** (34) Enthesis (5)
Enthesopathy (8) Factors (5) **Female** (12) Fibrocartilage (2)
Fossils (2) Functional (2) Hip (2) Humerus (4) Hunter-gatherers (2) Labor (2) Lifestyle (4)
Male (19) Manual (2) **Markers** (16) Micro-computed (2)
Middle (12) MSM (4) Muscle (4)
Musculoskeletal (21) Occupations (7)
Osteoarthritis (2) Physical (5) Physiological (2) Radius (2) Rheumatic (5) Rheumatoid (2)
Sex (6) Size (4) Skeletal (3) Socio-economic (2) Socioeconomic (2)
Spondyloarthropathy (4) Status (2) **Stress** (18) Tibia (2) Tomography (2)
Trabecular (2) Ulna (2) Workload (2) Young (6)

FIG. 4.7   The most frequently used keywords in the 56 articles identified as part of the journal search. This frequency cloud was generated using the open-source application 'TagCrowd' (Steinbock, 2008), with the maximum number of words set to $n=150$ and the minimum frequency per word set to $n=2$. Larger fonts and darker words represent greater frequency of words, with associated value positioned in brackets to the right of each word.

The reason bioarchaeological studies observe the entire skeleton (where possible) when trying to discern changes associated with activity, rather than singling out a specific joint, is because, when exploring behaviour in past populations, the aim is often to infer general patterns of activity—such as those related to sex, social class, or subsistence patterns—rather than define specific activities involving an isolated joint or entheseal complex. Additionally, osseous changes at the entheses and joints are also frequently used as differential diagnosis criteria of diseases that have no relation to activity, such as with osteoarthritis and various spondyloarthropathies. To accomplish analyses of general activity patterns, an assessment of the overall musculoskeletal system is necessary (Jurmain, 1999; Jurmain et al., 2012; Maat et al., 1995; Rogers and Waldron, 1995; Samsel et al., 2014; Waldron, 2012). Further, osseous changes are often recorded aiming to develop new methods to assess those same changes—entheseal or joint changes—which are subsequently used as proxies for activity and occupation. The focus is necessarily placed on the entire skeleton to objectively explore osseous changes while controlling for sex, age at death, and morphological variation (Alves-Cardoso and Henderson, 2013; Henderson et al., 2013b, 2016; Perréard Lopreno et al., 2013).

Although these studies do not solely focus on the shoulder, the "top ten" articles are representative of how osseous changes in the joints—including the shoulder—and associated entheses are assessed to infer activity patterns in human remains. Most of the studies used entheseal changes to infer activity, though others also combined entheseal changes with an analysis of degenerative/pathological changes in the joint. All studies surveyed followed the conventional methodological approach when assessing activity patterns in both unidentified and identified human remains. That is, the entheseal changes and joint alterations were visually assessed with

TABLE 4.3 Top 10 most frequently cited articles (ordered by total number of citations).

Title	Authors	Year	Source title	Cited by[a]	Author keywords
The effect of age, sex, and physical activity on entheseal morphology in a contemporary Italian skeletal collection	Milella et al.	2012	American Journal of Physical Anthropology	71	Ageing; entheseal changes; entheses; markers of activity
Enthesopathies and activity patterns in the Early Medieval Great Moravian population: Evidence of division of labour	Havelková et al.	2011	International Journal of Osteoarchaeology	59	Early Middle Ages; Enthesopathies; Enthesopathy; Great Moravia; MSM; Occupational stress markers; Sexual division of labour
Sex differences in activity-related osseous change in the spine and the gendered division of labor at Ensay and Wharram Percy, UK	Sofaer Derevenski	2000	American Journal of Physical Anthropology	58	Gender; Lifestyle; Osseous modification; Spine
Recording specific entheseal changes of fibrocartilaginous entheses: Initial tests using the Coimbra method	Henderson et al.	2013	International Journal of Osteoarchaeology	36	Ageing; Enthesis; Enthesopathy; Methodology; Musculoskeletal stress markers (MSM); SIMON Collection, Geneva
The categorisation of occupation in identified skeletal collections: A source of bias?	Alves-Cardoso and Henderson	2013	International Journal of Osteoarchaeology	31	Ageing; Coimbra identified collection; Entheseal changes (EC); Fibrocartilaginous entheses; Luis Lopes skeletal collection; Portugal; Socio-economic status
Activity patterns in New Kingdom Nubia: An examination of entheseal remodelling and osteoarthritis at Tombos	Schrader	2012	American Journal of Physical Anthropology	31	Egypt; imperialism; musculoskeletal stress markers; Nile valley
The new 'Coimbra Method': A biologically appropriate method for recording specific features of fibrocartilaginous entheseal changes	Henderson et al.	2016	International Journal of Osteoarchaeology	25	Activity markers; entheses; enthesis; musculoskeletal stress markers (MSM)
Activity reconstruction of post-medieval Dutch rural villagers from upper limb osteoarthritis and entheseal changes	Palmer et al.	2016	International Journal of Osteoarchaeology	25	Activity patterns; Dairy farming; Entheseal changes; Historic cemetery; Palaeopathology; Skeletal activity markers; The Netherlands
Exploring the relationship between entheseal changes and physical activity: A multivariate study	Milella et al.	2015	American Journal of Physical Anthropology	25	Entheseal robusticity; Multivariate study; Occupations

the assumption that they were, at least to some extent, related to activity (Alves-Cardoso and Henderson, 2013; Havelková et al., 2011; Henderson, et al., 2013a, 2016; Maat et al., 1995; Milella et al., 2012, 2015; Schrader, 2012; Sofaer Derevenski, 2000).

Furthermore, the topics addressed in the "top ten" articles are diverse, though many investigated the relationship between activity patterns and osseous changes through case studies of a particular site or time period. One study explored the effects of multiple variables, such as age, sex, and physical activity, on entheseal morphology in a contemporary sample (Milella et al., 2012). Others addressed the evidence of sexual division of labour in different archaeological contexts. For example, Havelková et al. (2011) assessed the prevalence of entheseal changes in two skeletal populations from early medieval Great Moravia, the Mikulčice site. These authors aimed to test different living and occupational conditions, focusing on high- and low-status populations and suggesting that the latter were exposed to a higher degree of physical stress. Their results confirmed this hypothesis for male individuals, but not in the female portion of their assemblage. Mikulčice females associated with the high-status site (castle) had more entheseal changes than females from the low-status site (Mikulčice hinterland)—the explanation offered was that Mikulčice castle females were not of privileged social class. Another example is the work of Sofaer Derevenski (2000), which analysed osseous changes in the vertebrae of individuals from Ensay and Wharram Percy (UK) to infer possible sex-associated differences in activities performed at these sites. The results were as the authors predicted—males and females had contrasting levels of activity-related changes: the male and female differences at Ensay were related to different types of activities, while the patterns observed in male and females in the Wharram Percy assemblage suggested similar patterns in gendered activities and lifestyles.

Other studies investigated activity patterns using a combination of entheseal changes and other degenerative and/or pathological changes to the joint. For instance, Schrader (2012) used entheseal remodelling and osteoarthritis to assess activity patterns in New Kingdom Nubia within a context of Egyptian imperial expansion. The results led to the conclusion that the site analysed—Tombos—served as an administrative centre since the remains showed low entheseal remodelling scores and little evidence of osteoarthritis—interpreted to indicate low levels of physical activity. Another study by Palmer et al. (2016) observed the same osseous changes (i.e., entheseal and degenerative joint changes) to assess the activities in an assemblage of postmedieval rural Dutch villagers, comparing individuals of low and high social status. These authors found that individuals from higher social classes had fewer and less marked osseous changes than their lower class counterparts. They also found that the correlation between entheseal changes and osteoarthritis was low, highlighting their complex aetiologies. In another study, Maat et al. (1995) studied degenerative changes in vertebrae aiming to contribute to the knowledge of conditions such as vertebral osteophytosis, vertebral osteoarthritis, and diffuse idiopathic skeletal hyperostosis (DISH). Although this study primarily focused on sex and age differences in disease prevalence, the authors noted that even in a group of individuals living a "physically moderate life" (i.e., not overly strenuous nor overly sedentary), peripheral joint osteoarthritis co-occurred with cases of vertebral osteoarthritis more than osteophytosis, and suggested that activity-related factors appeared to play somewhat more of a role than did weight bearing—even in other weight-bearing joints like the hip or knee.

It is also worth noting a subset of papers that outlined the inherent bias and variability of interpreting activity patterns when the known occupation at death is used to categorise activity patterns (Alves-Cardoso and Henderson, 2013; Milella et al., 2015). The paper by Alves-Cardoso and Henderson (2013) tested how the method for categorising occupation affected the interpretation of entheseal changes and their relation with activity. Their results showed that the associations that could be made between entheseal changes and occupations depended on the way in which various occupations were categorised. For instance, differences could arise if individual occupations were considered versus broader occupational classes such as industrial/administrative/service work, or depending on which specific activities were considered 'heavy' or 'light' (and how the boundaries of these categories were defined in the first place). Milella et al. (2015) explored the patterns of entheseal changes in documented occupations. Without prior categorisation of the sample group by occupation, they used multivariate analysis of the observed patterns of osseous changes to cluster occupation into three classes related to farming, physically demanding but generalised occupation, and nonphysically demanding occupations. One of the major accomplishments of this research was demonstrating the value of research designs devoid of a priori assumptions of how different groups might cluster when testing activity-related hypothesis. Both the Alves-Cardoso and Henderson (2013) and Milella et al. (2015) studies have also demonstrated that, even when the occupation of an individual is known, the connections between activity and osseous changes are not straightforward (heavy activity does not necessarily equal robust/rugose bony changes).

As a final observation, two papers address new approaches to the study of entheseal changes, describing improved methods for recording entheseal changes in fibrocartilaginous entheses, hoping to contribute to a better understanding of fibrocartilaginous entheseal changes and their association with bone and entheses morphology, sex, age, and activity (Henderson, et al., 2013b, 2016). The method discussed in these papers has become known as 'The New Coimbra Method' and is described as a biologically appropriate method for assessing fibrocartilaginous entheseal changes, without previous assumptions of the association of entheseal changes and activity (Henderson, et al., 2013b, 2016).

Although most bioarchaeological studies examining activity-related changes tend to focus on the whole skeleton, some have specifically addressed the shoulder joint complex. Though these studies did not necessarily analyse solely activity-related issues, their insights are relevant to understanding how the bones of the shoulder respond to various stressors. One example of bioarchaeological work focusing on the shoulder is that of Roberts et al. (2007). These authors sought to confirm whether the range of clinically recognised pathological changes associated with rotator cuff disease (RCD)—a musculoskeletal condition where there is a tear in the tendon of one or more of the four rotator cuff muscles—could be identified in skeletal remains. They also aimed to determine a possible sequence for the pattern of osseous changes to the cuff insertions and joint structures to illustrate the way in which this disease may worsen with age (Roberts et al., 2007). The authors investigated osseous changes occurring at both the articular surfaces and the nonarticular insertion points of the shoulder in 86 skeletons from archaeological sites throughout Britain. In their study, Roberts and co-authors assessed the prevalence of osseous changes at the cuff insertions based on anatomical side, sex, and area of the shoulder. They also analysed whether entheseal changes associated with

RCD affected a particular side or sex more frequently, and how RCD progressed across the shoulder, highlighting potential connections between specific anatomical areas of the shoulder and osseous changes. The authors proposed that observing specific anatomical areas of the shoulder could provide useful clinical information about the relationship between various pathological/entheseal changes in the shoulder, which would help researchers to better understand the multifactorial aetiology of RCD. Regarding anatomical side and biological sex, the authors found that differences in side-specific prevalence of RCD were not significant and observed only slight prevalence differences in pathological changes at the subacromion, GHJ, and bicipital sulcus when the assemblage was divided by sex (Roberts et al., 2007). Roberts et al. (2007) also observed a relatively high frequency of osseous changes on the lesser tubercle of the humerus, including marked surface/marginal osteophyte, pitting, and/or insertion contour change. They noted that changes were observed on 38% (36/96) of the lesser tubercles compared to only 16% (14/85) of greater tubercles. Changes co-occurring at the subacromion and superior cuff tendon insertions on the humerus (thought to be indicative of subacromial impingement) appeared only moderately associated, while lesions on both the lesser tubercle and bicipital groove co-occurred more frequently (Roberts et al., 2007). Though subacromial impingement is often described in the clinical literature as one of the primary catalysts for the onset of RCD, Roberts and colleagues suggest these findings may indicate alternative pathways of shoulder joint breakdown not involving impingement of the *supraspinatus* beneath the acromion. The high prevalence of pathological change at the *subscapularis* insertion observed in this study supports a multifactorial and progressive model of RCD pathogenesis, as reported more recently in clinical literature, and also highlights the different areas of the shoulder in which pathological changes may appear.

Unfortunately, Roberts et al. (2007) did not consider the prevalence data within the social or historical contexts from which the skeletal material originated or made any comparisons between the different time periods surveyed. Though their primary aim was to identify patterns of degeneration in the shoulder—rather than analyse how the observed patterns of change related to specific activities/occupations, developed with age, or varied between populations—some discussion about how degeneration patterns might have differed temporally or geographically and why these differences might exist could have enriched the skeletal analysis. For example, comparison between the subsets of the assemblage could have illuminated variations between the sites that may have indicated a difference in aetiological factors influencing disease progression, alternate cultural activities, or even potential changes in the pathogenesis of RCD over time.

Another set of studies specific to the shoulder joint complex and associated entheses were performed by Miles (1996, 1998, 1999a,b, 2000). These studies include the observation of osseous changes related to subacromial impingement in two skeletal assemblages—one from the site of Ensay, located in Scotland (Miles, 1996, 2000) and another from the Spitalfields collection in London (Miles, 1998, 1999a,b, 2000). The 1998 article, however, is a case study of only two individuals. This study on the ossification of the coracoacromial ligament is of little value beyond describing the pathogenesis of this condition and will not be discussed further. In the remaining studies, Miles investigated how the observed pathological changes correlate to sex (Miles, 1996, 1999a,b, 2000), anatomical side (Miles, 1996, 2000), and increasing age (Miles, 1996, 1999a,b, 2000), and attempted to highlight potential relationships between activity and pathological changes in the joint structures/stabiliser insertions

(Miles, 1996, 2000). Three articles (Miles, 1999a,b, 2000) also divided skeletons into graded categories based on the individual's age and the degree of expression of the pathological alterations in an attempt to discern a possible progression for the observed changes. Four investigations (Miles, 1996, 1999a,b, 2000) used large skeletal samples, which allowed for an understanding of the variation in pathological changes within and between assemblages. These studies are summarised as follows.

The 1996 investigation analysed the osseous changes occurring with RCD and, in particular, the entrapment of the supraspinatus tendon between the inferior acromion and the superior humerus (subacromial impingement). Miles noted that the majority of scapulae from Ensay, Scotland, had more pronounced changes in their right acromion, with more frequent/larger areas of eburnation, larger osteophytic projections, and more pronounced/raised subacromial faceting. Miles also described both the greater and lesser humeral tubercles as being affected by similar destructive processes—though there were several cases where the surface of the former was considerably less degraded than the latter. Throughout the analysis Miles referenced a range of clinical, biomechanical, and anatomical literature, and presented diagrams of the shoulder architecture to illustrate how various movements affect the joint structures and lead to pathological change. The insight these supports provide in terms of both the anatomical distribution of pathological changes as well as the degree to which those changes can occur in the shoulder highlights how important these modern clinical sources are for understanding joint breakdown and entheseal changes in an archaeological context.

Miles (1999a) study focused on the acromioclavicular joints (ACJ) of skeletons from the crypt at Christ Church, Spitalfields, with the aim of illustrating how this joint breaks down in a way that generally worsens with increased age. Individuals were divided into 10-year age categories (26–35y, 36–45y, 46–55y, 65y +), with a roughly even sex distribution in each age category. As previously, Miles (1999a) made reference to studies of cadaveric material to support his descriptions of the osseous changes and illustrate how particular findings in the soft tissue are reflected in the skeleton. Miles found that breakdown of the joint surface began at a relatively early age, with all individuals aged 36 years and older showing changes that could have been classified clinically as osteoarthritis. Given that symptoms of shoulder dysfunction and pain are often observed to develop later in life in clinical studies, this finding suggests that skeletal changes in the ACJ may develop before the emergence of any clinically detrimental consequences or that the disease behaved differently in these individuals compared to modern contexts. Miles also found more expressed examples of pathological change present in males, suggesting this pattern may be due to a tendency to perform heavier physical activities in many societies as well as their heavier musculature in general. That said, he does not suggest any *specific* activity-related factors responsible for the changes in the ACJ, nor does he explicitly state that activity was the sole factor in the joint facet breakdown.

In both the later studies comparing skeletons from Ensay to those from Spitalfields (Miles, 1999b, 2000), Miles sought to determine any differences in the level of disease prevalence/expression in the shoulder joints and in the demography of the affected individuals at the two sites. He divided the assemblages into age categories to assess how increasing age affected the degenerative changes noted in the shoulder. This division highlighted differences in the age at which particular types of change developed in the shoulders of individuals from either site, with more marked pathological changes appearing in younger individuals at Ensay. The youngest examples of what Miles considered to be true humeral impingement and probable

RC breakdown (indicated by osseous changes in the humeral tubercles) were observed in a 35-year-old male and a 40-year-old individual of indeterminate sex. In contrast, deterioration of the greater or lesser tubercles did not occur in the Spitalfields assemblage until the individuals were roughly a decade older than at Ensay. From these findings (Miles, 1999b, 2000), he suggested that differences in the age of onset of pathological changes and earlier transition between phases of degenerative change may indicate that the Ensay individuals led a generally more physically rigorous lifestyle and performed heavier activities more regularly than the Spitalfields individuals. However, Miles (1999b, 2000) does caution that this hypothesis is difficult to test from the osteological material alone. He also discusses the generally inconsistent way in which occupational records were kept, citing this as a confounding factor for determining the relative effects of specific activities on the bodies of individuals in the assemblage (Miles, 2000). Though higher levels/more pronounced pathological changes were noted in younger individuals at Ensay, Miles observed that including the older individuals from both sites did result in the overall proportion of Spitalfields individuals with pathological changes in the shoulder becoming equivalent to those observed at Ensay (Miles, 2000). From these findings, Miles suggested that the more comparable total prevalence for each site is a function of the greater number of old individuals at Spitalfields compared to Ensay. This observation also illustrates the multifactorial nature of RCD and impingement, with increased age-related joint breakdown perhaps playing a larger role than—and/or obfuscating the effects of—activity-related factors.

    Though there are some shortcomings in the methodology, data presentation, and terminology employed in all of Miles' studies, the articles represent in-depth examinations of shoulder pathology and activity in an archaeological context. Despite the issues, the thorough descriptions of shoulder joint anatomy and biomechanics in each study, as well as the comprehensive discussion of osseous changes, and the emphasis on the compounding/progressive way in which joint breakdown occurs in the shoulder, highlight the importance of this clinical information for facilitating a robust assessment of pathological changes in archaeological material. Additionally, throughout these investigations, Miles was generally careful not to infer beyond the limits of the data or make concrete statements regarding aetiology as he explored the factors that may have resulted in the observed osseous changes. This approach not only reflects the multifactorial nature of osseous changes in the shoulder joint, but also highlights the difficulties associated with establishing direct, causal relationships between observed pathological conditions and facets of the once-living individual's life.

    This Scopus search of papers published between 1990 and 2020 offers only a partial view on the studies addressing activity, activity-related changes, and past populations. The search results will of course vary depending on the database queried and the search criteria used, meaning that different search parameters may produce a slightly different selection of literature. However, the authors felt the criteria used here were broad enough to include a number of texts representative of the types of bioarchaeological studies that have been performed, as well as to illustrate how the methodology has changed over time. Though more studies certainly exist, the range of articles analysed offers an overarching view of studies addressing activity, and activity-related changes based on the analysis of osseous changes to the joints and associated entheses. Unfortunately, most of the studies surveyed presented an account of the overall skeleton to infer activity rather than specifically addressing the shoulder joint and its associated entheses.

4.7 Synthesising clinical and bioarchaeological studies

The "total skeleton" approach in bioarchaeological studies has shown that regardless of the joint analysed, age at death is a major confounding variable in activity-related studies of archaeological material. As highlighted in both the clinical and archaeological literature, age-related breakdown of the shoulder stabilisers may lead to increased enthesopathic and degenerative change regardless of the activities being performed, thus obscuring any between-group differences that may have previously been present (Pandey and Jaap Willems, 2015; Rhode and Rhode, 2015). Consequently, the age at death of an individual is an important factor that all studies aiming to infer activity from osseous changes must consider (Alves-Cardoso and Henderson, 2013; Henderson, et al., 2013b; Milella et al., 2012; Miles, 1996, 1999a, 2000; Niinimäki and Sotos, 2013). The difficulties are not merely associated with the methodological limitations of age-at-death assessment in human osteological remains—i.e., issues converting subjective assessments of bone tissue development and maturity into chronological age (Buckberry and Chamberlain, 2002; Clark et al., 2020; Cunha et al., 2009)—but also with the fact that skeletons of older individuals may exhibit osseous changes from both pathological changes, as well as the cumulative years of normal efforts. These factors could confound interpretations of activity. Age must also be considered when exploring differences related to biological sex. Although sexual differences in the prevalence of entheseal or joint changes are found in some studies and eagerly assessed as a division of labour along gender lines, any differences between sexes also need to address the impact of age on the development of bone changes (Cardoso and Henderson, 2010). It is not sufficient to compare male and female differences within or between assemblages without considering this variable. Ignoring the effects of age can lead to biases. For example, positive association of the presence of entheseal changes between sexes, without controlling for age, may obscure the fact that sex subsamples (e.g. females) may be composed of older individuals. It is not that females were working harder or were performing more demanding activities, but simply that they were older.

Bone changes may correlate with a lifetime of activity, and therefore, bioarchaeologists need to consider that people may change occupations over time meaning that any activities associated with that occupation may change. Some alterations in the entheses and/or joint itself may also be unrelated to activity, but instead, could result from traumatic events or underlying pathological conditions capable of mimicking the types of osseous changes that might otherwise be attributed to activity (Alves-Cardoso, 2019; Henderson, et al., 2013b). The same critical approach needs to be considered when exploring populations from different temporal and geographical contexts, as variation in the inherent robusticity and/or morphology of different skeletal assemblages may skew the interpretation of activity (Nolte and Wilczak, 2013; Palmer et al., 2016). Lastly, one needs to remember that the types of bony alterations used to infer activity have complex aetiologies and, therefore, osseous changes at joint and entheseal sites should ideally be analysed without a priori assumptions about their connection to certain activities. Only by exploring the specific cultural, socioeconomic, temporal, and geographic context of the remains can any suggestion about the relationship between activity and entheseal/joint changes in the shoulder be made (Acosta et al., 2017; Godde et al., 2018; Milella et al., 2015; Myszka et al., 2019; Niinimäki and Sotos, 2013; Nikita et al., 2019; Palmer and Waters-Rist, 2019).

Unlike many archaeological investigations, the goal of contemporary clinical and biomechanical studies is seldom to determine what activities an individual was performing based on the changes observed in their joints. On the contrary, there is often prior knowledge of what activities the patients were performing or what incident might have incited some degree of shoulder dysfunction. With this information, these clinical studies can assess how particular activity-related factors may have contributed to the development of these observed joint changes and then address issues of patient discomfort, determine preventative measures to limit joint injuries, or design rehabilitation programmes. However, despite the range of studies that have investigated the relationship between activity and shoulder joint changes, methodological differences between them—including the use of a range of classification systems for similar conditions (e.g., rotator cuff tendonitis), different imaging modalities to assess disease presence/absence, and the use of variable diagnostic criteria (see Walker-Bone and Cooper, 2005 for detailed discussion)—make it difficult to determine patterns of cause and effect with a high level of consistency (Walker-Bone and Cooper, 2005).

Establishing direct connections between activity and osseous changes in the shoulder becomes especially challenging when considering how factors such as age, biological sex, abnormal joint morphology, preexisting pathological conditions—such as DISH or acromegaly—and socioeconomic status may impact the ways in which musculoskeletal diseases or enthesopathies develop (Herberts et al., 1984; Resnick and Niwayama, 1983; Rhode and Rhode, 2015; Villotte and Knüsel, 2013). Given these numerous confounding factors, understanding patient history and exposure to risk factors of tendon injury is an essential part of clinical investigations into the relationship between activity and changes in the shoulder (Rhode and Rhode, 2015). Unfortunately, archaeologists seldom—if ever—have access to the personal histories of the individuals they study. Because of this generally incomplete picture, archaeologists must be wary of suggesting how *specific* occupations or activities affected the bodies of *individuals* based on their shoulder joints alone. Nonetheless, because there is evidence showing some connection between certain strenuous upper limb activities and changes in the shoulder joint, observing differences in prevalence and distribution of those joint changes between age- and sex-matched sections of archaeological assemblages may allow inferences to be made about how changes in the shoulder may reflect broad (group-level) differences in habitual activities.

It is important to emphasise that many of the bioarchaeological studies assessing the ways in which activity patterns may lead to skeletal changes tend to take a holistic approach—analysing the entire skeleton, rather than solely changes in the shoulder. However, the few studies surveyed here that focused specifically on the shoulder joint complex offer some important conclusions about the aetiological factors influencing changes in this joint. These studies are summarised in Table 4.4.

4.8 Conclusions and future directions

As many bioarchaeological and clinical studies have shown, correlations between entheseal changes and activity patterns are not straightforward. It goes without saying that entheseal changes and joint degeneration in the shoulder are affected by multiple risk factors related to both occupational activity and intrinsic biological mechanisms (Punnett and Wegman, 2004). Age and sex appear to be the main factors associated with the development

TABLE 4.4 Summary of the key bioarchaeological studies analysing changes in the shoulder.

Reference	Site	Time period(s)	Sample size	Diagnostic criteria	Key findings and conclusions
Miles (1996)	Ensay, Scotland	AD1500–AD1800	4 males, 8 females	Miles' own criteria, including subacromial/humeral facets, eburnation, new bone growth, roughness, pitting Minimum changes required for positive diagnosis not clearly defined	- Findings suggested that degenerative changes were mostly age related - Miles also argued that, due to their less robust musculoskeletal systems, women were less adapted to the strenuous physical activity the islanders performed on a regular basis, which may have led to the more pronounced examples of pathological change observed in the sample
Miles (1999a)	Christ Church Spitalfields, London	19th century	71 males, 73 females	4-Grade system for grading ACJ changes: 0 – surface smooth; 1 – largely smooth with small pits; 2 – as in grade 1 but with slightly raised ACJ facet; 3 – confluent pits with areas of erosion, enlarged margins with osteophytes; 3A – as in grade 3 but at least half of surface pitted/eroded; 4 – as in grade 3 but erosion/new bone distorts whole surface, possible surface flattening, eburnation and facet breakdown.	- Pathological changes again showed a strong association with increasing age—though changes did develop early (at around 35 years of age) they became more expressed with increased age - Most pronounced pathological changes (stage 4) were found in males, which Miles suggested may be related to their heavier musculature and them having performed more physically demanding activities on a regular basis
Miles (1999b)	Ensay, Scotland; Christ Church Spitalfields, London	AD1500–AD1800; 19th century	- Ensay (n = 120 scapulae) - Spitalfields (n = 99 scapulae) - All individuals under 55 years old at death	Miles' own system: 0 – epiphyseal union with cupped surface and minor trabeculae; 1 – marked surface new bone, acromial flattening with trabeculae; 2 – as in grade 1 with more defined flattening (facet formation), some pressure flattening on superior humerus; 3 – as in grade 2 with addition of pressure facets, pitting and/or erosion on superior humerus *Grades 0–2 are considered 'preimpingement' and grade 3 is defined as 'impingement'	- Documented the morphology of the acromial undersurface, finding the increased expression of pathological changes on the subacromion to be associated with age - Findings also suggested that the difference in age distribution of changes found between the populations (i.e., more marked changes in younger individuals at Ensay) might indicate possible associations with activity

Continued

TABLE 4.4 Summary of the key bioarchaeological studies analysing changes in the shoulder.—cont'd

Reference	Site	Time period(s)	Sample size	Diagnostic criteria	Key findings and conclusions
Miles (2000)	Ensay, Scotland; Christ Church Spitalfields, London	AD1500 –AD1800; 19th century	- Ensay (n = 92 scapulae) Spitalfields (n = 163 scapulae) - These individuals were added to the samples from Miles (1999b) in the tables for a total of n = 212 at Ensay and n = 262 at Spitalfields	Same as Miles (1999b) with the addition of the following categories: 4 – formation of poorly defined, facet-like areas on subacromial surface, erosion on greater and lesser tuberosities, osteophytes; 5 – raised subacromial facet and humeral contour changes *Categories 0–2 defined as 'preimpingement' and categories 3–5 defined as 'impingement'	- Findings emphasised the significant association between increasing age and the development of degenerative changes in the shoulder joint - Confirmed findings of previous study (Miles, 1999b), showing the early breakdown of shoulder joint in individuals from Ensay, which may reflect their heavier physical lifestyle compared to more sedentary Londoners - Found that nearly all individuals from both samples showed evidence of pathological change after age 65 years - Miles proposed possible association with handedness following observations of generally more pronounced changes on the right side of paired shoulder joints
Roberts et al. (2007)	Various UK sites (n = 7)	Iron age— 18th/19th century	~86 individuals total	Rogers and Waldron's (1995) criteria for diagnosing degenerative joint disease	- Investigated osseous changes occurring at both the articular surfaces and the nonarticular insertion points of the shoulder, assessing osseous changes at the cuff insertions based on anatomical side, sex category, and specific area of the shoulder - Found strong associations between increased age and disease prevalence/ expression - Did not find any differences in disease prevalence/ expression by side in paired shoulder joints - Only found slight differences in prevalence at a few anatomical points when sample was divided by biological sex - Observed different patterns of degenerative change—possibly related to different aetiological factors (i.e. age-related versus trauma-related tendon tears)

of shoulder pathology, with the former being of greater impact than the latter, but other factors like preexisting diseases or socioeconomic status may play a role as well (Milella et al., 2012; Punnett and Wegman, 2004). These confounding factors make drawing connections to the activities an individual performed during life difficult, particularly in bioarchaeological contexts, as the deceased individuals presenting with evidence of pathological changes are undoubtedly unable to provide any patient feedback (unlike in clinical cases). Therefore, to predict the activities a person habitually performed during life based on the analysis of their skeletal remains alone will always carry a level of interpretative error. Having said this, a diligent assessment of human remains, with careful consideration of the context (temporal, geographical, socioeconomic, etc.) from which the remains originate, may yield information on overall patterns of differential biomechanical and/or pathological changes due to activity. This assessment requires careful consideration of a range of factors known to influence the presence of musculoskeletal bone changes (including both entheseal changes and degenerative joint alterations)—age at death and preexisting diseases such as other spondyloarthropathies, for example, are two additional factors that must be considered in addition to activity.

Additionally, it is important to remember that individuals likely did not perform the same activities for the whole of their lifetimes, a fact that further confounds attempts to link musculoskeletal changes to specific activities. Considering that some of the main goals in bioarchaeological investigations are to describe/assess behaviour, determine differences in social or cultural practices, and assess changes in subsistence patterns, the classification of general activities under the notion of a particular "occupation" adds an extra layer of difficulty. Furthermore, the classification of "occupations" may also be ambiguous with regard to the tasks performed, while other classifications may be subject to hierarchies within a particular occupation. For instance, the term shoemaker may refer to both an apprentice, as well as a master shoemaker—individuals who would potentially be performing different types of activities and whose joints could be exposed to different stressors. Of course, some information in the records may contain errors or be biased towards recording or ignoring particular segments of the population (i.e., women, 'unskilled' workers, itinerant/temporary workers). Analysing osseous changes in the entire skeleton where possible—rather than just a single joint—is a more well-rounded approach when observing the effects of activity in the skeleton, but even then, the myriad confounding factors must be considered. All of these limitations are becoming more apparent to those exploring collections with identified human skeletal remains (Henderson and Cardoso, 2018). Unfortunately, bioarchaeologists may often be tempted to create narratives about past individuals based on changes observed in the skeletal remains (Jurmain et al., 2012). While hypothesising a general level of activity as one of the factors leading to joint changes may be more palatable on an assemblage level—when there is sufficient contextual information to bolster those assertions—ascribing certain activities as the cause of individual skeletal changes without specific supporting evidence is much less tenable (Jurmain et al., 2012; Waldron, 2012). This concept has been illustrated by examples presented throughout this chapter (e.g., Alves-Cardoso and Henderson, 2013; Havelková et al., 2013; Milella et al., 2012, 2015; Sofaer Derevenski, 2000).

It is certainly enticing to be able to describe the daily lives and individual experiences of the once-living people studied in this discipline; however, this temptation to extend beyond the limits of the data that bioarchaeologists have access to is something to be avoided during investigations of the past. Bioarchaeologists must understand that any conclusions to be drawn about the effects of activity on the shoulder—and the body in general—are almost always going to be limited to inferences at the group level. Additionally, the biographical information available in archaeological contexts does not provide anywhere near the entire picture of an individual's life history—it is more of a snapshot at the time of death. In this sense, although collections with available documentary information about the skeletons and good temporal/geographical/socioeconomic context can help to better understand the observed changes in the skeleton and test hypotheses relating osseous changes to activities, it will always be necessary to exercise caution when making interpretations.

This chapter has highlighted the ways in which the effects of activity have been interpreted from osseous changes in the shoulder in both bioarchaeological and clinical contexts and also emphasised the various intrinsic and extrinsic factors that affect how changes may develop in this joint complex. This array of factors must be considered when attempting to interpret the relationship between activity patterns of once-living individuals and the entheseal or pathological changes observed in archaeological skeletal remains. Most important to comprehend is the fact that inferring activity from osseous changes in human remains is not an easy task, and bioarchaeologists must avoid associating joint and entheseal changes solely with activity. Understanding the context from which the remains originate, carefully considering the confounding factors that affect osseous changes and limiting oneself to the observed data is essential. Though it may sadly limit what bioarchaeologists can say about the things people were doing in the past, acknowledging the many confounding factors when addressing relationships between activity and human remains is the most robust and methodologically sound approach.

References

Acosta, M. A., Henderson, C. Y., & Cunha, E. (2017). The effect of terrain on entheseal changes in the lower limbs. *International Journal of Osteoarchaeology, 27*(5), 828–838. https://doi.org/10.1002/oa.2597.

Alves-Cardoso, F. (2019). "Not of one's body": The creation of identified skeletal collections with Portuguese human remains. In K. Squires, D. Errickson, & N. Márquez-Grant (Eds.), *Ethical approaches to human remains: A global challenge in bioarchaeology and forensic anthropology* (pp. 503–518). Cham, Switzerland: Springer.

Alves-Cardoso, F., & Henderson, C. (2013). The categorisation of occupation in identified skeletal collections: A source of bias? *International Journal of Osteoarchaeology, 23*(2), 186–196. https://doi.org/10.1002/oa.2285.

Apostolakos, J., Durant, T. J., Dwyer, C. R., Russell, R. P., Weinreb, J. H., Alaee, F., et al. (2014). The enthesis: A review of the tendon-to-bone insertion. *Muscles, Ligaments and Tendons Journal, 4*(3), 333–342.

Arai, R., Nimura, A., Yamaguchi, K., Yoshimura, H., Sugaya, H., Saji, T., et al. (2014). The anatomy of the coracohumeral ligament and its relation to the subscapularis muscle. *Journal of Shoulder and Elbow Surgery, 23*(10), 1575–1581. https://doi.org/10.1016/j.jse.2014.02.009.

Becker, S. K. (2019). Osteoarthritis, entheses, and long bone cross-sectional geometry in the Andes: Usage, history, and future directions. *International Journal of Paleopathology, 29*, 45–53. https://doi.org/10.1016/j.ijpp.2019.08.005.

Beltran, J., Bencardino, J., Padron, M., Shankman, S., Beltran, L., & Ozkarahan, G. (2002). The middle glenohumeral ligament: Normal anatomy, variants and pathology. *Skeletal Radiology, 31*(5), 253–262. https://doi.org/10.1007/s00256-002-0492-1.

Bencardino, J. T., Garcia, A. I., & Palmer, W. E. (2003). Magnetic resonance imaging of the shoulder: Rotator cuff. *Topics in Magnetic Resonance Imaging, 14*(1), 51–67.

Benjamin, M., Evans, E. J., & Copp, L. (1986). The histology of tendon attachments to bone in man. *Journal of Anatomy, 149*, 89–100.

Benjamin, M., Kumai, T., Milz, S., Boszczyk, B. M., Boszczyk, A. A., & Ralphs, J. R. (2002). The skeletal attachment of tendons—Tendon "entheses". *Comparative Biochemistry and Physiology. Part A, Molecular & Integrative Physiology, 133*(4), 931–945. https://doi.org/10.1016/s1095-6433(02)00138-1.

Benjamin, M., Toumi, H., Ralphs, J. R., Bydder, G., Best, T. M., & Milz, S. (2006). Where tendons and ligaments meet bone: Attachment sites ('entheses') in relation to exercise and/or mechanical load. *Journal of Anatomy, 208*(4), 471–490. https://doi.org/10.1111/j.1469-7580.2006.00540.x.

Bindé, M., Cochard, D., & Knüsel, C. J. (2019). Exploring life patterns using entheseal changes in equids: Application of a new method on unworked specimens. *International Journal of Osteoarchaeology, 29*(6), 947–960. https://doi.org/10.1002/oa.2809.

Bitter, N. L., Clisby, E. F., Jones, M. A., Magarey, M. E., Jaberzadeh, S., & Sandow, M. J. (2007). Relative contributions of infraspinatus and deltoid during external rotation in healthy shoulders. *Journal of Shoulder and Elbow Surgery, 16*, 563–568.

Buckberry, J. L., & Chamberlain, A. T. (2002). Age estimation from the auricular surface of the ilium: A revised method. *American Journal of Physical Anthropology, 119*(3), 231–239. https://doi.org/10.1002/ajpa.10130.

Buikstra, J. E., & Ubelaker, D. H. (1994). In *Standards for data collection from human skeletal remains: Proceedings of a seminar at The Field Museum of Natural History*. Fayetteville, AK: Arkansas Archaeological Survey.

Calce, S. E., Kurki, H. K., Weston, D. A., & Gould, L. (2018). The relationship of age, activity, and body size on osteoarthritis in weight-bearing skeletal regions. *International Journal of Paleopathology, 22*, 45–53. https://doi.org/10.1016/j.ijpp.2018.04.001.

Cardoso, F. A., & Henderson, C. Y. (2010). Enthesopathy formation in the humerus: Data from known age-at-death and known occupation skeletal collections. *American Journal of Physical Anthropology, 141*(4), 550–560. https://doi.org/10.1002/ajpa.21171.

Chourasia, A. O., Buhr, K. A., Rabago, D. P., Kijowski, R., Lee, K. S., Ryan, M. P., et al. (2013). Relationships between biomechanics, tendon pathology, and function in individuals with lateral epicondylosis. *Journal of Orthopaedic and Sports Physical Therapy, 43*, 368–378. https://doi.org/10.2519/jospt.2013.4411.

Clark, J. M., & Harryman, D. T. (1992). Tendons, ligaments, and capsule of the rotator cuff. Gross and microscopic anatomy. *The Journal of Bone and Joint Surgery. American Volume, 74*(5), 713–725.

Clark, M. A., Simon, A., & Hubbe, M. (2020). Aging methods and age-at-death distributions: Does transition analysis call for a re-examination of bioarchaeological data? *International Journal of Osteoarchaeology, 30*, 206–217. https://doi.org/10.1002/oa.2848.

Cunha, E., & Umbelino, C. (1995). What can bones tell about labour and occupation. The analysis of skeletal markers of occupational stress in the Identified Skeletal Collection of the Anthropological Museum of the University of Coimbra (preliminary results). *Antropologia Portuguesa, 13*, 49–68.

Cunha, E., Baccino, E., Martrille, L., Ramsthaler, F., Prieto, J., Schuliar, Y., et al. (2009). The problem of aging human remains and living individuals: A review. *Forensic Science International, 193*(1–3), 1–13. https://doi.org/10.1016/j.forsciint.2009.09.008. Epub 2009 Oct 29.

Curtis, A. S., Burbank, K. M., Tierney, J. J., Scheller, A. D., & Curran, A. R. (2006). The insertional footprint of the rotator cuff: An anatomic study. *Arthroscopy: The Journal of Arthroscopic and Related Surgery, 22*, 603–609.

Cutts, S., Gangoo, S., Modi, N., & Pasapula, C. (2020). Tennis elbow: A clinical review article. *Journal of Orthopaedics, 17*, 203–207. https://doi.org/10.1016/j.jor.2019.08.005.

Day, A., Taylor, N. F., & Green, R. A. (2012). The stabilizing role of the rotator cuff at the shoulder—Responses to external perturbations. *Clinical Biomechanics (Bristol, Avon), 27*(6), 551–556. https://doi.org/10.1016/j.clinbiomech.2012.02.003.

Edwards, T. B., Walch, G., Nové-Josserand, L., Gerber, C., Warner, J. J. P., Iannotti, J. P., et al. (2005). Anterior superior rotator cuff tears: Repairable and irreparable tears. In *Complex and Revision Problems in Shoulder Surgery* (2nd, pp. 107–128). Philadelphia, PA: Lippincott Williams & Wilkins.

Farid, N., Bruce, D., & Chung, C. B. (2008). Miscellaneous conditions of the shoulder: Anatomical, clinical, and pictorial review emphasizing potential pitfalls in imaging diagnosis. *European Journal of Radiology, 68*(1), 88–105. https://doi.org/10.1016/j.ejrad.2008.02.029.

Felstead, A. J., & Ricketts, D. (2017). Biomechanics of the shoulder and elbow. *Orthopaedics and Traumatology, 31*(5), 300–305. https://doi.org/10.1016/j.mporth.2017.07.004.

Forde, M. S., & Buchholz, B. (2004). Task content and physical ergonomic risk factors in construction ironwork. *International Journal of Industrial Ergonomics, 34*(4), 319–333. https://doi.org/10.1016/j.ergon.2004.04.011.

Gellhorn, A. C., Gillenwater, C., & Mourad, P. D. (2015). Intense focused ultrasound stimulation of the rotator cuff: Evaluation of the source of pain in rotator cuff tears and tendinopathy. *Ultrasound in Medicine & Biology, 41*(9), 2412–2419. https://doi.org/10.1016/j.ultrasmedbio.2015.05.005.

Godde, K., Wilson Taylor, R. J., & Gutierrez, C. (2018). Entheseal changes and demographic/health indicators in the upper extremity of modern Americans: Associations with age and physical activity. *International Journal of Osteoarchaeology, 28*(3), 285–293. https://doi.org/10.1002/oa.2653.

Hanvold, T. N., Wærsted, M., Mengshoel, A. M., Bjertness, E., & Veiersted, K. B. (2015). Work with prolonged arm elevation as a risk factor for shoulder pain: A longitudinal study among young adults. *Applied Ergonomics, 47,* 43–51. https://doi.org/10.1016/j.apergo.2014.08.019.

Havelková, P., Hladík, M., & Velemínský, P. (2013). Entheseal changes: Do they reflect socioeconomic status in the early medieval central European population? (Mikulčice – Klášteřisko, Great Moravian Empire, 9th–10th century). *International Journal of Osteoarchaeology, 23*(2), 237–251.

Havelková, P., Villotte, S., Velemínský, P., Poláček, L., & Dobisíková, M. (2011). Enthesopathies and activity patterns in the early medieval great Moravian population: Evidence of division of labour. *International Journal of Osteoarchaeology, 21*(4), 487–504. https://doi.org/10.1002/oa.1164.

Henderson, C. Y., & Alves-Cardoso, F. (2013). Special issue entheseal changes and occupation: Technical and theoretical advances and their applications. *International Journal of Osteoarchaeology, 23*(2), 127–134. https://doi.org/10.1002/oa.2298.

Henderson, C. Y., & Cardoso, F. A. (Eds.). (2018). *Identified skeletal collections: The testing ground of anthropology?* Oxford, UK: Archaeopress Publishing Limited.

Henderson, C. Y., Craps, D. D., Caffell, A. C., Millard, A. R., & Gowland, R. (2013a). Occupational mobility in 19th century rural England: The interpretation of entheseal changes. *International Journal of Osteoarchaeology, 23*(2), 197–210. https://doi.org/10.1002/oa.2286.

Henderson, C. Y., Mariotti, V., Pany-Kucera, D., Villotte, S., & Wilczak, C. (2013b). Recording specific entheseal changes of fibrocartilaginous entheses: Initial tests using the Coimbra method. *International Journal of Osteoarchaeology, 23*(2), 152–162. https://doi.org/10.1002/oa.2287.

Henderson, C. Y., Mariotti, V., Pany-Kucera, D., Villotte, S., & Wilczak, C. (2016). The new 'Coimbra method': A biologically appropriate method for recording specific features of fibrocartilaginous entheseal changes. *International Journal of Osteoarchaeology, 26*(5), 925–932. https://doi.org/10.1002/oa.2477.

Herberts, P., Kadefors, R., Högfors, C., & Sigholm, G. (1984). Shoulder pain and heavy manual labor. *Clinical Orthopaedics and Related Research, 191,* 166–178.

Higgins, L. D., & Warner, J. J. P. (2005). Massive tears of the posterosuperior rotator cuff. In J. J. P. Warner, J. P. Iannotti, & E. L. Flatow (Eds.), *Complex and revision problems in shoulder surgery* (pp. 129–159). Philadelphia, PA: Lippincott Williams & Williams.

Huri, G., Kaymakoglu, M., & Garbis, N. (2019). Rotator cable and rotator interval: Anatomy, biomechanics and clinical importance. *EFORT Open Reviews, 4,* 56–62.

Ibáñez-Gimeno, P., Galtés, I., Jordana, X., Manyosa, J., & Malgosa, A. (2015). Activity-related sexual dimorphism in Alaskan foragers from Point Hope: Evidences from the upper limb. *Anthropologischer Anzeiger, 72*(4), 473–489. https://doi.org/10.1127/anthranz/2015/0505.

Itoi, E., Berglund, L. J., Grabowski, J. J., Schultz, F. M., Growney, E. S., Morrey, B. F., et al. (1995). Tensile properties of the supraspinatus tendon. *Journal of Orthopaedic Research, 13*(4), 578–584. https://doi.org/10.1002/jor.1100130413.

Jones, G. T., Pallawatte, N., El-Metwally, A., Macfarlane, G. J., Reid, D. M., & Dick, F. (2007). *Associations between work-related exposure and the occurrence of rotator cuff disease and/or biceps tendinitis: A reference document.*

Jurmain, R. (1999). *Stories from the skeleton. Behavioral reconstruction in human osteology.* Gordon and Breach Publishers.

Jurmain, R., Cardoso, F. A., Henderson, C., & Villotte, S. (2012). Bioarchaeology's Holy Grail: The reconstruction of activity. In *A Companion to Paleopathology* (pp. 531–552). Chichester, UK: Wiley Blackwell. https://onlinelibrary.wiley.com/doi/abs/10.1002/9781444345940.ch29.

Kennedy, K. A. R. (1998). Markers of occupational stress: Conspectus and prognosis of research. *International Journal of Osteoarchaeology, 8*(5), 305–310.

Kikukawa, K., Ide, J., Kikuchi, K., Morita, M., Mizuta, H., & Ogata, H. (2014). Hypertrophic changes of the teres minor muscle in rotator cuff tears: Quantitative evaluation by magnetic resonance imaging. *Journal of Shoulder and Elbow Surgery, 23,* 1800–1805.

Krupp, R. J., Kevern, M. A., Gaines, M. D., Kotara, S., & Singleton, S. B. (2009). Long head of the biceps tendon pain: Differential diagnosis and treatment. *The Journal of Orthopaedic and Sports Physical Therapy, 39*(2), 55–70. https://doi.org/10.2519/jospt.2009.2802.

Larsen, C. S. (2015). *Bioarchaeology: Interpreting behavior from the human skeleton (second).* Cambridge, UK: University of Cambridge Press.

Linaker, C. H., & Walker-Bone, K. (2015). Shoulder disorders and occupation. *Best Practice & Research. Clinical Rheumatology, 29*(3), 405–423. https://doi.org/10.1016/j.berh.2015.04.001.

Longo, U. G., Berton, A., Marinozzi, A., Maffulli, N., & Denaro, V. (2012). Subscapularis tears. *Medicine and Sport Science, 57*, 114–121. https://doi.org/10.1159/000328886.

Lugo, R., Kung, P., & Ma, C. B. (2008). Shoulder biomechanics. *European Journal of Radiology, 68*(1), 16–24. https://doi.org/10.1016/j.ejrad.2008.02.051.

Maat, G. J. R., Mastwijk, R. W., & van der Velde, E. A. (1995). Skeletal distribution of degenerative changes in vertebral osteophytosis, vertebral osteoarthritis and DISH. *International Journal of Osteoarchaeology, 5*(3), 289–298. https://doi.org/10.1002/oa.1390050308.

Mackinnon, S. E., & Novak, C. B. (1997). Repetitive strain in the workplace. *The Journal of Hand Surgery, 22*(1), 2–18. https://doi.org/10.1016/S0363-5023(05)80174-1.

Marmery, H. (2013). Imaging the shoulder. *Imaging, 22*(1), 20110061. https://doi.org/10.1259/imaging.20110061.

Matsuhashi, T., Hooke, A. W., Zhao, K. D., Sperling, J. W., Steinmann, S. P., & An, K.-N. (2013). Effect of humeral head rotation on bony glenohumeral stability. *Clinical Biomechanics (Bristol, Avon), 28*(9–10), 961–966. https://doi.org/10.1016/j.clinbiomech.2013.09.011.

McGeough, J. A. (2013). *The engineering of human joint replacements* (1st ed.). Chichester, UK: John Wiley & Sons.

Mehta, S., Gimbel, J. A., & Soslowsky, L. J. (2003). Etiologic and pathogenetic factors for rotator cuff tendinopathy. *Clinics in Sports Medicine, 22*(4), 791–812. https://doi.org/10.1016/s0278-5919(03)00012-7.

Milella, M., Giovanna Belcastro, M., Zollikofer, C. P. E., & Mariotti, V. (2012). The effect of age, sex, and physical activity on entheseal morphology in a contemporary Italian skeletal collection. *American Journal of Physical Anthropology, 148*(3), 379–388. https://doi.org/10.1002/ajpa.22060.

Melis, B., DeFranco, M. J., Lädermann, A., Barthelemy, R., & Walch, G. (2011). The teres minor muscle in rotator cuff tendon tears. *Skeletal Radiology, 40*, 1335–1344.

Milella, M., Alves Cardoso, F., Assis, S., Perréard Lopreno, G., & Speith, N. (2015). Exploring the relationship between entheseal changes and physical activity: A multivariate study. *American Journal of Physical Anthropology, 156*(2), 215–223. https://doi.org/10.1002/ajpa.22640.

Miles, A. E. W. (1996). Humeral impingement on the Acromion in a Scottish Island Population of c. 1600 AD. *International Journal of Osteoarchaeology, 6*, 259–288. https://doi.org/10.1002/(SICI)1099-1212(199606)6:3<259::AID-OA270>3.0.CO;2-5.

Miles, A. E. W. (1998). New light on the acromial attachment of the human coraco acromial ligament. *International Journal of Osteoarchaeology, 8*, 274–279. https://doi.org/10.1002/(SICI)1099-1212(199807/08)8:4<274::AID-OA426>3.0.CO;2-O.

Miles, A. E. W. (1999a). A five-grade categorization of age-related change in the acromio-clavicular joint derived from the skeletal remains of early 19th century Londoners of known sex and age. *International Journal of Osteoarchaeology, 9*, 83–101. https://doi.org/10.1002/(SICI)1099-1212(199903/04)9:2<83::AID-OA461>3.0.CO;2-V.

Miles, A. E. W. (1999b). Observations on the undersurface of the skeletalized human acromion in two populations. *International Journal of Osteoarchaeology, 9*, 131–145. https://doi.org/10.1002/(SICI)1099-1212(199903/04)9:2<131::AID-OA462>3.0.CO;2-P.

Miles, A. E. W. (2000). Two shoulder-joint dislocations in early 19th century Londoners. *International Journal of Osteoarchaeology, 10*, 125–134. https://doi.org/10.1002/(SICI)1099-1212(200003/04)10:2<125::AID-OA515>3.0.CO;2-%23.

Millar, N. L., Silbernagel, K. G., Thorborg, K., Kirwan, P., Galatz, L., Abrams, G., et al. (2021). Tendinopathy. *Nature Reviews. Disease Primers, 7*, 1. https://doi.org/10.1038/s41572-020-00234-1.

Molnar, P. (2006). Tracing prehistoric activities: Musculoskeletal stress marker analysis of a stone-age population on the Island of Gotland in the Baltic Sea. *American Journal of Physical Anthropology, 129*(1), 12–23.

Morag, Y., Jamadar, D. A., Miller, B., Dong, Q., & Jacobson, J. A. (2011). The subscapularis: Anatomy, injury, and imaging. *Skeletal Radiology, 40*, 255–269.

Mura, N., O'Driscoll, S. W., Zobitz, M. E., Heers, G., Jenkyn, T. R., Chou, S.-M., et al. (2003). The effect of infraspinatus disruption on glenohumeral torque and superior migration of the humeral head: A biomechanical study. *Journal of Shoulder and Elbow Surgery, 12*, 179–184.

Myszka, A., Krenz-Niedbała, M., Tomczyk, J., & Zalewska, M. (2019). Osteoarthritis: A problematic disease in past human populations. A dependence between entheseal changes, body size, age, sex, and osteoarthritic changes development. *Anatomical Record, 303*, 2357–2371. https://doi.org/10.1002/ar.24316.

Niinimäki, S., & Sotos, L. B. (2013). The relationship between intensity of physical activity and entheseal changes on the lower limb. *International Journal of Osteoarchaeology, 23*(2), 221–228. https://doi.org/10.1002/oa.2295.

Nikita, E., Xanthopoulou, P., Bertsatos, A., Chovalopoulou, M.-E., & Hafez, I. (2019). A three-dimensional digital microscopic investigation of entheseal changes as skeletal activity markers. *American Journal of Physical Anthropology, 169*(4), 704–713. https://doi.org/10.1002/ajpa.23850.

Nolte, M., & Wilczak, C. (2013). Three-dimensional surface area of the distal biceps enthesis, relationship to body size, sex, age and secular changes in a 20th century American sample. *International Journal of Osteoarchaeology, 23*(2), 163–174. https://doi.org/10.1002/oa.2292.

Oyama, S., Myers, J. B., Blackburn, J. T., & Colman, E. C. (2010). Changes in infraspinatus cross-sectional area and shoulder range of motion with repetitive eccentric external rotator contraction. *Clinical Biomechanics, 26*, 130–135.

Palmer, J. L. A., & Waters-Rist, A. L. (2019). Acts of life: Assessing entheseal change as an indicator of social differentiation in postmedieval Aalst (Belgium). *International Journal of Osteoarchaeology, 29*(2), 303–313. https://doi.org/10.1002/oa.2740.

Palmer, J. L. A., Hoogland, M. H. L., & Waters-Rist, A. L. (2016). Activity reconstruction of post-medieval Dutch rural villagers from upper limb osteoarthritis and entheseal changes. *International Journal of Osteoarchaeology, 26*(1), 78–92. https://doi.org/10.1002/oa.2397.

Pandey, V., & Jaap Willems, W. (2015). Rotator cuff tear: A detailed update. *Asia-Pacific Journal of Sports Medicine, Arthroscopy, Rehabilitation and Technology, 2*(1), 1–14. https://doi.org/10.1016/j.asmart.2014.11.003.

Passanante, G. J., Skalski, M. R., Patel, D. B., White, E. A., Schein, A. J., Gottsegen, C. J., et al. (2017). Inferior glenohumeral ligament (IGHL) complex: Anatomy, injuries, imaging features, and treatment options. *Emergency Radiology, 24*(1), 65–71. https://doi.org/10.1007/s10140-016-1431-0.

Perréard Lopreno, G., Alves-Cardoso, F., Assis, S., Milella, M., & Speith, N. (2013). Categorization of occupation in documented skeletal collections: Its relevance for the interpretation of activity-related osseous changes. *International Journal of Osteoarchaeology, 23*(2), 175–185. https://doi.org/10.1002/oa.2301.

Peterson, J., & Hawkey, D. E. (1998). Preface. *International Journal of Osteoarchaeology, 8*(5), 303–304. https://doi.org/10.1002/(SICI)1099-1212(1998090)8:5<303::AID-OA450>3.0.CO;2-N.

Prescher, A. (2000). Anatomical basics, variations, and degenerative changes of the shoulder joint and shoulder girdle. *European Journal of Radiology, 35*(2), 88–102. https://doi.org/10.1016/S0720-048X(00)00225-4.

Punnett, L., & Herbert, R. (2000). Work-related musculoskeletal disorders: Is there a gender differential, and if so, what does it mean? In M. B. Goldman, & M. C. Hatch (Eds.), *Women and health* (pp. 474–492). San Diego, CA: Academic Press.

Punnett, L., & Wegman, D. H. (2004). Work-related musculoskeletal disorders: The epidemiologic evidence and the debate. *Journal of Electromyography and Kinesiology, 14*(1), 13–23. https://doi.org/10.1016/j.jelekin.2003.09.015.

Rathi, S., Taylor, N. F., & Green, R. A. (2017). The upper and lower segments of subscapularis muscle have different roles in glenohumeral joint functioning. *Journal of Biomechanics, 63*, 92–97.

Resnick, D., & Niwayama, G. (1983). Entheses and enthesopathy. Anatomical, pathological, and radiological correlation. *Radiology, 146*(1), 1–9. https://doi.org/10.1148/radiology.146.1.6849029.

Rhode, B. A., & Rhode, W. S. (2015). Occupational risk factors for shoulder tendon disorders 2015 update. *MOJ Orthopedics & Rheumatology, 3*(4), 00104. https://doi.org/10.15406/mojor.2015.03.00104.

Riggin, C. N., Morris, T. R., & Soslowsky, L. J. (2015). Tendinopathy II: Etiology, pathology, and healing of tendon injury and disease. In M. E. Gomes, R. L. Reis, & M. T. Rodrigues (Eds.), *Tendon regeneration: Understanding tissue physiology and development to engineer functional substitutes* (pp. 149–183). Amsterdam, NL: Academic Press.

Roberts, A. M., Peters, T. J., & Brown, K. R. (2007). New light on old shoulders: Palaeopathological patterns of arthropathy and enthesopathy in the shoulder complex. *Journal of Anatomy, 211*, 485–492. https://doi.org/10.1111/j.1469-7580.2007.00789.x.

Rogers, J., & Waldron, T. (1995). *A field guide to joint disease in archaeology*. Chichester: Wiley.

Salmi, A.-K., & Niinimäki, S. (2016). Entheseal changes and pathological lesions in draught reindeer skeletons—Four case studies from present-day Siberia. *International Journal of Paleopathology, 14*, 91–99. https://doi.org/10.1016/j.ijpp.2016.05.012.

Samsel, M., Kacki, S., & Villotte, S. (2014). Palaeopathological diagnosis of spondyloarthropathies: Insights from the biomedical literature. *International Journal of Paleopathology, 7*, 70–75. https://doi.org/10.1016/j.ijpp.2014.07.002.

Schlecht, S. H. (2012). Understanding entheses: Bridging the gap between clinical and anthropological perspectives. *The Anatomical Record, 295*(8), 1239–1251. https://doi.org/10.1002/ar.22516.

Schrader, S. A. (2012). Activity patterns in New Kingdom Nubia: An examination of entheseal remodeling and osteoarthritis at Tombos. *American Journal of Physical Anthropology, 149*(1), 60–70. https://doi.org/10.1002/ajpa.22094.

Seitz, A. L., McClure, P. W., Finucane, S., Boardman, N. D., & Michener, L. A. (2011). Mechanisms of rotator cuff tendinopathy: Intrinsic, extrinsic, or both? *Clinical Biomechanics, 26*(1), 1–12. https://doi.org/10.1016/j.clinbiomech.2010.08.001.

Shanahan, E. M., & Sladek, R. (2011). Shoulder pain at the workplace. *Best Practice & Research. Clinical Rheumatology, 25*(1), 59–68. https://doi.org/10.1016/j.berh.2011.01.008.

Shindle, M. K., Chen, C. C. T., Robertson, C., DiTullio, A. E., Paulus, M. C., Clinton, C. M., et al. (2011). Full-thickness supraspinatus tears are associated with more synovial inflammation and tissue degeneration than partial-thickness tears. *Journal of Shoulder and Elbow Surgery, 20*(6), 917–927. https://doi.org/10.1016/j.jse.2011.02.015.

Simons, S. M., Dixon, J. B., Kruse, D., & Grayzel, J. (2019). Presentation and diagnosis of rotator cuff tears. In *UpToDate*. https://www.uptodate.com/contents/presentation-and-diagnosis-of-rotator-cuff-tears.

Snedeker, J. G., & Foolen, J. (2017). Tendon injury and repair—A perspective on the basic mechanisms of tendon disease and future clinical therapy. *Acta Biomaterialia, 63*, 18–36. https://doi.org/10.1016/j.actbio.2017.08.032.

Sofaer Derevenski, J. R. (2000). Sex differences in activity-related osseous change in the spine and the gendered division of labor at Ensay and Wharram Percy, UK. *American Journal of Physical Anthropology, 111*(3), 333–354. https://doi.org/10.1002/(SICI)1096-8644(200003)111:3<333::AID-AJPA4>3.0.CO;2-K.

Steinbock, D. (2008). *Tag crowd*. http://www.tagcrowd.com.

Thorpe, C. T., Birch, H. L., Clegg, P. D., Screen, H. R. C., Gomes, M., Reis, R., et al. (2015). Tendon physiology and mechanical behavior: Structure-function relationships. In *Tendon regeneration understanding tissue physiology and development to engineer functional substitutes* (pp. 3–39). Amsterdam, NL: Academic Press. https://doi.org/10.1016/B978-0-12-801590-2.00001-6.

Tytherleigh-Strong, G., Hirahara, A., & Miniaci, A. (2001). Rotator cuff disease. *Current Opinion in Rheumatology, 13*(2), 135–145. https://doi.org/10.1097/00002281-200103000-00007.

van der Windt, D. A., Thomas, E., Pope, D. P., de Winter, A. F., Macfarlane, G. J., Bouter, L. M., et al. (2000). Occupational risk factors for shoulder pain: A systematic review. *Occupational and Environmental Medicine, 57*(7), 433–442. https://doi.org/10.1136/oem.57.7.433.

Villotte, S., & Knüsel, C. J. (2013). Understanding entheseal changes: Definition and life course changes. *International Journal of Osteoarchaeology, 23*(2), 135–146. https://doi.org/10.1002/oa.2289.

Villotte, S., Castex, D., Couallier, V., Dutour, O., Knüsel, C. J., & Henry-Gambier, D. (2010). Enthesopathies as occupational stress markers: Evidence from the upper limb. *American Journal of Physical Anthropology, 142*(2), 224–234. https://doi.org/10.1002/ajpa.21217.

Villotte, S., Assis, S., Cardoso, F. A., Henderson, C. Y., Mariotti, V., Milella, M., et al. (2016). In search of consensus: Terminology for entheseal changes (EC). *International Journal of Paleopathology, 13*, 49–55. https://doi.org/10.1016/j.ijpp.2016.01.003.

Villotte, S., & Knüsel, C. J. (2014). "I sing of arms and of a man…": Medial epicondylosis and the sexual division of labour in prehistoric Europe. *Journal of Archaeological Science, 43*, 168–174. https://doi.org/10.1016/j.jas.2013.12.009.

Walch, G., Nove-Josserand, L., Levigne, C., & Renaud, E. (1994). Tears of the supraspinatus tendon associated with "hidden" lesions of the rotator interval. *Journal of Shoulder and Elbow Surgery, 3*(6), 353–360. https://doi.org/10.1016/S1058-2746(09)80020-7.

Waldron, T. (2009). *Palaeopathology*. Cambridge: Cambridge University Press.

Waldron, T. (2012). Joint disease. In *A companion to paleopathology* (pp. 513–530). Chichester, UK: Wiley Blackwell. https://onlinelibrary.wiley.com/doi/abs/10.1002/9781444345940.ch28.

Walker-Bone, K., & Cooper, C. (2005). Hard work never hurt anyone: Or did it? A review of occupational associations with soft tissue musculoskeletal disorders of the neck and upper limb. *Annals of the Rheumatic Diseases, 64*(8), 1112–1117. https://doi.org/10.1136/ard.2004.026484.

Wang, J. H.-C. (2006). Mechanobiology of tendon. *Journal of Biomechanics, 39*(9), 1563–1582. https://doi.org/10.1016/j.jbiomech.2005.05.011.

Wanner, I. S., Sierra Sosa, T., Alt, K. W., & Tiesler Blos, V. (2007). Lifestyle, occupation, and whole bone morphology of the pre-Hispanic Maya coastal population from Xcambó, Yucatan, Mexico. *International Journal of Osteoarchaeology, 17*, 253–268. https://doi.org/10.1002/oa.873.

Weiss, E., & Jurmain, R. (2007). Osteoarthritis revisited: A contemporary review of aetiology. *International Journal of Osteoarchaeology, 17*, 437–450.

Wilk, K. E., Arrigo, C. A., & Andrews, J. R. (1997). Current concepts: The stabilizing structures of the glenohumeral joint. *Journal of Orthopaedic & Sports Physical Therapy, 25*(6), 364–379. https://doi.org/10.2519/jospt.1997.25.6.364.

Yamaguchi, K., Ditsios, K., Middleton, W. D., Hildebolt, C. F., Galatz, L. M., & Teefey, S. A. (2006). The demographic and morphological features of rotator cuff disease. A comparison of asymptomatic and symptomatic shoulders. *The Journal of Bone and Joint Surgery. American Volume, 88*(8), 1699–1704. https://doi.org/10.2106/JBJS.E.00835.

Yamamoto, N., & Itoi, E. (2015). A review of biomechanics of the shoulder and biomechanical concepts of rotator cuff repair. *Asia-Pacific Journal of Sports Medicine, Arthroscopy, Rehabilitation and Technology, 2*(1), 27–30. https://doi.org/10.1016/j.asmart.2014.11.004.

Archery and the arm

Jessica Ryan-Despraz

Laboratory of Prehistoric Archaeology and Anthropology, Department F.-A. Forel for Environmental and Aquatic Sciences, Section of Earth and Environmental Sciences, University of Geneva, Geneva, Switzerland

5.1 Introduction

Current research in biological anthropology aims to identify activity and behaviour from the skeleton as a means for understanding past populations. While much work is yet to be done, advancements in this field would prove invaluable to domains such as archaeology, prehistory, and evolutionary biology that hope to understand our collective pasts, how we have evolved, and our daily lives.

One example of one such application is the identification of specialised archer activity. Archaeologically, the first confirmed bows date to the Mesolithic period, most notably from Holmegaard (Denmark), though evidence from arrowheads indicates that they could have existed already during the Upper Palaeolithic (Becker, 1945; Cattelain, 1997; Guilaine and Zammit, 2008; Junkmanns, 2001). The two primary functions of archery were hunting and warfare, the former of which likely appeared first. Archaeologists have identified examples of individuals suffering from likely arrow wounds dating from as early as the Mesolithic, such as at the cemeteries at Vassil'evka 3 and Voloshkii (Ukraine) and the Schela Cladovei cave (Romania) (Boroneanț et al., 1999; Dolukhanov, 2004; Guilaine and Zammit, 2008; Thorpe, 2003). Yet, even with these examples, instances of arrow wounds that could potentially imply the presence of warfare are minimal before the Neolithic period; therefore, it remains likely that the original function of archery was for hunting. Numerous Mesolithic cave paintings depicting a prevalence of hunting scenes also provide evidence for this (Dams, 1984). In fact, the first confirmed instances of warfare occurred during the Bronze Age, raising questions as to the presence and evolution of warfare and archery during the preceding Neolithic period (Ryan-Despraz, 2021). In terms of prehistory, determining which populations and/or individuals utilised regular archery could therefore contribute to the understanding of hunting strategies and the origin of warfare.

Archery has been prominent throughout history as an activity used and adapted by cultures on all continents. To name a few, it has been an important part of the hunting culture in many Native American tribes; the warrior culture of the Mongol armies under Genghis Khan; and as a weapon, hunting tool, and overall cultural staple for the English through-out the second millennium CE (Blitz, 1988; Raphael, 2009; Rogers, 2011). The ability to trace archery on the human skeleton would therefore prove valuable not only for identifying these individuals and understanding their daily lives but also contextualising them within a broader cultural scope.

However, one main issue is that there are no confirmed skeletons belonging to a special-ised, daily archer to which we can compare ancient specimens. There are many assumed collections (e.g. the *Mary Rose* and Towton skeletal collections in England), but, the identities of these individuals remain unconfirmed, thus making comparative studies problematic. For this reason, at this point in time, researchers require data from biomechanics and modern archers in order to create a theoretical model of what an archer's skeletal adaptations may look like. This chapter will therefore include an analysis of the basic biomechanics of archery, common bone adaptation responses with a specific regard to archery (including degenerative joint disease, entheseal changes, and cross-sectional bone geometry), and common injuries and muscle activations identified in modern archery (see Table 5.1 for a list of common abbre-viations). With regard to common injuries, those with the potential to leave evidence on the bone will be most important to this perspective. Lastly, this chapter will look at other studies from biological anthropology that have examined likely archer remains and outline the com-mon and significant findings. This will hopefully provide a helpful starting point for future studies looking to identify archer activity on the human skeleton.

5.2 Basic biomechanics: The kinesiology of archery

Theoretically, archery is interesting to explore in terms of activity identification because of its seemingly asymmetric muscular activation—one arm pulls the string while another one holds the bow. However, when looking at the biomechanics involved, this interpreta-tion might not be so simple. Since muscles are activated in groups, the same muscles are often activated in both arms, while it is the magnitude of force on each arm generated that varies.

Overall, archery is a static activity that requires strength in the trunk and arms (Kaynaroğlu and Kiliç, 2012). Bone diaphyses are especially capable of modifying their mass based on

TABLE 5.1   A list of common abbreviations used throughout this chapter.

Abbreviation	Definition
DJD	Degenerative joint disease
EC	Entheseal change
CSBG	Cross-sectional bone geometry
DISH	Diffuse idiopathic skeletal hyperostosis
RCS	Rotator cuff syndrome
RCD	Rotator cuff disease

biomechanical stress; therefore, activities such as archery that require a higher level of muscular development are best for trying to interpret activity from bone (Lanyon, 1992; Rubin et al., 1990; Shaw et al., 2012).

Due to being a static activity (isometric muscle contraction), posture, stance, and muscle contraction timing are important for archery mastery. In order to maximise efficiency, the optimal shooting position creates a line allowing for the forces to pass most efficiently through the arms and body (Ahmad et al., 2014). Creating such a line means that the forces pass through the bones rather than the muscles, thus minimising fatigue, which by extension also minimises the chance of injury and aids each shot in being more precise (Ahmad et al., 2014; Janson, 2004; Larven, 2007).

In terms of stance and posture, modern archery identifies the ideal position as having straight knees and spine with the feet shoulder width apart and facing forward (Kosar and Demirel, 2004). To hold this position, the erector spinae muscles maintain isometric contraction of the vertebral column, and the trunk flexor muscles (*m. rectus abdominis, m. external abdominal obliques, m. internal abdominal obliques*) prevent hyperextension while maintaining balance, which is especially important during the force transfer experienced during the arrow release (Kosar and Demirel, 2004). Additionally, the *m. transverse abdominals, m. multifidus,* and the *m. internal oblique abdominals* also stabilise the lower back during trunk rotation, which is important for an archer's posture (Littke, 2004a). These are all examples of isometric contraction. Also, the head should always be turned towards the target. Since archery is an asymmetric activity, the head will always be turned in the same direction, towards the bow arm. The primary muscles activated in neck rotation are the *m. splenius* (capitus and cervicus), *m. suboccipitals, m. semispinalis,* and the *m. sternocleidomastoid* (Hall, 2003). Kaynaroğlu and Kiliç (2012) speculate that this constant and asymmetric turning of the head constitutes the likely reason behind most neck pain in modern archers.

5.2.1 Draw arm

The movement of the draw arm largely requires a high level of scapular adduction, or pulling the scapula towards the midline, which also extends the glenohumeral joint horizontally. This involves both concentric and isometric contraction of the *m. minor rhomboid, m. major rhomboid,* and the middle and inferior *m. trapezius* as well as activation from the posterior *m. deltoid, m. infraspinatus,* and *m. teres minor* (Ertan, 2006; Kosar and Demirel, 2004). During the draw, the scapula also experiences an upward rotation (*m. serratus anterior* and the inferior *m. trapezius*) and the arm abducts to 90 degree (*m. deltoid* and *m. supraspinatus*) while rotating towards the interior (anterior *m. deltoid, m. infraspinatus, m. subscapularis,* and the *m. latissimus dorsi*). Once the string is fully drawn, the arm is in a flexed and semi-pronated position, which implicates the *m. biceps br, m. brachioradialis, m. brachialis, m. pronator teres,* and the *m. pronator quadratus* (Kosar and Demirel, 2004). Throughout the draw motion, there is also slight wrist and metacarpophalangeal joint extension (*m. extensor carpi radialis brevis, m. extensor carpi radialis longus, m. extensor carpi ulnaris, m. extensor digitorum, m. extensor indicis,* and the *m. extensor digiti minimi*). The most common finger position uses the second, third, and fourth fingers and involves the flexion of the distal interphalangeal joints (*m. flexor digitorum profundus* and the *m. flexor digitorum superficialis*) (Kosar and Demirel, 2004). It is also important to point out that the study from Clarys et al. (1990) found no differences between the types of grip used and the muscles activated, which indicates that the general principles of biomechanics are valid across several archery schools. When the arrow is released, the *m. extensor digitorum* is among the most highly activated muscles, and its antagonist muscle is the *m. flexor digitorum superficialis*.

5.2.2 Bow arm

Moving on to the bow arm, two primary forces derive from this position. The first is the bow's weight applying a downward force on the arm and the second comes from pulling the string. The latter creates a compression force into the elbow and shoulder as well as both a traction in the draw arm and a compression between the upper arm and shoulder (Axford, 1995). In terms of movement, the primary motion is extension and horizontal abduction. Throughout the draw, the bow arm needs to hold steady, making shoulder stability vital for a successful shot. At the shoulder joint, and as seen with the draw arm, when the arm is raised horizontally to 90 degree, this activates scapular contraction (shoulder adduction) with the glenohumeral joint also performing horizontal extension and internal rotation (Kosar and Demirel, 2004). Since holding the bow also rotates the scapula upwards, this will activate the *m. serratus anterior*. Another very important joint is the elbow, which holds a semi-pronated position while extended (*m. pronator teres, m. pronator quadratus, m. triceps br*, and the *m. anconeus*). This joint is also responsible for resisting the external forces that come from holding the bow, which further implicates the wrist as an important joint in the transfer and support of force and biomechanical stress (Kosar and Demirel, 2004). At the wrist, the radiocarpal joint is also in extension (*m. flexor carpi ulnaris* and the *m. flexor carpi radialis*) and while the hand should be relaxed, the fingers are still required to contract around the bow (*m. flexor digitorum superficialis, m. flexor digitorum profundus, m. lumbricalis, m. interossei palmaris, m. interossei dorsalis, m. flexor digiti minimi brevis*, and the *m. opponens digiti minimi*). Table 5.2 presents a summary of this information.

5.3 Common bone adaptation responses to archery

As these collected works attest, principles of entheseal changes could be applied to the skeleton to identify repeated activities performed during life. Just as one can observe modern professional athletes and guess his or her sport (e.g. a rugby player vs. a marathon runner), theoretically it could be possible to observe one's skeleton and evaluate his or her repeated physical activities. This is possible because just as things like physical activity and posture influence muscular development, muscular development also influences bone development and morphology (Clarke, 2008; Knudson, 2007; Pearson and Lieberman, 2004; Wolff, 1869). Therefore, by studying the changes that appear at the meeting points between muscles and bones (entheses), one could potentially deduce at what point certain muscles were activated during life. To do this, researchers tend to look at three primary factors: degenerative joint disease (DJD), entheseal changes (EC), and cross-sectional bone geometry (CSBG) (Jurmain et al., 2012). However, it is important to clarify that this field of activity reconstruction from the bone is still ongoing. Specifically with regard to CSBG and EC, researchers are

TABLE 5.2 A summary of the primary muscle activation in archery.

Muscles	Function
Trunk and Neck	
Trunk flexors (*m. rectus abdominis, m. external abdominal obliques, m. internal abdominal obliques*)	Maintain posture (stability)
Lower back stabilisers (*m. transverse abdominals, m. multifudus, m. internal oblique abdominals*)	
Neck rotation (*m. splenius* (capitus and cervicus), *m. suboccipitals, m. semispinalis, m. sternocleidomastoid*)	Head position
Draw Arm	
m. minor rhomboid, m. major rhomboid, middle and inferior *m. trapezius*, posterior *m. deltoid, m. infraspinatus, m. teres minor*	Scapular adduction
m. serratus anterior, inferior *m. trapezius*	Upward rotation of the scapula
m. deltoid, m. supraspinatus	Arm abduction to 90 degree
Anterior *m. deltoid, m. infraspinatus, m. subscapularis, m. latissimus dorsi*	Arm rotation towards the interior
m. biceps br, m. brachioradialis, m. brachialis, m. pronator teres, m. pronator quadratus	Arm in a flexed and semi-pronated position
m. extensor carpi radialis brevis, m. extensor carpi radialis longus, m. extensor carpi ulnaris, m. extensor digitorum, m. extensor indicis, m. extensor digiti minimi	Wrist and metacarpophalangeal joint extension
m. flexor digitorum profundus, m. flexor digitorum superficialis	Flexion of the distal interphalangeal joints (second, third, and fourth digits)
Bow Arm	
m. minor rhomboid, m. major rhomboid, middle and inferior *m. trapezius*, posterior *m. deltoid, m. infraspinatus, m. teres minor*	Scapular adduction
m. deltoid, m. supraspinatus, m. infraspinatus, m. teres minor, m. pectoralis major, m. latissimus dorsi, m. teres major, m. subscapularis	Horizontal extension and internal rotation of the glenohumeral joint
m. serratus anterior, inferior *m. trapezius*	Upward rotation of the scapula
m. pronator teres, m. pronator quadratus, m. triceps br, m. anconeus	Semi-pronation of the elbow while extended
m. flexor carpi ulnaris, m. flexor carpi radialis	Extension of radiocarpal joint
m. flexor digitorum superficialis, m. flexor digitorum profundus, m. lumbricalis, m. interossei palmaris, m. interossei dorsalis, m. flexor digiti minimi brevis, m. opponens digiti minimi	Finger contraction

still working to understand the link between subchondral bone remodelling and the surface appearance for muscle insertions and the extent to which mechanical stress influences such development. Therefore the following bone adaptations potentially linked to specialised activity should be treated as guidelines rather than strict, black-and-white rules for identifying activity from the skeleton.

5.3.1 Degenerative joint disease (DJD)

DJD is also commonly referred to as osteoarthritis, but this is an example of misleading nomenclature because the "-itis" suffix denotes inflammation, but its primary characteristic is degenerative. For this reason, DJD is a more accurate term, referring to the degeneration of articular cartilage that results in direct contact between bones (Rogers and Waldron, 1995; Schwartz, 2007; Weiss and Jurmain, 2007). On bone, this appears as eburnation on the articular surface (a primary identifier) and secondary identifiers include osteophytes along the margin of the articular surface, bone proliferation and erosion on the joint surface, and changes in the form of the joint contour (Fig. 5.1) (Rogers and Waldron, 1995; Waldron, 2009). It is among the most common joint diseases and due to its degenerative nature, it is most common in older individuals (up to 90% of octogenarians) as a result of ageing and the body's decreased ability to regenerate cellular matrices (Mann and Hunt, 2005); this is another reason why older individuals should be avoided in studies looking at entheseal changes (EC) and bone modifications linked to activity. While DJD occurs with ageing, it can also be exacerbated by other pathologies as well as genetic predisposition.

However, while its onset and severity can vary depending on such factors, none of this changes the fact that it is still a result of biomechanical functions (Chen et al., 2017; Goodman and Martin, 2002; Rogers and Waldron, 1995; Salter, 2002; White, 2000). The direct bone-on-bone contact is what creates osteological markers; if the individual practised minimal movement, the bones would not rub together to leave traces on bone. Therefore it is still highly linked to biomechanics and its presence in younger individuals can be especially telling in terms of repetitive activity. Indeed, Weiss and Jurmain (2007) and Jurmain et al. (2012) identified a correlation between the appearance of DJD and physical activity.

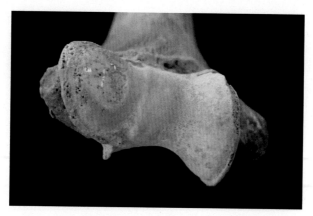

FIG. 5.1 DJD on the distal humerus, including eburnation on the capitulum (image not to scale). *Image courtesy of J. Ryan-Despraz.*

However, the possibility of secondary DJD should also be considered. Primary DJD is the type previously mentioned, driven by factors such as age, mechanics, and genetics. Secondary DJD can derive from trauma, metabolic disorders, and endocrine disturbances, to name a few (Salter, 2002). This means that external factors, such as traumas, could hasten and augment the severity and onset of DJD. For example, a rotator cuff injury can often lead to joint instability, and this additional movement and stress at the articulation could accelerate the cartilage degeneration that would lead to bone-on-bone contact. For this reason, the presence of DJD should also be examined with respect to any other EC or bone modifications in the same biomechanical region as they could be related, or not. The asymmetric appearance of DJD is one indicator that it is more closely tied to biomechanics than some other pathology or genetic predisposition. In terms of archery, this could be an important distinction as the intensity of muscle activation for each arm is not always symmetric. Asymmetric DJD together with entheseal changes could serve as an excellent indicator for repetitive physical activity.

5.3.2 Entheseal changes (EC)

Entheses are the insertions for tendons, muscles, ligaments, and articular capsules on bone and studying them can be useful for identifying certain diseases, most notably metabolic and joint, as well as muscular development (Benjamin et al., 2002, 2004; Benjamin and McGonagle, 2001). The muscle attachments to bone are either fibrous or fibrocartilaginous. Fibrous insertions usually involve larger muscles whose fibres insert directly into the bone and fibrocartilaginous entheses have tendons that transition first into cartilage before attaching to bone (Benjamin et al., 1986, 2002; Benjamin and Ralphs, 1995). The cartilage makes these insertions more flexible, thus they appear more commonly around joints with higher mobility, such as the rotator cuff. However, this also makes them more susceptible to biomechanical stress and injuries (Apostolakos et al., 2014; Benjamin et al., 1986; Selvanetti et al., 1997). Villotte (2009) provides a summary of the primary characteristics for both types of enthesis. A normal and healthy fibrocartilaginous enthesis is smooth, almost resembling an articular surface whereas a normal fibrous enthesis can also present slight colour and texture changes (Benjamin and McGonagle, 2001; Villotte et al., 2016), and any change to this composition is an entheseal change (EC). Some literature has also referred to these as 'musculoskeletal stress markers'; however, 'EC' has become the preferred term because it does not assume a cause (Schlecht, 2012; Villotte et al., 2016).

Villotte et al. (2016, p. 52) defined the three types of EC as mineralised tissue formation, surface discontinuity, and loss of the original bone morphology. This corresponds to previous observations by Mariotti et al. (2004, 2007) that bone has three responses to biomechanical stress: robusticity (usually characterised by size, shape, colour, and texture), proliferation (such as the formation of a bony spur), and erosion (including porosities) (Fig. 5.2). Each of these three characteristics is associated with muscular development; however, robusticity and surface irregularities are more minor and do not imply the possible presence of a pathology, whereas bone proliferation and erosion could signify a level of use and activation that became pathological (hence the common term "enthesopathy") (Benjamin et al., 2002; Sudoł-Szopińska et al., 2015). Factors that can cause bone proliferation and erosion include mechanical loading, haemorrhagic processes, infections, metabolic diseases, traumas, and tumours (Martin-Francés et al., 2013). Because so many different problems can lead to EC, linking

FIG. 5.2 Examples of: (A) robusticity (colour and texture change), (B) osteophytic development, (C) osteolytic development (cortical defect). *Images courtesy of J. Ryan-Despraz.*

them to biomechanical stress (and therefore activity) needs to be handled carefully. In this sense and as with DJD, age is a very important consideration because it is a main factor in the appearance of EC (Alves Cardoso and Henderson, 2010; Pearson and Lieberman, 2004). This means that studies of EC are not feasible on older individuals because it is no longer possible to say whether an EC is linked to mechanical loading or is rather an effect of biological bone ageing. In addition to age, steps also need to be taken to identify diffuse idiopathic skeletal hyperostosis (DISH) and spondylarthropathies so that they can be ruled out (or not) as a potential cause for the presence of EC (Villotte and Knüsel, 2013).

For the identification of archery, EC are important to consider with respect to the biomechanics. For example, Pederzini et al. (2012) found that repetitive movement and mechanical stress most commonly manifest on bone in the form of hypertrophy, including traction spurs and osteophytes, as well as loose bone fragments (such as with osteochondritis dissecans) and other osteochondral defects. As with all activities, it is necessary to understand how all muscles work together to perform the specific functions that will allow for in-depth skeletal interpretations. Therefore the EC most relevant to study when attempting to identify archery on the skeleton are those linked to the specific movements involved in archery, particularly those defined in Section 5.2. At the same time, it would also be helpful to analyse the entheses of those muscles not previously identified as largely activated by archers in order to see if there are developmental differences. Such comparisons could be helpful in assessing and differentiating between both specialised archer activity as well as other activities that might resemble archery from a biomechanical perspective.

5.3.3 Cross-sectional bone geometry (CSBG)

Lastly, CSBG examines a bone's cross-section (Fig. 5.3) in order to understand how instances of loading may have led to modifications affecting the bone's overall strength (Cowgill, 2018). As with EC, the basic idea is that bones are a living tissue capable of adapting to internal and external forces, such as biomechanical loading. With EC, one looks at surface adaptations with regard to entheses, and with CSBG, one considers the overall structural integrity of the bone as it acts like a lever, thus adapting its internal structure in order to support loading (Becker, 2019; Runestad et al., 1993). CSBG therefore aims to examine a long bone's cross-section as a means for interpreting its strength and resistance. For example, loading has been shown to influence cortical bone thickness and directionality changes in bone apposition as well as morphological modifications (Becker, 2019;

FIG. 5.3   Cross-section of a distal tibia. *Image courtesy of J. Ryan-Despraz.*

Ruff et al., 2006). However, and as with DJD and EC, factors other than biomechanics can also influence CSBG, for example hormones, age, nutrition, and genetics (Cowgill, 2018; Jurmain et al., 2012; Lieberman et al., 2004).

In terms of archery, these techniques could be applied to the humerus in order to better understand how loading could influence the bone's internal structure. Rhodes and Knüsel (2005) examined three medieval English collections: two with individuals killed violently (blade wounds from Towton and Fishergate) and one comparative group from Fishergate without blade wounds. Based on the archaeological and historical context of England for each of these collections, the group from Towton were likely archers, whereas the groups from Fishergate were not (Rhodes and Knüsel, 2005). The main findings from this study noted that diaphyseal robusticity varied between collections and diaphyseal morphology varied both between right and left sides of the Towton collection as well as between the collections with blade wounds and the comparative group. Specifically, the Towton group exhibited an increased resistance to medial-lateral bending. Additionally, findings of bilateral asymmetry indicated developmental differences likely linked to activity both between groups as well as within the same group.

When looking at potential bone adaptations linked to archery, and indeed other activities, it is important to consider it in terms of specialisation. Studies, such as those by Clarys et al. (1990), Kian et al. (2013), and Nishizono et al. (1987), have found developmental and biomechanical differences in archers based on their experience level, which could in turn reflect on how the bone adapts to loading constraints. For this reason, it is important to remember that activity reconstruction is possible because an individual performed the same movement or set of movements repeatedly and regularly, thus allowing the bone to develop and conform according to its environment. These methods are therefore not fit to analyse activities that could not be considered "specialised". However, again, it is necessary to clarify that researchers are still working to understand how and to what extent physical activity influences subchondral bone as well as how this can be defined in biological anthropology. For now, the primary takeaway is that physical activity, and more specifically repeated and strenuous activity, influences bone morphological adaptation.

5.4 Injuries in modern archers

Since it is not possible to discuss injuries and occurrences with archaeological collections, this section will gather evidence for injuries from modern, documented competitive archers and clinical reports. When attempting to deduce activity and behaviour from the skeleton, archery is interesting from a muscular development perspective, but it is more difficult to assess in terms of common injuries. It is a relatively safe sport, with minimal instances of trauma and the traumas that do occur tend to be superficial or not affecting the bone (such as finger blisters and contusions caused by the bow string) (Åman et al., 2016; Lapostolle et al., 2004). Of the injuries and issues that are present, most studies agree that the most common problem derives from overuse (Chen et al., 2005; Ergen et al., 2004; Ertan, 2006; Hildenbrand and Rayan, 2010; Kaynaroğlu and Kiliç, 2012; Lapostolle, 2004; Prine et al., 2016). Overuse occurs when movement continues after the muscles have reached a point of fatigue, which leads to microtears that, when not given time to heal, can lead to tendinosis or other muscle injuries (Sheibani-Rad et al., 2013). These issues can then create inflammation that leads to bone degeneration (Walz et al., 2010).

In reference to overuse injuries in archery, most researchers agree that the most common injury among modern archers is to the draw arm shoulder (see Tables 5.3 and 5.4). This mostly includes rotator cuff impingement and tendinosis of the *m. supraspinatus, m. subscapularis,* and the *m. biceps br.* The three most common secondary injuries are to the fingers, draw arm wrist, and bow arm elbow. The bow arm will tend to see the appearance of lateral and medial epicondylosis, though archers have also reported overuse injuries resulting in tendinosis of the *m. biceps br* and the *m. triceps br.* Table 5.3 summarises these findings.

TABLE 5.3 A summary of some common injuries in archery.

Draw arm	Bow arm
Shoulder tendinitis, mainly the *m. supraspinatus* but also the *m. infraspinatus, m. teres minor,* and *m. subscapularis* (the muscles of the rotator cuff) (Kaynaroğlu and Kiliç, 2012; Mann and Littke, 1989)	*Lateral epicondylitis* (Frostick et al., 1999; Niestroj et al., 2017)
Shoulder impingement (Kaynaroğlu and Kiliç, 2012; Mann and Littke, 1989; Prine et al., 2016)	*Osteochondritis dissecans* (Andrews and Whiteside, 1993)
Biceps tendinitis (Kaynaroğlu and Kiliç, 2012; Mann and Littke, 1989)	*Osseous hypertrophy at the elbow* (Andrews and Whiteside, 1993)
	De Quervain's (Kaynaroğlu and Kiliç, 2012)

TABLE 5.4 Common injuries in modern archers as found in various studies.

Primary injuries	Secondary Injuries	Source
Wrist, knee	Shoulder, waist, hands, upper arms	Chen et al. (2005)
Draw arm: finger blisters, shoulder, wrist	N/A	Ertan (2006)
Shoulder, fingers	Spine, elbow	FITA Medical Committee (Ergen et al., 2004; Lapostolle et al., 2004)
Extensor muscle fatigue	N/A	Frostick et al. (1999)
Posterior instability, recurrent dislocation	N/A	Fukuda and Neer (1988)
Draw arm shoulder (⅔ of injuries on draw arm)	Neck, back, bow arm: shoulder and elbow; draw arm: hand, wrist, elbow, Achilles tendon	Gopal Adkitte et al. (2016)
Shoulder instability, upper extremities	Chest, neck, back	Hildenbrand and Rayan (2010)
Novice archers: abrasions and contusions of bow arm; archers > 35 years: rotator cuff tears, lateral epicondylitis, medial elbow tendinopathy, de Quervain tendinopathy, metacarpal fractures, overuse injuries: tendinitis, muscle fatigue, shoulder instability; tendinitis at biceps and triceps tendons, anterior impingement with tendinitis of *m. biceps br* longhead and/or *m. supraspinatus*, epicondylitis and epitrochleitis	N/A	Kaynaroğlu and Kiliç (2012)
Epicondylitis (bow arm)	Epitrochleitis (draw arm), biceps tendinitis at radial tuberosity (draw arm), triceps tendinitis at olecranon (bow arm)	Lapostolle et al. (2004)
Acromio-humeral space (impingement), overuse	N/A	Littke (2004b)
Supraspinatus impingement tendinitis, infraspinatus/teres minor traction tendinitis	N/A	Mann and Littke (1989)
Epiphyseal injury of coracoid process	N/A	Naraen et al. (1999)
Draw shoulder (especially impingement and tendinosis)	Bow elbow (especially lateral epicondylitis)	Niestroj et al. (2017)
Lacerations (arrow mishandling)	Contusions (lash of string)	Palsbo (2012)

Continued

TABLE 5.4 Common injuries in modern archers as found in various studies—cont'd

Primary injuries	Secondary Injuries	Source
Overuse shoulder injuries (70%)	Back injuries (31.7%), elbow (18.8%), wrist (15.8%), fingers (12.9%), forearm (11.9%)	Prine et al. (2016)
Bilateral medial epicondylitis, median nerve compression at wrist, de Quervain's tenosynovitis, median nerve compression at elbow	N/A	Rayan (1992)
Shoulder	Upper back muscles	Renfro and Fleck (1991)
Inflamed epicondyle muscles (especially *m. ext. carpi ulnaris* of bow arm)	Finger pain (draw arm)	Sessa (1994)
Rotator cuff impingement and tears, lateral and medial epicondylitis, traumatic synovitis (elbow), de Quervain's tenosynovitis (wrist), ext. tendon synovitis (wrist)	Scapular dyskinesis, posterior shoulder dislocation, lacerations and puncture wounds (from arrows)	Singh and Lhee (2016)
Bow arm lateral epicondylitis	N/A	Whiteside and Andrews (1989)

From Ryan-Despraz, J., 2021. The application of biomechanics and bone morphology to interpret specialized activity and social stratification: The case of bell beaker archery (PhD thesis). University of Geneva, Geneva, Switzerland.

In terms of injury location, the FITA Medical Committee performed a survey of senior (experienced) archers and again found the majority of problems to be with the draw arm shoulder, which accounted for 49% of cases studied (Ergen et al., 2004). The next most common injury locations were the fingers (12.3%), elbow (11.4%), and lower back (10%). Table 5.4 summarises additional, various findings from several researchers concerning primary injury locations in archers.

While traumatic musculoskeletal injuries are rare in archery, the overuse aspect of muscle activation is interesting in terms of repetitive activity. This level of physical activity is exactly the type of development biological anthropologists might look for when examining EC. Some injuries resulting from overuse include tendinitis, muscle fatigue, and shoulder instability. Additionally, many sources cited impingement combined with m. *biceps br* and *m. supraspinatus* tendinitis as an injury seen in modern archers, and one common way to avoid this problem is to strengthen the entire rotator cuff (Kaynaroğlu and Kiliç, 2012), which could in turn create an EC.

5.4.1 How common injuries might manifest on bone

This section will examine the osteological markers that could form as a result of the aforementioned injuries common in modern archers. This includes general overuse, rotator cuff impingement, osteochondritis dissecans, lateral epicondylosis, and medial epicondylosis.

5.4.1.1 Overuse

Overuse refers to muscle fatigue and can therefore result in a number of injuries, most notably tendinopathies, stress fractures, and osteochondritis dissecans to name a few (Aicale et al., 2018).

Beginning with tendinopathies, studies have calculated that they make up 30%–50% of sports-related injuries (Kaux et al., 2011; Scott and Ashe, 2006). There are two primary tendon injuries. The first is a tendon rupture which results from a mechanical overload, meaning the tendon was not physiologically capable of performing the given task (Lipman et al., 2018). The second is repetitive microtrauma of the tendon which is linked to muscle fatigue. This means that the action in and of itself is within an individual's physiological limits, but the microtears that result from overuse were not given time to repair, thus augmenting to the point of tendinopathy (Lipman et al., 2018; Nakama et al., 2007). In sports science, tendon rupture and the accumulation of microtraumas is the difference between an injury due to lifting too much weight one time versus injuries caused by lifting smaller weights too often. As with epicondylosis, the primary identification of tendinopathies on the skeleton will be through the presence of enthesopathies, including an erosion surface at the insertion as well as possible enthesophytes (osteophytic development). Some studies have discussed the probability that traction forces, such as what might occur with a tendon rupture, will cause the appearance of enthesophytes or osteophytes on the bone (Boutsiadis et al., 2012; Rogers et al., 1997). Smillie (1970) examined the tibiae of football players and hypothesised that inflammation and repetitive strain lead to such formations. The idea is that when tendons pull on bones, they also gradually pull the bone away from the original surface creating a spur, which would also add stability as the insertion surface becomes larger (i.e. three-dimensional) and thus possibly able to support more biomechanical loading. As an additional note, calcifying tendinitis occurs when calcium hydroxyapatite forms in the tendon, which can then lead to an erosion of the cortical bone at the insertion zone, though this is rare (Chan et al., 2004). This appears most commonly at the shoulder and could be the result of degeneration, metaplasia, or necrosis of the soft tissue, or repetitive trauma, and the presence of pain would limit mobility (Ebenbichler et al., 1999; Uhthoff and Loehr, 1998). Causes for cortical bone erosion include inflammation, vascularisation, and muscle traction (Dürr et al., 1997; Hayes et al., 1987).

Stress fractures are one of the most common injuries resulting from intense physical activity and are frequently studied in athletes and soldiers, most often in the lower limbs (Başbozkurt et al., 2012; Finestone and Milgrom, 2012a,b). Stress fractures consist of fatigue microdamage that results from stress and overuse, leading the body to correct this damage by creating an increase in calcium movement, which then accumulates in the afflicted areas (Mann et al., 2012). Stress fractures occur because the bone is unable to handle the loads placed on it, for this reason it is most commonly seen in athletes and untrained individuals beginning physical activity (this is also one likely reason for the aforementioned differences observed between professional and novice archers) (Mann et al., 2012). Bones continuously undergo remodelling and physical activity is capable of aiding this process by strengthening them; however, excessive remodelling can lead to an imbalance (Martin-Francés et al., 2013). Signs of a stress fracture include cortical radiolucent lines and callous formations, and if left untreated they can develop into a transverse fracture (Martin-Francés et al., 2013). Arangio et al. (1998) proposed four primary factors that cause stress fractures: load direction (horizontal force can be worse than vertical), the quantity of the load (heavier vs. lighter), load location (some areas are naturally weaker than others), and muscle reaction when subjected to external forces. Zwas et al. (1987) classified four stages of stress fractures ranging from mild to severe when examining bones; however due to their morphologies, other maladies such as tumours, osteomyelitis, entrapment syndromes, compartment syndromes, and periostitis must be ruled out before making a diagnosis (Başbozkurt et al., 2012).

Osteochondritis dissecans occurs when part of an articulation does not receive enough blood flow, causing a part of the bone tissue to die and break off (this is most common at the knee, though it is also seen on the capitulum of the humerus (Fig. 5.4), talus, femoral head, glenoid and humeral head, and scaphoid) (Roberts and Manchester, 2007; Waldron, 2009). The main causes are trauma, including the microtraumas resulting from overuse and compressive strain, and it often appears in younger athletes (Kajiyama et al., 2017; Waldron, 2009). It is possible that the bone fragment reattaches and heals, leaving a scar; however if this does not happen, the appearance on the bone will be as a defined, circular area with a porous surface, usually about 10mm in diameter. Additionally, one side effect of living with osteochondritis can be the development of DJD (Waldron, 2009).

5.4.1.2 *Rotator cuff disease and impingement*

Rotator cuff impingement is part of a larger problem commonly known as Rotator Cuff Syndrome (RCS) or Rotator Cuff Disease (RCD), and also includes bursitis and tendinitis, which could lead to the development of DJD of the glenohumeral joint and rotator cuff arthropathy (Varacallo and Mair, 2018). While both the femoroacetabular and glenohumeral articulations are ball-and-socket joints, the latter is much less pronounced and requires the muscles of the rotator cuff to hold the humerus in position (Waldron, 2009). This makes a highly activated joint more unstable and the role and health of the muscles and tendons invaluable to proper movement and stability. RCD thus occurs most often from overuse, though other pathologies, such as os acromiale, can influence its appearance (Waldron, 2009).

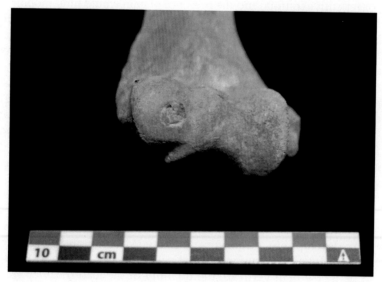

FIG. 5.4   Osteochondritis on the capitulum of the humerus. *Image courtesy of J. Ryan-Despraz.*

In RCD, modifications to the rotator cuff region can include the acromion, acromioclavicular joint, coracoid process, and the bicipital groove. The muscles of the rotator cuff are the *m. supraspinatus, m. infraspinatus, m. teres minor,* and the *m. subscapularis,* and it is important to examine the insertions for these muscles when looking for signs of RCD. Waldron (2009) identifies the following osteological features common to RCD (Fig. 5.5):

- Entheseal changes of the rotator cuff muscles, including pitting on the insertion surface and enthesophytes on or around the enthesis.
- Changes in the enthesis contour.

When considering RCD in terms of impingement, it is also important to understand the role of the *m. deltoid.* The deltoids pull the humerus up into the glenohumeral joint, and to avoid pulling too high, muscles of the rotator cuff also pull down on the humerus to hold it in a central, functional position. However, due to factors such as muscle fatigue and overuse, sometimes the deltoids manage to raise the humerus too high. This causes the muscles and tendons of the rotator cuff, most notably the *m. subscapularis,* to be pinched and rub against the bottom of the acromion, which can lead to tears. This is why impingement is most common in overhead throwing athletes and other activities that raise the arm, such as archery. Problems can be augmented if the acromion has any type of bone deformation, such as a spur, that both makes the space between it and the tendon smaller as well as presents a surface that is more likely to damage the tendon. Examining the bottom of the acromion is therefore one way to identify rotator cuff impingement and therefore RCD. Waldron (2009, p. 42) provides the following criteria for identifying rotator cuff impingement on the skeleton:

- Eburnation on the top of the humeral head
- Eburnation on the bottom of the acromion

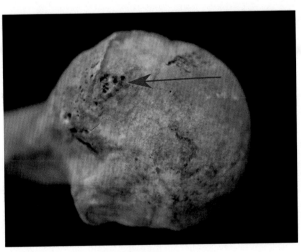

FIG. 5.5 Example EC of the rotator cuff, including an erosion surface and a pronounce contour (including colour change) (image not to scale). *Image courtesy of J. Ryan-Despraz.*

5.4.1.3 Lateral and medial Epicondylosis

In many sports, including archery, epicondylosis remains a common problem with regard to injury. Linguistically, it is also commonly referred to as epicondylitis; however, this implies an inflammation (− itis) that is not necessarily there (this is the same linguistic problem as with "osteoarthritis"). Because it is a degenerative pathology, epicondylosis is therefore a more accurate term, especially when details of the problem are unknown. Epicondylalgia also usually refers to the same pathology, though linguistically it emphasises the presence of pain (− algia) and therefore is more common in living patients (Waugh, 2005). Basically, all of these terms refer to the same underlying problem—overuse of the flexor or extensor tendons leading to issues at the epicondyles of the elbow—just with linguistic nuances that draw attention to varying aspects of the overlying symptoms (Ahmad et al., 2013).

Lateral epicondylosis affects the tendon for the common extensor muscles which originates on the lateral epicondyle of the humerus. The common extensor muscles include the *m. extensor carpi radialis brevis, m. extensor digitorum, m. extensor digiti minimi*, and the *m. extensor carpi ulnaris*; however, studies have shown that these muscles merge, forming one tendon ("enthesis organ") that then attaches to the lateral epicondyle (Milz et al., 2004). Lateral epicondylosis therefore results from an overuse of these muscles, whose main job is extending the wrist as well as forearm pronation and supination (Guadilla et al., 2016). Identifying lateral epicondylosis on the skeleton will therefore require the presence of entheseal changes at the insertion of this tendon (Fig. 5.6). This pathology is also linked to osteochondritis dissecans of the radiocapitellar joint and entheseal changes of the ulnar collateral ligament (Connell et al., 2001; Guadilla et al., 2016).

Medial epicondylosis appears at the medial epicondyle of the humerus at the origin for the common flexor tendon which, as with the common extensors, acts as an "enthesis organ" for the *m. flexor carpi ulnaris, m. palmaris longus, m. flexor carpi radialis, m. pronator teres*, and the *m. flexor digitorum superficialis*. The movements associated with these muscles, besides wrist

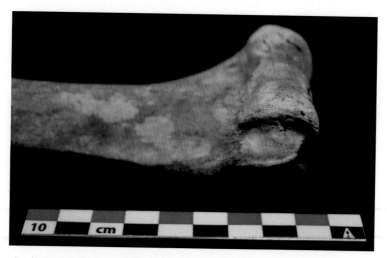

FIG. 5.6   Example of EC (osteophytic development at the enthesis contour) of the lateral epicondyle. *Image courtesy of J. Ryan-Despraz.*

flexion, are forearm pronation and a general overload of valgus stress (Amin et al., 2015). Prolonged and untreated overuse of these muscles leads to degeneration, which could become an EC at the tendon insertion zone on the medial epicondyle.

5.5 Studies in biological anthropology

Identifying specialised archery activity in an archaeological context could be valuable to prehistoric and historic interpretations, therefore several studies have examined the skeletons of suspected archers (suspected due to the presence of certain burial goods linked to archery, such as wristguards and arrowheads, however confirming the individuals' identities is not currently possible). A summary of these studies and their findings are outlined in Table 5.5 and discussed in this section.

The findings of Dutour (1986) specify particular instances of EC at the insertions for specific muscles, which are also muscles highly activated in archery. Dutour (1986) noted the additional appearance of osteolytic development specifically on the right coronoid process, which when recognising that the majority of individuals are right-handed, would be the draw arm. The author also observed a highly developed *m. triceps br* and lesions on the distal face of the olecranon fossa on the left arm, which would likely be the bow arm for many archers. These two features found together correspond with expectations of biomechanical activation of the elbow in archery. Overall, this study puts archaeologists one step closer to more accurate interpretations of daily life during the Saharan Neolithic.

Merbs (1983) noticed the appearance of osteoarthritis (published under this term, but this chapter uses the term DJD) in articulations of the humerus, mainly at the glenoid cavity and the humeroradial joint. Here it would be important to confirm the ages of the individuals to exclude the possibility that such development is linked to age. Also, it would be worthwhile to note the symmetry (or lack thereof), especially at the glenoid cavity. Both arms experience stress, but the bow arm is additionally subject to compression forces deriving from the bow. The findings of this study improved not only the understanding of daily life for this Canadian Inuit population, but it was also able to link specialised activity to the idea of probable gender roles.

Molnar (2006) found highly developed rotator cuffs (meaning irregular surfaces for the insertions of the *m. supraspinatus*, *m. infraspinatus*, *m. teres minor*, and the *m. subscapularis*) with a slight difference favouring right-sided development (the draw arm for a right-handed archer). This was coupled with greater development (irregular insertion surfaces) of the deltoids and the *m. triceps br* on the left (bow arm for a right-handed archer). In terms of archery, this corresponds to the expected biomechanical stresses placed on the draw arm shoulder and the bow arm elbow. From an archaeological perspective, this study aimed to enhance interpretations of common prehistoric activities for the region (archery, harpooning, and kayaking), which then has the potential to illustrate not only prehistoric daily life, but local economics.

Stirland (2013) studied the skeletons from the sunken English warship, the *Mary Rose*. This ship surely had specialised archers on board, but the precise identities of the recovered remains cannot be confirmed. Overall, the author found higher than average rates of os acromiale (mostly bilateral, but still a very slight preference for the left side), a bone modification that appears when the distal acromion does not fuse to the rest of the scapula. The problem

TABLE 5.5 A summary of the osteological findings for various anthropological studies examining suspected and likely archers.

Collection	Osteological findings	Source
Neolithic Sahara	Osteophytes on right *m. biceps br*	Dutour (1986)
	Hypertrophy of right *m. teres minor*	
	Developed left *m. triceps br*	
	Osteolytic band on right coronoid process	
	Lesion on left distal face of olecranon fossa	
20th century Hudson Bay Inuit (Canada)	Osteoarthritis of the glenoid cavity	Merbs (1983)
	Osteoarthritis of the humeroradial joint	
Middle Neolithic, Gotland (Sweden)	High rotator cuff development on both sides, but slightly more on right	Molnar (2006)
	Greater *m. deltoid* and *m. triceps br* development on left	
15th century warship (England)	Os acromiale (bilateral, slight left preference)	Stirland (2013)
	Attachment area for humeral abduction bigger on left	
	Greater shoulder dimension for bow arm	
Middle Neolithic (France)	More robust ulna (right)	Thomas (2014)
	No significant stature difference between suspected and non-suspected archers	
	More robust radius	
	Overall more robust	
	Greater clavicle midshaft circumference	
	Anterior-posterior flattening of the left ulna	
	Archers have more enthesopathies (specifically in the left upper limb)	
	No significant differences in lower limbs	
Hungarian conquest period (10th century)	Minimal degeneration	Tihanyi et al. (2015)
	Hypertrophies are the dominant lesions	
	Markers are bilateral, severity is asymmetric	
	High-frequency enthesopathy sites: costoclavicular, *m. deltoid*, *m. trapezius*, *m. teres major*, *m. pectoralis major*, *m. latissimus dorsi*, common flexor and extensor origins, *m. biceps br*, *m. margo interosseous*	
Bell Beaker inhumations (Central Europe)	Overall robusticity (large and irregular surfaces), especially for insertions of the rotator cuff muscles, *m. pectoralis major*, *m. latissimus dorsi*, and *m. deltoid*	Ryan et al. (2018)
	Enthesopathies of the *m. trapezius* insertion	
	Likely DJD of the glenoid cavity	
	Osteolytes at the costoclavicular ligament insertion	
	Likely DJD of the elbow	

From Ryan-Despraz, J., 2021. The application of biomechanics and bone morphology to interpret specialized activity and social stratification: The case of bell beaker archery (PhD thesis). University of Geneva, Geneva, Switzerland.

with this pathology is that it appears in 3–6% of modern populations, indicating that it could be genetic, as well as traumatic and mechanical. Indeed, Hunt and Bullen (2007) and Rovesta et al. (2017) found that os acromiale appearance corresponds to both sex (being more common in females) and ancestry, though admittedly both trauma and mechanical stress could also contribute to the problem. Additionally, it is necessary to note that natural fusion of this bone occurs between the ages of 20 and 25 years, therefore once again, age is an important consideration when examining this pathology. From a muscular standpoint, this part of the acromion contains both an origin for the deltoid muscles and an insertion for the *m. trapezius*, two muscles highly activated in archery. For this reason, its appearance in larger-than-average numbers in the *Mary Rose* collection could link it to specialised archery. However, in an examination of living humans, Park et al. (1994) concluded that os acromiale is a painful problem that can limit range of motion, putting into question one's capacity for specialised archery with such a pathology. This skeletal collection from a known warship is valuable to archaeological and anthropological interpretations because it contained at least some certain specialised archers (and warriors); researchers simply cannot confirm if these individuals were the recovered skeletons. Osteological analyses from such a relatively well-known context are important for comparative analyses with other collections.

Thomas (2014) found that the group of suspected archers was overall more robust (larger, more irregular surfaces), especially in the radius, ulna, and clavicle (midshaft circumference). The author also found anterior-posterior flattening of the left ulna (bow arm for a right-handed archer), which could, together with the larger forearm bones, potentially be linked to the biomechanical stresses of holding a bow. In general, the suspected archers had more EC than the non-suspected archers, especially in the left upper limb. Lastly, Thomas (2014) noted no differences in the lower limbs for the suspected archers. Seeing as legs are not actively involved in archery, this is unsurprising and an indication that the other skeletal modifications are possibly linked to archery. Overall, this study managed to link certain specific bone markers to the presence of arrowheads in the burial context.

Tihanyi et al. (2015) found minimal cases of degeneration on the skeletons, with hypertrophies being the most common lesions. Interestingly, the appearance of EC was bilateral, indicating both sides had prominent muscle activation; however, the severity was asymmetric. This is a notable remark because it does correspond with findings that muscle activation in archery is less asymmetric than one might imagine. For example, the draw arm activates more prominently the *m. biceps br*, but as an antagonist muscle the *m. triceps br* would also be stimulated, albeit to a much lesser degree. On the other hand, the *m. triceps br* is a principally activated muscle of the bow arm, though as an antagonist muscle the *m. biceps br* is again mildly stimulated. This would therefore allow for the prediction that, for example, both the *m. biceps br* and the *m. triceps br* would be developed, but the *m. biceps br* more so on the draw arm and the *m. triceps br* more so on the bow arm. Lastly, the areas identified as demonstrating the most instances of EC were the costoclavicular insertion (a ligament involved in movement of the pectoral girdle), *m. deltoid*, *m. trapezius*, *m. teres major*, *m. pectoralis major*, *m. latissimus dorsi*, the common flexor and extensor origins, *m. biceps br*, and the *m. margo interosseous*. With the exception of the *m. margo interosseous*, each of these muscles has been identified as being highly activated in archery. However, this study did not describe any changes of the rotator cuff. Archaeologically, this research worked to draw a parallel between the grave context and the possible archery specialisation of the interred individual during the Hungarian Conquest Period.

Lastly, Ryan et al. (2018) found an overall greater robusticity (size and rate of irregular surfaces) in the group of suspected archers, similar to the findings of Molnar (2006), Thomas (2014), and Tihanyi et al. (2015). This especially included development of the rotator cuff region, *m. pectoralis major*, *m. latissimus dorsi*, and the *m. deltoid* as well as an erosion surface at the insertion for the costoclavicular ligament. Ryan et al. (2018) also found probable instances of DJD at the elbow joint for some of the suspected archers. However, in this case, it is worth noting that extreme cases of DJD would be both painful and limit mobility, therefore raising the question as to an individual's capacity for performing a strenuous activity such as archery with this condition. One archaeological goal of this study was to examine the possible existence (origin?) of warfare during the end of the Neolithic, and the ability to identify specialised archery could be an important step in this process.

Each of these studies contributes to modern interpretations of daily life in the analysed populations. This in turn impacts how researchers understand concepts such as labour practices, gender identity, warfare and hunting (and by extension local economics), and social organisation throughout prehistory and history. Archaeologically speaking, the continued exploration of these topics will further illustrate the lives and evolution of past populations.

All of these findings (summarised in Box 5.1) correspond with those of the common injuries seen in modern archers and the hypothesised biomechanics.

5.6 Conclusion

While identifying activity from the skeleton remains a promising field of study, it is important to recognise its current limits. The first issue to address is the actual identification of an activity because there are several actions that could yield similar results on the skeleton. For this reason, it is prudent to work with individuals linked to a particular context. This means testing a specific hypothesis based on a skeleton's associated goods and environment rather than simply looking at any skeleton and attempting to identify any activity. Second, it is important that the anthropological community improve the global understanding of how biomechanics and muscle activation actually influence bone morphology. This means practical experiments aimed at improving our grasp on sports science and all the ways in which movement can affect bone development, and in particular the relationship between

---

Box 5.1

- EC of the rotator cuff (*m. supraspinatus*, *m. infraspinatus*, *m. teres minor*, and the *m. subscapularis*), *m. biceps br.*, *m. triceps br.*, *m. teres major*, *m. deltoid*, *m. trapezius*, *m. pectoralis major*, *m. latissimus dorsi*, *m. margo interosseous*, and the origins for the common flexors and extensors.

- DJD and other lesions of the glenoid cavity and around the elbow joint.
- Os acromiale.
- Overall larger bone dimensions and irregular surfaces.

the subchondral bone and the visible entheseal surface. Specifically, in terms of biomechanics, what type and extent of movement constitutes the array of factors leading to bone erosion, bone proliferation, and robusticity? And, at what point might such development imply an individual's inability to continue performing a particular action?

One of the biggest problems when looking at EC is identifying whether or not a surface irregularity is truly linked to muscle use or rather some other condition. In order to control for this as best as possible, a general skeletal analysis should be performed in an attempt to identify other existing pathologies that could also potentially leave traces on bone (such as DISH). One of the most important criteria to control for is therefore age because it is a primary driver for EC and other pathologies such as DJD. Because it is currently not possible to tell whether bone modifications are the result of age or biomechanics, the skeletons of senior adults should not be used for activity identification using these techniques.

Biological anthropology would also benefit from new methods of bone analysis that focus on quantitative and replicable data as well as the application of computational science tools, such as machine learning algorithms aimed at organising, creating, and predicting osteological models. Advanced photographic and scanning capabilities are slowly becoming a part of laboratories and research institutions everywhere. Indeed, modern medical facilities are starting to experiment with using machine learning as a means for identifying diseases based on images of blood and tissue samples from patients (Cifuentes-Alcobendas and Domínguez-Rodrigo, 2019; Domínguez-Rodrigo and Baquedano, 2018; Esteva et al., 2019). The field of biological anthropology would greatly benefit from methodologies that rely less heavily on simple visual observations and more on the quantitative data that can be provided by technology, such as high-precision scanners. This would increase repeatability and reliability across the field as well as facilitate more complex data analyses, such as predictive models, and data sharing between researchers.

Lastly, it is important to address the limits linked to bone preservation. Even with modern skeletons, it is rare to be able to measure and score every possible feature, and for older and even prehistoric skeletons a complete dataset is impossible. Therefore researchers would also benefit from improved methods of data analysis with missing data, which becomes especially problematic when analysing questions pertaining to such aspects as potential behaviours and population comparisons. This is another reason why it is important to focus on the techniques of data analysis in the field of biological anthropology.

Despite these potential limits to modern osteological analyses of identifying behaviour (including archery) on the skeleton, this area continues to be a promising and intriguing field of study. The capacity to improve our understanding both of past daily life and how movement science influences bone development has the potential to contribute valuable insights into numerous fields, such as anthropology, archaeology, history, kinesiology, and sports medicine.

Acknowledgments

This work derives from a PhD dissertation (Ryan-Despraz, 2021), therefore the author would like to thank her supervisors, Marie Besse, Jocelyne Desideri, and Sébastien Villotte, for their support as well as the Swiss National Science Foundation (FNS, grant number 175169) and the Laboratory of Prehistoric Archaeology and Anthropology at the University of Geneva for project funding. An additional thanks goes to Marie Besse, Jocelyne Desideri, and the Laboratory of Prehistoric Archaeology and Anthropology at the University of Geneva for allowing access to the skeletal collections for the photos included in this chapter.

References

Ahmad, Z., Siddiqui, N., Malik, S. S., Abdus-Samee, M., Tytherleigh-Strong, G., & Rushton, N. (2013). Lateral epicondylitis: A review of pathology and management. *The Bone & Joint Journal, 95-B,* 1158–1164.

Ahmad, Z., Taha, Z., Arif Hassan, H., Azrul Hisham, M., Hadi Johari, N., & Kadirgama, K. (2014). Biomechanics measurements in archery. *Journal of Mechanical Engineering Science, 6,* 762–771.

Aicale, R., Tarantino, D., & Maffulli, N. (2018). Overuse injuries in sport: A comprehensive overview. *Journal of Orthopaedic Surgery and Research, 13,* 309.

Alves Cardoso, F., & Henderson, C. Y. (2010). Enthesopathy formation in the humerus: Data from known age-at-death and known occupation skeletal collections. *American Journal of Physical Anthropology, 141,* 550–560.

Åman, M., Forssblad, M., & Henriksson-Larsén, K. (2016). Incidence and severity of reported acute sports injuries in 35 sports using insurance registry data. *Scandinavian Journal of Medicine & Science in Sports, 26,* 451–462.

Amin, N. H., Kumar, N. S., & Schickendantz, M. S. (2015). Medial epicondylitis: Evaluation and management. *Journal of the American Academy of Orphopaedic Surgeons, 23,* 348–355.

Andrews, J. R., & Whiteside, J. A. (1993). Common elbow problems in the athlete. *Journal of Orthopaedic & Sports Physical Therapy, 17,* 289–295.

Apostolakos, J., Durant, T. J., Dwyer, C. R., Russell, R. P., Weinreb, J. H., Alaee, F., et al. (2014). The enthesis: A review of the tendon-to-bone insertion. *Muscles, Ligaments and Tendons Journal, 4,* 333–342.

Arangio, G. A., Beam, H., Kowalczyk, G., & Salathé, E. P. (1998). Analysis of stress in the metatarsals. *Foot and Ankle Surgery, 4,* 123–128.

Axford, R. (1995). *Archery anatomy.* Souvenir Press.

Başbozkurt, M., Demiralp, B., & Atilla, H. A. (2012). Stress fractures in military personnel of Turkish army. In *Sports injuries: Prevention, diagnosis, treatment and rehabilitation* (pp. 853–857). Berlin, Heidelberg: Springer-Verlag.

Becker, C. J. (1945). *En 8000-aarig Stenalderboplads i Holmegaards Mose* (pp. 61–72). Fra Nationalmuseets Arbejdsmark.

Becker, S. K. (2019). Osteoarthritis, entheses, and long bone cross-sectional geometry in the Andes: Usage, history, and future directions. *International Journal of Paleopathology, 29,* 45–53. https://doi.org/10.1016/j.ijpp.2019.08.005.

Benjamin, M., Evans, E. J., & Copp, L. (1986). The histology of tendon attachments to bone in man. *Journal of Anatomy, 149,* 89–100.

Benjamin, M., Kumai, T., Milz, S., Boszczyk, B. M., Boszczyk, A. A., & Ralphs, J. R. (2002). The skeletal attachment of tendons—tendon 'entheses'. *Comparative Biochemistry and Physiology Part A: Molecular & Integrative Physiology, 133,* 931–945.

Benjamin, M., & McGonagle, D. (2001). The anatomical basis for disease localisation in seronegative spondyloarthropathy at entheses and related sites. *The Journal of Anatomy, 199,* 503–526.

Benjamin, M., Moriggl, B., Brenner, E., Emery, P., McGonagle, D., & Redman, S. (2004). The "enthesis organ" concept: Why enthesopathies may not present as focal insertional disorders. *Arthritis and Rheumatism, 50,* 3306–3313.

Benjamin, M., & Ralphs, J. R. (1995). Functional and developmental anatomy of tendons and ligaments. In S. L. Gordon, S. J. Blair, & L. J. Fine (Eds.), *Repetitive motion disorders of the upper extremity* (pp. 185–203). Rosemont, Illinois: American Academy of Orthopaedic Surgeons.

Blitz, J. H. (1988). Adoption of the bow in prehistoric North America. *North American Archaeologist, 9,* 123–145.

Boroneanţ, V., Bonsall, C., McSweeney, K., Payton, R., & Macklin, M. (1999). A mesolithic burial area at Schela Cladovei, Romania. In A. Thévenin (Ed.), *L'Europe Des Derniers Chasseurs: Épipaléolithique et Mésolithique. (Actes Du 5e Colloque International UISPP, Commission XII, Grenoble, 18–23 Septembre 1995).* Éditions du Comité des Travaux Historiques et Scientifiques, Paris (pp. 385–390).

Boutsiadis, A., Karataglis, D., & Papadopoulos, P. (2012). Pathology of rotator cuff tears. In *Sports injuries: Prevention, diagnosis, treatment and rehabilitation* (pp. 81–86). Berlin, Heidelberg: Springer-Verlag.

Cattelain, P. (1997). Hunting during the upper paleolithic: Bow, spearthrower, or both? In H. Knecht (Ed.), *Projectile technology* (pp. 213–240). New York: Plenum Press.

Chan, R., Kim, D. H., Millett, P. J., & Weissman, B. N. (2004). Calcifying tendinitis of the rotator cuff with cortical bone erosion. *Skeletal Radiology, 33,* 596–599.

Chen, S.-K., Cheng, Y.-M., Lin, Y.-C., Hong, Y.-J., Huang, P-J., & Chou, P.-H. (2005). Investigation of management models in elite athlete injuries. *Kaohsiung Journal of Medical Sciences, 21,* 220–227.

Chen, D., Shen, J., Zhao, W., Wang, T., Han, L., Hamilton, J. L., et al. (2017). Osteoarthritis: Toward a comprehensive understanding of pathological mechanism. *Bone Research, 5,* 16044.

Cifuentes-Alcobendas, G., & Domínguez-Rodrigo, M. (2019). Deep learning and taphonomy: High accuracy in the classification of cut marks made on fleshed and defleshed bones using convolutional neural networks. *Scientific Reports, 9*, 1–12. https://doi.org/10.1038/s41598-019-55439-6.

Clarke, B. (2008). Normal bone anatomy and physiology. *Clinical Journal of the American Society of Nephrology, 3*, S131–S139.

Clarys, J. P., Cabri, J., Bollens, E., Sleeckx, R., Taeymans, J., Vermeiren, M., et al. (1990). Muscular activity of different shooting distances, different release techniques, and different performance levels, with and without stabilizers, in target archery. *Journal of Sports Sciences, 9*, 235–257.

Connell, D., Burke, F., Coombes, P., McNealy, S., Freeman, D., Pryde, D., et al. (2001). Sonographic examination of lateral epicondylitis. *American Journal of Roentgenology, 176*, 777–782.

Cowgill, L. (2018). Cross-sectional geometry. In W. Trevathan (Ed.), *The international encyclopedia of biological anthropology* (pp. 1–3). John Wiley & Sons Inc. https://doi.org/10.1002/9781118584538.ieba0113.

Dams, L. (1984). *Les Peintures Rupestres du Levant Espagnol*. Paris: Picard.

Dolukhanov, P. M. (2004). War and peace in prehistoric Eastern Europe. In J. Carman, & A. Harding (Eds.), *Ancient warfare: Archaeological perspectives* (pp. 73–87). Sutton publishing.

Domínguez-Rodrigo, M., & Baquedano, E. (2018). Distinguishing butchery cut marks from crocodile bite marks through machine learning methods. *Scientific Reports, 8*, 1–8. https://doi.org/10.1038/s41598-018-24071-1.

Dürr, H. R., Lienemann, A., Silbernagl, H., Nerlich, A., & Refior, H. J. (1997). Acute calcific tendinitis of the pectoralis major insertion associated with cortical bone erosion. *European Radiology, 7*, 1215–1217. https://doi.org/10.1007/s003300050277.

Dutour, O. (1986). Enthesopathies (lesions of muscular insertions) as indicators of the activities of Neolithic Saharan populations. *American Journal of Physical Anthropology, 71*, 221–224.

Ebenbichler, G. R., Erdogmus, C. B., Resch, K. L., Funovics, M. A., Kainberger, F., Barisani, G., et al. (1999). Ultrasound therapy for calcific tendinitis of the shoulder. *New England Journal of Medicine, 340*, 1533–1538. https://doi.org/10.1056/NEJM199905203402002.

Ergen, E., Çirçi, E., Lapostolle, J.-C., & Hibner, K. (2004). FITA medical committee archery injuries survey (seniors). In E. Ergen, & K. Hibner (Eds.), *Sports science and medicine in archery* (pp. 59–64). FITA: Fédération Internationale de Tir à l'Arc.

Ertan, H. (2006). Injury patterns among Turkish archers. *The Shield-Research Journal of Physical Education & Sports Science, 1*, 19–29.

Esteva, A., Robicquet, A., Ramsundar, B., Kuleshov, V., DePristo, M., Chou, K., et al. (2019). A guide to deep learning in healthcare. *Nature Medicine, 25*, 24–29. https://doi.org/10.1038/s41591-018-0316-z.

Finestone, A. S., & Milgrom, C. (2012a). Diagnosis and treatment of stress fractures. In *Sports injuries: Prevention, diagnosis, treatment and rehabilitation* (pp. 775–785). Berlin, Heidelberg: Springer-Verlag.

Finestone, A. S., & Milgrom, C. (2012b). Epidemiology and anatomy of stress fractures. In *Sports injuries: Prevention, diagnosis, treatment and rehabilitation* (pp. 769–773). Berlin, Heidelberg: Springer-Verlag.

Frostick, S. P., Mohammad, M., & Ritchie, D. A. (1999). Sports injuries of the elbow. *British Journal of Sports Medicine, 33*, 301–311.

Fukuda, H., & Neer, C. S. I. (1988). Archer's shoulder: Recurrent posterior subluxation and dislocation of the shoulder in two archers. *Orthopedics, 11*, 171–174.

Goodman, A. H., & Martin, D. L. (2002). Reconstructing health profiles from skeletal remains. In R. H. Steckel, & J. C. Rose (Eds.), *The backbone of history. Health and nutrition in the Western Hemisphere* (pp. 11–60). Cambridge: Cambridge University Press.

Gopal Adkitte, R., Shah, S., Jain, S., Walla, S., Chopra, N., & Kumar, H. (2016). Common injuries amongst Indian elite archers: A prospective study. *Saudi Journal of Sports Medicine, 16*, 210–213. https://doi.org/10.4103/1319-6308.187559.

Guadilla, J., Lopez-Vidriero, E., Lopez-Vidriero, R., Padilla, S., Delgado, D., Arriaza, R., et al. (2016). PRP in lateral elbow pain. In L. A. Pederzini, D. Eygendaal, & M. Denti (Eds.), *Elbow and sport, ESSKA (European society of sports traumatology, knee surgery & arthroscopy)* (pp. 109–124). Springer.

Guilaine, J., & Zammit, J. (2008). *The origins of war: Violence in prehistory*. John Wiley & Sons.

Hall, S. (2003). *Basic biomechanics* (4th ed.). McGraw-Hill.

Hayes, C. W., Rosenthal, D. I., Plata, M. J., & Hudson, T. M. (1987). Calcific tendinitis in unusual sites associated with cortical bone erosion. *American Journal of Roentgenology, 149*, 967–970. https://doi.org/10.2214/ajr.149.5.967.

Hildenbrand, J. C., & Rayan, G. M. (2010). Archery. In D. Caine, P. Harmer, & M. A. Schiff (Eds.), *Epidemiology of injury in olympic sports, the encyclopedia of sports medicine* (pp. 18–25). Wiley-Blackwell.

Hunt, D. R., & Bullen, L. (2007). The frequency of os acromiale in the Robert J. Terry collection. *International Journal of Osteoarchaeology, 17*, 309–317.

Janson, L. (2004). Muscles in archery. In E. Ergen, & K. Hibner (Eds.), *Sports medicine and science in archery* (pp. 13–14). FITA: Fédération Internationale de Tir à l'Arc.

Junkmanns, J. (2001). *Arc et Flèche: Fabrication et Utilisation au Néolithique.* Bienne: Éditions Musée Schwab.

Jurmain, R., Alves Cardoso, F., Henderson, C., & Villotte, S. (2012). Bioarchaeology's holy grail: The reconstruction of activity. In A. L. Grauer (Ed.), *A companion to paleopathology* Blackwell Publishing Ltd.

Kajiyama, S., Muroi, S., Sugaya, H., Takahashi, N., Matsuki, K., Kawai, N., et al. (2017). Osteochondritis dissecans of the humeral capitellum in young athletes: Comparison between baseball players and gymnasts. *The Orthopaedic Journal of Sports Medicine, 5*, 1–5.

Kaux, J.-F., Forthomme, B., Goff, C. L., Crielaard, J.-M., & Croisier, J.-L. (2011). Current opinions on tendinopathy. *Journal of Sports Science and Medicine, 10*, 238–253.

Kaynaroğlu, V., & Kiliç, Y. A. (2012). Archery-related sports injuries. In M. N. Doral, & J. Karlsson (Eds.), *Sports injuries: Prevention, diagnosis, treatment and rehabilitation* (pp. 1081–1086). Berlin, Heidelberg: Springer-Verlag.

Kian, A., Ghomshe, F. T., & Norang, Z. (2013). Comparing the ability of controlling the bow hand during aiming phase between two elite and beginner female compound archers: A case study. *European Journal of Experimental Biology, 3*.

Knudson, D. (2007). *Fundamentals of biomechanics* (2nd ed.). US: Springer. https://doi.org/10.1007/978-0-387-49312-1.

Kosar, N. S., & Demirel, H. A. (2004). Kinesiological analysis of archery. In E. Ergen, & K. Hibner (Eds.), *Sports medicine and science in archery* (pp. 3–12). FITA: Fédération Internationale de Tir à l'Arc.

Lanyon, L. E. (1992). Control of bone architecture by functional load bearing. *Journal of Bone and Mineral Research, 7*, S369–S375.

Lapostolle, J.-C. (2004). Elbow pathologies in archery. In E. Ergen, & K. Hibner (Eds.), *Sports medicine and science in archery* (pp. 70–81). FITA: Fédération Internationale de Tir à l'Arc.

Lapostolle, J.-C., Hibner, K., & Ergen, E. (2004). FITA medical committee archery injuries survey (juniors). In E. Ergen, & K. Hibner (Eds.), *Sports medicine and science in archery* (pp. 65–69). FITA: Fédération Internationale de Tir à l'Arc.

Larven, J. (2007). *Shooting technique biomechanics* (pp. 1–28). Archery Australia.

Lieberman, D. E., Polk, J. D., & Demes, B. (2004). Predicting long bone loading from cross-sectional geometry. *American Journal of Physical Anthropology, 123*, 156–171. https://doi.org/10.1002/ajpa.10316.

Lipman, K., Wang, C., Ting, K., Soo, C., & Zheng, Z. (2018). Tendinopathy: Injury, repair, and current exploration. *Drug Design, Development and Therapy, 12*, 591–603. https://doi.org/10.2147/DDDT.S154660.

Littke, N. (2004a). Back pain and the archer. In E. Ergen, & K. Hibner (Eds.), *Sports medicine and science in archery* (pp. 105–110). FITA: Fédération Internationale de Tir à l'Arc.

Littke, N. (2004b). Shoulder injuries: A rehab perspective. In E. Ergen, & K. Hibner (Eds.), *Sports medicine and science in archery* (pp. 82–86). FITA: Fédération Internationale de Tir à l'Arc.

Mann, G., Constantini, N., Nyska, M., Dolev, E., Barchilon, V., Shabat, S., et al. (2012). Stress fractures: Overview. In *Sports injuries: Prevention, diagnosis, treatment and rehabilitation* (pp. 787–813). Berlin, Heidelberg: Springer-Verlag.

Mann, R. W., & Hunt, D. R. (2005). *Photographic regional atlas of bone disease: A guide to pathologic and normal variation in the human skeleton* (3rd ed.). Springfield: Charles C. Thomas.

Mann, D., & Littke, N. (1989). Shoulder injuries in archery. *Canadian Journal of Sport Science, 14*, 85–92.

Mariotti, V., Facchini, F., & Giovanna Belcastro, M. (2004). Enthesopathies-proposal of a standardized scoring method and applications. *Collegium Antropologicum, 28*, 145–159. https://doi.org/572.781:572.08:616.71.

Mariotti, V., Facchini, F., & Giovanna Belcastro, M. (2007). The study of entheses: Proposal of a standardised scoring method for twenty-three entheses of the postcranial skeleton. *Collegium Antropologicum, 31*, 291–313.

Martin-Francés, L., Martinon-Torres, M., Gracia-Téllez, A., & Bermúdez de Castro, J. M. (2013). Evidence of stress fracture in a homo antecessor metatarsal from Gran Dolina site (Atapuerca, Spain). *International Journal of Osteoarchaeology, 25*, 564–573. https://doi.org/10.1002/oa.2310.

Merbs, C. F. (1983). *Patterns of activity-induced pathology in a Canadian Inuit population. Vol. 119* (pp. 120–128). Archaeological Survey Canada.

Milz, S., Tischer, T., Buettner, A., Schieker, M., Maier, M., Redman, S., et al. (2004). Molecular composition and pathology of entheses on the medial and lateral epicondyles of the humerus: A structural basis for epicondylitis. *Annals of the Rheumatic Diseases, 63*, 1015–1021.

Molnar, P. (2006). Tracing prehistoric activities: Musculoskeletal stress marker analysis of a stone-age population on the island of Gotland in the Baltic Sea. *American Journal of Physical Anthropology, 129*, 12–23.

Nakama, L. H., King, K. B., Abrahamsson, S., & Rempel, D. M. (2007). Effect of repetition rate on the formation of microtears in tendon in an in vivo cyclical loading model. *Journal of Orthopaedic Research, 25*, 1176–1184. https://doi.org/10.1002/jor.20408.

Naraen, A., Giannikas, K. A., & Livesley, P. J. (1999). Overuse epiphyseal injury of the coracoid process as a result of archery. *International Journal of Sports Medicine, 20*, 53–55. https://doi.org/10.1055/s-2007-971092.

Niestroj, C. K., Schöffl, V., & Küpper, T. (2017). Acute and overuse injuries in elite archers. *The Journal of Sports Medicine and Physical Fitness, 58*, 1063–1070.

Nishizono, H., Shibayama, H., Izuta, T., & Saito, K. (1987). Analysis of archery shooting techniques by means of electromyography. In *International society of biomechanics in sports, symposium V. Presented at the symposium V, Athens* (pp. 364–372).

Palsbo, S. E. (2012). Epidemiology of recreational archery injuries: Implications for archery ranges and injury prevention. *The Journal of Sports Medicine and Physical Fitness, 52*, 293–299.

Park, J. G., Lee, J. K., & Phelps, C. T. (1994). Os acromiale associated with rotator cuff impingement: MR imaging of the shoulder. *Radiology, 193*, 255–257.

Pearson, O. M., & Lieberman, D. E. (2004). The aging of Wolff's "law": Ontogeny and responses to mechanical loading in cortical bone. *Yearbook of Physical Anthropology, 47*, 63–99.

Pederzini, L. A., Tosi, M., Prandini, M., Nicoletta, F., & Barberio, V. I. (2012). Sports-related elbow problems. In M. N. Doral, & J. Karlsson (Eds.), *Sports injuries: Prevention, diagnosis, treatment and rehabilitation*. Berlin, Heidelberg: Springer-Verlag.

Prine, B., Prine, A., Leavitt, T., Wasser, J., & Vincent, H. (2016). Prevalence and characteristics of archery-related injuries in archers: 3083 board #148 June 3. *Medicine & Science in Sports & Exercise, 48*, 874.

Raphael, K. (2009). Mongol siege warfare on the banks of the euphrates and the question of gunpowder (1260-1312). *Journal of the Royal Asiatic Society, 19*, 355–370.

Rayan, G. M. (1992). Archery-related injuries of the hand, forearm, and elbow. *Southern Medical Journal, 85*, 961–964.

Renfro, G. J., & Fleck, S. J. (1991). *Preventing rotator cuff injury and reaching optimal athletic performance in archery through resistance exercise* (pp. 194–195). The US Archer.

Rhodes, J., & Knüsel, C. (2005). Activity-related skeletal change in medieval humeri: Cross-sectional and architectural alterations. *American Journal of Physical Anthropology, 128*, 536–546.

Roberts, C., & Manchester, K. (2007). *The archaeology of disease* (3rd ed.). Cornell University Press.

Rogers, C. J. (2011). The development of the longbow in late medieval England and 'technological determinism'. *Journal of Medieval History, 37*, 321–341. https://doi.org/10.1016/j.jmedhist.2011.06.002.

Rogers, J., Shepstone, L., & Dieppe, P. (1997). Bone formers: Osteophyte and enthesophyte formation are positively associated. *Annals of the Rheumatic Diseases, 56*, 85–90.

Rogers, J., & Waldron, T. (1995). *A field guide to joint disease in archaeology*. West Sussex: John Wiley & Sons.

Rovesta, C., Marongiu, M. C., Corradini, A., Torricelli, P., & Ligabue, G. (2017). Os acromiale: Frequency and a review of 726 shoulder MRI. *Musculoskeletal Surgery, 101*, 201–205.

Rubin, C. T., McLeod, K. J., & Bain, S. D. (1990). Functional strains and cortical bone adaptation: Epigenetic assurance of skeletal integrity. *Journal of Biomechanics, 23*, 43–54.

Ruff, C., Holt, B., & Trinkaus, E. (2006). Who's afraid of the big bad Wolff?: "Wolff's law" and bone functional adaptation. *American Journal of Physical Anthropology, 129*, 484–498. https://doi.org/10.1002/ajpa.20371.

Runestad, J. Q., Ruff, C. B., Nieh, J. C., Thorington, R. W., & Teaford, M. F. (1993). Radiographic estimation of long bone cross-sectional geometric properties. *American Journal of Physical Anthropology, 90*, 207–213. https://doi.org/10.1002/ajpa.1330900207.

Ryan, J., Desideri, J., & Besse, M. (2018). Bell beaker archers: Warriors or an ideology? *Journal of Neolithic Archaeology, 20*, 97–122.

Ryan-Despraz, J. (2021). *The application of biomechanics and bone morphology to interpret specialized activity and social stratification: The case of bell beaker archery* (PhD thesis). Geneva, Switzerland: University of Geneva.

Salter, D. M. (2002). Degenerative joint disease. *Current Diagnostic Pathology, 8*, 11–18.

Schlecht, S. H. (2012). Understanding entheses: Bridging the gap between clinical and anthropological perspectives. *The Anatomical Record, 295*, 1239–1251. https://doi.org/10.1002/ar.22516.

Schwartz, J. H. (2007). *Skeleton keys: An introduction to human skeletal morphology, development, and analysis* (2nd ed.). New York: Oxford University Press.

Scott, A., & Ashe, M. (2006). Common tendinopathies in the upper and lower extremities. *Current Sports Medicine Reports, 5*, 233–241. https://doi.org/10.1097/01.CSMR.0000306421.85919.9c.

Selvanetti, A., Cipolla, M., & Puddu, G. (1997). Overuse tendon injuries: Basic science and classification. *Operative Techniques in Sports Medicine, 5,* 110–117.

Sessa, E. (1994). Etude de la biomécanique du tir à l'arc: Incidence sur la pathologie et la rééducation. *Annales de kinésithérapie, 21,* 435–442.

Shaw, C., Hofmann, C., Petraglia, M., Stock, J., & Gottschall, J. (2012). Neandertal humeri may reflect adaptation to scraping tasks, but not spear thrusting. *PLoS One, 7,* e40349.

Sheibani-Rad, S., Wolfe, S., & Jupiter, J. (2013). Hand disorders in musicians: The orthopaedic surgeon's role. *The Bone & Joint Journal, 95,* 1–5.

Singh, A., & Lhee, S.-H. (2016). Injuries in archers. *Saudi Journal of Sports Medicine, 16,* 168.

Smillie, I. S. (1970). *Injuries of the knee joint.* Edinburgh: Churchill Livingstone.

Stirland, A. J. (2013). *The men of the Mary Rose: Raising the dead.* Gloucestershire: The History Press.

Sudoł-Szopińska, I., Kwiatkowska, B., Prochorec-Sobieszek, M., & Maśliński, W. (2015). Enthesopathies and enthesitis. Part 1. Etiopathogenesis. *Journal of Ultrasonography, 15,* 72–84.

Thomas, A. (2014). Bioarchaeology of the middle neolithic: Evidence for archery among early European farmers. *American Journal of Physical Anthropology, 154,* 279–290.

Thorpe, I. J. N. (2003). Anthropology, archaeology, and the origin of warfare. *World Archaeology, 35,* 145–165.

Tihanyi, B., Bereczki, Z., Molnár, E., Berthon, W., Révész, L., Dutour, O., et al. (2015). Investigation of Hungarian Conquest Period (10th c. AD) archery on the basis of activity-induced stress markers on the skeleton- -preliminary results. *Acta Biologica Szegediensis, 59,* 65–77.

Uhthoff, H., & Loehr, J. (1998). The shoulder. In C. Rockwood, & F. Matsen (Eds.), *Calcifying tendinitis* (pp. 989–1008). Philadelphia: Saunders.

Varacallo, M., & Mair, S. D. (2018). Rotator cuff syndrome. In *StatPearls.* Treasure Island (FL): StatPearls Publishing.

Villotte, S. (2009). *Enthésopathies et activités des hommes préhistoriques—Recherche méthodologique et application aux fossiles européens du Paléolithique supérieur et du Mésolithique, BAR International Series 1992.* Oxford: Archaeopress.

Villotte, S., Assis, S., Alves Cardoso, F., Henderson, C., Mariotti, V., Milella, M., et al. (2016). In search of consensus: Terminology for entheseal changes (EC). *International Journal of Paleopathology, 13,* 49–55.

Villotte, S., & Knüsel, C. J. (2013). Understanding entheseal changes: Definition and life course changes. *International Journal of Osteoarchaeology, 23,* 135–146. https://doi.org/10.1002/oa.2289.

Waldron, T. (2009). *Palaeopathology.* Cambridge University Press.

Walz, D. M., Newman, J. S., Konin, G. P., & Ross, G. (2010). Epicondylitis: Pathogenesis, imaging, and treatment. *RadioGraphics, 30,* 167–184.

Waugh, E. J. (2005). Lateral epicondylalgia or epicondylitis: What's in a name? *Journal of Orthopaedic & Sports Physical Therapy, 35,* 200–202.

Weiss, E., & Jurmain, R. (2007). Osteoarthritis revisited: A contemporary review of aetiology. *International Journal of Osteoarchaeology, 17,* 437–450.

White, T. D. (2000). *Human osteology* (2nd ed.). Academic Press.

Whiteside, J. A., & Andrews, J. R. (1989). Common elbow problems in the recreational athlete. *The Journal of Musculoskeletal Medicine, 6,* 17–34.

Wolff, J. (1869). Über die Bedeutung der Architektur des spongiösen Substanz. *Zentralblatt für die medizinischen Wissenschaft VI,* 223–234.

Zwas, S. T., Elkanovitch, R., & Frank, G. (1987). Interpretation and classification of bone scintigraphic findings in stress fractures. *The Journal of Nuclear Medicine, 28,* 452–457.

6

Tool use and the hand

*Christopher J. Dunmore[a], Fotios Alexandros Karakostis[b],
Timo van Leeuwen[c], Szu-Ching Lu[d,e], and Tomos Proffitt[f]*

[a]Skeletal Biology Research Centre, School of Anthropology and Conservation, University of Kent, Canterbury, United Kingdom, [b]DFG (*Deutsche Forschungsgemeinschaft*) Center for Advanced Studies "Words, Bones, Genes, Tools," Eberhard Karls University of Tübingen, Tübingen, Germany, [c]Department of Development and Regeneration, KU Leuven, KULAK, Kortrijk, Belgium, [d]Laboratory for Innovation in Autism, University of Strathclyde, Glasgow, United Kingdom, [e]School of Education, University of Strathclyde, Glasgow, United Kingdom, [f]Technological Primate Research Group, Max Planck Institute for Evolutionary Anthropology, Leipzig, Germany

Comprising over a quarter of the total number of bones in the human skeleton, the biomechanically complex joints formed by the short and long bones of our hands provide a wealth of information. As the subject of this volume, it may come as no surprise to the reader that this information can be used to infer human behaviour, and for the hands this phrase could not be more apt. Unlike most other living primates, our hands are not principally used for locomotion but instead for manipulation. It might be thought that this is a consequence of being a bipedal ape but, although true to a large extent, this idea belies the complexity and diversity of manual behaviours that we use to interact with our surroundings. Simply put, an idea will stay just that until we use our hands to make it a reality. Most technology requires manual interaction; you are likely using your hands to either hold this book or to digitally access this chapter right now. Indeed, the earliest technology preserved in the archaeological record, stone tools, represents our first record of complex material culture, the result of hands performing distinctly human behaviours.

While entire volumes exist that cover the morphology, behaviour, and evolution of primate hands in more detail (Kivell et al., 2016; Lewis, 1989; Napier, 1993; Preuschoft and Chivers, 1993), here we provide an overview of what the hand bones can tell us about manual behaviour using emerging methodologies and evidence. This chapter explores which behaviours are truly unique to the modern human hand among our living relatives and how these behaviours can be inferred in the archaeological record, with the ultimate goal of understanding how they evolved, as well as what this means for our hands today.

6.1 What behaviours are unique to the human hand?

When we ask what behaviours are unique to the human hand, the first answer will likely be an opposable thumb that allows us to create and use tools. The idea that this trait is one of the defining features of our species is an old one (Huxley, 1863). However, tool use (Beck, 1980; but see St Amant and Horton, 2008) has been documented in many animals that do not possess hands, from dolphins (*Tursiops* sp.; Krützen et al., 2005) to crows (*Corvus moned-uloides*; Taylor et al., 2007), and has yet to be recorded in some species that have opposable digits. Chameleons and birds both have digits that oppose each other around tree branches (Abourachid et al., 2017; Anderson and Higham, 2014). Arboreal mammals, especially primates, were known to use an opposable thumb for the same purpose over a century ago (Wood-Jones, 1917). More recently it has been demonstrated that opposable thumbs are not restricted to arboreal primates (Jolly, 1970), as they are present in most monkeys and all apes (Fragaszy and Crast, 2016), although only some of these species use their hands for tool use (Carpenter, 1887; Urbani, 1998; Visalberghi, 1990). The fact that humans commonly exhibit a level of dexterity and technology not found in birds, other primates, or even other apes cannot, therefore, be solely attributed to their opposable thumb. This behavioural distinction is the result of grips habitually used by humans in complex manipulation, facilitated by the anatomy of the modern human hand.

Primate grips can be broadly classified as power or precision grips. Power grips do not typically recruit the thumb to a large extent and often involve the palm, such as a diagonal power grip used by chimpanzees (*Pan troglodytes*) or bonobos (*Pan paniscus*) when climbing trees (Marzke and Wullstein, 1996; Napier, 1960; Neufuss et al., 2017; Samuel et al., 2018). An exception is the uniquely human power 'squeeze' grip, in which the thumb directs the force of this grip around a cylindrical object, such as a hammer (Marzke et al., 1992). Precision grips primarily involve the fingers and the thumb, rather than the palm, and are used when manipulating small objects, such as a pen (Marzke and Wullstein, 1996; Napier, 1960).

When grasping small objects, great apes frequently use precision grips, opposing their thumb to the side of the index finger in 'pad-to-side' grips (Christel, 1993; Marzke and Wullstein, 1996, Fig. 6.1C). Historically, this type of thumb opposition was thought of as less dexterous than that of humans (Day and Napier, 1963; Napier, 1960). However, further research has shown that these 'pad-to-side' grips are just one example of the many grips used in proficient great ape manipulation, that all oppose the pad, side, or tip of the thumb, to the side or tip of another digit (Bardo et al., 2016, 2017; Christel, 1993; Marzke and Wullstein, 1996; Neufuss et al., 2019). Several chimpanzee populations (McGrew, 2013; Whiten et al., 1999; Whiten and Boesch, 2001) use these grips in a wide variety of, primarily organic, tool use, including grooming (Mcgrew and Tutin, 1973; Tutin et al., 2001) and subsistence activities, such as termite fishing or nut cracking (Boesch et al., 2020; Boesch and Boesch, 1993). Humans also use 'pad-to-side' grips (Key et al., 2018; Fig. 6.1A) but unlike great apes, we also employ habitual 'pad-to-pad' grips during manipulation. 'Pad-to-pad' grips require

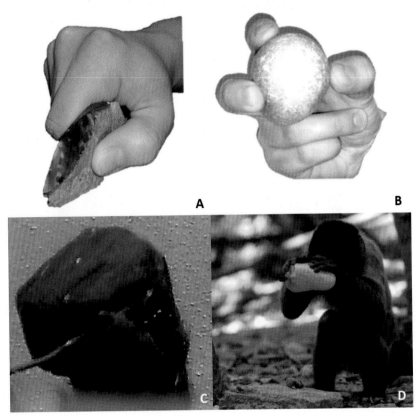

FIG. 6.1 Primate grips. (A) A human holding a stone flake in a 'pad-to-side' grip about to cut an object. (B) A human holding a hammerstone in a 'Three-jaw chuck pad-to-pad grip' used during knapping, note that most of the hammerstone surface is exposed. (C) A bonobo holding a stick in a 'pad-to-side' grip, note the thumb would contact the long palm if the fingers were extended rather than flexed. (D) A capuchin using a symmetrical bimanual power grip to pound a nut on an anvil with a hammerstone. *(C) Image courtesy of Ameline Bardo.*

spreading the thumb radially,[a] away from the palm, while simultaneously flexing it and turning the thumb pad to face, and fully oppose the finger pads (Marzke and Wullstein, 1996; Napier, 1960; Fig. 6.1B). The extant great ape thumb is not long enough to fully oppose the pad of other digits in 'pad-to-pad' grips of small objects (Christel, 1993; Figs. 6.1 and 6.2), whereas a shorter palm and fingers, rather than a long thumb, facilitates this grip in humans (Almécija et al., 2015; *contra* Susman, 1994).

Human 'pad-to-pad' grips are thought to underlie much of our dexterity. These grips use a small contact area to hold an object without encompassing it, exposing much of the object surface for use with another, such as in the production of stone tools, or for visual inspection

---

[a] Radial and ulnar are used in this chapter as directional terms as these forearm bones consistently articulate with one side of the hand. The medial or lateral side of the hand changes when it is palm-up (supinated, as in anatomical position) or palm-down (pronated).

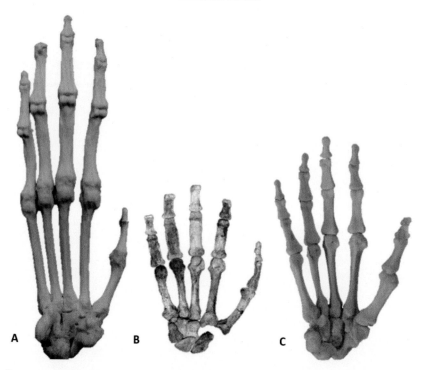

FIG. 6.2 Great ape and hominin right hand skeletons in palmar view. (A) An articulated orangutan hand skeleton, note the length of the metacarpals in the palm as well as the length and curvature of the phalanges. (B) The almost complete right hand of *Australopithecus sediba*; the distal phalanges of the fingers are missing. (C) A modern human hand, note the shorter metacarpals than the orangutan and the straighter phalanges relative to both other hands depicted. *(B) Image courtesy of Peter Schmid and adapted from Kivell, T.L., Kibii, J.M., Churchill, S.E., Schmid, P., & Berger, L.R. (2011). Australopithecus Sediba hand demonstrates mosaic evolution of locomotor and manipulative abilities. Science, 333(6048), 1411–1417.*

(Marzke, 1997, 2013; Fig 6.1B). It must be remembered that object inspection is crucial to complex manipulation, as it is as much a mental exercise as a manual one (Coudé et al., 2019). The 'pad-to-pad' grip also provides the space for other digit pads to contact an object at the same time, applying force from a different angle. This may further secure an object in a static grip or form a transitionary grip in which an object is moved from one finger pad to another during 'in-hand' or unimanal manipulation. Many other grips may be part of 'in-hand' manipulation, but they do not expose as much of the object or facilitate as much independent movement of the digits as 'pad-to-pad' grips. These sequential in-hand grips, 'precision handling', are finer, more elaborate, and more frequent in humans than in other apes (Christel, 1993; Fragaszy and Crast, 2016; Landsmeer, 1962). Similarly, though bimanual manipulation has been observed in baboons (Vauclair et al., 2005), capuchins (Meunier and Vauclair, 2007), and several great apes (Bardo et al., 2015; Byrne et al., 2001; Forrester et al.,

2016), bimanual movements are more frequent in humans and more often incorporate asymmetric movements, where each hand performs a different function (Bardo et al., 2017; Mutalib et al., 2019).

Unlike apes, several monkey species have a thumb-to-finger length ratio closer to humans (AlMécija et al., 2015; Liu et al., 2016), which may facilitate tool use. For example, this ratio in long tailed macaques (*Macaca fascicularis*) has been related to their unimanual use of hammerstones to crack open hard shelled marine resources on stationary anvils (Gumert et al., 2009; Gumert and Malaivijitnond, 2012; Malaivijitnond et al., 2007; Marzke, 2013). Moreover, certain monkey species can also use some form of 'pad-to-pad' grip and bimanual coordination (Fragaszy and Crast, 2016; Macfarlane and Graziano, 2009; Tanaka, 1998). Various groups of bearded capuchins monkeys employ bimanual grips on hammerstones to crack open a range of encased foods (*Sapajus libidinosus*; Barrett et al., 2018; Canale et al., 2009; Falótico and Ottoni, 2016; Ferreira et al., 2010; Fragaszy et al., 2004; Ottoni and Izar, 2008; Spagnoletti et al., 2011; Fig 6.1D). However, the series of habitual 'pad-to-pad' grips that constitute complex human manipulation are thought to be more forceful (Marzke, 1997, 2013). For example, the 'three-jaw chuck pad-to-pad' grip (Marzke and Steven Shackley, 1986; Fig. 6.1B) can effectively secure a stone between just the thumb, index, and middle finger pads when it strikes another stone hard enough to fracture it (Key and Dunmore, 2015; Williams-Hatala et al., 2018). This type of forceful 'pad-to-pad' precision grip has not yet been observed in other species (Marzke et al., 2015) and is made possible by human hand anatomy.

Forceful opposition of the thumb to the fingers requires manual anatomy that balances mobile digits with force-generating capacity. Human thumb (thenar) musculature is large relative to that of other apes (Marzke et al., 1999) but is also thought to limit flexion of the first metacarpophalangeal joint (Barakat et al., 2013; Dunmore et al., 2020; Tuttle, 1969). In particular, the relatively large human *opponens pollicis* muscle abducts the thumb and, unlike in other apes, secondarily flexes it (Marzke et al., 1998, 1999). In this opposed position, the absolutely broad human apical tuft that underlies the large thumb pad (Shrewsbury et al., 2003) is thought to better distribute and transmit the force of a strong grip, to increasingly proximal joints (Rolian et al., 2011) and eventually the large, relatively flat trapezio-metacarpal joint (Marzke et al., 2010). The flatness of this joint in humans is also thought to permit medial axial rotation of the first metacarpal, aiding thumb opposition, but this joint retains a level curvature that prevents dislocation of the thumb joints under high forces, which can be up to 12 times the force applied at the tip of thumb (Cooney 3rd and Chao, 1977; Marzke, 2013; Marzke et al., 2010). On the other side of a human forceful 'pad-to-pad' grip, the relatively short fingers face the thumb due to metacarpal torsion (Almécija et al., 2015; Drapeau, 2015) and the hypothenar musculature of the little finger 'cups' the hand, facilitating in-hand manipulation (Marzke, 1997, 2013). Likely due to its musculature's role in these forceful 'pad-to-pad' grips and 'in-hand' manipulation, the human little finger is also crucial for stone tool production (Domalain et al., 2017; Key et al., 2018, 2019; Marzke, 1997, 2013).

In sum, human dexterity is not simply the result of a long opposable thumb unique to our species. Rather, the pads of our ancestrally short fingers are opposed by the full pad of our thumb. To achieve this opposition the thumb is simultaneously abducted, medially rotated as well as flexed, and maintains a balance of mobility and strength. Habitual sequences of this forceful 'pad-to-pad' opposition in precision grips with multiple digits, constitute precise in-hand manipulation of objects, as well as complex bimanual manipulation, unique to us among extant species. The inference of this behaviour, and its evolution, is the aim of much osteological research but its result, complex technology, has also been the subject of much research.

6.2 Hominin hand morphology and tools in the past

The most direct evidence for the breadth of manual behaviours that past populations and fossil hominins could practise are the hand bones themselves and the technology they created. The size, shape, and robusticity of bones that comprise their external morphology are a major factor in determining whether a hand can perform certain behaviours. Hominin fossils were traditionally qualitatively assessed (e.g. Napier, 1962; Weidenreich, 1941) and subsequently their linear dimensions and the angles between them were measured (e.g. Alba et al., 2003; Bush et al., 1982). Many modern analyses also measure external fossil shape with landmarks and associated geometric morphometric techniques (e.g. Bardo et al., 2020; Galletta et al., 2019). Regardless of the measurement technique employed, aspects of fossil shape are considered 'primitive' if they more resemble those of modern primates and are considered more 'derived' if they more resemble modern humans. In each case, the observed manual behaviour is functionally related to the bone shape(s) of the living species and inferred for the similarly shaped fossil species. This comparative morphology approach to functional interpretation can be become complex due to the different mosaics of 'primitive' and 'derived' traits seen in fossil hominin hands (Kivell et al., 2011, 2015) and is further complicated by low numbers of relatively complete and undamaged fossil hands (Kivell, 2015; *see below*). Though a full review of fossil hominin manual morphology is beyond the scope of this chapter, below we broadly characterise the external morphology, grasping capabilities and associated early technology of hominin species.

The oldest fossil hominin hand bones yet described are attributed to *Orrorin tugenensis*, dated to ~6 million years ago (Ma). They comprise a curved finger bone (the proximal phalanx; PP; BAR 349′00) similar to those of extant arboreal primates (Senut et al., 2001) and a human-like first distal phalanx (DP1; BAR 1901′01; Gommery and Senut, 2006). While the ranges of proximal phalangeal curvature in extant and extinct primate taxa certainly overlap, proximal phalanges are straighter in modern humans compared to arboreal extant great apes (Kivell et al., 2018). Curved fingers, such as those found in *Orrorin* (Senut et al., 2001), are thought to reduce bending stresses during arboreal climbing (Nguyen et al., 2014; Richmond, 2007) and are thought to develop through ontogeny in climbing primates (Richmond, 2007; but see Wallace et al., 2020). The DP1 possesses a broad apical tuft, unlike the smooth thin rod-like morphology of the extant ape DP1, which provides more surface area for the thumb pad that is useful for manipulating objects in human 'pad-to-pad' precision grips. Combined with an asymmetric insertion of the *flexor pollicis longus* (FPL) muscle, which causes the distal thumb to pronate during flexion, this morphology has been used to infer 'pad-to-pad' precision grips in this species (Almécija et al., 2010; Gommery and Senut, 2006). Thus, these two finger bones evidence a hand useful for manipulation and in the trees, early in the hominin lineage.

The oldest relatively complete fossil hominin hand (ARA-VP-6/500) is assigned to *Ardipithecus ramidus* (~4.4 Ma). Though its metacarpals are almost as short as those of *Homo*, this hand has a finger-to-thumb length ratio similar to extant apes due to its long, curved fingers and short thumb (Almécija et al., 2015; Lovejoy et al., 2009; Prang et al., 2021; Simpson et al., 2019; White et al., 2015). While these proportions would have precluded forceful 'pad-to-pad' grasping, the relatively gracile first metacarpal (Mc1) has a radio-ulnarly broad sellar base, indicating a slightly greater capacity for thumb opposition than in extant great apes. The McP and interphalangeal joint shape in *Ar. ramidus* is more similar to those of suspensory apes than of humans or extant monkeys, leading some researchers to interpret this hand as suspensory (Prang et al., 2021). However, small facets on the second to fourth metacarpals (Mc2–4), as well as the capitate and hamate, suggest that *Ar. ramidus* did not have the strong flexor tendons that extant apes use when vertically climbing or during arboreal suspension (Lovejoy et al., 2009). Instead, these long fingers are interpreted by Lovejoy and colleagues as advantageous for grasping branches during 'careful climbing' and the capitate head would have allowed hyperdorsiflexion of the wrist during palmigrade arboreal clambering (Lovejoy et al., 2009; Simpson et al., 2019; White et al., 2015).

In the genus of *Australopithecus*, the skeletal morphology of the hand becomes increasingly human-like. Within this genus, the less derived hand of *Australopithecus anamensis* (4.2–3.8 Ma) is only known from a curved but broken PP (KNM-KP 30503) and a capitate (KNM-KP 31724) found at Kanapoi, Kenya (Ward et al., 2001). This capitate has a Mc2 facet that faces radially, parasagittal in dorsal view, as in extant great apes where it is thought to better transmit axial forces from the fingers during power grasping (Leakey et al., 1998; Macho et al., 2010; Ward et al., 2001). At 3.7 million years ago (Ma) StW 573 or 'Little Foot' is attributed to early *Australopithecus africanus* or *Australopithecus prometheus* (Clarke, 1999, 2013). This specimen is an almost complete skeleton that includes the oldest complete and associated hominin hand skeleton in South Africa, but the hand remains largely undescribed (Clarke, 1999, 2013; Kivell et al., 2020).

The many isolated hand fossils of *Australopithecus afarensis* (3.85–2.95 Ma) and *Au. africanus* (3–2 Ma), which have been described as "strikingly similar" (Ward, 2013, p. 240), are known from several sites in eastern and South Africa respectively (Bush et al., 1982; Kivell et al., 2020; Pickering et al., 2018; Ricklan, 1987; Ward et al., 2012). While there has been considerable debate regarding the hand proportions of both species, there is consensus that their finger-to-thumb length ratios were more human-like than *Ardipithecus*, and that they were capable of some form of 'pad-to-pad' precision grips (Alba et al., 2003; Almécija and Alba, 2014; Feix et al., 2015; Green and Gordon, 2008; Kivell et al., 2020; Marzke, 1983; Pickering et al., 2018; Prang et al., 2021; Ricklan, 1987; Rolian and Gordon, 2013, 2014). A greater emphasis on thumb loading relative to *Ardipithecus* is reflected in an oblique, or disto-radially facing, capitate facet for the Mc2 in these species (KNM-WT 22944, AL 288-1w, AL 333–40, TM 1526; Macho et al., 2010; McHenry, 1983; Tocheri et al., 2008). Similarly, the DP1 of *Au. afarensis* and *Au. africanus* both have broad apical tufts as well as an insertion for the FPL muscle associated with forceful manipulation, though this may be slightly more marked in the former species (AL 333–159, StW 294; Almécija and Alba, 2014; Kivell et al., 2020; Ward et al., 2012). Yet, these austalopith precision grips were unlikely to be the same as in modern humans. For example, a lack of radial expansion in the *Au. afarensis* distal Mc1 articular surface (AL333w-39; Galletta et al., 2019) and the ape-like curvature of the proximal Mc1 articular surface in

Au. africanus (StW418; Marchi et al., 2017) would have limited thumb opposition in these species. A biomechanical model, incorporating *opponens pollicis* and its interaction with Mc1 and trapezial shape, also demonstrated weaker thumb opposition in *Australopithecus* relative to later hominins (Karakostis et al., 2021). Further, the relative Mc1 robusticity of *Au. africanus*, either measured as relative breadth or in terms of bending rigidity, is similar to great apes, implying that these grips were not as forceful as those of modern humans (Dunmore et al., 2020b; Green and Gordon, 2008). The proximal phalangeal curvature of *Au. africanus*, and especially *Au. afarensis*, combined with well-developed flexor sheath ridges and ape-like lunate morphology in the latter (AL444-3 & KNM-WT22944j), likely reflect an ape-like facility for arboreal power grasping in the hands of these species (Bush et al., 1982; Kivell et al., 2020; Pickering et al., 2018; Stern and Susman, 1983; Ward, 2013; Ward et al., 2012). Thus, these species probably used their curved fingers to power grasp branches when climbing trees and were likely also capable of a type of 'pad-to-pad' manipulation, though not one quite the same as that of modern humans.

As australopithecine hand morphology is more derived than earlier hominins, it is perhaps unsurprising that the earliest currently known stone tools, or lithics, found at Lomekwi 3 in Kenya, also date to this period at 3.3 Ma (Harmand et al., 2015; but see Archer et al., 2020). These tools consist of large sharp-edged flakes detached from larger stone cores as they strike a stone anvil, or as the core is struck by a hammerstone on an anvil, known as passive hammer or bi-polar knapping, respectively (Harmand et al., 2015; Inizan et al., 1999; Lewis and Harmand, 2016). Smaller versions of these flakes are produced unintentionally by capuchins when they pulverise quartzite cobbles into dust (Proffitt et al., 2016). While we do not know if it was *Kenyanthropus platyops* (for which we have no hand bones, Leakey et al., 2001) or *Au. afarensis* that made these flakes, this hominin was already capable of bimanual manipulation required for bi-polar knapping (Harmand et al., 2015; Lewis and Harmand, 2016). Potential cut marks on fossil bones at roughly the same time at Dikika, Ethiopia (3.4 Ma; McPherron et al., 2010; but see Domínguez-Rodrigo et al., 2010, 2012; Sahle et al., 2017) also evidence forceful precision grips and asymmetric bimanual manipulation required to use these tools to butcher a carcass (Key et al., 2018; Williams-Hatala et al., 2018), a behaviour which cannot be inferred from fossil hands alone. While organic tool production and use does not preserve in the Plio-Pleistocene record, it was also likely an important component of the early hominin manual behaviour repertoire (Panger et al., 2003), as it is for orangutans (*Pongo* sp.; Meulman and van Schaik, 2013; Bardo et al., 2017) and chimpanzees, as well as other species, today (Marzke et al., 2015; McGrew, 2013). The manipulative capabilities that facilitated the production and use of these organic tools and other non-lithic behaviours are therefore inferred from the preserved fossil hands.

The associated hand of *Australopithecus sediba* (MH2) dated to 2 Ma in Malapa, South Africa, possesses hand proportions much closer to those of modern humans, further facilitating 'pad-to-pad' precision grips (Almécija et al., 2015; Kivell et al., 2011, 2018; Fig. 6.2). Moreover, its robust fifth metacarpal (Mc5) with an asymmetric head and well- developed enthesis for *opponens digiti minimi* suggest a strongly flexed abducted fifth finger, a position similar to that of modern humans during manipulation (Key et al., 2019; Kivell et al., 2018; Marzke, 1997, 2013). A reciprocally asymmetric Mc2 head is found in *Au. afarensis* (Drapeau et al., 2005) and *Au. sediba*, but a proximo-distal Mc2-capitate articulation likely limited pronation of the second finger in the latter species (Kivell et al., 2018). Though the Mc1 is absolutely gracile, it is more robust relative to the rest of the hand than in *Au. africanus* and its distal palmar beak implies the presence of two sesamoid bones within strong flexor tendons (Dunmore et al., 2020b; Kivell et al., 2018). When size is controlled for, the modelled torque of the *opponens pollicis* during thumb opposition is higher than that of other *Australopithecus* species in *Au. sediba*, though still much lower than that of fossil *Homo* species (Karakostis et al., 2021). This potentially more manipulative hand was also capable of arboreal power grips, however, as evidenced by proximal phalangeal curvature and well-developed flexor sheath ridges of both the proximal and, unlike other australopiths, intermediate phalanges (Kivell et al., 2011, 2018).

The stone tools at Olduvai, Tanzania were the first to be closely associated with a fossil hominin hand, the juvenile type specimen of *Homo habilis* (OH7; 1.8 Ma; Napier, 1962; Susman and Creel, 1979). However, some authors have questioned this attribution (Moyà-Solà et al., 2008) and hand fossils that date to around 2.5–1.5 Ma in eastern Africa, and later in South Africa, are problematic as they generally can be attributed to either early *Homo* or *Paranthropus*, which were sympatric genera (Tocheri et al., 2008). Regardless of its attribution, OH7 does have a Mc1-trapezium articulation that is at least as flat as that of *Homo neanderthalensis*, which would be less stable than those of modern humans but also capable of transmitting large loads from the thumb (Marzke et al., 2010; Trinkaus, 1989). The broad DP1 apical tuft of OH7 also suggests a manipulative hand (Napier, 1962), though this bone lacks the distinctive insertion for the FPL seen in earlier and later hominins (Almécija et al., 2010). Indeed, the OH7 proximal phalanges possess a median bar and are more curved than *Au. sediba* and *Au. africanus*, (Kivell et al., 2018, 2020) suggesting this hand was capable of arboreal power grasping (Susman and Creel, 1979). Similarly, recently described *Paranthropus boisei* hand bones (KNM-ER 47000; 1.5 Ma) possess hand proportions that would facilitate precision grips, but even more curved proximal phalanges than OH7, similar to *Au. afarensis*, with well-developed flexor sheath ridges and a gracile Mc1, suggest manipulation may have been less important in this hand (Richmond et al., 2020). The hand material variably assigned to *Paranthropus robustus* or early *Homo* at Swartkrans in South Africa appears more derived than that of the east African material (Constantino and Wood, 2004; Susman, 1989; Susman et al., 2001). The curvature of the proximal phalanges (SKX 15468; SKX 5018) is within the range humans (Kivell et al., 2018; Susman, 1989) and the DP1 apical tufts of some specimens (TM1517k; SKX5016) are as broad as that of OH7 (Almécija et al., 2010; Kivell et al., 2018; Susman, 1988). Further, SKX5020 is an Mc1 wide enough to support this broad DP1 morphology (Susman, 1988) and has been modelled with an *opponens pollicis* that could have generated similar force to that of a modern human during thumb opposition (Karakostis et al., 2021), suggesting a more manipulative hand in this species than in remains attributed to *P. boisei*. Conversely, the proximal phalanges are described as having an "ape-like shape" (Susman et al., 2001, p. 615) and SK84 is a much smaller Mc1, with a more curved (ape-like) trapezial facet (Marchi et al., 2017; Trinkaus and Long, 1990). Thus, while there is significant morphological variation and taxonomic ambiguity between 2.5 and 1.5 Ma, some of these hand remains possess derived features associated with increased thumb loading and manipulation coincident with the proliferation of the Oldowan stone tool industry in Africa.

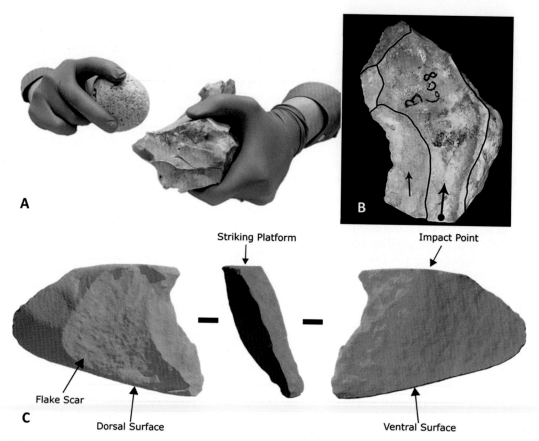

FIG. 6.3 Knapping. (A) Human using a hammerstone, held in a 'three-jaw chuck pad-to-pad grip', to strike a flint core, held in '5-finger-cradle-power-grip'. This asymmetric bimanual strike causes a conchoidal fracture in the core to detach a flake, leaving a flake scar on the core. (B) An example of an Oldowan core with several flake scars. The black arrows denote the direction of the hammerstone impact with which the flakes were struck. (C) A digital model of a simple flake detachment, showing the core *(left)*, the struck flake in profile *(middle)* and its exposed ventral surface *(right)*.

While often associated with *Homo*, Oldowan stone tools are just as likely to have been made by *Paranthropus* (Susman, 1988; Tocheri et al., 2008). In either case, by 2.6 Ma hominins seemed to have fully mastered the fine manipulative postures (Semaw et al., 2003) required to habitually detach sharp flakes from a handheld stone core (freehand knapping; Fig. 6.3), the basis of Oldowan technology (Braun et al., 2019; Semaw et al., 1997, 2003; Stout et al., 2010). Modern replication experiments have demonstrated that freehand knapping requires a forceful 'three-jaw chuck pad-to-pad' precision grip on the hammerstone (Figs. 6.1B and 6.3; Marzke and Steven Shackley, 1986; Williams et al., 2012) before it can be used to strike the core, itself secured by forceful precision grips (Key et al., 2019; Key and Dunmore, 2015). Localised and intense battering marks on Oldowan hammerstones evidence their precise manipulation, so that only the convex surface, exposed in 'pad-to-pad' grips, struck the core (Mora and de la Torre, 2005). This strike is a forceful and precise asymmetric bimanual action (Geribàs et al., 2010; Marzke et al., 1998). The refitted stone tools from Lokalalei 2C, Kenya demonstrate that by 2.3 Ma hominins could modulate the hammerstone velocity and strike cores in an optimal location for detaching a flake (Delagnes and Roche, 2005). When captive bonobos (*P. paniscus*) were taught to knap they could not produce Oldowan tools, due to an inability to direct precise hammerstone blows (Toth et al., 2006), as they lacked the types of precision grips required for the effective manipulation of both the hammerstone and core (Pelegrin, 2005) because of their hand morphology (Almécija et al., 2015; *see above*). Thus, the habitual production of stone tools was facilitated by the precision grips made possible by the hand morphology of *Homo* or *Paranthropus*, or both, between 2.6 and 1.5 Ma.

Also found in this time period is a proximal phalanx (OH86; 1.8 Ma), likely from the fifth ray, that has been tentatively attributed to *Homo erectus sensu lato* because it is relatively straight, similar to that of modern humans (Domínguez-Rodrigo et al., 2015). The *H. erectus* juvenile skeleton, KNM-WT 15000 (1.6 Ma), also preserves a relatively straight proximal phalanx, as well as two possible Mc1 shafts (-BV; -BU) which are robust but only tentatively attributed to this specimen and potentially not hominin, making further manual behaviour inferences difficult (Richmond et al., 2020; Tocheri et al., 2008; Walker et al., 1993). However, a large third metacarpal (Mc3) from Kaitio, Kenya, also attributed to *H. erectus* sensu *lato* (KNM-WT 51260; 1.4 Ma), is modern human-like in length and, crucially, preserves a distinct styloid process (Ward et al., 2014). The styloid process, shared with modern humans to the exclusion of great apes, is thought to provide greater resistance against hyperextension of the metacarpal, and palmar dislocation of its base, during heavy loading of a flexed third finger, such as during stone tool production (Marzke and Marzke, 1987).

Outside of Africa, the post-crania at Dmanisi, Georgia, represent early *Homo* but preserve just two distal phalanges of hand, again making functional inferences challenging (D2679/D3480; 1.8 Ma; Lordkipanidze et al., 2007). Three localities in Atapuerca, Spain comprise the remainder of the Early, and much of the Middle, Pleistocene manual fossil hominin record. Similar to OH86, a straight and relatively robust fifth proximal phalanx (ATE9-2) from Sima de la Elepante is dated to 1.2 Ma. Similarly, straight proximal phalanges, with less developed flexor-sheath ridges than OH7, are dated to 700–900 thousand years ago (ka) at the TD-6 horizon of Grand Dolina (Duval et al., 2018; Lorenzo et al., 1999). These remains, along with a well preserved capitate that lacks the facet for a Mc3 styloid process, have been attributed by some researchers to *Homo antecessor* (Lorenzo et al., 1999). Lastly, a remarkable sample of 519 hand bones (131 carpals, 63 metacarpals, and 325 phalanges) date to ∼430 ka at Sima de los Huesos (SH; Arsuaga et al., 2015). Originally attributed to *Homo heidlebergensis* (García and Arsuaga, 2011), these SH hominins are now recognised to be morphologically and genetically more similar to early *Homo neanderthalensis* (Arsuaga et al., 2014; Meyer et al., 2016). Long palmar tubercles of the carpals, especially that of the trapezium, would have formed a deep carpal tunnel for large flexor tendons, and suggest manipulative power-grasping in SH hominins, though the pisiform was pea-shaped, like modern humans to the exclusion of extant apes. The large *opponens pollicis* Mc1 attachment, combined with a sellar facet for the Mc1 on the trapezium, the potential for Mc2 rotation, broad distal trochleae of the intermediate phalanges and broad apical tufts of the distal phalanges, also provide evidence of forceful precision 'pad-to-pad' grip capability in the SH hominins (Arsuaga et al., 2015).

While the hands of Pleistocene *Homo* described above appear increasingly modern human-like, other Pleistocene hominins exhibit much more primitive morphology. *Homo naledi* is found in the Dinaledi (Berger et al., 2015) and Lesedi chambers (Hawks et al., 2017) of the Rising Star cave system, South Africa, which contain over 1500 postcranial elements representing at least 18 individuals (300 ka; Dirks et al., 2017). Proximal phalanges as curved as OH7 and *A. afarensis*, and uniquely curved intermediate phalanges with a median bar, suggest arboreal power-grasping in *H. naledi* that was distinct from that of other hominins (Kivell et al., 2015). However, the proportions of the nearly complete and associated Hand 1 indicate a relatively long thumb that lies at the top of the modern human range of variation (Kivell et al., 2015). Further the large, relatively robust Mc1 head is radially extended and ultimately supports a broad DP1 (Galletta et al., 2019; Kivell et al., 2015). Though the Mc3 lacks a styloid process, the thumb morphology combined with a mildly sellar Mc5-hamate articulation, which would facilitate supination of the fifth finger, would facilitate forceful 'pad-to-pad' precision grips in *H. naledi* (Kivell et al., 2015).

In eastern Asia, *H. erectus* manual remains are represented by a partial lunate from Zhoukoudian cave, China, dated to ~500–400 ka though a modern description has yet to be published (Boaz et al., 2004; Shen et al., 2001; Weidenreich, 1941). In southeastern Asia, the manual remains of at least six *Homo floresiensis* individuals have been found on the island of Flores, Indonesia, dated to 100–60 ka (Larson et al., 2009; Orr et al., 2013; Sutikna et al., 2016; Tocheri et al., 2007). Much like *H. naledi*, despite this recent geological age, the *H. floresiensis* wrist lacks all the derived features of modern human and Neanderthal hands thought to better deal with oblique or radial load from the thumb, including a palmarly expanded trapezoid, a Mc3 styloid process facet and a more oblique Mc2 facet on the capitate (Orr et al., 2013; Tocheri et al., 2007). Well-defined flexor sheath ridges on *H. floresiensis* proximal phalanges combined with this wrist morphology suggest a hand more adapted for axial loads from power-grasping. However, these phalanges are relatively straight and a distinct insertion for the FPL on the DP1 suggests that the thumb may have been used to create the lithic tools found at this site (Larson et al., 2009).

Recent late Pleistocene finds hint at more manual morphology to be found. For instance, the hand of *Homo luzonensis*, found in the Philippines, is currently only known from an intermediate phalanx with modern-human like dimensions but also described as curved with well-developed flexor sheath ridges (CCH-2; 67 ka; Détroit et al., 2019). The only other manual element is a distal phalanx (CCH-5) which has unique proportions among hominins (Détroit et al., 2019). One final late Pleistocene hominin group includes a single distal phalanx found in Denisova, dated to around 75–50 ka (Douka et al., 2019), that is genetically distinct from both *H. sapiens* and *H. neanderthalensis* (Reich et al., 2010).

Neanderthals present the most well-preserved hominin hand bone remains in the fossil record, represented by several relatively complete manual skeletons from Europe and the Near East (Kivell, 2015; Tocheri et al., 2008; Trinkaus, 1983). Their manual morphology presents most of the derived traits observed in modern humans, including a relatively long and robust thumb with well-defined attachment areas for the muscles *opponens pollicis* and *first dorsal interosseous* (Maki and Trinkaus, 2011; Niewoehner, 2006; Karakostis et al., 2018b; see below). Similarly, Neanderthal metacarpal shafts are robust, their proximal and intermediate phalanges exhibit lower curvature relative to older hominins, and their distal phalanges bear wide apical tufts as well as protruding entheses for *flexor pollicis longus*. The Mc3 styloid process is also present (but see below). Their wrist bone morphology is also closer to modern humans than earlier hominin species, showing traits such as a boot-shaped trapezoid, bearing a Mc2 articulation with relatively proximo-distal orientation, and a facet for the scaphoid that is relatively small and rectangular (Niewoehner, 2006; Tocheri et al., 2008; Trinkaus, 1983; Trinkaus and Villemeur, 1991). Like in modern humans, the Neanderthal capitate exhibits a broad neck and capitate-hamate joint (Marzke and Shackley, 1986; Tocheri et al., 2008). Though it is grossly distinct from great apes, the Neanderthal capitate bears a relatively proximo-distally oriented Mc2 articulation, a trait that is often contrasted with modern humans (Niewoehner, 2006; Niewoehner et al., 1997). The pisiform also has the characteristic derived 'pea-shape', in contrast with its 'rod-shape' in earlier hominins (Niewoehner, 2006; Tocheri et al., 2008; Trinkaus, 1983).

There are several differences observed between the hand of Neanderthals and that of modern humans, which had formerly led researchers to assume a more limited manual dexterity in Neanderthals (Musgrave, 1971). Primarily, the Neanderthal thumb is differently proportioned, consisting of a relatively longer distal phalanx and a relatively shorter proximal phalanx than in modern humans (Almécija et al., 2015; Musgrave, 1971; Trinkaus, 1983; Trinkaus and Villemeur, 1991). At the same time, the thumb's trapezio-metacarpal joint is flatter and less palmarly extended, suggesting a less stable basal thumb joint (Marzke et al., 2010; Niewoehner, 2006). When all of the joints of the Neanderthal trapezium and the Mc1 are considered together, they evidence a thumb that would have been more suited to loading in extension and adduction, such as when gripping a hafted tool, than the thumb of modern humans (Bardo et al., 2020). There are also several subtle differences in the orientation and morphology of wrist joints (for more details see Niewoehner, 2006) and a smaller Mc3 styloid process in Neanderthals that may have adversely affected the hand's resistance to radial biomechanical loading relative to modern humans. However, these differences are subtle and there is considerable intra-specific variation and gross morphological overlap between these species (Bardo et al., 2020), suggesting a similar manual dexterity in Neanderthals and humans (Bardo et al., 2020; Feix et al., 2015; Niewoehner et al., 2003; Tocheri et al., 2008). This viewpoint has been supported by virtual biomechanical simulations revealing a similar precision-grasping capacity between the two species (Feix et al., 2015; Niewoehner et al., 2003).

Moving from pre-history to the historical period, it is unfortunate that intra-specific variation in modern *Homo sapiens* external hand morphology is rarely the focus of osteoarchaeological research (Karakostis et al., 2016) and is an area in which far more research could be done. However, some work has linked the robusticity of modern human bones with labor intensive occupations in past populations (Becker, 2016; Cope et al., 2005) and further work has begun to infer manual behaviours from soft tissue traces, pathology and internal bone morphology in archaeological populations (see Sections 6.4 and 6.5).

The above broad characterizations of external fossil hominin hand morphology show a general chronological trend away from hands with a facility for efficient arboreal power-grasping to those that could more heavily load an increasingly mobile thumb during forceful precision 'pad-to-pad' manipulation. However, derived features of the hominin hand evolved in a piecemeal manner, often co-occurring with phylogenetically primitive traits in different mosaic morphologies (Richmond et al., 2016). These external traits provide the richest source of data regarding the breadth of fossil hominin hand capabilities, yet this data *alone* cannot tell us which of these possible hand postures were habitually used by hominins during their life.

6.3 Soft tissues

While bone morphology imposes limits on joint movements, it is important to remember that this manual behaviour is a result of soft tissues in vivo, including muscular force and how this is mediated via connective tissues. As bone is the substrate that allows for muscle to execute its function, bone morphology shapes to facilitate that function. The more habitual the function, the more likely it is to be represented in the bone morphology. It is this recipro-cal interaction that allows us to infer musculoskeletal function from bone, and by extension, behaviour in the absence of soft tissue.

Skeletal muscle attaches to bone via connective tissue in a variety of ways and each of these interactions result in a variety of bone morphologies since repetitive muscle contrac-tions affect bone reformation. Connective tissues consist of collagenous and elastic fibres, among other materials, allowing it to withstand tensile stresses, and different tissue types are characterised by specific fibre type composition, as well as fibre orientation. Multiple types of connective tissue are involved in the interaction of muscle and bone, be it direct interaction or with the intervention of other connective tissues such as tendons. Both bone and muscle are sheathed in connective tissue, periosteum, and fascia respectively, and their direct inter-action is usually reflected on bone by smooth fossae due to the uniform distribution of mus-cle forces. Indirect interaction usually occurs through tendinous tissue consisting of a dense arrangement of, relatively, parallel collagen fibres that have visco-elastic properties. This al-lows for not only the direct transmission of muscle forces, but also for energy storage caus-ing tendons to function as springs during repeated muscle contractions, lowering metabolic expenditure (Bennett et al., 1986; Cavagna et al., 1977). Tendon fibres may occur in parallel bundles that attach to a single bony structure, forming tubercles and other bony protrusions (for example the hamulus of hamate bone on which many structures including the *flexor carpi ulnaris* insert), or they can form a sheet structure that directs forces along a line, resulting in ridges along the bone surface. The architecture of bone corresponds to muscle architecture, and when certain bony landmarks are relatively prominent, it is important to understand to which muscle structure and function it relates.

The primate hand is operated by approximately 30 named muscles although the exact number varies due to both interspecific and intraspecific variation. While hand function var-ies considerably, from the brachiating gibbons, to the semi-arboreal, knuckle-walking African apes, and primarily manipulative modern humans, overall forelimb musculature is remark-ably consistent across most primates (Aversi-Ferreira et al., 2010; Diogo et al., 2017; Thorpe et al., 1999; Tuttle, 1969; van Leeuwen et al., 2018; Vanhoof et al., 2020). One of the few ex-ceptions, most notable in the context of the hand, is the presence of both a separate long thumb flexor and extensor in humans. The muscles that affect the hand can be classified into functional groups based on their dominant movement in 3D space, for example wrist and digit flexors, extensors, radial and ulnar deviators, as well as forearm rotators, pronators, and supinators. Muscle nomenclature frequently refers to both the function and the structure that is acted upon for example the *flexor pollicis longus*, the long thumb flexor muscle. Muscles that act upon the hand but originate from within the forearm are referred to as extrinsic hand muscles, while intrinsic hand muscles are situated entirely within the hand.

Force generation is the main functionality of any muscle tendon structure and morphological adaptations to force generation can be observed on many levels. At the microscopic level, different internal muscle morphologies produce different force outputs, balancing biomechanical requirements. For example, the specific orientation of muscle fibres in pennate muscles produce relatively high force output while taking up relatively little space, at the expense of effective contractive length and range of motion (Gans, 1982). Similarly, the higher proportion of type II, fast-twitch muscle fibres in chimpanzee muscles has been associated with their increased force generating capacity compared to humans, who possess more relatively slow, but more efficient type I muscle fibres (O'Neill et al., 2017).

On a macroscopic level, a simple adaptation is just increasing muscle mass or volume. Having more muscle tissue means having more contractile elements, resulting in more force generating capacity. A commonly used standardised metric for mass/volume is Physiological Cross-Sectional Area (PCSA), as it accounts for muscle fibre orientation. An example of increasing muscle volume is that due to the importance of grasping, i.e. wrist and digital flexion, in many primate locomotor types. The PCSA ratio of forearm flexor to extensor muscle mass is relatively high, around 3:1, in most hominoids (van Leeuwen et al., 2018). Modern humans are, however, characterised by a flexor to extensor ratio of close to 1.5:1, quite plausibly due to the lack of locomotor function. The trouble with simply increasing PCSA is that it usually increases muscle size. Space within the body is a limited resource, so this leads to a trade-off between muscle size and available space. For example, the strongly contracted human thenar muscle belly is thought to limit metacarpophalangeal (McP) joint flexion of the thumb (Barakat et al., 2013; Dunmore et al., 2020a; Tuttle, 1969). Underuse of the possible range of movement at McP joints has been hypothesised to underlie the much higher rates of osteoarthritis (OA) in human hands compared to that of chimpanzees (Alexander, 1994; Bain et al., 2015; Jurmain, 1989, 2000; Murai et al., 2018; Rothschild and Rühli, 2005a,b). On the macroscopic level, a structural solution to this issue of limited space is the optimal use of moment arms (Marzke et al., 1999). By distancing the line of muscle/tendon action from the joint rotation centre, the muscle/tendon moment arm increases (Fig. 6.4) and a torque is created that equals the muscle force multiplied by that moment arm. For example, tendon bowstringing in the fingers when phalangeal flexor sheaths break, or are otherwise removed, creates a pathologically high level of torque (Lu et al., 2015; Fig. 6.4). Recent experimentally validated models of manual movements in the extant non-human primate hand, which incorporate tendon paths, also highlight the relative roles of muscular and gravitational forces during certain hand postures (Lu et al., 2018; Synek et al., 2019b, 2020). Changes in muscle force generating capacity and attachment location may be well interpretable from bone as landmarks such as tubercles are expected to reflect these parameters in their location and relative size. Hand function, however, is not only dictated by individual muscle action.

Musculoskeletal function is a complex network of intermuscular synergies. Concerted contraction of muscles with deviating primary functions may result in a combined, unique

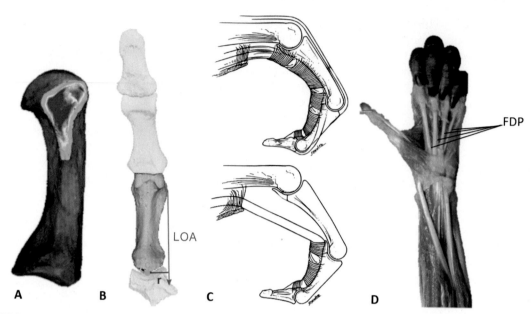

FIG. 6.4 Muscle and bone interaction. (A) A virtual delineation of the muscle attachment site (enthesis) of *opponens pollicis* on a thumb metacarpal (following the protocols of the V.E.R.A. method), this muscle acts to flex and abduct the thumb during human opposition. (B) A depiction of the joint moment arm length (*r*) of muscle *opponens pollicis* for abduction, defined by the trapezio-metacarpal's joint's centre (c) and the muscle's line of action (LOA). (C) The difference in the moment arms between a healthy *(top)* and a pathologically 'bowstrung' tendon *(bottom)* of flexor digitorum profundus in a human (FDP). (D) The tendons of FDP in a partially dissected, but otherwise intact, bonobo hand. Note the lack of space in the palm for intrinsic muscles and the necessity for extrinsic flexors to act on these large tendons. *(A) From Karakostis, F. A., & Lorenzo, C. (2016). Morphometric patterns among the 3D surface areas of human hand Entheses. American Journal of Physical Anthropology, 160(4), 694–707; Karakostis, F. A., Centeno, H. R., Francken, M., Hotz, G., Rademaker, K., & Harvati, K. (2020). Biocultural evidence of precise manual activities in an Early Holocene individual of the high-altitude Peruvian Andes. American Journal of Physical Anthropology, 174(1), 35–48. (B) From Karakostis, F. A., Centeno, H. R., Francken, M., Hotz, G., Rademaker, K., & Harvati, K. (2020). Biocultural evidence of precise manual activities in an Early Holocene individual of the high-altitude Peruvian Andes. American Journal of Physical Anthropology, 174(1), 35–48. (C) From Ryzewicz, M., & Wolf, J. M. (2006). Trigger Digits: Principles, Management, and Complications. The Journal of Hand Surgery, 31(1), 135–146. https://doi.org/10.1016/j.jhsa.2005.10.013.*

motion of the hand, while antagonistic muscles may work against each other to carefully balance out dexterous movements. Therefore, anatomical structures that inherently facilitate such concerted action can significantly impact the capabilities of the complete anatomical structure. For example, both humans and bonobos have a functional tendon insertion for the long thumb flexor, but in bonobos, this tendon inserts directly onto the flexor tendon of the index finger, making it functionally part of the common digit flexor (*flexor digitorum profundus*), while in humans, this tendon originates from an individual muscle belly (*flexor pollicis longus*; Diogo et al., 2017; van Leeuwen et al., 2018). This type of functional coupling, or muscle individualization, depending on your perspective, will have a significant impact on the overall capabilities and dexterity of the primate hand, but is also likely to play a role in the ability for controlled in-hand manipulation in humans. Unfortunately, such nuanced interactions between muscles are hard to reclaim from hard tissue morphology alone but should definitely be considered in behavioural reconstructions.

The more tractable but also under considered elements of soft tissue that mediate joint movement, and therefore behaviour, are ligaments and joints. Where muscle tissue is responsible for the active function of the hand, passive soft tissue, and specifically ligament structures, can significantly affect joint capabilities. Ligaments are bands of connective tissue between bones that attach to them at bony landmarks, such as ridges or protrusions. These non-contractile tissues that are only functional when tensed, specifically structured to resist tensile stresses, and stabilise all joint types by passive restriction of motion. For example, the anterior oblique ligament attaches to the volar beak of the first metacarpal, where it stabilises the trapeziometacarpal (TMc) joint of the thumb during extension. Furthermore, joint functionality is dictated by its congruency (Rafferty, 1990), and as such by its cartilage layer. Cartilage is not strictly uniformly distributed within synovial joints (Dourthe et al., 2019) and without representative cartilage information the exact congruency of the joint is uncertain. The importance of these passive structures in defining joint range of motion should not be understated, as it has been indicated that bone morphology alone could not account for thumb range of motion in the bonobo basal thumb joint (Marzke et al., 2010; Rafferty, 1990; Tocheri et al., 2003; van Leeuwen et al., 2019). Thus, to infer fossil manual behaviour we need as much information about these soft tissues, including ligaments, joint cartilage, and muscles, as possible.

6.4 Soft tissue traces in the archaeological record

Among the most frequently studied traces of soft tissue in the fossil record are muscle attachment sites, or entheses, which are defined as the areas of the bone surface where muscles or ligaments attach (Benjamin et al., 1986; Benjamin and Ralphs, 1998; Fig. 6.4A). They comprise the only preserved direct remains of the musculotendinous unit in the fossil and osteoarchaeological records (Schrader, 2018). Depending on the nature of their insertion, entheses are broadly divided into fibrous and fibrocartilaginous (Benjamin et al., 2006). The latter correspond to muscle/tendon attachments onto the bone via uncalcified and calcified fibrocartilage layers, usually located near bone epiphyses. By contrast, fibrous entheses represent direct attachment of muscles via fibres, typically spreading along long bone diaphyses (Benjamin et al., 2006). The concept of using entheses as markers of physical activity relies on the bone's capacity to adapt to biomechanical pressure throughout life (Rauch, 2005; Ruff et al., 2006). Consequently, the interface of bone and muscle at entheses is expected to respond to repetitive biomechanical forces applied on its surface during habitual muscle recruitment (Rossetti et al., 2017). Essentially, a larger entheseal surface area on the bone is thought to allow a muscle to pull harder on its bony substrate (Deymier-Black et al., 2015; Rossetti et al., 2017), and the occurrence of entheseal bone changes related to repetitive biomechanical stress has been experimentally as well as histologically demonstrated (Benjamin et al., 2006; Karakostis et al., 2019a; Rossetti et al., 2017; Vickerton et al., 2014).

Several studies have questioned the functional significance of entheses based on the low precision of various visual scoring systems (see Wilczak et al., 2017) and on a former lack of experimental evidence associating entheseal morphology with physical activity (Rabey et al., 2015; Zumwalt, 2006). "Validated Entheses-based Reconstruction of Activity" (V.E.R.A.) is a method introduced in 2016 (Karakostis and Lorenzo, 2016), which has been recently tested using laboratory studies conducted following blind analytical procedures (e.g. Karakostis et al., 2019a,c). The latter have provided strong experimental evidence that physical activity directly affects entheseal patterns (e.g. Karakostis et al., 2019a,c). V.E.R.A. relies on a comprehensive and detailed protocol for delineating and measuring the 3D areas of entheseal bone surfaces (based on the criteria of surface elevation, complexity, and coloration; Fig. 6.4A), followed by the application of multivariate statistical procedures. Therefore, instead of comparing each single entheseal structure across different individuals and groups, this new method's multivariate approach aims at identifying group correlations among different entheseal areas that reflect the recruitment of frequent muscle synergy groups (Karakostis et al., 2019a,c; Karakostis and Lorenzo, 2016). Such synergies are better discerned from group correlations among different muscle entheses, rather than from direct inter-individual comparisons of a single entheseal structure, as its attaching muscle can be used for a diverse range of hand movements (Karakostis et al., 2019c). A detailed step-by-step outline of V.E.R.A.'s protocol is provided in a recent literature review focusing on this approach (Karakostis and Harvati, 2021).

Previous research has suggested that changes in skeletal morphology are dependent on muscular and gravitational loads (Oxnard, 1991, 2004). The hand possesses the smallest proportion of body weight in the entire human skeleton (0.66% in males and 0.50% in females; see Tözeren, 1999), while the weight of its bulky extrinsic muscles is mainly applied on the forearm and not the hand itself (Benjamin et al., 2008; Fig. 6.4). Therefore, all other things being equal, significant muscle recruitment would be expected to have a greater influence on hand bones than more weight-bearing skeletal elements, such as the lower limbs (see Karakostis et al., 2019b). More traditional methods also support the unique influence of musculature on the hand. An 18th–19th century English population was found to display a different pattern of manual entheseal bilateral asymmetry than that found in the humeri of the same individuals (Cashmore and Zakrzewski, 2013). Entheseal development has also been related to manual behaviours in past populations (Becker, 2016; Cope et al., 2005). A recent study applying the V.E.R.A. method on the hand entheses of human skeletons with life-long and detailed occupational documentation (19th century Basel-Spitalfriedhof collection, Switzerland) identified distinctive differences between individuals with sustained high-force grip professions (mainly lifelong construction workers) and those with less strenuous professions, such as tailors and painters (Karakostis et al., 2017). The relatively larger entheses of the high-force grip professions were those associated with the fifth digit, thumb, and wrist, known to be important in human power grips (Karakostis et al., 2017; Marzke et al., 1998). Conversely, the less strenuous professions had relatively larger entheses for intrinsic thumb muscles and on the index finger, that have been associated with human precision grasping (Karakostis et al., 2017). It should be noted that another study questioned the ability to directly infer muscle architecture based on the raw size of single hand entheses (Williams-Hatala et al., 2016). However, the cadaveric sample used in that research comprised only old adults (likely due to ethical considerations) and so this correlation was unlikely to be found because bone's responsiveness to forces greatly declines with biological age (Maïmoun and Sultan, 2011). Also, muscle and bone are known to react differently to biomechanical stress (Bennell et al., 1997; Brumitt and Cuddeford, 2015) and, thus, the lack of a direct association between the two tissues does not negate the experimentally documented effects of activity-induced repetitive muscle forces on entheseal bone morphology (Deymier et al., 2019; Karakostis et al., 2019a,c; Schlecht et al., 2019).

In fossil hominin hands, several studies have focused on the entheses of thumb muscles that play a fundamental role for human-like precision grasping and stone tool-related activities (Hamrick et al., 1998; Marzke et al., 1998), such as *opponens pollicis* (associated with thumb opposition; Fig. 6.4A and B) or *flexor pollicis longus* (associated with flexion of the distal thumb; Almécija et al., 2010; Karakostis et al., 2018b; Maki and Trinkaus, 2011; Marzke, 2013; Shrewsbury et al., 2003). The latter muscle attaches onto the distinctive arch-shaped - or "gabled" - ridge on the palmar surface of the distal pollical phalanx (Almécija et al., 2010; Shrewsbury et al., 2003). The existence of this pronounced structure in hominins has been directly associated with the palmar pulp's full compartmentalization, which also involves the presence of the ungual fossa, spines, as well as broad apical tuft (Almécija et al., 2010; Kivell et al., 2020). Importantly, previous cadaveric research on different primate genera has demonstrated the importance of the *flexor pollicis longus* enthesis (and its resulting moment arm) for human-like manipulation efficiency (Shrewsbury et al., 2003), while the greater bony protrusion of the same enthesis (i.e. the gabled ridge) in various early hominin species (such as *Orrorin tugenensis* or *Australopithecus sediba*) has been linked with early indications of human-like precision grasping (Almécija et al., 2010; Gommery and Senut, 2006; Kivell et al., 2011, 2020; Marzke, 2013, see above). This is arguably due to this muscle's central role in sustaining and manipulating objects within the hand using the thumb's pulp, a fundamental requirement for biomechanically efficient grasping in humans (Kivell et al., 2015, 2020).

Maki and Trinkaus (2011) measured the radio-ulnar breadth of the first metacarpal as a proxy of radial projection of the *opponens pollicis* enthesis, relying on digital photographs of thumb metacarpals. Their results indicated relatively high mechanical effectiveness in Neanderthals and Middle Palaeolithic modern humans, which may be associated with lifestyle changes in modern humans during the Upper Palaeolithic. Those authors' interpretations relied on the biomechanical principle that greater entheseal extension is bound to increase an attaching muscle's moment arm and, by definition, its force-producing capacity (Karakostis et al., 2018a,b; Maki and Trinkaus, 2011; Shrewsbury et al., 2003). This principle has been statistically supported in two recent biomechanical and geometric morphometric studies, which found that greater 3D projection of the *opponens pollicis* enthesis is strongly correlated with longer moment arms (and thus, force producing capacity) for both flexion and abduction at the thumb's trapezio-metacarpal joint (Karakostis et al., 2020, 2021). It is worth noting that, even though the potential influence of the missing soft tissue morphology is largely unknown, a muscle's force-producing capacity is shown to be substantially affected by its moment arm, even when comparing specimens with identical muscle architecture and strength (Ward et al., 2010).

Niewoehner (2001, 2006) noted the overall high robusticity of entheses in the hand and wrist bones of Neanderthal specimens, associating it directly with increased overall muscle use and the habitual performance of transverse power-grasping movements. By contrast, an application of the V.E.R.A. approach found that Neanderthals exhibit a distinctive precision-grasping entheseal pattern involving muscles of the thumb and the index finger, which was only present in documented lifelong precision workers (Karakostis et al., 2018b). Similarly, experimental evidence also supports the fundamental contribution of thumb-index precision-grasping to the reproduction of lithic industries associated with Neanderthals (Key and Lycett, 2018; also see discussion in Karakostis et al., 2018b). Moreover, in that previous study, early modern humans showed both power- and precision- grasping entheseal patterns, potentially reflecting the emergence of greater occupational variability in Upper Palaeolithic modern humans (Karakostis et al., 2018b). It must be noted that, as indicated by recent research identifying differences in bone trabecular morphology between two Neanderthal specimens (Dunmore et al., 2020b), the exact nature of physical activities would also be expected to vary among Neanderthal groups relying on different subsistence strategies. This hypothesis can also be supported by the observed diversity of Neanderthal stone tool kits (Hardy et al., 2013).

In sum, recent methodological improvements, such as the 3D multi-entheseal approach for reconstructing muscle synergies (i.e. V.E.R.A.), have the potential to reveal and further refine inferred fossil hominin manual behaviour, from the most common traces of soft tissue in the fossil record; entheses. Yet, records of habitual manual behaviour are not restricted only to entheses but also reside with the bone itself.

6.5 Internal bone morphology

Just as osteological traces are thought to record soft tissue morphology, and thus behaviour, bone itself adapts to habitual and significant loads via a process known as bone functional adaptation (Currey, 2012; Ruff et al., 2006). Not to be confused with evolutionary adaptations, bone functional adaptation refers to changes in morphology that allow a bone to better resist loads experienced during life. This process is referred to by some authors as modelling and others as remodelling (see Chapter 2 in this volume; Barak, 2020), and the relative frequency or strength of load required to produce it is still debated (Judex and Carlson, 2009; Robling, 2009). Yet, the mechanisms by which these forces are transduced into cellular signals, that subsequently alter bone form, are becoming increasingly well-known (Chen et al., 2010).

Perhaps, the most intuitive form of bone functional adaptation, occurs in the diaphyses of long bones, where forces that do not align closely with the long axis of the bone, promote deposition of cortical bone in that plane to better resist them. This change in cortical thickness can be observed in a 2D slice orthogonal to the long axis of the bone (Fig. 6.5D and E), and its contribution to the bone's stiffness can be measured via moments of inertia, commonly referred to as cross-sectional geometry (Ruff and Hayes, 1983; Ruff et al., 2006). For example, African apes habitually knuckle-walk, which strongly loads their hands in compression. These species' third and fourth metacarpals (Mc3 and Mc4) are significantly stiffer than those of orangutans and humans, which do not practise this form of locomotion (Marchi, 2005). Similarly, habitual and forceful 'pad-to-pad' opposition of the thumb seen during human manipulation (Key and Dunmore, 2015; Williams-Hatala et al., 2018) has yet to be reported in wild great ape hand use. Consistent with these behavioural differences, humans have a stiffer first metacarpal (Mc1) relative to the other metacarpals (Wong et al., 2018) unlike other great apes (Dunmore et al., 2020b). Further, the abduction of the thumb during 'pad-to-pad' opposition is reflected by a human Mc1 that is stiffest in the radio-ulnar plane (Dunmore et al., 2020b) and pronation of the index finger during human manipulation is also consistent with an Mc2 that has greater bending strength in a dorso-radial to palmo-ulnar plane (Lazenby, 1998). Within humans, some studies have reported a lack of correlation between known occupation and the cross-sectional properties of the Mc2 in an 18th Century

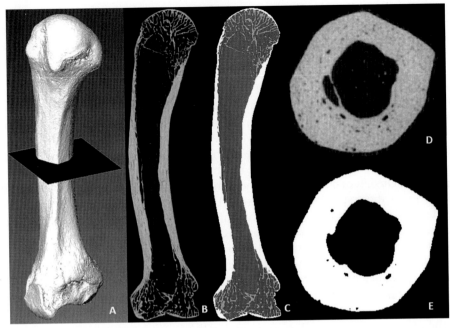

FIG. 6.5 Internal bone morphology of human right second metacarpal. (A) Outer surface with mid-shaft cross-section orthogonal to the long-axis of the bone. (B) Parasagittal cross-section of a CT-scan of this bone. (C) The same scan but with trabeculae highlighted in grey and cortical bone in white. (D) The mid-shaft cross-section in superior view. (E) The cross-section binarized for cross-sectional geometry analysis.

English population (Mays, 2001). Conversely, Hagihara (2021) found that the Mc3 was dorso-palmarly stiffer in the hands of a Jomon period population (3500 BCE–500 BCE) relative to modern Japanese people, linking this to more strenuous power-grasping behaviours in the archaeological population. Future analyses of manual cross-sectional geometry in more human populations will likely provide a clearer picture of how these properties are linked to manual behaviours.

While cross-sectional differences can be readily observed and related to directional forces, they also necessarily conflate cortical thickness and bone form to some extent. Factoring out differences in form has demonstrated that the human Mc3 has a significantly thinner cortical diaphysis relative to orangutans and chimpanzees (Tsegai et al., 2017), which is consistent with lower loading of the primarily manipulative human hand. Though this abstraction of cortical thickness is harder to relate to real-world forces acting on a bone than cross-sectional geometry, it is simpler to relate to bone functional adaptation, works in 3D and can be applied to non-diaphyseal cortical bone, such as the carpals (Tsegai et al., 2017).

Occurring throughout the carpals and inside the epiphyses, and pseudo-epiphyses, of metacarpals and phalanges, trabecular bone also undergoes functional adaptation (Fig. 6.5A–C). The trabecular struts that comprise this tissue have been experimentally shown to increase in in thickness (Tb.Th), and number (Tb.N), as well as change orientation (Barak et al., 2011; Pontzer et al., 2006) to better transmit habitual loads to more proximal elements (Currey, 2002). Similar to cross-sectional geometry, the resultant increase in trabecular bone volume, calculated as Bone Volume/Total Volume (BV/TV), can be directly related to trabecular stiffness, accounting for up to 89%, as measured by Young's modulus (Stauber et al., 2006).

During hand use, loads are transmitted from the base of the fingers to the metacarpal head. Where fingers are habitually used in similar positions, this directional load will be greatest on a particular region of the metacarpal head, promoting increased BV/TV in this region. BV/TV is highest in the dorsal aspect of the Mc3 head in African apes, consistent with fingers loaded in hyper-extension during knuckle walking (Chirchir et al., 2017; Dunmore et al., 2019; Tsegai et al., 2013; but see Barak et al., 2017). Further, the type of knuckle-walking practised by gorillas, to the exclusion of chimpanzees and bonobos, also appears to be reflected in BV/TV distribution across these species' Mc2–5 heads (Dunmore et al., 2019, 2020b). Conversely, the BV/TV is highest in disto-palmar aspects of orangutan Mc2–5 heads, reflecting fingers habitually loaded in an extended to flexed position during hook and 'double-lock' branch grasps (Chirchir et al., 2017; Dunmore et al., 2019; Tsegai et al., 2013). Humans also habitually flex their fingers, which is consistent with the highest BV/TV in the disto-palmar aspect of the Mc3 head (Barak et al., 2017; Chirchir et al., 2017; Tsegai et al., 2013). However, the primarily manipulative human hand is distinguished from the branch grasping of orangutans when BV/TV distribution across Mc2–5 is analysed (Dunmore et al., 2020b; Stephens et al., 2018). Further, the presumably low loading associated with manipulation, relative to locomotion, is also consistent with lower BV/TV throughout the human Mc3 head and radial wrist bones compared to other great apes (Barak et al., 2017; Chirchir et al., 2017; Schilling et al., 2014; Tsegai et al., 2013).

Trabecular evidence consistent with human manipulation has also been found in the Mc1 of the thumb. A slightly higher BV/TV in the right Mc1, not found in chimpanzees, is consistent with the ~90% frequency of right-handed manipulation in our species (Faurie et al., 2005; Lazenby et al., 2011; Stephens et al., 2016). Similarly, a radio-palmar BV/TV concentration human Mc1 head and base is consistent with 'pad-to-pad' thumb opposition (Dunmore et al., 2020a; Stephens et al., 2018). In particular, the TMc joint is a frequent location of OA in humans (Caspi et al., 2001; Gillis et al., 2011; Haugen et al., 2011; Polatsch and Paksima, 2006; Turker and Thirkannad, 2011). Indeed, the severity of OA in many hand joints has been associated with external robusticity and intensive manual behaviours in the Tiwanaku culture (500-1100 CE; Becker, 2016) and has been strongly correlated with robusticity of the Mc1 at the TMc joint in the Bronze age population of Tell Abraq (2300 BCE; Cope et al., 2005). OA of the TMc joint initially develops in the radio-palmar aspect of the Mc1 base (Koff et al., 2003). This region that most closely contacts the trapezium during 'pad-to-pad' thumb opposition (D'Agostino et al., 2017) and is reinforced by the highest BV/TV of the human Mc1 base (Dunmore et al., 2020a; Stephens et al., 2018). The human trabecular distribution of the Mc1 is significantly different from greater disto-ulnar BV/TV in the Mc1 head and palmar Mc1 base found in great apes, consistent with their observed habitual 'pad-to-side' precision grips (Dunmore et al., 2020a,b).

This association between observed hand use and BV/TV distribution in living species allows the inference of in vivo behaviours in fossil hominins, including fossilised modern humans. High BV/TV in the disto-ulnar Mc3 head and palmar Mc1 base have been used to infer human-like manipulation and possibly tool-use in *Au. africanus* (Skinner et al., 2015). Similarly, BV/TV distribution consistent with orang-like power grasping in the Mc2–5 but human-like manipulation of the Mc1 have been used to infer a unique combination locomotor and manipulative uses of the *Au. sediba* hand (Dunmore et al., 2020b) consistent with the interpretation of its external morphology (Kivell et al., 2011, 2018; see above). Conversely, the power grasping BV/TV pattern in the Mc2–5, and variable Mc1 BV/TV of Neanderthals, may initially appear at odds with broadly modern human-like external manual morphology seen in *H. neanderthalensis* (Dunmore et al., 2020b; Niewoehner, 2006; see above). However, this inferred manual behaviour is consistent with the frequent use of power 'squeeze' grips (Marzke et al., 1992) on hafted tools, associated with this species in the archaeological record (Anderson-Gerfaud, 1990; Dunmore et al., 2020b; Niewoehner, 2006). Fossil modern humans have a similar BV/TV distribution to their modern counterparts (Dunmore et al., 2020b), but their BV/TV is generally higher (Stephens et al., 2018), consistent with more forceful modern-human-like manipulation. This pattern is replicated in thicker Mc3 diaphyseal cortical bone (Tsegai et al., 2017) as well as higher BV/TV in other skeletal elements (Chirchir et al., 2015; Ryan and Shaw, 2015) and likely reflects lower force requirements of our modern lives.

Aside from BV/TV increases, trabeculae adapt to habitual loads by becoming oriented in their direction (Barak et al., 2011; Pontzer et al., 2006). The relative number of trabeculae oriented in a similar axis, is therefore thought to reflect the level of habitual loading in that direction, measured as degree of anisotropy (DA) which accounts for ~10% of trabecular bone stiffness (Maquer et al., 2015). Unfortunately, while there is some evidence that more arboreal apes have lower Mc3 head DA (Tsegai et al., 2013), most studies have not found a clear functional association between DA and primate hand use (Chirchir et al., 2017; Dunmore et al., 2019, 2020a; Lazenby et al., 2011; Ragni, 2020). Analysing the principal direction of trabecular orientation (PTO), however, demonstrates that suspensory orangutans and gibbons have primarily proximo-distally oriented trabeculae throughout the third finger and metacarpal, consistent with hook-like power grasping, whereas quadrupedal macaques demonstrated a primarily palmo-dorsal alignment consistent with loading during palmigrady (Matarazzo, 2015). Correspondingly, chimpanzees shift from proximo-distal to palmo-dorsal PTO in the Mc3 base with the adoption of knuckle-walking during ontogeny (Ragni, 2020). This palmo-dorsal PTO, also found in the adult chimpanzee Mc3 head (Barak et al., 2017; Matarazzo, 2015), is significantly different from modern humans (Barak et al., 2017) and is concomitant with a mixture of PTO's in the whole third finger, which are also consistent with knuckle-walking (Matarazzo, 2015). Therefore PTO, rather than DA, holds promise for functional interpretation of fossil hand use from trabecular bone.

Emerging techniques, like PTO, will provide increasingly clear and direct functional links between fossil hominin hand morphology and the in vivo manual behaviours that produced it. Skeletonisation, is another new approach that reduces the trabecular network to lines of uniform thickness, its pure topology (Reznikov et al., 2020). This technique has the potential to separate out the contribution of Tb.N from other factors like Tb.Th to the more gross measure of BV/TV, while the novel application of finite element analysis (FEA) has already begun to ground different internal morphology in the biomechanics of hand bones (Nguyen et al., 2014; Synek et al., 2019a). Where the external morphology of the hand bones largely delimits the range of movements of which the hand is capable, the internal, and trace soft, morphology discussed above can tell us what that fossil hand did (Ruff and Runestad, 1992). Understanding significant manual behaviours in past populations, is not only important for understanding our own archaeological and evolutionary history, but also provides a holistic basis to understand how we use our hands now.

6.6 Conclusions

This chapter has given a brief overview of what habitual behaviours are unique to the modern human hand and how these are used to create technology that is qualitatively more complex than that of other living primates. This context allows us to interpret some of the manual behaviours of fossil hominins that made similar lithic technology which survives in the archaeological record. Fossil anatomy provides insight into the broader repertoire of hand postures and behaviours of which these hands were capable. Yet, it is a better understanding of extant primate soft tissue anatomy, its record in fossil bone external morphology, and signals in internal bone structure, that allow us to infer which of these manual capabilities were habitually employed by fossil hominins. Together, these approaches elucidate how our dexterous hands evolved and how they were used by past populations. This provides a context for how we understand the biomechanics of our hands today, and in turn, how we can treat the aspects of their pathology that relate to our evolutionary past.

Glossary

AL Refers to fossils attributed to the genus *Australopithecus,* housed in the Kenya National Museum and found in Kenya.

ARA-VP Refers to fossils attributed to the genus *Ardipithecus* found in the Aramis vertebrate palaeontology localities, Ethiopia.

ATE Refers to fossils attributed to the genus *Homo* and found in pit of elephants (Sima de la Elepante), Atapuerca, Spain.

BAR Refers to fossils attributed to the genus *Orrorin,* found in Baringo county, Kenya.

Bipolar knapping A knapping technique by which a core is held and stabilised against an anvil and struck from above in a perpendicular manner with a hammerstone to initiate a fracture.

BV/TV Bone volume/Total volume, a measure of trabecular bone density.

CCH Refers to fossils attributed to the species *Homo luzonensis* and found in Callao Cave, the Philippines.

Conchoidal Fracture The fracture type, which can be initiated in brittle stone that possesses a high degree of isotropy.

Core A piece of stone from which flakes have been detached. Cores normally possess one or more flake scars, a flaking surface (where flakes are detached) and a knapping platform (where the core is struck).

DA Degree of Anisotropy, a measure of trabecular alignment.

Dorsal surface The exterior surface of the cores flaking surface preserved on a flake. This surface often preserves either the unmodified (cortical) cores surface, or one or more flake scars from previous flake detachments from the core.

DP1 Distal phalanx of the first digit or thumb, the bone that underlies the thumb pad.

FEA Finite element modelling, an analysis technique for solving engineering problems virtually.

Flake The detached stone from a core, possessing an impact point on a striking platform and a clear dorsal and ventral surface.

FPL *Flexor pollicis longus*, a distinct muscle that flexes the thumb in humans (and gibbons).

Freehand knapping A knapping technique by which the core is held in one hand and the hammerstone in another.

Hammerstone Typically, a rounded stone held in one or two hands used to impact a core in order to initiate a conchoidal fracture to detach flakes.

ka Kilo-annum or thousands of years ago.

Knapping The process of intentionally shaping various types of stone through either direct or indirect percussion or pressure to produce different stone tools.

KNM-ER Refers to fossils various attributed to the genera *Australopithecus, Paranthropus and Homo,* housed in the Kenya National Museum and found in localities East of lake Rudolf (now Turkana), Kenya.

KNM-KP Refers to fossils attributed to the genus *Australopithecus,* housed in the Kenya National Museum and found in Kanapoi locality, Kenya.

KNM-WT Refers to fossils various attributed to the genera *Australopithecus, Paranthropus and Homo,* housed in the Kenya National Museum and found in localities in the West of the Turkana basin, Kenya.

Ma Mega-annum or millions of years ago.

Mc Metacarpal, the palm bones of the hand that connects the wrist to the fingers, Mc1 refers to the thumb metacarpal and Mc5 refers to the little finger metacarpal.

McP joint Metacarpophalangeal joint, the joint between each distal metacarpal and their proximal phalanx at the knuckle.

MH Refers to fossils attributed to the species *Australopithecus sediba* and found in Malapa, South Africa.

OA Osteoarthritis, a joint disease in which cartilage breaks down, resulting damage to other joint tissues, restricting joint range of motion and reducing strength (McDonough and Jette, 2010).

OH Refers to fossils various attributed to the genera *Australopithecus, Paranthropus and Homo,* found in localities in Olduvai gorge, Tanzania.

Passive hammer knapping A knapping technique in which a core is held in one or two hands and is struck against a stationary rock.

PCSA Physiological Cross sectional Area, a measure of muscle size.

PP Proximal phalanx, the finger bone immediately distal to the metacarpal.

PTO Principle trabecular orientation, a measure of the direction trabeculae are aligned in.

SH Refers to fossils attributed to the genus *Homo* and found in pit of bones (Sima de los Huesos), Atapuerca, Spain.

SKX Refers to various fossils attributed to the genera *Australopithecus, Paranthropus and Homo,* and found in Swartkrans, South Africa.

StW Refers to various fossils attributed to the genera *Australopithecus, Paranthropus and Homo, housed in the University of the Witwatersrand,* found in Stekfontein localities in South Africa.

TD-6 Refers to fossils variously attributed to the species *Homo heidlebergensis* and *H. anetecessor* and found in level 6 of the Dolina trench (Trinchera Dolina), Atapuerca, Spain.

TM Refers to various fossils attributed to the genera *Australopithecus, Paranthropus and Homo,* housed in the Transvaal (Now Ditsong) Museum and found in localities in South Africa.

TMc joint Trapizeiometacarpal joint, the joint between the proximal first metacarpal and the trapezium where the radial palm meets the wrist.

Ventral surface The interior surface of the core formed due to a flake fracture and preserved on a flake. This surface is often smooth and convex, sometimes possessing ripples along its surface and a bulb of percussion.

References

Abourachid, A., Fabre, A.-C., Cornette, R., & Höfling, E. (2017). Foot shape in arboreal birds: Two morphological patterns for the same pincer-like tool. *Journal of Anatomy, 231*(1), 1–11.

Alba, D. M., Moyà-Solà, S., & Köhler, M. (2003). Morphological affinities of the Australopithecus afarensis hand on the basis of manual proportions and relative thumb length. *Journal of Human Evolution, 44*(2), 225–254.

Alexander, C. J. (1994). Utilisation of joint movement range in arboreal Primates compared with human subjects: An evolutionary frame for primary osteoarthritis. *Annals of the Rheumatic Diseases, 53*(11), 720–725.

Almécija, S., & Alba, D. M. (2014). On manual proportions and pad-to-pad precision grasping in Australopithecus afarensis. *Journal of Human Evolution, 73*, 88–92.

Almécija, S., Moyà-Solà, S., & Alba, D. M. (2010). Early origin for human-like precision grasping: A comparative study of Pollical distal phalanges in fossil hominins. *PLoS One, 5*(7), e11727.

Almécija, S., Smaers, J. B., & Jungers, W. L. (2015). The evolution of human and ape hand proportions. *Nature Communications, 6*(1), 7717.

Amant, R. S., & Horton, T. E. (2008). Revisiting the definition of animal tool use. *Animal Behaviour, 75*(4), 1199–1208.

Anderson, C. V., & Higham, T. E. (2014). Chameleon anatomy. In K. A. Tolley, & A. Herrel (Eds.), *The biology of chameleons* (pp. 7–56). University of California Press.

Anderson-Gerfaud, P. C. (1990). Aspects of behaviour in the middle Palaeolithic: Functional analysis of stone tools from Southwest France. In P. Mellars (Ed.), *The emergence of modern humans: An archaeological perspective* (pp. 389–418). Ithaca: Cornell University Press.

Archer, W., Aldeias, V., & McPherron, S. P. (2020). What is 'in situ'? A reply to Harmand et al. (2015). *Journal of Human Evolution, 142*, 102740.

Arsuaga, J. L., Carretero, J.-M., Lorenzo, C., Gómez-Olivencia, A., Pablos, A., Rodríguez, L., et al. (2015). Postcranial morphology of the middle Pleistocene humans from Sima de Los Huesos, Spain. *Proceedings of the National Academy of Sciences, 112*(37), 11524–11529.

Arsuaga, J. L., Martínez, I., Arnold, L. J., Aranburu, A., Gracia-Téllez, A., Sharp, W. D., et al. (2014). Neandertal roots: Cranial and chronological evidence from Sima de Los Huesos. *Science, 344*(6190), 1358–1363.

Aversi-Ferreira, T. A., Diogo, R., Potau, J. M., Bello, G., Pastor, J. F., & Ashraf Aziz, M. (2010). Comparative anatomical study of the forearm extensor muscles of Cebus Libidinosus (Rylands et al., 2000; Primates, Cebidae), modern humans, and other Primates, with comments on primate evolution, phylogeny, and Manipulatory behavior. *The Anatomical Record, 293*(12), 2056–2070.

Bain, G. I., Polites, N., Higgs, B. G., Heptinstall, R. J., & McGrath, A. M. (2015). The functional range of motion of the finger joints. *Journal of Hand Surgery (European Volume), 40*(4), 406–411.

Barak, M. M. (2020). Bone modeling or bone remodeling: That is the question. *American Journal of Physical Anthropology, 172*(2), 153–155.

Barak, M. M., Lieberman, D. E., & Hublin, J.-J. (2011). A Wolff in Sheep's clothing: Trabecular bone adaptation in response to changes in joint loading orientation. *Bone, 49*(6), 1141–1151.

Barak, M. M., Sherratt, E., & Lieberman, D. E. (2017). Using principal trabecular orientation to differentiate joint loading orientation in the 3rd metacarpal heads of humans and chimpanzees. *Journal of Human Evolution, 113*, 173–182.

Barakat, M. J., Field, J., & Taylor, J. (2013). The range of movement of the thumb. *The Hand, 8*(2), 179–182.

Bardo, A., Borel, A., Meunier, H., Guéry, J.-P., & Pouydebat, E. (2016). Behavioral and functional strategies during tool use tasks in bonobos. *American Journal of Physical Anthropology, 161*(1), 125–140.

Bardo, A., Cornette, R., Borel, A., & Pouydebat, E. (2017). Manual function and performance in humans, gorillas, and orangutans during the same tool use task. *American Journal of Physical Anthropology, 164*(4), 821–836.

Bardo, A., Moncel, M. H., Dunmore, C. J., Kivell, T. L., Pouydebat, E., & Cornette, R. (2020). The implications of thumb movements for Neanderthal and modern human manipulation. *Scientific Reports, 10*(1), 1–12.

Bardo, A., Pouydebat, E., & Meunier, H. (2015). Do bimanual coordination, tool use, and body posture contribute equally to hand preferences in bonobos? *Journal of Human Evolution, 82*, 159–169.

Barrett, B. J., Monteza-Moreno, C. M., Dogandžić, T., Zwyns, N., Ibáñez, A., & Crofoot, M. C. (2018). Habitual stone-tool-aided extractive foraging in White-faced capuchins, Cebus Capucinus. *Royal Society Open Science, 5*(8), 181002.

Beck, B. B. (1980). *Animal tool behavior: The use and manufacture of tools by animals.* Garland STPM Pub.

Becker, S. K. (2016). Skeletal evidence of craft production from the Ch'iji Jawira site in Tiwanaku, Bolivia. *Journal of Archaeological Science: Reports, 9*, 405–415.

Benjamin, M., Evans, E. J., & Copp, L. (1986). The histology of tendon attachments to bone in man. *Journal of Anatomy, 149*, 89–100.

Benjamin, M., Kaiser, E., & Milz, S. (2008). Structure-function relationships in tendons: A review. *Journal of Anatomy, 212*(3), 211–228.

Benjamin, M., & Ralphs, J. R. (1998). Fibrocartilage in tendons and ligaments — An adaptation to compressive load. *Journal of Anatomy, 193*(4), 481–494.

Benjamin, M., Toumi, H., Ralphs, J. R., Bydder, G., Best, T. M., & Milz, S. (2006). Where tendons and ligaments meet bone: Attachment sites ("Entheses") in relation to exercise and/or mechanical load. *Journal of Anatomy, 208*(4), 471–490.

Bennell, K. L., Malcolm, S. A., Khan, K. M., Thomas, S. A., Reid, S. J., Brukner, P. D., et al. (1997). Bone mass and bone turnover in power athletes, endurance athletes, and controls: A 12-month longitudinal study. *Bone, 20*(5), 477–484.

Bennett, M. B., Ker, R. F., Imery, N. J., & Alexander, R. M. N. (1986). Mechanical properties of various mammalian tendons. *Journal of Zoology, 209*(4), 537–548.

Berger, L. R., Hawks, J., de Ruiter, D. J., Churchill, S. E., Schmid, P., Delezene, L. K., et al. (2015). Homo Naledi, a new species of the genus Homo from the Dinaledi Chamber, South Africa. *ELife, 4*, e09560. edited by J. Krause and N. J. Conard.

Boaz, N. T., Ciochon, R. L., Xu, Q., & Liu, J. (2004). Mapping and taphonomic analysis of the *Homo erectus* loci at locality 1 Zhoukoudian, China. *Journal of Human Evolution, 46*(5), 519–549.

Boesch, C., & Boesch, H. (1993). Different hand postures for pounding nuts with natural hammers by wild chimpanzees. In H. Preuschoft, & D. J. Chivers (Eds.), *Hands of primates* (pp. 31–43). Vienna: Springer Vienna.

Boesch, C., Kalan, A. K., Mundry, R., Arandjelovic, M., Pika, S., Dieguez, P., et al. (2020). Chimpanzee ethnography reveals unexpected cultural diversity. *Nature Human Behaviour*, 1–7.

Braun, D. R., Aldeias, V., Will Archer, J., Arrowsmith, R., Baraki, N., Campisano, C. J., et al. (2019). Earliest known Oldowan artifacts at> 2.58 ma from Ledi-Geraru, Ethiopia, highlight early technological diversity. *Proceedings of the National Academy of Sciences, 116*(24), 11712–11717.

Brumitt, J., & Cuddeford, T. (2015). Current concepts of muscle and tendon adaptation to strength and conditioning. *International Journal of Sports Physical Therapy, 10*(6), 748–759.

Bush, M. E., Owen Lovejoy, C., Johanson, D. C., & Coppens, Y. (1982). Hominid carpal, metacarpal, and phalangeal bones recovered from the Hadar formation: 1974–1977 collections. *American Journal of Physical Anthropology, 57*(4), 651–677.

Byrne, R. W., Corp, N., & Byrne, J. M. (2001). Manual dexterity in the Gorilla: Bimanual and digit role differentiation in a natural task. *Animal Cognition, 4*(3–4), 347–361.

Canale, G. R., Guidorizzi, C. E., Kierulff, M. C. M., & Gatto, C. A. F. R. (2009). First record of tool use by wild populations of the yellow-breasted capuchin monkey (Cebus Xanthosternos) and new Records for the Bearded Capuchin (Cebus Libidinosus). *American Journal of Primatology, 71*(5), 366–372.

Carpenter, A. (1887). Monkeys Opening Oysters. *Nature, 36*, 53.

Caslumore, L. A., & Zakrzewski, S. R. (2013). Assessment of musculoskeletal stress marker development in the hand. *International Journal of Osteoarchaeology, 23*(3), 334–347.

Caspi, D., Flusser, G., Farber, I., Ribak, J., Leibovitz, A., Habot, B., et al. (2001). Clinical, radiologic, demographic, and occupational aspects of hand osteoarthritis in the elderly. *Seminars in Arthritis and Rheumatism, 30*(5), 321–331.

Cavagna, G. A., Heglund, N. C., & Taylor, C. R. (1977). Mechanical work in terrestrial locomotion: Two basic mechanisms for minimizing energy expenditure. *American Journal of Physiology-Regulatory, Integrative and Comparative Physiology, 233*(5), R243–R261.

Chen, J.-H., Liu, C., You, L., & Simmons, C. A. (2010). Boning up on Wolff's law: Mechanical regulation of the cells that make and maintain bone. *Journal of Biomechanics, 43*(1), 108–118.

Chirchir, H., Kivell, T. L., Ruff, C. B., Hublin, J.-J., Carlson, K. J., Zipfel, B., et al. (2015). Recent origin of low trabecular bone density in modern humans. *Proceedings of the National Academy of Sciences, 112*(2), 366–371.

Chirchir, H., Zeininger, A., Nakatsukasa, M., Ketcham, R. A., & Richmond, B. G. (2017). Does trabecular bone structure within the metacarpal heads of Primates vary with hand posture? *Comptes Rendus Palevol, 16*(5), 533–544.

Christel, M. (1993). Grasping techniques and hand preferences in Hominoidea. In H. Preuschoft, & D. J. Chivers (Eds.), *Hands of primates* (pp. 91–108). Vienna: Springer.

Clarke, R. J. (1999). Discovery of complete arm and hand of the 3.3 million-year-old Australopithecus skeleton from Sterkfontein. *South African Journal of Science, 95*, 477–480.

Clarke, R. (2013). Australopithecus from Sterkfontein Caves, South Africa. In K. E. Reed, J. G. Fleagle, & R. E. Leakey (Eds.), *The Paleobiology of Australopithecus, vertebrate Paleobiology and Paleoanthropology* (pp. 105–123). Dordrecht: Springer Netherlands.

Constantino, P., & Wood, B. (2004). Paranthropus paleobiology. *Biological Sciences Faculty Research*.

Cooney, W. P., 3rd, & Chao, E. Y. (1977). Biomechanical analysis of static forces in the thumb during hand function. *Journal of Bone and Joint Surgery, 59*(1), 27–36.

Cope, J. M., Berryman, A. C., Martin, D. L., & Potts, D. D. (2005). Robusticity and osteoarthritis at the trapeziometacarpal joint in a bronze age population from tell Abraq, United Arab Emirates. *American Journal of Physical Anthropology, 126*(4), 391–400.

Coudé, G., Toschi, G., Festante, F., Bimbi, M., Bonaiuto, J., & Ferrari, P. F. (2019). Grasping neurons in the ventral premotor cortex of macaques are modulated by social goals. *Journal of Cognitive Neuroscience, 31*(2), 299–313.

Currey, J. D. (2002). *Bones: Structure and mechanics*. Princeton University Press.

Currey, J. D. (2012). The structure and mechanics of bone. *Journal of Materials Science, 47*(1), 41–54.

D'Agostino, P., Dourthe, B., Faes Kerkhof, G., Van Lenthe, H., Stockmans, F., & Vereecke, E. E. (2017). In vivo biomechanical behavior of the Trapeziometacarpal joint in healthy and osteoarthritic subjects. *Clinical biomechanics, 49*, 119–127.

Day, M. H., & Napier, J. (1963). The functional significance of the deep head of flexor Pollicis brevis in Primates. *Folia Primatologica, 1*(2), 122–134.

Delagnes, A., & Roche, H. (2005). Late Pliocene hominid knapping skills: The case of Lokalalei 2C, West Turkana, Kenya. *Journal of Human Evolution, 48*, 435–472.

Détroit, F., Mijares, A. S., Corny, J., Daver, G., Zanolli, C., Dizon, E., et al. (2019). A new species of Homo from the late Pleistocene of the Philippines. *Nature, 568*(7751), 181–186.

Deymier, A. C., Schwartz, A. G., Cai, Z., Daulton, T. L., Pasteris, J. D., Genin, G. M., et al. (2019). The multiscale structural and mechanical effects of mouse supraspinatus muscle unloading on the mature Enthesis. *Acta Biomaterialia, 83*, 302–313.

Deymier-Black, A. C., Pasteris, J. D., Genin, G. M., & Thomopoulos, S. (2015). Allometry of the tendon Enthesis: Mechanisms of load transfer between tendon and bone. *Journal of Biomechanical Engineering, 137*(11).

Diogo, R., Shearer, B., Potau, J. M., Pastor, J. F., de Paz, F. J., Arias-Martorell, J., et al. (2017). *Photographic and descriptive musculoskeletal atlas of bonobos*. Cham: Springer International Publishing.

Dirks, P. H. G. M., Roberts, E. M., Hilbert-Wolf, H., Kramers, J. D., Hawks, J., Dosseto, A., et al. (2017). The age of Homo Naledi and associated sediments in the rising star cave, South Africa. *ELife, 6*, e24231. edited by G. H. Perry.

Domalain, M., Bertin, A., & Daver, G. (2017). Was Australopithecus afarensis able to make the Lomekwian stone tools? Towards a realistic biomechanical simulation of hand force capability in fossil hominins and new insights on the role of the fifth digit. *Comptes Rendus Palevol, 16*(5), 572–584.

Domínguez-Rodrigo, M., Pickering, T. R., Almécija, S., Heaton, J. L., Baquedano, E., Mabulla, A., et al. (2015). Earliest modern human-like hand bone from a new >1.84-million-year-old site at Olduvai in Tanzania. *Nature Communications, 6*(1), 7987.

Domínguez-Rodrigo, M., Pickering, T. R., & Bunn, H. T. (2010). Configurational approach to identifying the earliest hominin butchers. *PNAS, 107*.

Domínguez-Rodrigo, M., Pickering, T. R., & Bunn, H. T. (2012). Experimental study of cut Marks made with rocks unmodified by human flaking and its bearing on claims of ~3.4-million-year-old butchery evidence from Dikika, Ethiopia. *Journal of Archaeological Science, 39*(2), 205–214.

Douka, K., Slon, V., Jacobs, Z., Ramsey, C. B., Shunkov, M. V., Derevianko, A. P., et al. (2019). Age estimates for hominin fossils and the onset of the upper Palaeolithic at Denisova cave. *Nature, 565*(7741), 640–644.

Dourthe, B., Nickmanesh, R., Wilson, D. R., D'Agostino, P., Patwa, A. N., Grinstaff, M. W., et al. (2019). Assessment of healthy Trapeziometacarpal cartilage properties using indentation testing and contrast-enhanced computed tomography. *Clinical biomechanics, 61*, 181–189.

Drapeau, M. S. M. (2015). Metacarpal torsion in apes, humans, and early *Australopithecus*: implications for manipulatory abilities. *PeerJ, (3)*, e1311.

Drapeau, M. S. M., Ward, C. V., Kimbel, W. H., Johanson, D. C., & Rak, Y. (2005). Associated cranial and forelimb remains attributed to Australopithecus afarensis from Hadar, Ethiopia. *Journal of Human Evolution, 48*(6), 593–642.

Dunmore, C. J., Bardo, A., Skinner, M. M., & Kivell, T. L. (2020a). Trabecular variation in the first metacarpal and manipulation in hominids. *American Journal of Physical Anthropology, 171*(2), 219–241.

Dunmore, C. J., Kivell, T. L., Bardo, A., & Skinner, M. M. (2019). Metacarpal trabecular bone varies with distinct hand-positions used in hominid locomotion. *Journal of Anatomy, 235*(1), 45–66.

Dunmore, C. J., Skinner, M. M., Bardo, A., Berger, L. R., Hublin, J.-J., Pahr, D. H., et al. (2020b). The position of Australopithecus Sediba within fossil hominin hand use diversity. *Nature Ecology & Evolution, 4*(7), 911–918.

Duval, M., Grün, R., Parés, J. M., Martín-Francés, L., Campaña, I., Rosell, J., et al. (2018). The first direct ESR dating of a hominin tooth from Atapuerca gran Dolina TD-6 (Spain) supports the antiquity of Homo antecessor. *Quaternary Geochronology, 47*, 120–137.

Falótico, T., & Ottoni, E. B. (2016). The manifold use of pounding stone tools by wild capuchin monkeys of Serra Da Capivara National Park, Brazil. *Behaviour, 153*(4), 421–442.

Faurie, C., Schiefenhövel, W., le Bomin, S., Billiard, S., & Raymond, M. (2005). Variation in the frequency of left-handedness in traditional societies. *Current Anthropology, 46*(1), 142–147.

Feix, T., Kivell, T. L., Pouydebat, E., & Dollar, A. M. (2015). Estimating thumb-index finger precision grip and manipulation potential in extant and fossil primates. *Journal of the Royal Society Interface, 12*(106), 20150176.

Ferreira, R. G., Emidio, R. A., & Jerusalinsky, L. (2010). Three stones for three seeds: Natural occurrence of selective tool use by capuchins (Cebus libidinosus) based on an analysis of the weight of stones found at nutting sites. *American Journal of Primatology, 72*(3), 270–275.

Forrester, G. S., Rawlings, B., & Davila-Ross, M. (2016). An analysis of bimanual actions in natural feeding of semiwild chimpanzees. *American Journal of Physical Anthropology, 159*(1), 85–92.

Fragaszy, D. M., & Crast, J. (2016). Functions of the hand in Primates. In T. L. Kivell, P. Lemelin, B. G. Richmond, & D. Schmitt (Eds.), *The evolution of the primate hand: Anatomical, developmental, functional, and paleontological evidence, developments in primatology: Progress and prospects* (pp. 313–344). New York, NY: Springer.

Fragaszy, D., Izar, P., Visalberghi, E., Ottoni, E. B., Gomes, M., & de Oliveira. (2004). Wild capuchin monkeys (Cebus Libidinosus) use anvils and stone pounding tools. *American Journal of Primatology, 64*(4), 359–366.

Galletta, L., Stephens, N. B., Bardo, A., Kivell, T. L., & Marchi, D. (2019). Three-dimensional geometric morphometric analysis of the first metacarpal distal articular surface in humans, great apes and fossil hominins. *Journal of Human Evolution, 132*, 119–136.

Gans, C. (1982). FIBER architecture and muscle function. *Exercise and Sport Sciences Reviews, 10*(1), 160–207.

García, N., & Arsuaga, J. L. (2011). The Sima de Los Huesos (Burgos, northern Spain): Palaeoenvironment and habitats of Homo Heidelbergensis during the middle Pleistocene. *Quaternary Science Reviews, 30*(11), 1413–1419.

Geribàs, N., Mosquera, M., & Vergès, J. M. (2010). What novice Knappers have to learn to become expert stone toolmakers. *Journal of Archaeological Science, 37*(11), 2857–2870.

Gillis, J., Calder, K., & Williams, J. (2011). Review of thumb carpometacarpal arthritis classification, treatment and outcomes. *Canadian Journal of Plastic Surgery, 19*(4), 134–138.

Gommery, D., & Senut, B. (2006). La phalange distale du pouce d'Orrorin tugenensis (Miocène supérieur du Kenya). *Geobios, 39*(3), 372–384.

Green, D. J., & Gordon, A. D. (2008). Metacarpal proportions in Australopithecus Africanus. *Journal of Human Evolution, 54*(6), 705–719.

Gumert, M. D., Kluck, M., & Malaivijitnond, S. (2009). The physical characteristics and usage patterns of stone axe and pounding hammers used by Long-tailed macaques in the Andaman Sea region of Thailand. *American Journal of Primatology, 71*(7), 594–608.

Gumert, M. D., & Malaivijitnond, S. (2012). Marine prey processed with stone tools by Burmese Long-tailed macaques (Macaca Fascicularis Aurea) in intertidal habitats. *American Journal of Physical Anthropology, 149*(3), 447–457.

Hagihara, Y. (2021). Dorso-palmar elongation of the diaphysis of the third metacarpal bone in prehistoric Jomon people. *Anatomical Science International, 96*(1), 119–131.

Hamrick, M. W., Churchill, S. E., Schmitt, D., & Hylander, W. L. (1998). EMG of the human flexor Pollicis longus muscle: Implications for the evolution of hominid tool use. *Journal of Human Evolution, 34*(2), 123–136.

Hardy, B. L., Moncel, M.-H., Daujeard, C., Fernandes, P., Béarez, P., Desclaux, E., et al. (2013). Impossible Neanderthals? Making string, throwing projectiles and catching small game during marine isotope stage 4 (Abri Du Maras, France). *Quaternary Science Reviews, 82*, 23–40.

Harmand, S., Lewis, J. E., Feibel, C. S., Lepre, C. J., Prat, S., Lenoble, A., et al. (2015). 3.3-million-year-old stone tools from Lomekwi 3, West Turkana, Kenya. *Nature, 521*(7552), 310–315.

Haugen, I. K., Englund, M., Aliabadi, P., Niu, J., Clancy, M., Kvien, T. K., et al. (2011). Prevalence, incidence and progression of hand osteoarthritis in the general population: The Framingham osteoarthritis study. *Annals of the Rheumatic Diseases, 70*(9), 1581–1586.

Hawks, J., Elliott, M., Schmid, P., Churchill, S. E., de Ruiter, D. J., Roberts, E. M., et al. (2017). New fossil remains of Homo Naledi from the Lesedi Chamber, South Africa. *ELife, 6*, e24232. edited by G.H. Perry.

Huxley, T. H. (1863). *Evidence as to Man's place in nature.* Williams and Norgate.

Inizan, M.-L., Reduron-Ballinger, M., & Roche, H. (1999). *Technology and terminology of knapped stone: Followed by a multilingual vocabulary Arabic, English, French, German, Greek, Italian, Portuguese, Spanish. Vol. 5.* (Cercle de Recherches et d'Etudes Préhistoriques).

Jolly, C. J. (1970). The seed-eaters: A new model of hominid differentiation based on a baboon analogy. *Man, 5*(1), 5–26.

Judex, S., & Carlson, K. J. (2009). Is Bone's response to mechanical signals dominated by gravitational loading? *Medicine & Science in Sports & Exercise, 41*(11), 2037–2043.

Jurmain, R. (1989). Trauma, degenerative disease, and other pathologies among the Gombe chimpanzees. *American Journal of Physical Anthropology, 80*(2), 229–237.

Jurmain, R. (2000). Degenerative joint disease in African great apes: An evolutionary perspective. *Journal of Human Evolution, 39*(2), 185–203.

Karakostis, F. A., Centeno, H. R., Francken, M., Hotz, G., Rademaker, K., & Harvati, K. (2020). Biocultural evidence of precise manual activities in an Early Holocene individual of the high-altitude Peruvian Andes. *American Journal of Physical Anthropology, 174*(1), 35–48.

Karakostis, F. A., Haeufle, D., Anastopoulou, I., Moraitis, K., Hotz, G., Tourloukis, V., et al. (2021). Biomechanics of the human thumb and the evolution of dexterity. *Current Biology, 31*(6), 289–291.

Karakostis, F. A., & Harvati, K. (2021). New horizons in reconstructing past human behavior: Introducing the "Tübingen University Validated Entheses-based Reconstruction of Activity" method. *Evolutionary Anthropology.* https://doi.org/10.1002/evan.21892.

Karakostis, F. A., Hotz, G., Scherf, H., Wahl, J., & Harvati, K. (2017). Occupational manual activity is reflected on the patterns among hand Entheses. *American Journal of Physical Anthropology, 164*(1), 30–40.

Karakostis, F. A., Hotz, G., Scherf, H., Wahl, J., & Harvati, K. (2018a). A repeatable geometric morphometric approach to the analysis of hand Entheseal three-dimensional form. *American Journal of Physical Anthropology, 166*(1), 246–260.

Karakostis, F. A., Hotz, G., Tourloukis, V., & Harvati, K. (2018b). Evidence for precision grasping in Neandertal daily activities. *Science Advances, 4*(9), eaat2369.

Karakostis, F. A., Jeffery, N., & Harvati, K. (2019a). Experimental proof that multivariate patterns among muscle attachments (Entheses) can reflect repetitive muscle use. *Scientific Reports, 9*(1), 16577.

Karakostis, F. A., Le Quéré, E., Vanna, V., & Moraitis, K. (2016). Assessing the effect of manual physical activity on proximal hand phalanges using Hellenistic and modern skeletal samples from Greece. *Homo, 67*(2), 110–124.

Karakostis, F. A., & Lorenzo, C. (2016). Morphometric patterns among the 3D surface areas of human hand Entheses. *American Journal of Physical Anthropology, 160*(4), 694–707.

Karakostis, F. A., Vlachodimitropoulos, D., Piagkou, M., Scherf, H., Harvati, K., & Moraitis, K. (2019b). Is bone elevation in hand muscle attachments associated with biomechanical stress? A histological approach to an anthropological question. *The Anatomical Record, 302*(7), 1093–1103.

Karakostis, F. A., Wallace, I. J., Konow, N., & Harvati, K. (2019c). Experimental evidence that physical activity affects the multivariate associations among muscle attachments (Entheses). *Journal of Experimental Biology, 222*(23).

Key, A. J. M., & Dunmore, C. J. (2015). The evolution of the hominin thumb and the influence exerted by the non-dominant hand during stone tool production. *Journal of Human Evolution, 78*, 60–69.

Key, A. J. M., Dunmore, C. J., & Marzke, M. W. (2019). The unexpected importance of the fifth digit during stone tool production. *Scientific Reports, 9*(1), 16724.

Key, A. J. M., & Lycett, S. J. (2018). Investigating interrelationships between lower Palaeolithic stone tool effectiveness and tool user biometric variation: Implications for technological and evolutionary changes. *Archaeological and Anthropological Sciences, 10*(5), 989–1006.

Key, A., Merritt, S. R., & Kivell, T. L. (2018). Hand grip diversity and frequency during the use of lower Palaeolithic stone cutting-tools. *Journal of Human Evolution, 125*, 137–158.

Kivell, T. L. (2015). Evidence in hand: Recent discoveries and the early evolution of human manual manipulation. *Philosophical Transactions of the Royal Society B: Biological Sciences, 370*(1682), 20150105.

Kivell, T. L., Churchill, S. E., Kibii, J. M., Schmid, P., & Berger, L. R. (2018). The hand of Australopithecus Sediba. *Paleoanthropology, 52* (Special Issue: Australopithecus sediba).

Kivell, T. L., Deane, A. S., Tocheri, M. W., Orr, C. M., Schmid, P., Hawks, J., et al. (2015). The hand of Homo Naledi. *Nature Communications, 6*(1), 8431.

Kivell, T. L., Kibii, J. M., Churchill, S. E., Schmid, P., & Berger, L. R. (2011). Australopithecus Sediba hand demonstrates mosaic evolution of locomotor and manipulative abilities. *Science, 333*(6048), 1411–1417.

Kivell, T. L., Lemelin, P., Richmond, B. G., & Schmitt, D. (2016). *The evolution of the primate hand: Anatomical, developmental, functional, and paleontological evidence.* Springer.

Kivell, T. L., Ostrofsky, K., Richmond, B. G., & Drapeau, M. (2020). Metacarpals & manual phalanges. In B. Zipfel, B. G. Richmond, & C. V. Ward (Eds.), *Hominin postcranial remains from Sterkfontein, South Africa, 1936-1995, human evolution series* (pp. 105–141). Oxford University Press.

Koff, M. F., Ugwonali, O. F., Strauch, R. J., Rosenwasser, M. P., Ateshian, G. A., & Mow, V. C. (2003). Sequential Wear patterns of the articular cartilage of the thumb carpometacarpal joint in Osteoarthritis1 1No benefits in any form have been received or Will be received from a commercial party related directly or indirectly to the subject of this article. *The Journal of Hand Surgery, 28*(4), 597–604.

Krützen, M., Mann, J., Heithaus, M. R., Connor, R. C., Bejder, L., & Sherwin, W. B. (2005). Cultural transmission of tool use in bottlenose dolphins. *Proceedings of the National Academy of Sciences, 102*(25), 8939–8943.

Landsmeer, J. M. F. (1962). Power grip and precision handling. *Annals of the Rheumatic Diseases, 21*(2), 164–170.

Larson, S. G., Jungers, W. L., Tocheri, M. W., Orr, C. M., Morwood, M. J., Sutikna, T., et al. (2009). Descriptions of the upper limb skeleton of Homo Floresiensis. *Journal of Human Evolution, 57*(5), 555–570.

Lazenby, R. A. (1998). Second metacarpal cross-sectional geometry: Rehabilitating a circular argument. *American Journal of Human Biology, 10*(6), 747–756.

Lazenby, R. A., Skinner, M. M., Hublin, J.-J., & Boesch, C. (2011). Metacarpal trabecular architecture variation in the chimpanzee (Pan troglodytes): Evidence for locomotion and tool-use? *American Journal of Physical Anthropology, 144*(2), 215–225.

Leakey, M. G., Feibel, C. S., McDougall, I., Ward, C., & Walker, A. (1998). New specimens and confirmation of an early age for Australopithecus Anamensis. *Nature, 393*(6680), 62–66.

Leakey, M. G., Spoor, F., Brown, F. H., Gathogo, P. N., Kiarie, C., Leakey, L. N., et al. (2001). New hominin genus from eastern Africa shows diverse middle Pliocene lineages. *Nature, 410*(6827), 433–440.

Lewis, O. J. (1989). *Functional morphology of the evolving hand and foot*. Oxford University Press.

Lewis, J. E., & Harmand, S. (2016). An earlier origin for stone tool making: Implications for cognitive evolution and the transition to Homo. *Philosophical Transactions of the Royal Society B, 371*(1698), 20150233.

Liu, M.-J., Xiong, C.-H., & Di, H. (2016). Assessing the manipulative potentials of monkeys, apes and humans from hand proportions: Implications for hand evolution. *Proceedings of the Royal Society B: Biological Sciences, 283*(1843), 20161923.

Lordkipanidze, D., Jashashvili, T., Vekua, A., Marcia, S., de León, P., Christoph, P. E., et al. (2007). Postcranial evidence from early Homo from Dmanisi, Georgia. *Nature, 449*(7160), 305–310.

Lorenzo, C., Arsuaga, J. L., & Carretero, J. M. (1999). Hand and foot remains from the gran Dolina early Pleistocene site (sierra de Atapuerca, Spain). *Journal of Human Evolution, 37*(3–4), 501–522.

Lovejoy, C. O., Simpson, S. W., White, T. D., Asfaw, B., & Suwa, G. (2009). Careful climbing in the Miocene: The forelimbs of Ardipithecus Ramidus and humans are primitive. *Science, 326*(5949), 70. 70e1–8.

Lu, S.-C., Vereecke, E. E., Synek, A., Pahr, D. H., & Kivell, T. L. (2018). A novel experimental design for the measurement of metacarpal bone loading and deformation and fingertip force. *PeerJ, 6*, e5480.

Lu, S.-C., Yang, T.-H., Li-Chieh Kuo, I., Jou, M., Sun, Y.-N., & Fong-Chin, S. (2015). Effects of different extents of pulley release on tendon excursion efficiency and tendon moment arms. *Journal of Orthopaedic Research, 33*(2), 224–228.

Macfarlane, N. B., & Graziano, M. S. (2009). Diversity of grip in *Macaca mulatta. Experimental Brain Research, 197*(3), 255–268.

Macho, G. A., Spears, I. R., Leakey, M. G., McColl, D. J., Jiang, Y., Abel, R., et al. (2010). An exploratory study on the combined effects of external and internal morphology on load dissipation in primate capitates: Its potential for an understanding of the positional and locomotor repertoire of early hominins. *Folia Primatologica, 81*(5), 292–304.

Maïmoun, L., & Sultan, C. (2011). Effects of physical activity on bone remodeling. *Metabolism, 60*(3), 373–388.

Maki, J., & Trinkaus, E. (2011). Opponens pollicis mechanical effectiveness in Neandertals and early modern humans. *PaleoAnthropology, 2011*, 62–71.

Malaivijitnond, S., Lekprayoon, C., Tandavanittj, N., Panha, S., Cheewatham, C., & Hamada, Y. (2007). Stone-tool usage by Thai Long-tailed macaques (Macaca Fascicularis). *American Journal of Primatology, 69*(2), 227–233.

Maquer, G., Musy, S. N., Wandel, J., Gross, T., & Zysset, P. K. (2015). Bone volume fraction and fabric anisotropy are better determinants of trabecular bone stiffness than other morphological variables. *Journal of Bone and Mineral Research, 30*(6), 1000–1008.

Marchi, D. (2005). The cross-sectional geometry of the hand and foot bones of the Hominoidea and its relationship to locomotor behavior. *Journal of Human Evolution, 49*(6), 743–761.

Marchi, D., Proctor, D. J., Huston, E., Nicholas, C. L., & Fischer, F. (2017). Morphological correlates of the first metacarpal proximal articular surface with manipulative capabilities in apes, humans and south African early hominins. *Comptes Rendus Palevol, 16*(5), 645–654.

Marzke, M. W. (1983). Joint functions and grips of the Australopithecus afarensis hand, with special reference to the region of the capitate. *Journal of Human Evolution, 12*(2), 197–211.

Marzke, M. W. (1997). Precision grips, hand morphology, and tools. *American Journal of Physical Anthropology, 102*(1), 91–110.

Marzke, M. W. (2013). Tool making, hand morphology and fossil hominins. *Philosophical Transactions of the Royal Society B: Biological Sciences, 368*(1630), 20120414.

Marzke, M. W., Marchant, L. F., McGrew, W. C., & Reece, S. P. (2015). Grips and hand movements of chimpanzees during feeding in Mahale Mountains National Park, Tanzania. *American Journal of Physical Anthropology, 156*(3), 317–326.

Marzke, M. W., & Marzke, R. F. (1987). The third metacarpal styloid process in humans: Origin and functions. *American Journal of Physical Anthropology, 73*(4), 415–431.

Marzke, M. W., Marzke, R. F., Linscheid, R. L., Smutz, P., Steinberg, B., Reece, S., et al. (1999). Chimpanzee thumb muscle cross sections, moment arms and potential torques, and comparisons with humans. *American Journal of Physical Anthropology, 110*(2), 163–178.

Marzke, M. W., & Shackley, M. S. (1986). Hominid hand use in the Pliocene and Pleistocene: Evidence from experimental archaeology and comparative morphology. *Journal of Human Evolution, 15*(6), 439–460.

Marzke, M. W., Tocheri, M. W., Steinberg, B., Femiani, J. D., Reece, S. P., Linscheid, R. L., et al. (2010). Comparative 3D quantitative analyses of Trapeziometacarpal joint surface curvatures among living catarrhines and fossil hominins. *American Journal of Physical Anthropology, 141*(1), 38–51.

Marzke, M. W., Toth, N., Schick, K., Reece, S., Steinberg, B., Hunt, K., et al. (1998). EMG study of hand muscle recruitment during hard hammer percussion manufacture of Oldowan tools. *American Journal of Physical Anthropology, 105*(3), 315–332.

Marzke, M. W., & Wullstein, K. L. (1996). Chimpanzee and human grips: A new classification with a focus on evolutionary morphology. *International Journal of Primatology, 17*(1), 117–139.

Marzke, M. W., Wullstein, K. L., & Viegas, S. F. (1992). Evolution of the power ("squeeze") grip and its morphological correlates in hominids. *American Journal of Physical Anthropology, 89*(3), 283–298.

Matarazzo, S. A. (2015). Trabecular architecture of the manual elements reflects locomotor patterns in Primates. *PLoS One, 10*(3), e0120436.

Mays, S. (2001). Effects of age and occupation on cortical bone in a group of 18th-19th century British men. *American Journal of Physical Anthropology, 116*, 34–44.

McDonough, C. M., & Jette, A. M. (2010). The contribution of osteoarthritis to functional limitations and disability. *Clinics in Geriatric Medicine, 26*(3), 387–399.

McGrew, W. C. (2013). Is primate tool use special? Chimpanzee and new Caledonian crow compared. *Philosophical Transactions of the Royal Society B: Biological Sciences, 368*(1630), 20120422.

Mcgrew, W. C., & Tutin, C. E. G. (1973). Chimpanzee tool use in dental grooming. *Nature, 241*(5390), 477–478.

McHenry, H. M. (1983). The capitate of Australopithecus afarensis and A. Africanus. *American Journal of Physical Anthropology, 62*(2), 187–198.

McPherron, S. P., Alemseged, Z., Marean, C. W., Wynn, J. G., Reed, D., Geraads, D., et al. (2010). Evidence for stone-tool-assisted consumption of animal tissues before 3.39 million years ago at Dikika, Ethiopia. *Nature, 466*(7308), 857–860.

Meulman, E. J. M., & van Schaik, C. P. (2013). Orangutan tool use and the evolution of technology. In C. Sanz, J. Call, & C. Boesch (Eds.), *Tool use in animals* (pp. 176–202). Cambridge: Cambridge University Press.

Meunier, H., & Vauclair, J. (2007). Hand preferences on Unimanual and bimanual tasks in White-faced capuchins (Cebus Capucinus). *American Journal of Primatology, 69*(9), 1064–1069.

Meyer, M., Arsuaga, J.-L., de Filippo, C., Nagel, S., Aximu-Petri, A., Nickel, B., et al. (2016). Nuclear DNA sequences from the middle Pleistocene Sima de Los Huesos hominins. *Nature, 531*(7595), 504–507.

Mora, R., & de la Torre, I. (2005). Percussion tools in Olduvai beds I and II (Tanzania): Implications for early human activities. *Journal of Anthropological Archaeology, 24*(2), 179–192.

Moyà-Solà, S., Köhler, M., Alba, D. M., & Almécija, S. (2008). Taxonomic attribution of the Olduvai hominid 7 manual remains and the functional interpretation of hand morphology in robust australopithecines. *Folia Primatologica, 79*(4), 215–250.

Murai, T., Uchiyama, S., Nakamura, K., Ido, Y., Hata, Y., & Kato, H. (2018). Functional range of motion in the metacarpophalangeal joints of the hand measured by single Axis electric goniometers. *Journal of Orthopaedic Science, 23*(3), 504–510.

Musgrave, J. H. (1971). How Dextrous was Neanderthal man ? *Nature, 233*(5321), 538–541.

Mutalib, S. A., Mace, M., & Burdet, E. (2019). Bimanual coordination during a physically coupled task in unilateral spastic cerebral palsy children. *Journal of Neuroengineering and Rehabilitation*, 16(1), 1.

Napier, J. R. (1960). Studies of the hands of living Primates. *Proceedings of the Zoological Society of London*, 134(4), 647–657.

Napier, J. (1962). Fossil hand bones from Olduvai Gorge. *Nature*, 196(4853), 409–411.

Napier, J. R. (1993). In R. H. Tuttle (Ed.), *Hands*. Princeton University Press.

Neufuss, J., Robbins, M. M., Baeumer, J., Humle, T., & Kivell, T. L. (2017). Comparison of hand use and forelimb posture during vertical climbing in mountain gorillas (Gorilla Beringei Beringei) and chimpanzees (Pan troglodytes). *American Journal of Physical Anthropology*, 164(4), 651–664.

Neufuss, J., Robbins, M. M., Baeumer, J., Humle, T., & Kivell, T. L. (2019). Manual skills for food processing by mountain gorillas (Gorilla Beringei Beringei) in Bwindi impenetrable National Park, Uganda. *Biological Journal of the Linnean Society*, 127(3), 543–562.

Nguyen, H. N., Pahr, D. H., Gross, T., Skinner, M. M., & Kivell, T. L. (2014). Micro-finite element (MFE) modeling of the Siamang (Symphalangus Syndactylus) third proximal phalanx: The functional role of curvature and the flexor sheath ridge. *Journal of Human Evolution*, 67, 60–75.

Niewoehner, W. A. (2001). Behavioral inferences from the Skhul/Qafzeh early modern human hand remains. *Proceedings of the National Academy of Sciences*, 98(6), 2979–2984.

Niewoehner, W. A. (2006). Neanderthal hands in their proper perspective. In J.-J. Hublin, K. Harvati, & T. Harrison (Eds.), *Neanderthals revisited: New approaches and perspectives* (pp. 157–190). Dordrecht: Springer Netherlands.

Niewoehner, W. A., Bergstrom, A., Eichele, D., Zuroff, M., & Clark, J. T. (2003). Manual dexterity in Neanderthals. *Nature*, 422(6930), 395.

Niewoehner, W. A., Weaver, A. H., & Trinkaus, E. (1997). Neandertal Capitate-Metacarpal Articular Morphology. *American Journal of Physical Anthropology*, 103(2), 219–233.

O'Neill, M. C., Umberger, B. R., Holowka, N. B., Larson, S. G., & Reiser, P. J. (2017). Chimpanzee super strength and human skeletal muscle evolution. *Proceedings of the National Academy of Sciences*, 114(28), 7343–7348.

Orr, C. M., Tocheri, M. W., Burnett, S. E., Awe, R. D., Wahyu Saptomo, E., Sutikna, T., et al. (2013). New wrist bones of Homo Floresiensis from Liang Bua (Flores, Indonesia). *Journal of Human Evolution*, 64(2), 109–129.

Ottoni, E. B., & Izar, P. (2008). Capuchin monkey tool use. Overview and implications. *Evolutionary Anthropology: Issues, News, and Reviews*, 17(4), 171–178.

Oxnard, C. E. (1991). *Mechanical stress and strain at a point: Implications for biomorphometric and biomechanical studies of bone form and architecture. Vol. 4* (pp. 57–109).

Oxnard, C. E. (2004). Thoughts on bone biomechanics. *Folia Primatologica*, 75(4), 189–201.

Panger, M. A., Brooks, A. S., Richmond, B. G., & Wood, B. (2003). Older than the Oldowan? Rethinking the emergence of hominin tool use. *Evolutionary Anthropology: Issues, News, and Reviews*, 11(6), 235–245.

Pelegrin, J. (2005). Remarks about archaeological techniques and methods of knapping: Elements of a cognitive approach to stone knapping. In V. Roux, & B. Bril (Eds.), *Stone knapping: The necessary conditions for a uniquely hominid behaviour. McDonald Institute Monograph Series, Cambridge* (pp. 23–33). Cambridge: McDonald Institute monograph series.

Pickering, T. R., Heaton, J. L., Clarke, R. J., & Stratford, D. (2018). Hominin hand bone fossils from Sterkfontein caves, South Africa (1998–2003 excavations). *Journal of Human Evolution*, 118, 89–102.

Polatsch, D. B., & Paksima, N. (2006). Basal joint arthritis: Diagnosis and treatment. *Bulletin of the NYU Hospital for Joint Diseases*, 64(3–4), 178.

Pontzer, H., Lieberman, D. E., Momin, E., Devlin, M. J., Polk, J. D., Hallgrímsson, B., et al. (2006). Trabecular bone in the bird knee responds with high sensitivity to changes in load orientation. *Journal of Experimental Biology*, 209(1), 57–65.

Prang, T. C., Ramirez, K., Grabowski, M., & Williams, S. A. (2021). Ardipithecus hand provides evidence that humans and chimpanzees evolved from an ancestor with suspensory adaptations. *Science. Advances*, 7(9), eabf2474.

Preuschoft, H., & Chivers, D. J. (1993). *Hands of Primates*. Springer Science & Business Media.

Proffitt, T., Luncz, L. V., Falótico, T., Ottoni, E. B., de la Torre, I., & Haslam, M. (2016). Wild monkeys flake stone tools. *Nature*, 539(7627), 85–88.

Rabey, K. N., Green, D. J., Taylor, A. B., Begun, D. R., Richmond, B. G., & McFarlin, S. C. (2015). Locomotor activity influences muscle architecture and bone growth but not muscle attachment site morphology. *Journal of Human Evolution*, 78, 91–102.

Rafferty, K. F. (1990). *The functional and phylogenetic significance of the carpometacarpal joint of the thumb in anthropoid primates.*

Ragni, A. J. (2020). Trabecular architecture of the capitate and third metacarpal through ontogeny in chimpanzees (Pan troglodytes) and gorillas (Gorilla Gorilla). *Journal of Human Evolution, 138*, 102702.

Rauch, F. (2005). Bone growth in length and width: The yin and Yang of bone stability. *Journal of Musculoskeletal & Neuronal Interactions, 5*(3), 194–201.

Reich, D., Green, R. E., Kircher, M., Krause, J., Patterson, N., Durand, E. Y., et al. (2010). Genetic history of an archaic hominin group from Denisova Cave in Siberia. *Nature, 468*(7327), 1053–1060.

Reznikov, N., Alsheghri, A. A., Piché, N., Gendron, M., Desrosiers, C., Morozova, I., et al. (2020). Altered topological blueprint of trabecular bone associates with skeletal pathology in humans. *Bone Reports, 12*, 100264.

Richmond, B. G. (2007). Biomechanics of phalangeal curvature. *Journal of Human Evolution, 53*(6), 678–690.

Richmond, B. G., Green, D. J., Lague, M. R., Chirchir, H., Behrensmeyer, A. K., Bobe, R., et al. (2020). The upper limb of Paranthropus Boisei from Ileret, Kenya. *Journal of Human Evolution, 141*, 102727.

Richmond, B. G., Roach, N. T., & Ostrofsky, K. R. (2016). Evolution of the early hominin hand. In T. L. Kivell, P. Lemelin, B. G. Richmond, & D. Schmitt (Eds.), *The evolution of the primate hand: Anatomical, developmental, functional, and paleontological evidence, developments in primatology: Progress and prospects* (pp. 515–543). New York, NY: Springer.

Ricklan, D. E. (1987). Functional anatomy of the hand of Australopithecus Africanus. *Journal of Human Evolution, 16*(7), 643–664.

Robling, A. G. (2009). Is Bone's response to mechanical signals dominated by muscle forces? *Medicine and Science in Sports and Exercise, 41*(11), 2044–2049.

Rolian, C., & Gordon, A. D. (2013). Reassessing manual proportions in Australopithecus afarensis. *American Journal of Physical Anthropology, 152*(3), 393–406.

Rolian, C., & Gordon, A. D. (2014). Response to Almécija and Alba (2014) - on manual proportions in Australopithecus afarensis. *Journal of Human Evolution, 73*, 93–97.

Rolian, C., Lieberman, D. E., & Zermeno, J. P. (2011). Hand biomechanics during simulated stone tool use. *Journal of Human Evolution, 61*(1), 26–41.

Rossetti, L., Kuntz, L. A., Kunold, E., Schock, J., Müller, K. W., Grabmayr, H., et al. (2017). The microstructure and micromechanics of the Tendon-Bone Insertion. *Nature Materials, 16*(6), 664–670.

Rothschild, B. M., & Rühli, F. J. (2005a). Comparison of arthritis characteristics in lowland Gorilla Gorilla and mountain Gorilla Beringei. *American Journal of Primatology, 66*(3), 205–218.

Rothschild, B. M., & Rühli, F. J. (2005b). Etiology of reactive arthritis in Pan Paniscus, P. Troglodytes Troglodytes, and P. Troglodytes Schweinfurthii. *American Journal of Primatology, 66*(3), 219–231.

Ruff, C. B., & Hayes, W. C. (1983). Cross-sectional geometry of Pecos Pueblo femora and tibiae—A biomechanical investigation: I. method and general patterns of variation. *American Journal of Physical Anthropology, 60*(3), 359–381.

Ruff, C., Holt, B., & Trinkaus, E. (2006). Who's afraid of the big bad Wolff?: "Wolff's law" and bone functional adaptation. *American Journal of Physical Anthropology, 129*(4), 484–498.

Ruff, C. B., & Runestad, J. A. (1992). Primate Limb Bone Structural Adaptations. *Annual Review of Anthropology, 21*, 407–433.

Ryan, T. M., & Shaw, C. N. (2015). Gracility of the modern *Homo sapiens* skeleton is the result of decreased biomechanical loading. *Proceedings of the National Academy of Sciences, 112*(2), 372–377.

Ryzewicz, M., & Wolf, J. M. (2006). Trigger digits: Principles, management, and complications. *The Journal of Hand Surgery, 31*(1), 135–146. https://doi.org/10.1016/j.jhsa.2005.10.013.

Sahle, Y., El Zaatari, S., & White, T. D. (2017). Hominid butchers and biting crocodiles in the African Plio-Pleistocene. *Proceedings of the National Academy of Sciences, 114*(50), 13164–13169.

Samuel, D. S., Nauwelaerts, S., Stevens, J. M. G., & Kivell, T. L. (2018). Hand pressures during arboreal locomotion in captive bonobos (Pan Paniscus). *Journal of Experimental Biology, 221*(8).

Schilling, A.-M., Tofanelli, S., Hublin, J.-J., & Kivell, T. L. (2014). Trabecular bone structure in the primate wrist. *Journal of Morphology, 275*(5), 572–585.

Schlecht, S. H., Martin, C. T., Ochocki, D. N., Nolan, B. T., Wojtys, E. M., & Ashton-Miller, J. A. (2019). Morphology of mouse anterior cruciate ligament-complex changes following exercise during pubertal growth. *Journal of Orthopaedic Research, 37*(9), 1910–1919.

Schrader, S. (2018). *Activity, diet and social practice: Addressing everyday life in human skeletal remains.* Springer.

Semaw, S., Renne, P., Harris, J. W. K., Feibel, C. S., Bernor, R. L., Fesseha, N., et al. (1997). 2.5-million-year-old stone tools from Gona, Ethiopia. *Nature, 385*(6614), 333–336.

Semaw, S., Rogers, M. J., Quade, J., Renne, P. R., Butler, R. F., Domínguez-Rodrigo, M., et al. (2003). 2.6-million-year-old stone tools and associated bones from OGS-6 and OGS-7, Gona, Afar, Ethiopia. *Journal of Human Evolution, 45*(2), 169–177.

Senut, B., Pickford, M., Gommery, D., Mein, P., Cheboi, K., & Coppens, Y. (2001). First hominid from the Miocene (Lukeino formation, Kenya). *Comptes Rendus de l'Académie Des Sciences - Series IIA - Earth and Planetary Science, 332*(2), 137–144.

Shen, G., Ku, T., Cheng, H., Edwards, R., Yuan, Z., & Wang, Q. (2001). High-precision U-series dating of locality 1 at Zhoukoudian, China. *Journal of Human Evolution, 41*, 679–688.

Shrewsbury, M. M., Marzke, M. W., Linscheid, R. L., & Reece, S. P. (2003). Comparative morphology of the Pollical distal phalanx. *American Journal of Physical Anthropology, 121*(1), 30–47.

Simpson, S. W., Levin, N. E., Quade, J., Rogers, M. J., & Semaw, S. (2019). Ardipithecus Ramidus Postcrania from the Gona project area, Afar regional state, Ethiopia. *Journal of Human Evolution, 129*, 1–45.

Skinner, M. M., Stephens, N. B., Tsegai, Z. J., Foote, A. C., Huynh Nguyen, N., Gross, T., et al. (2015). Human-like hand use in Australopithecus Africanus. *Science, 347*(6220), 395–399.

Spagnoletti, N., Visalberghi, E., Ottoni, E., Izar, P., & Fragaszy, D. (2011). Stone tool use by adult wild bearded capuchin monkeys (Cebus Libidinosus). Frequency, efficiency and tool selectivity. *Journal of Human Evolution, 61*(1), 97–107.

Stauber, M., Laurent Rapillard, G., van Lenthe, H., Zysset, P., & Müller, R. (2006). Importance of individual rods and plates in the assessment of bone quality and their contribution to bone stiffness. *Journal of Bone and Mineral Research, 21*(4), 586–595.

Stephens, N. B., Kivell, T. L., Gross, T., Pahr, D. H., Lazenby, R. A., Hublin, J.-J., et al. (2016). Trabecular architecture in the thumb of Pan and Homo: Implications for investigating hand use, loading, and hand preference in the fossil record. *American Journal of Physical Anthropology, 161*(4), 603–619.

Stephens, N. B., Kivell, T. L., Pahr, D. H., Hublin, J.-J., & Skinner, M. M. (2018). Trabecular bone patterning across the human hand. *Journal of Human Evolution, 123*, 1–23.

Stern, J. T., & Susman, R. L. (1983). The locomotor anatomy of Australopithecus afarensis. *American Journal of Physical Anthropology, 60*(3), 279–317.

Stout, D., Semaw, S., Rogers, M. J., & Cauche, D. (2010). Technological variation in the earliest Oldowan from Gona, Afar, Ethiopia. *Journal of Human Evolution, 58*(6), 474–491.

Susman, R. L. (1988). Hand of Paranthropus Robustus from member 1, Swartkrans: Fossil evidence for tool behavior. *Science, 240*(4853), 781–784.

Susman, R. L. (1989). New hominid fossils from the Swartkrans formation (1979–1986 excavations): Postcranial specimens. *American Journal of Physical Anthropology, 79*(4), 451–474.

Susman, R. L. (1994). Fossil evidence for early hominid tool use. *Science, 265*(5178), 1570–1573.

Susman, R. L., & Creel, N. (1979). Functional and morphological affinities of the subadult hand (O.H. 7) from Olduvai Gorge. *American Journal of Physical Anthropology, 51*(3), 311–331.

Susman, R. L., de Ruiter, D., & Brain, C. K. (2001). Recently identified postcranial remains of Paranthropus and early Homo from Swartkrans cave, South Africa. *Journal of Human Evolution, 41*(6), 607–629.

Sutikna, T., Tocheri, M. W., Morwood, M. J., Saptomo, E. W., Awe, R. D., Wasisto, S., et al. (2016). Revised stratigraphy and chronology for Homo floresiensis at Liang Bua in Indonesia. *Nature, 532*(7599), 366–369.

Synek, A., Dunmore, C. J., Kivell, T. L., Skinner, M. M., & Pahr, D. H. (2019a). Inverse Remodelling algorithm identifies habitual manual activities of Primates based on metacarpal bone architecture. *Biomechanics and Modeling in Mechanobiology, 18*(2), 399–410.

Synek, A., Szu-Ching, L., Nauwelaerts, S., Pahr, D. H., & Kivell, T. L. (2020). Metacarpophalangeal joint loads during Bonobo locomotion: Model predictions versus proxies. *Journal of the Royal Society Interface, 17*(164), 20200032.

Synek, A., Szu-Ching, L., Vereecke, E. E., Nauwelaerts, S., Kivell, T. L., & Pahr, D. H. (2019b). Musculoskeletal models of a human and Bonobo finger: Parameter identification and comparison to in vitro experiments. *PeerJ, 7*, e7470.

Tanaka, I. C. H. I. R. O. U. (1998). Social diffusion of modified louse egg-handling techniques during grooming in free-ranging Japanese macaques. *Animal Behaviour, 56*(5), 1229–1236.

Taylor, A. H., Hunt, G. R., Holzhaider, J. C., & Gray, R. D. (2007). Spontaneous Metatool use by new Caledonian crows. *Current Biology, 17*(17), 1504–1507.

Thorpe, S. K. S., Crompton, R. H., Günther, M. M., Ker, R. F., & McNeill Alexander, R. (1999). Dimensions and moment arms of the hind- and forelimb muscles of common chimpanzees (Pan troglodytes). *American Journal of Physical Anthropology, 110*(2), 179–199.

Tocheri, M. W., Marzke, M. W., Liu, D., Bae, M., Jones, G. P., Williams, R. C., et al. (2003). Functional capabilities of modern and fossil hominid hands: Three-dimensional analysis of Trapezia. *American Journal of Physical Anthropology, 122*(2), 101–112.

Tocheri, M. W., Orr, C. M., Jacofsky, M. C., & Marzke, M. W. (2008). The evolutionary history of the hominin hand since the last common ancestor of Pan and Homo. *Journal of Anatomy, 212*(4), 544–562.

Tocheri, M. W., Orr, C. M., Larson, S. G., Sutikna, T., Jatmiko, E. W., Saptomo, R. A., et al. (2007). The primitive wrist of Homo Floresiensis and its implications for hominin evolution. *Science, 317*(5845), 1743–1745.

Toth, N., Schick, K., & Semaw, S. (2006). A comparative study of the stone tool-making skills of Pan, Australopithecus, and Homo sapiens. In N. Toth, & K. Schick (Eds.), *The Oldowan: Case studies into the earliest stone age* (pp. 155–222). Stone Age Institute Press.

Tözeren, A. (1999). *Human body dynamics: Classical mechanics and human movement.* Springer Science & Business Media.

Trinkaus, E. (1983). *The Mousterian legacy: Human biocultural change in the upper pleistocene.* B.A.R.

Trinkaus, E. (1989). Olduvai hominid 7 Trapezial metacarpal 1 articular morphology: Contrasts with recent humans. *American Journal of Physical Anthropology, 80*(4), 411–416.

Trinkaus, E., & Long, J. C. (1990). Species attribution of the Swartkrans member 1 first metacarpals: SK 84 and SKX 5020. *American Journal of Physical Anthropology, 83*(4), 419–424.

Trinkaus, E., & Villemeur, I. (1991). Mechanical advantages of the Neandertal thumb in flexion: A test of an hypothesis. *American Journal of Physical Anthropology, 84*(3), 249–260.

Tsegai, Z. J., Kivell, T. L., Thomas Gross, N., Nguyen, H., Pahr, D. H., Smaers, J. B., et al. (2013). Trabecular bone structure correlates with hand posture and use in hominoids. *PLoS One, 8*(11), e78781.

Tsegai, Z. J., Stephens, N. B., Treece, G. M., Skinner, M. M., Kivell, T. L., & Gee, A. H. (2017). Cortical bone mapping: An application to hand and foot bones in hominoids. *Comptes Rendus Palevol, 16*(5), 690–701.

Turker, T., & Thirkannad, S. (2011). Trapezio-metacarpal arthritis: The Price of an opposable thumb! *Indian Journal of Plastic Surgery, 44*(2), 308–316.

Tutin, C. E. G., Boesch, W. C., McGrew, R. W., Wrangham, S., Goodall, R., & Whiten, and Nishida. (2001). Charting cultural variation in chimpanzees. *Behaviour, 138*(11–12), 1481–1516.

Tuttle, R. H. (1969a). Quantitative and functional studies on the hands of the Anthropoidea. I. the Hominoidea. *Journal of Morphology, 128*(3), 309–363.

Urbani, B. (1998). An early report on tool use by Neotropical Primates. *Neotropical Primates, 6*(4), 123–124.

van Leeuwen, T., Vanhoof, M. J. M., Kerkhof, F. D., Stevens, J. M. G., & Vereecke, E. E. (2018). Insights into the musculature of the Bonobo hand. *Journal of Anatomy, 233*(3), 328–340.

van Leeuwen, T., Vanneste, M., Kerkhof, F. D., D'agostino, P., Vanhoof, M. J. M., Stevens, J. M. G., et al. (2019). Mobility and structural constraints of the Bonobo Trapeziometacarpal joint. *Biological Journal of the Linnean Society, 127*(3), 681–693.

Vanhoof, M. J. M., van Leeuwen, T., & Vereecke, E. E. (2020). The forearm and hand musculature of semi-terrestrial rhesus macaques (*Macaca Mulatta*) and arboreal gibbons (Fam. Hylobatidae). Part I. description and comparison of the muscle configuration. *Journal of Anatomy.*

Vauclair, J., Meguerditchian, A., & Hopkins, W. D. (2005). Hand preferences for Unimanual and coordinated bimanual tasks in baboons (Papio Anubis). *Cognitive Brain Research, 25*(1), 210–216.

Vickerton, P., Jarvis, J. C., Gallagher, J. A., Akhtar, R., Sutherland, H., & Jeffery, N. (2014). Morphological and histological adaptation of muscle and bone to loading induced by repetitive activation of muscle. *Proceedings of the Royal Society B: Biological Sciences, 281*(1788), 20140784.

Visalberghi, E. (1990). Tool use in Cebus. *Folia Primatologica, 54*(3–4), 146–154.

Walker, A., Leakey, R., & Leakey, R. E. (1993). *The Nariokotome Homo erectus skeleton.* Harvard University Press.

Wallace, I. J., Loring Burgess, M., & Patel, B. A. (2020). Phalangeal curvature in a chimpanzee raised like a human: Implications for inferring Arboreality in fossil hominins. *Proceedings of the National Academy of Sciences, 117*(21), 11223–11225.

Ward, C. V. (2013). Postural and locomotor adaptations of australopithecus species. In K. E. Reed, J. G. Fleagle, & R. E. Leakey (Eds.), *The paleobiology of australopithecus, vertebrate paleobiology and paleoanthropology* (pp. 235–245). Dordrecht: Springer Netherlands.

Ward, C. V., Kimbel, W. H., Harmon, E. H., & Johanson, D. C. (2012). New postcranial fossils of Australopithecus afarensis from Hadar, Ethiopia (1990–2007). *Journal of Human Evolution, 63*(1), 1–51.

Ward, C. V., Leakey, M. G., & Walker, A. (2001). Morphology of Australopithecus Anamensis from Kanapoi and Allia bay, Kenya. *Journal of Human Evolution, 41*(4), 255–368.

Ward, C. V., Tocheri, M. W., Michael Plavcan, J., Brown, F. H., & Manthi, F. K. (2014). Early Pleistocene third metacarpal from Kenya and the evolution of modern human-like hand morphology. *Proceedings of the National Academy of Sciences, 111*(1), 121–124.

Ward, S. R., Winters, T. M., & Blemker, S. S. (2010). The architectural Design of the Gluteal Muscle Group: Implications for movement and rehabilitation. *Journal of Orthopaedic & Sports Physical Therapy, 40*(2), 95–102.

Weidenreich, F. (1941). The extremity bones of *Sinanthropus pekinensis*. *Palaeontologica Sinica New Series D, 5*, 1–151.

White, T. D., Owen Lovejoy, C., Asfaw, B., Carlson, J. P., & Suwa, G. (2015). Neither chimpanzee nor human, Ardipithecus reveals the surprising ancestry of both. *Proceedings of the National Academy of Sciences, 112*(16), 4877–4884.

Whiten, A., & Boesch, C. (2001). The cultures of chimpanzees. *Scientific American, 284*(1), 60–67.

Whiten, A., Goodall, J., McGrew, W. C., Nishida, T., Reynolds, V., Sugiyama, Y., et al. (1999). Cultures in chimpanzees. *Nature, 399*(6737), 682–685.

Wilczak, C. A., Mariotti, V., Pany-Kucera, D., Villotte, S., & Henderson, C. Y. (2017). Training and Interobserver reliability in qualitative scoring of skeletal samples. *Journal of Archaeological Science: Reports, 11*, 69–79.

Williams, E. M., Gordon, A. D., & Richmond, B. G. (2012). Hand pressure distribution during Oldowan stone tool production. *Journal of Human Evolution, 62*(4), 520–532.

Williams-Hatala, E. M., Hatala, K. G., Gordon, M. K., Key, A., Kasper, M., & Kivell, T. L. (2018). The manual pressures of stone tool behaviors and their implications for the evolution of the human hand. *Journal of Human Evolution, 119*, 14–26.

Williams-Hatala, E. M., Hatala, K. G., Hiles, S., & Rabey, K. N. (2016). Morphology of muscle attachment sites in the modern human hand does not reflect muscle architecture. *Scientific Reports, 6*(1), 28353.

Wong, A. L., Meals, C. G., & Ruff, C. B. (2018). Computed tomographic analysis of the internal structure of the metacarpals and its implications for hand use, pathology, and surgical intervention. *Anatomical Science International, 93*(2), 231–237.

Wood-Jones, F. (1917). *Arboreal man*. Creative Media Partners LLC.

Zumwalt, A. (2006). The effect of endurance exercise on the morphology of muscle attachment sites. *Journal of Experimental Biology, 209*(3), 444–454.

7

Behaviour and the bones of the thorax and spine

Kimberly A. Plomp

School of Archaeology, University of the Philippines, Diliman, Quezon City, Manila, Philippines

7.1 Introduction

Together the spine and thorax house and protect many of the major organs critical for survival. The thorax, the region between your neck and abdomen, encircles the heart and lungs, required for circulation and respiration. The spine protects the spinal cord (technically an organ), which transmits messages from your brain to the rest of your body, making it vital for all bodily functions. The spine also has several important anatomical adaptations that allow us to stand upright and walk on two legs, a locomotor behaviour called bipedalism that is a key defining feature of our lineage. Since protecting vital organs and enabling movement are the ultimate functions of the bones that make up the spine and thorax, any modifications, either evolutionary, biological, or cultural, to these bones can have an immense impact on the function of important organs and the overall skeleton. As such, identifying and interpreting morphology and/or skeletal changes in the bones of the spine and thorax can provide important information on the lives of past peoples, such as cultural practices, violence, medical interventions, and, potentially, daily activities (although this is controversial). Skeletal changes may also tell us about important shifts in hominin evolution that led to anatomically modern humans.

This chapter provides a detailed overview of the types of information that can be gleaned from analysing the bones of the spine and thorax in archaeological and palaeoanthropological skeletal remains, and how this helps bioarchaeologists infer behaviour in the past. It starts with a brief outline of the human bones in this region and moves on to discuss how understanding why these bones have the morphology and characteristics they do is important for our locomotion and understanding human evolution. Next, it focuses on bioarchaeological analyses of skeletal indicators of behaviour that we can interpret to help us understand the lives of past peoples. It does so by covering various topics organised into three sections—corseting and binding, activity-related palaeopathology of the spine, and violence and ritual behaviour. This is not an exhaustive list of what bioarchaeologists can identify in the thorax and spine of human remains, but it provides insight into the types of approaches and hypotheses developed in bioarchaeology to try to infer behaviour in the past from skeletal remains.

7.2 The human spine

The spine consists of individual vertebrae that articulate with each other to create the vertebral column. The typical number of vertebrae in a spine varies between species. This number is known as the vertebral formula and reflects the number of vertebrae in each region of the spine. In humans, the vertebral formula generally includes 7 cervical vertebrae (all mammals except sloths and manatees have 7 cervical vertebrae, even giraffes!), 12 thoracic, 5 lumbar, 5 sacral, and 3–5 coccygeal vertebrae. Each human pre-sacral vertebra (i.e. cervical, thoracic, and lumbar), with the exception of the first two cervical vertebrae, has somewhat similar anatomy, although there are differences between regions. Specifically, they have one vertebral body, two laminae, two pedicles, two transverse processes, one spinous process, and four zygapophyseal facets (two superior and two inferior) (Fig. 7.1) (Been et al., 2010; El-Khoury and Whitten, 1993; Latimer and Ward, 1993; Russo, 2010; Sanders and Bodenbender, 1994; Shapiro, 1993; Williams and Russo, 2015). The zygapophyseal joints are synovial joints (i.e. moveable) that articulate each vertebra to their inferiorly or superiorly adjoining vertebrae (Latimer and Ward, 1993; Russo, 2010; Shapiro, 1993; Williams and Russo, 2015). All pre-sacral vertebrae, again except for the first two cervical vertebrae, also articulate to one another at the vertebral bodies through fibrocartilaginous joints created with the intervertebral discs (Taylor, 1975). The discs allow for some movement and flexibility in the spine and withstand the compressive loading associated with holding the body upright (Davis, 1961; Hernandez et al., 2009; Latimer and Ward, 1993; Rose, 1975; Shapiro, 1991, 1993). The thoracic vertebrae also have synovial joints that articulate with the ribs, called costovertebral joints.

As noted, the first two cervical vertebrae differ from the others in their morphology. The first cervical vertebra is called the atlas, a name which stems from the Titan of Greek mythology who carried the earth on his shoulders. The vertebra was given this name because it articulates with the occipital condyles on the base of the cranium, essentially 'carrying' the

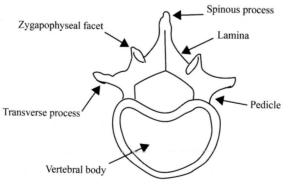

FIG. 7.1   A typical human thoracic or lumbar vertebra.

skull, as Atlas does the earth. The second cervical vertebra is called the axis because it has a pillar-like protuberance called the dens (or odontoid process) that articulates with the atlas through a pivot synovial joint called the median atlantoaxial joint, which allows the atlas to rotate, like on an axis. The morphology of the atlas and dens enables cervical rotation of the head (e.g. moving your head from side to side).

7.3 The human thorax

The bones of the human thorax include the thoracic vertebrae, ribs, and the sternum. The ribs articulate to the thoracic vertebrae, with two ribs (one right and one left) for each vertebra. Thus humans typically have 24 ribs, although, congenitally, the number can vary (for more on supernumerary ribs, see Aly et al., 2016; Bots et al., 2011; Eaves-Johnson, 2010). Of the 24 ribs, 18 articulate to cartilage at their anterior/ventral end, which acts as a bridge articulating the ribs to the sternum. The lower six ribs are called 'floating ribs' because they do not attach to the sternum (Jellema et al., 1993). The shape of the ribcage has been described as 'barrel' shaped in humans and gibbons, compared to the 'funnel' shape of the ribcage in other non-human apes (Bastir et al., 2013). The sternum comprises three individual bones: the manubrium, sternal body, and xiphoid process, which can fuse together in some older individuals. The manubrium connects the thorax to the shoulder girdle through its articulation with the clavicles. The manubrium also articulates with the first ribs, while the remaining non-floating ribs articulate to the sternal body through the costal cartilage (Bastir et al., 2013).

7.4 Locomotion

One of the defining features of our lineage, the hominins, is our unique form of obligate bipedalism. Humans, and all hominins that came before us, walk upright on two legs. We carry our centre of gravity over our hips and knees, unlike other bipedal animals such as birds or kangaroos, whose centre of gravity pulls their chests forward and down. This unique form of bipedalism is possible in humans because of a number of functional and morphological adaptations throughout our skeleton, from the shape of our cranium right down to our feet. This chapter focuses on the spine and thorax, and as such, the evolutionary adaptations of the vertebrae, ribs, and sternum related to bipedalism. In this section, we will focus on adaptations of the spine because its function is the most well understood to date.

The shape of the human spine is unique and has many adaptations related to bipedalism. Our spines have a sinuous shape, which is created through the curvatures of the four regions of the spinal column (Fig. 7.2) (Abitbol, 1995; Been et al., 2010, 2017; Keith, 1923; Schultz, 1961; Shapiro, 1993; Ward and Latimer, 2005; Whitcome et al., 2007). Starting at the neck and moving down, the cervical spine has a lordotic curve, which means that it curves forward towards the anterior of the body; the thoracic region has the opposite curvature shape, known as kyphotic curve; and the lumbar spine again moves to a lordotic curvature. These curvatures are a result of changes in the shapes of the intervertebral discs and vertebral bodies. In the cervical spine, the intervertebral discs are dorsally wedged

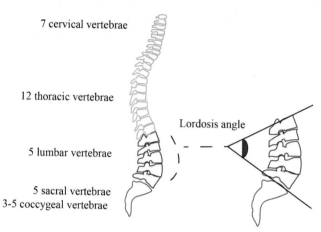

FIG. 7.2 The five regions of the human spine that create the curves unique to the human spine, including the lumbar lordotic angle.

(i.e. the discs are craniocaudally shorter on the dorsal end than the ventral end), which causes the lordosis (Been et al., 2010), while the thoracic kyphosis is created through the ventral wedging of the vertebral bodies (i.e. the thoracic vertebrae are craniocaudally shorter on the ventral end than the dorsal end) (Latimer and Ward, 1993). The lumbar spine is similar to the cervical region except that its lordosis is created through dorsal wedging of both the intervertebral discs and the vertebral bodies (i.e. the lumbar vertebrae and discs are craniocaudally shorter on the dorsal end than the ventral end) (Been et al., 2010). As for the sacrum and coccyx, they typically have a kyphotic curve that is amplified by the ventral tilting of the sacral promontory, that is the cranial end of the sacrum (Been et al., 2019).

The curves of the spine, especially the lumbar lordosis, bring the body's centre of gravity over the hips, which allows us to balance our upper body over our legs and hips during bipedal locomotion (Latimer and Ward, 1993; Whitcome et al., 2007). In contrast, a chimpanzee, for example, has a 'C' shaped spine rather than a sinuous shaped spine and carries its centre of gravity more forward in the thorax, pulling its body ventrally and making it difficult to stand upright for long periods of time. To reiterate, the curves in the human spine are created through wedging of the vertebral bodies and/or intervertebral discs, attributes that are morphologically identifiable and distinguishable among humans demonstrating obligate bipedalism compared to other facultatively biped non-human primates. In addition to wedging of their bodies, a number of other traits distinguish human vertebrae from those of non-human apes, and thus, possibly relate to bipedalism. Specifically, compared to non-human apes, all healthy human vertebrae tend to have craniocaudally taller bodies, mediolaterally wider vertebral foramina, dorsoventrally longer laminae, and shorter spinous processes with flatter tips (see Fig. 7.1 for vertebral elements) (Gómez-Olivencia et al., 2013; Hernandez et al., 2009; Latimer and Ward, 1993; Plomp et al., 2019; Sanders and Bodenbender, 1994; Ward, 1991). There are also many other traits that are unique to individual vertebrae in the human spine, a list of which, along with their potential functions, can be found in Plomp et al. (2019).

In fossil vertebrae, the presence of these traits, including the spinal curves, has been used to argue whether or not an extinct taxon was capable of bipedalism. This question is important in anthropology because identifying the oldest hominoid taxon capable of habitual bipedalism will provide a wealth of information about human evolution, such as when and where bipedalism evolved (allowing it to be connected to specific environments and pressures), what locomotor behaviour preceded it, and how the last common ancestor of humans and non-human great apes looked and acted.

Recently, researchers have used vertebral traits to argue that a Miocene ape taxon may have used habitual bipedalism. Böhme et al. (2019) analysed cervical and thoracic vertebrae, along with other skeletal elements, of a Miocene ape, *Danuvius guggenmosi*, and suggested that this ape possessed adaptive traits indicating that it was capable of a form of bipedalism, although not one identical to our own (Böhme et al., 2019). Specifically, they argued that the *D. guggenmosi* specimen exhibits a thorax that was broad on the mediolateral plane, similar to modern humans and gibbons, and had an orthograde (i.e. upright) spine. Böhme et al. (2019) based this interpretation on a combination of traits, including a low costal facet angle on the first thoracic vertebra and a dorsal orientation of the thoracic transverse processes, which is also found in modern human vertebrae (Plomp et al., 2019). In addition, they posited that since the spinous processes of the cervical and thoracic vertebrae changed in inclination from the upper spine to the lower thoracic spine, *D. guggenmosi* would have had both the cervical lordosis and a thoracic kyphosis seen in later bipedal hominins, including modern humans (Böhme et al., 2019). Lastly, they argued that the diaphragmatic thoracic vertebrae, which is the vertebra that transitions from the thoracic to the lumbar spine (i.e. has thoracic shaped superior zygapophyseal facets and lumbar shaped inferior zygapophyseal facets), was not the ultimate, or last, thoracic vertebrae in the *D. guggenmosi* spine. In modern humans, the diaphragmatic thoracic vertebrae is typically also the ultimate thoracic vertebrae, and if there were additional rib-bearing thoracic vertebrae lower than the diaphragmatic vertebra in *D. guggenmosi*, this would indicate that it had a longer lower spine (i.e. more vertebrae) than either modern humans or non-human apes. Böhme et al. (2019) argued that this longer lower spine would allow *D. guggenmosi* to carry its centre of gravity over its hips and knees, which is required for bipedalism. The authors considered these traits in combination with adaptations identified in other skeletal elements, such as the proximal tibia, that suggests *D. guggenmosi* had an extended hip and knees posture like later hominins (compared to bent hips and knees in non-human apes), to argue that *D. guggenmosi* would have been capable of habitual bipedal posture, along with arboreal suspension (similar to modern non-human apes) (Böhme et al., 2019). However, this interpretation has been contested (Williams et al., 2020) and debated (Böhme et al., 2020).

Regardless of the outcome of this argument, the *D. guggenmosi* example illustrates how important understanding the relationship between skeletal morphology and functional anatomy is for palaeoanthropology, and how skeletal traits in the spine and thorax can provide not only information on the locomotor behaviour of a long extinct ape, but also the evolution of human locomotion. If one day it is accepted that *D. guggenmosi* used habitual bipedalism along with arboreal locomotion, this would significantly push back the currently accepted timeline for the evolution of bipedalism in the hominoid lineage.

Building on the evolutionary evidence for early bipedalism, morphological studies have also found evidence that subtle shape variations in the human vertebrae can have a major impact on the health of the spine. Humans have far more spinal disease than any other primate (Filler, 2007; Jurmain, 1989; Lovell, 1990; Lowenstine et al., 2016) and for decades, scholars have suggested this is due to the stress placed on our spines when we move on two feet (Filler, 2007; Jurmain, 2000; Keith, 1923; Latimer, 2005; Merbs, 1996a; Plomp et al., 2015a,b). Based on this hypothesis, Plomp and colleagues used bioarchaeological and documented human remains to investigate the potential relationship between bipedalism and spinal health by analysing and comparing vertebral shape between humans with and without spinal pathologies to vertebrae of non-human apes and fossil hominins. They found that people with certain vertebral shape traits tend to have one of two spinal lesions, Schmorl's nodes or spondylolysis. Furthermore, individuals with Schmorl's nodes tended to have vertebrae that shared more shape similarities with vertebrae of chimpanzees and fossil hominins than did human vertebrae without Schmorl's nodes (Plomp et al., 2015a,b, 2019). They also found that people with spondylolysis tended to have vertebrae that shared fewer shape traits with chimpanzees than did healthy human vertebrae (Plomp et al., 2020). They interpreted the findings of these studies to indicate that human vertebral shape variation can be described as an evolutionary spectrum, with one end having ancestral shape traits that are closer to those seen in chimpanzees and fossil hominins and the opposite end having shape traits that are hyper-derived adaptations for bipedalism (Fig. 7.3) (Plomp et al., 2015b, 2019, 2020, 2022). They based this interpretation on the fact that humans and chimpanzees share a relatively recent common ancestor but do not share locomotor repertoires, and thus, any shape traits shared between them are likely plesiomorphic (ancestral) and those that differ are derived. Depending on where your vertebral shape sits on the evolutionary shape spectrum could

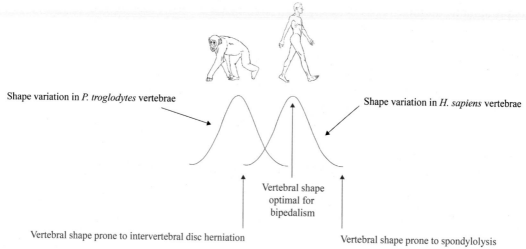

Shape variation in *P. troglodytes* vertebrae

Shape variation in *H. sapiens* vertebrae

Vertebral shape optimal for bipedalism

Vertebral shape prone to intervertebral disc herniation

Vertebral shape prone to spondylolysis

FIG. 7.3 An illustration of the Evolutionary Shape Hypothesis, where human vertebral shape can be visualised as a spectrum with one end being 'ancestral' and shares shape traits with chimpanzees, and the other end has hyper-derived adaptations for bipedalism. Where a person's vertebrae sit on this spectrum may influence their propensity to develop Schmorl's nodes or spondylolysis. *From Plomp, K. A., et al. (2022). Palaeopathology and evolutionary medicine: An integrated approach. Oxford, UK: Oxford University Press.*

have a significant impact on your spinal health, with individuals who have vertebrae on the ancestral end being more prone to Schmorl's nodes and those on the derived end, more prone to spondylolysis (Plomp et al., 2015a, 2019, 2020, 2022).

This section highlights the wealth of important information that can be gained from understanding, analysing, and comparing the skeletal variation of the human thorax and spine. Namely, the study by (Böhme et al., 2019) may enlighten us about details on the timing and history of the evolution of bipedalism in the hominoid lineage, while the research by Plomp et al. (2015a,b, 2019, 2020, 2022) can inform us of how and why humans continue to experience repercussions related to these evolutionary adaptations. This type of work helps biological anthropologists continue to consider how even subtle variations in human anatomy and morphology can help us better understand human evolution, as well as potential 'trade-offs' related to our evolutionary history.

7.5 Activity-related palaeopathology of the spine

As demonstrated in the previous section, palaeopathological investigations can provide important insight into human movements and experiences. The previous section outlines how palaeopathology has helped us understand evolutionary trade-offs related to locomotion. This section builds on this to discuss how palaeopathology has been used to attempt to help biological anthropologists better understand human activities. Attempting to infer activity, and specifically, subsistence-based activities, in archaeological populations has been a major research focus in bioarchaeology for decades. This is because the ability to identify not only whether certain activities, such as those related to subsistence, were performed in past populations, but also patterns that may indicate gendered or social differences in activities would provide an immense amount of valuable insight into life in the past. Thus, over the years, many bioarchaeologists have interpreted different frequency rates and distribution patterns of spondylosis, zygapophyseal osteoarthritis, and Schmorl's nodes, often in tandem with appendicular osteoarthritis, within and between skeletal populations to suggest gendered division of labour and/or occupational differences between communities separated by space and/or time (a small selection of papers: Bridges, 1994; Eng, 2016; Klaus et al., 2009; Lovell, 1994; Pickering, 1979; Stirland and Waldron, 1997). One example out of many such studies is was performed by Zhang et al. (2017) who analysed the occurrence of osteoarthritis (appendicular and zygapophyseal), Schmorl's nodes, and spondylosis in 167 skeletons from two archaeological sites (Xiaomintunn and Xin'anzhuang) in Yinxu, China, dated to the Late Shang Dynasty (1250–1046 BCE). They compared the frequency and distribution patterns of lesions between the two cemetery populations and between sexes. The authors found that individuals from one site, Xiaomintunn, exhibited more overall osteoarthritis than did the individuals from Xin'anzhuang, and that males of both populations were more frequently afflicted with all lesions in their upper body than were females. They interpreted these patterns to indicate that both sites had a gendered division of labour, with males tending to perform activities that required repetitive lifting and carrying heavy loads more often than females, and that more physically intensive activities were being performed at Xiaomintunn than Xin'anzhuang.

There is an important consideration to make here. While there is a relationship between repetitive activities and loading to the breakdown of joint components leading to osteoarthritis, Schmorl's nodes, and spondylosis, there are a number of other factors that need to be considered when using these pathologies to infer behaviour in the past. All three conditions are also highly correlated with increasing age and a number of other aetiological factors, including genetics, biochemistry, skeletal morphology, body weight, and diet (Dar et al., 2009, 2010; Herrero-Beaumont et al., 2009; Middleton and Fish, 2009; Sarzi-Puttini et al., 2005; Spector et al., 1996; Spector and MacGregor, 2004; Weiss and Jurmain, 2007). Additionally, there is evidence to suggest that the development of Schmorl's nodes may be highly influenced by vertebral shape and bipedalism (Plomp et al., 2015 a,b, 2019; see also Section 7.4). Thus, although Zhang et al. (2017) and the other studies referenced earlier have all used the presence of Schmorl's nodes, osteoarthritis, and/or spondylosis to infer that past individuals performed specific activities, the multifactorial aetiologies of these conditions confound their conclusions. Namely, since there can be many reasons aside from activity that lead to the development of these conditions, it is nearly impossible, at least to date, to distinguish patterns of differing frequency and distribution related to variations in activity and behaviour and those related to different gene pools, diets, and skeletal morphology. This not only makes it difficult to infer that people were performing specific activities, such as carrying heavy loads (e.g. Zhang et al., 2017), but also to say that general differences in frequencies or trends of pathologies within and between populations indicate levels of activity, divisions of labour, or differing occupations.

Spondylolysis, which is a cleft in the neural arch at the pars interarticularis, typically located in the lower lumbar vertebrae (Hu et al., 2008), is more reliably associated with activity than osteoarthritis or Schmorl's nodes and can also be identified in the archaeological record. Although its aetiology has been debated, it is now generally accepted to be a result of a fatigue fracture that occurs when the inferior articular process of a vertebra repeatedly impacts the pars interarticularis of the vertebrae below it (Hu et al., 2008; Mays, 2007). In fact, spondylolysis only occurs in humans and does not naturally develop in other animals, leading Merbs (1996a) to suggest that it is related to our bipedal locomotion. Modern clinical research has identified that spondylolysis is more likely to occur in young individuals who undergo extreme spinal extension, such as gymnasts (Soler and Calderón, 2000) and other athletes (Iwamoto et al., 2004), with one study finding that up to 47% of adolescent athletes who suffer from low back pain had spondylolysis (Micheli and Wood, 1995).

Considering the clinical association with spondylolysis and spinal movement and activity, the bioarchaeological studies that use the presence of the lesion to infer behaviour in past populations make compelling cases. For example, in his classic paper, Merbs (1996b) found that, in a sample of 373 individuals from archaeological Inuit populations from Alaska and Canada, 16 individuals had spondylolysis of the sacrum. Fifteen of these individuals were males and 14 were young adults. The author interpreted this frequency pattern to indicate that these young men were likely routinely participating in strenuous activities such as weightlifting, kayaking, harpooning, and wrestling. Similarly, Arriaza (1997) found an unusually high prevalence of spondylolysis in the lumbar spines of a burial population from Hyatt Site, Tumon Bay, Guam, dated to between 1200 and 1521 CE. Specifically, he found that of the 38 individuals with preserved spines, 8 (21%) had lumbar spondylolysis, 5 of whom were male and 3 female. Arriaza (1997) argued that the high prevalence of spondylolysis in this population could be explained by the way they constructed their homes. During this period, houses were built out of large stone pillars, and Arriaza proposed that moving these large stones from one site to another may have induced abnormal extension and torque in the lower spine, leading to microfractures developing on the pars interarticularis and eventually resulting in spondylolysis.

Although the link between activity and spinal movement with spondylolysis seems clear cut, it is more complicated than previously thought. Recent studies have found a strong correlation between vertebral morphology and the presence of spondylolysis. In particular, a clinical study undertaken by Roussouly et al. (2005) found that people with spondylolysis tended to have lumbar spines with larger lordotic angles (more pronounced lordotic curve) than those without the lesion. This corresponds with the later findings of Masharawi (2012), who found that individuals with spondylolysis tend to have final lumbar vertebrae with bodies with a greater degree of dorsal wedging, which would create larger lordotic angles. Using archaeological populations, Ward and Latimer (2005), Ward et al. (2007, 2010), and Plomp et al. (2020) found that other vertebral traits were also correlated with the presence of spondylolysis. All traits associated with spondylolysis can be argued to be evolutionary adaptations for bipedalism, which supports Merbs's (1996a) earlier claim that spondylolysis may be a consequence of walking on two legs. Overall, based on these studies, it is generally accepted now that spondylolysis is more likely to develop in people with vertebrae with hyper-derived shape traits related to bipedalism because they increase the direct contact between the inferior zygapophyseal facets of the superior vertebrae with the pars interarticularis of the inferior vertebrae, thus leading to a culmination of fatigue fractures (Masharawi et al., 2007; Plomp et al., 2020, 2022; Roussouly et al., 2005; Ward and Latimer, 2005; Ward et al., 2007, 2010).

The relationship between vertebral morphology, bipedalism, and spondylolysis does not refute the arguments that it is linked with activity and particular spinal movements but does add a caveat to the argument in that those activities may only increase the chances of developing spondylolysis when someone has lumbar vertebrae with those shape traits, or that these individuals may develop spondylolysis even without performing activities that require extensive spinal extension. The same can be said for the relationship between vertebral morphology and Schmorl's nodes discussed earlier (Plomp et al., 2015a,b, 2019). Ultimately, the main implication of these studies for using the presence of spondylolysis or other spinal pathologies to infer behaviour in the past is that one should consider other factors, such as individual vertebral morphology, before interpretations are made.

7.6 Corsetry and binding

Humans are similar to many other animals in that there are certain ways we signify and infer group affiliation and perceive health and evolutionary fitness. One way that humans do this is through clothing and accessories, where particular styles can indicate social status, age group, profession, and economic standing. What is signified by certain styles is culture dependent and changes over time, a phenomenon known as fashion. In bioarchaeology, we can only infer fashion in the past when associated materials, such as clothing, fabrics, jewellery, and other accessories, are preserved, or when said fashion leaves traces directly on human remains. Examples of this can be tattoos, as those seen on bog bodies and Ötzi and other mummies (Deter-Wolf et al., 2016; Friedman et al., 2018; Pabst et al., 2010), as well as when the shape of skeletal elements is altered due to cultural practices, or in the name of fashion, such as with some examples of intentional cranial modification (Pomeroy et al., 2010; Torres-Rouff, 2020) and foot binding (Berger et al., 2019; Zhao et al., 2020).

In bioarchaeology, we can find signs of a very popular fashion trend that swept through post-Medieval Europe and persists in some forms today—corseting. Between the 17th and 20th centuries, women and men began wearing corsets to form their torsos into 'desirable' shapes or to attempt to treat certain health conditions (Gibson, 2015, 2021; Stone, 2012). Corsets can be made of a variety of materials and can vary in stiffness and tightness, depending on the desired look, and can be worn as an undergarment, part of a dress or top, or as the main garment in an outfit (Bendall, 2014; Gibson, 2015). In the 1800s, the body, especially female, was corseted into an unnatural shape that was considered 'civilised' at the time, and in some circles today is still considered 'feminine' and 'sexy' (Gibson, 2015). In the extreme, however, where the corset synches in the waist in an exaggerated way, it can cause severe deformation to both the skeleton and internal organs (Gibson, 2015, 2021; Stone, 2012, 2020a,b). In the Victorian age, fashionable corsets were often incredibly restrictive of the human waist. For example, Gibson (2015) analysed collections of Victorian corsets housed at museums in London and Paris and determined that, on average, the corsets reduced women's waist circumference to 56.33 cm. In 1951 the average female waist in the UK was 70 cm (Gibson, 2015). This modification in waist size puts so much pressure on the thorax that the thoracic vertebrae, ribs, and sternum can all show temporary and permanent deformation in shape, depending on the duration and habit of binding.

The process of corseting acts to compress and confine the structures of the human torso. This compression places stress on the skeletal structures, which eventually will exhibit plastic deformation due to the strain (Zimmermann and Ritchie, 2015). When deformation related to corseting is sustained, it will modify the body into an unnatural shape, ultimately changing the shape of the underlying skeletal structures in ways that can be recognised by bioarchaeologists (Gibson, 2015; Stone, 2012, 2020a). If a restrictive corset was routinely placed on a child, this would restrict the growth of the bones of the thorax and result in deformation of the bones during skeletal development. Conversely, if corseting started during adulthood, after skeletal development was completed, the bones of the thorax could be altered through plastic responses to pressures acting on the bones. According to the principles in Young's modulus, bone bends to stresses applied, deforming when the strain is too great, and eventually will fail and fracture (Turner, 2006). This is likely the pattern observed in the corseted ribs of adults, whereby this long application of strain results in sustained deformation and may even cause fractured ribs (Brickley and Mays, 2019).

The way that the skeleton deforms varies depending on the shape of the corset itself, but there are several key deformations that are often associated with most restrictive corseting of the thorax and waist. Specifically, we see flattening and deviation at the spinal processes

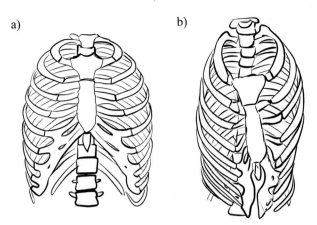

FIG. 7.4 (A) A typical human thorax and ribcage and (B) a human thorax that has been routinely corseted throughout life.

of thoracic vertebrae, stronger inward angulation at the rib angle, depression at the xiphoid process, and in some cases, inferior folding of the iliac crests (Fig. 7.4) (Gibson, 2015, 2021; Stone, 2012, 2020a). The pressure on the posterior elements of the spine causes the spinous processes of the thoracic vertebrae to be deformed in a way that they become more caudally and laterally oriented than those in non-corseted spines. Generally, healthy spinous processes project from the vertebra at a 45° angle, while Gibson (2015) found that corseted spines have spinous processes that project anywhere from a 10° to 30° angle. This causes enough pressure that one can lose the kyphotic curve in their thoracic spine, leading it to straighten (Gibson, 2015). Furthermore, the change in spinous process ordination results in a substantial decrease in the spacing between the processes, which is important for spinal movement and stability (Latimer and Ward, 1993; Shapiro, 1993), and can result in pseudo-facets forming where one spinous process directly contacts another. In addition, tight corsets place pressure on the ribs and sternum, which results in a deformation of the shape of the ribcage. Typically, a human ribcage is barrel shaped and is wider in their coronal plane than their sagittal plane. Also, healthy ribs are 'C' shaped and extend from the thoracic vertebrae in a lateral and caudal direction until the sternal ends articulates with the costocartilage, anteriorly. The sternum then projects out and upwards, creating space in the inferior thorax to house vital organs. As described by Gibson (2015, 2021), when restricted through habitual corset wearing, the ribcage becomes radially compressed and the ratio between the coronal and sagittal dimensions decreases, creating a more circular ribcage. The ribs lose their typical 'C' shape and no longer project inferiorly, but instead project anteriorly from the thorax. The sternum is compressed into the thorax cavity, and the floating ribs are pushed both up and inwards (Fig. 7.4) (Gibson, 2015, 2021). Also, depending on the shape of the corset, the pelvis can also be impacted, which can lead to difficulties during childbirth and increased maternal and infant deaths (Stone, 2009, 2012; Stone and Walrath, 2006). In Chapter 8 of this book, Decrausaz and Laudicina describe how corseting impacted the shape of the female pelvis and the implications this had for the health and longevity of people in the past.

Although restrictive corsetry for fashion is often viewed, rightfully so, as social violence against women (Stone, 2012, 2020a,b), the practice has also been used as a form of medical intervention and treatment (Ponseti, 1991; Ueyoshi and Shima, 1985). For centuries, variations of corsets, braces, and binding have been used to treat, albeit generally unsuccessfully, many spinal disorders, including spondylosis, spondylolisthesis, ankylosing spondylitis, and pain related to intervertebral disc degeneration (Turner, 1913; Ueyoshi and Shima, 1985). Such practices can sometimes leave the same traces on the skeleton as those made by fashionable corsets. For example, Moore and Buckberry (2016) investigated a case of an adult male from a 19th century cemetery population from Wolverhampton, UK, who exhibited lesions consistent with a chronic tuberculosis infection that affected the spine, called Pott's disease, as well as evidence suggesting that a corset was commonly worn. When a tuberculosis infection spreads to the spine, it can cause resorption and collapse of the vertebral bodies that result in a destabilisation of the spine and eventual pathological kyphosis (e.g. more pronounced than the healthy kyphotic curve) of the thoracic spine. This individual also had flatted spinous processes and abnormal angulation of the ribs, which correspond to skeletal changes associated with corseting (Gibson, 2021), leading the authors to suggest that an orthopaedic corset was used to try to correct the abnormal curvature of the spine caused by Pott's disease (Moore and Buckberry, 2016).

Another example of how skeletal changes can help bioarchaeologists infer medical interventions or treatments using restrictive materials was presented by Groves et al. (2003). They analysed the skeletal remains of a young woman found during excavations of Ripon Cathedral in North Yorkshire, UK, dated to the late 15th century. The authors described abnormal skeletal changes in the thorax, including flattening of the thoracic spinous processes, anterior bowing of the sternum, and abnormal curvature of the ribs with them being less laterally curved and more anteriorly projecting, resulting in a narrower ribcage. The authors provide differential diagnoses that include fashionable corset wearing and rickets but conclude that the best explanation for the combination of bone abnormalities is a developmental condition called *Pectus carinatum*. The aetiology of this condition is still unknown, but it has been suggested to result from premature fusion of the sternal ossification centres and hypergrowth of the costal cartilage, causing the ribs to 'buckle' and push the sternum forward. The condition is also known as pigeon or keel chest due to the narrow, forward projecting chest (Groves et al., 2003; Stevenson et al., 1993). However, Groves et al. (2003) note that *Pectus carinatum* would not explain the flattening of the spinous processes in the thoracic vertebrae and that this deformation could indicate an attempted treatment of the deformity. Since flattened spinous processes are seen when restrictive corsets are worn (Gibson, 2021), the authors suggest that a brace or restrictive strap may have been used to compress the chest so that it would eventually change into a more average shape. Thus, in this case study, the identification and description of bone changes in the thorax not only resulted in a diagnosis of a relatively rare developmental disorder, *Pectus carinatum*, but also in the identification and recognition of attempted treatment and medical intervention for the condition in the past (Groves et al., 2003).

The identification of such deformations in the thorax can help bioarchaeologists determine not only whether someone in the past typically wore restrictive clothing, or binding due to cultural practices or for health reasons, but when contextualised within one or more cemetery populations, this can provide unique insight into the social perceptions and behaviours surrounding corseting or other binding practices at the time. For example, one could analyse the skeletal remains of different cemetery populations through time and map if the prevalence of corset wearing or thorax binding changed with time, location, age, gender, and diseases, as well as identify trends in the materials used and their shapes. At the moment, this line of inquiry has only received a somewhat limited focus in bioarchaeology (Gibson, 2015, 2017, 2021; Stone, 2012, 2020a,b; Wescott et al., 2010), but as the discipline grows and we strengthen our understanding of the relationships between skeletal remains, biocultural remains, and social cultural behaviour, we could develop a reliable framework to identify and analyse such deformations in archaeological skeletons.

7.7 Violence and ritual behaviour

Identifying and understanding ritual and similar cultural practices has been a major focus of archaeological inquiry since the field began. Fortunately for this chapter, ritual behaviour can leave identifiable evidence on the spine and thorax of skeletons exhumed from archaeological contexts (e.g. Klaus et al., 2010; Selinsky, 2015; Tiesler and Olivier, 2020; Toyne, 2011). Interpersonal violence is not only related to ritual or warfare. Unfortunately, domestic violence and abuse have plagued human societies throughout history (Gaither, 2012; Gaither and Murphy, 2012; Wheeler et al., 2013). Based on clinical and forensic research, bioarchaeologists have identified specific patterns of trauma that can indicate interpersonal abuse in the past, such as blunt force trauma focused on the head, neck, and ribcage (Gowland, 2016; Redfern, 2017). Typically, this evidence is related to sharp and/or blunt force trauma and analysing patterns and distribution of such trauma can provide a wealth of information about both societal practices and the quality of life of people in the past. It is important to note that generally, we would consider the entirety of the human body for assessments of violence, but evidence of trauma on the thorax and skull has been used to infer particular modes of interpersonal violence, as will be discussed later. Also, many instances of trauma to the thorax will only affect the soft tissue and so will not be identifiable in the skeleton, making our estimation of this trauma likely lower than the actual prevalence. Investigating evidence of interpersonal trauma is such a recurring theme in bioarchaeology that this section can only discuss a portion of the studies published to date, and so will provide a brief overview to illustrate how such evidence on the thorax has been used in bioarchaeology to understand violence and ritual behaviour in the past.

Rib fractures are among the most commonly identified trauma in human skeletal remains (Jurmain, 1999; Lovell, 1997; Matos, 2009), and their occurrence has been suggested to potentially indicate strenuous activities (Brickley, 2006) or interpersonal violence (Lovell, 1997; Ross and Juarez, 2016). Barsness et al. (2003) undertook a review of clinical reporting of rib fractures in children and found that they were commonly the only evidence of abuse and that their presence could predict the occurrence of child abuse with an accuracy of 95%. Obviously it is critical to consider context when leaning towards an interpretation of abuse based on rib fractures, but if there is more than one fracture at different stages of healing in one individual, this strongly suggests ongoing interpersonal violence, likely domestic in origin (Barsness et al., 2003; Walker et al., 1997). This reasoning has been used in bioarchaeology to suggest possible cases of child abuse. In one such study, Tomasto-Cagigao (2009) analysed 198 skeletons from sites in the Nazca Valley of Peru dated from 3800 BCE to 900 CE. The sample included 54 juveniles, with two showing evidence of potential child abuse. One infant aged 6–12 months had a femoral fracture in the process of healing, and another child aged 8–16 months exhibited five rib fractures with evidence of ongoing healing at two different stages. Rib fractures rarely occur accidently at that age due to the flexibility of developing bone (Barsness et al., 2003; Walker et al., 1997), and thus, the presence of five fractures showing different stages of healing strongly suggest that this child was a victim of abuse (Tomasto-Cagigao, 2009). Although the ability to recognise abuse and domestic violence can help bioarchaeological interpretations of violence and social unrest in past populations, it is far more important in forensic anthropology, where identifying victims of abuse or episodes of acute or chronic abuse can have immeasurable importance in a forensic case (Bilo et al., 2010; Marks et al., 2009).

Evidence of sharp force trauma, such as cut marks, on the cervical vertebrae, clavicles, first two ribs, and the sternum has been used to infer human sacrifice in past cultures (Klaus et al., 2010; Toyne, 2011), and although this evidence is, in nature, indirect, it does often make a convincing argument for ritual behaviour. For example, Klaus et al. (2010) identified patterns of perimortem cut marks on the upper cervical vertebrae, clavicles, ribs, and sternum on 22 out of 29 individuals at archaeological site in Cerro Cerrillos, Peru, dated to between 900 and 1100 CE. Based on the pattern of cut marks identified on the remains, the authors interpreted that these 22 individuals were victims of ritual sacrifice (Klaus et al., 2010). Similarly, Toyne (2011) examined the skeletal remains of adult males and juveniles who had been interred at the Inca site of the Temple of the Sacred Stone at Túcume, Peru, dated to 1100–1532 CE. Of these individuals, 93% had evidence of perimortem cut marks on their cervical vertebrae, clavicles, and manubriums that were interpreted as indicating that these people had their throats cut, chests opened (likely to remove the heart), and head removed as part of ritualistic sacrifices (Toyne, 2011). There are several reasons to accept the author's interpretation. First, the inferred behaviour aligns with historic accounts of pre-contact Inca rituals (Cobo and Hamilton, 1990; Toyne, 2011). Second, the fact that all people at this sacred site were juveniles or adult males indicates that the assemblage was not random but instead chosen based on sex and age. And lastly, that almost all individuals showed similar patterns of trauma that are consistent with violent ritualistic acts (e.g. throat cutting, removal of organs, decapitation) support the suggestion that these people were victims of ritualistic human sacrifices (Toyne, 2011).

Traumatic lesions identified on bones of the spine and thorax, along with other postcranial trauma, have also been used to argue for the occurrence of non-ritualistic violence in past populations (Alfsdotter and Kjellström, 2019; Meyer et al., 2009; Murphy et al., 2010; Steadman, 2008). For example, Milella et al. (2021) examined the skeletal remains of 87 individuals that showed evidence of cranial and postcranial perimortem trauma from a site in Tuva, Southern Siberia, dating to the 2nd to 4th centuries. Of the traumatic lesions identified, mostly on males, they interpreted that chop marks on the vertebrae and slice marks on the cranium and cervical vertebrae (perhaps from throat slitting) indicate that this nomadic steppe population underwent a period of warfare that likely resulted in the violent deaths of these individuals (Milella et al., 2021). Since the head, neck, and thorax house the majority of organs required for survival, evidence of perimortem sharp force trauma to these regions, as identified by Milella et al. (2021), provides a fairly clear idea of the malicious intentions behind the trauma and thus, the presence of such trauma on archaeological remains can provide insight into social unrest, interpersonal violence, and/or episodes of warfare among and between past populations.

7.8 Conclusion and ways forward

This chapter has provided an overview of some of the ways that skeletal indicators of behaviour have been used in bioarchaeological analyses to understand various biological and social aspects of past humans and our fossil relatives. After reading this chapter, you will have a sense of the types of evidence of behaviour that can be present in the bones of the spine and thorax and how bioarchaeologists can interpret that evidence to infer different types of behaviours being performed by past peoples and their potential intentions (e.g. medical intervention, ritual, violence). Further, understanding human morphological variation and how we differ from our close relatives, both extinct and extant, can help us better understand the evolution of bipedalism and ultimately, modern humans. Thus, understanding both healthy and pathological skeletal variation of the human thorax is an important aspect of biological anthropology, including bioarchaeology and palaeoanthropology, that can enable us to identify, analyse, and interpret differences and changes in the bones of the thorax and spine to infer behaviour in the past.

There are many more examples for behaviour preserved in the human thorax than what can be covered in this chapter and even more avenues of research yet to be explored. For one, rib fractures are not always indicative of abuse or interpersonal violence, and clinically, stress fractures in the ribcage have been correlated with certain physical activities involving thorax rotation, such as rowing (Karlson, 1998; Madison et al., 2021), throwing (Coris and Higgins, 2005), and golfing (Lord et al., 1996). To my knowledge, few bioarchaeological analyses have interpreted rib fractures as evidence of repeated activities related to subsistence or specific movements, such as throwing or rowing (Carrasco Gamboa, 2019; Lovell and Dublenko, 1999), making this a promising area to investigate in the future. In addition, as discussed in this chapter, the link between particular activities and possible 'activity-related' pathologies in the spine and thorax is complicated, making reliable inferences about past behaviours difficult to reach. One reason the relationship is so difficult to pin down is that most pathologies have multiple aetiologies, some of which are unrelated to behaviour, such as genetic predispositions for arthritis. There have been some positive steps forward in understanding the relationships between shape, morphology, and pathology of the bones of the thorax and spine and their relation to locomotion, interpersonal violence, and even socially imposed beauty standards. However, there is still much left to do to help unravel why and how certain people experience pathological conditions in their spine and thorax while others do not will undoubtedly help us unravel the complex interaction between our biology and behaviour.

References

Abitbol, M. M. (1995). Lateral view of Australopithecus afarensis: Primitive aspects of bipedal positional behavior in the earliest hominids. *Journal of Human Evolution, 28,* 211e229.

Alfsdotter, C., & Kjellström, A. (2019). The Sandby borg massacre: Interpersonal violence and the demography of the dead. *European Journal of Archaeology, 22*(2), 210–231.

Aly, I., Chapman, J. R., Oskouian, R. J., Loukas, M., & Tubbs, R. S. (2016). Lumbar ribs: A comprehensive review. *Child's Nervous System, 32*(5), 781–785.

Arriaza, B. T. (1997). Spondylolysis in prehistoric human remains from Guam and its possible etiology. *American Journal of Physical Anthropology, 104*(3), 393–397.

Barsness, K. A., Cha, E. S., Bensard, D. D., Calkins, C. M., Partrick, D. A., Karrer, F. M., et al. (2003). The positive predictive value of rib fractures as an indicator of nonaccidental trauma in children. *Journal of Trauma and Acute Care Surgery, 54*(6), 1107–1110.

Bastir, M., García Martínez, D., Recheis, W., Barash, A., Coquerelle, M., Rios, L., et al. (2013). Differential growth and development of the upper and lower human thorax. *PLoS One, 8*(9), e75128.

Been, E., Gomez-Olivencia, A., Shefi, S., Soudack, M., Bastir, M., & Barash, A. (2017). Evolution of spinopelvic alignment in hominins. *The Anatomical Record, 300*, 900.

Been, E., Peleg, S., Marom, A., & Barash, A. (2010). Morphology and function of the lumbar spine of the Kebara 2 Neandertal. *American Journal of Physical Anthropology, 142*, 549e557.

Been, E., Simonovich, A., & Kalichman, L. (2019). Spinal posture and pathology in modern humans. In E. Been, A. Gomez-Olivencia, & P. A. Kramer (Eds.), *Spinal evolution* (pp. 310–320). New York: Springer.

Bendall, S. (2014). To write a distick upon it: Busks and the language of courtship and sexual desire in sixteenth-and seventeenth-century England. *Gender & History, 26*(2), 199–222.

Berger, E., Yang, L., & Ye, W. (2019). Foot binding in a Ming dynasty cemetery near Xi'an, China. *International Journal of Paleopathology, 24*, 79–88.

Bilo, R. A., Robben, S. G., & Rijn, R. R. V. (2010). General aspects of fractures in child abuse. In *Forensic aspects of pediatric fractures* (pp. 1–13). Berlin, Heidelberg: Springer.

Böhme, M., Spassov, N., DeSilva, J. M., & Begun, D. R. (2020). Reply to: Reevaluating bipedalism in Danuvius. *Nature, 586*(7827), E4–E5.

Böhme, M., Spassov, N., Fuss, J., Tröscher, A., Deane, A. S., Prieto, J., et al. (2019). A new Miocene ape and locomotion in the ancestor of great apes and humans. *Nature, 575*(7783), 489–493.

Bots, J., Wijnaendts, L. C., Delen, S., Van Dongen, S., Heikinheimo, K., & Galis, F. (2011). Analysis of cervical ribs in a series of human fetuses. *Journal of Anatomy, 219*(3), 403–409.

Brickley, M. (2006). Rib fractures in the archaeological record: A useful source of sociocultural information? *International Journal of Osteoarchaeology, 16*(1), 61–75.

Brickley, M. B., & Mays, S. (2019). Metabolic disease. In J. E. Buikstra (Ed.), *Ortner's identification of pathological conditions in human skeletal remains* (3rd ed., pp. 531–566). Academic Press.

Bridges, P. S. (1994). Vertebral arthritis and physical activities in the prehistoric southeastern United States. *American Journal of Physical Anthropology, 93*(1), 83–93.

Carrasco Gamboa, P. (2019). *The lives of the people from Banken 1.: A study based on muscular development and other activity markers*. Undergraduate Thesis Uppsala University. https://www.diva-portal.org/smash/record.jsf?pid=diva2%3A1348927&dswid=-3950.

Cobo, F. B., & Hamilton, R. (1990). *Inca religion and customs*. University of Texas Press.

Coris, E. E., & Higgins, H. W. (2005). First rib stress fractures in throwing athletes. *The American Journal of Sports Medicine, 33*(9), 1400–1404.

Dar, G., Masharawi, Y., Peleg, S., Steinberg, N., May, H., Medlej, B., et al. (2010). Schmorl's nodes distribution in the human spine and its possible etiology. *European Spine Journal, 19*(4), 670–675.

Dar, G., Peleg, S., Masharawi, Y., Steinberg, N., May, H., & Hershkovitz, I. (2009). Demographic aspects of Schmorl nodes: A skeletal study. *Spine, 34*(9), E312–E315.

Davis, P. R. (1961). Human lower lumbar vertebrae: Some mechanical and osteological considerations. *Journal of Anatomy, 95*, 337e344.

Deter-Wolf, A., Robitaille, B., Krutak, L., & Galliot, S. (2016). The world's oldest tattoos. *Journal of Archaeological Science: Reports, 5*, 19–24.

Eaves-Johnson, K. L. (2010). Supernumerary lumbar rib in human prehistory. *The FASEB Journal, 24*, 449.

El-Khoury, G. Y., & Whitten, C. G. (1993). Trauma to the upper thoracic spine: Anatomy, biomechanics, and unique imaging features. *American Journal of Roentgenology, 160*(1), 95–102.

Eng, J. T. (2016). A bioarchaeological study of osteoarthritis among populations of northern China and Mongolia during the bronze age to iron age transition to nomadic pastoralism. *Quaternary International, 405*, 172–185.

Filler, A. G. (2007). Emergence and optimization of upright posture among hominiform hominoids and the evolutionary pathophysiology of back pain. *Neurosurgery Focus, 23*, E4.

Friedman, R., Antoine, D., Talamo, S., Reimer, P. J., Taylor, J. H., Wills, B., et al. (2018). Natural mummies from Predynastic Egypt reveal the world's earliest figural tattoos. *Journal of Archaeological Science, 92*, 116–125.

Gaither, C. (2012). Cultural conflict and the impact on non-adults at Puruchuco-Huaquerones in Peru: The case for refinement of the methods used to analyze violence against children in the archeological record. *International Journal of Paleopathology, 2*(2–3), 69–77.

Gaither, C. M., & Murphy, M. S. (2012). Consequences of conquest? The analysis and interpretation of subadult trauma at Puruchuco-Huaquerones, Peru. *Journal of Archaeological Science, 39*(2), 467–478.

Gibson, R. (2015). Effects of long term corseting on the female skeleton: A preliminary morphological examination. *Nexus: The Canadian Student Journal of Anthropology, 23*(2), 45–60.

Gibson, R. (2017). *"To mold the wax of the woman": An examination of changes in skeletal morphology due to corseting.* American University.

Gibson, R. (2021). *The corseted skeleton; the bioarchaeology of binding.* London: Palgrave Mcmillan Cham.

Gómez-Olivencia, A., Been, E., Arsuaga, J. L., & Stock, J. T. (2013). The Neandertal vertebral column 1: The cervical spine. *Journal of Human Evolution, 64*(6), 608–630.

Gowland, R. L. (2016). Elder abuse: Evaluating the potentials and problems of diagnosis in the archaeological record. *International Journal of Osteoarchaeology, 26*(3), 514–523.

Groves, S., Roberts, C., Johnstone, C., Hall, R., & Dobney, K. (2003). A high status burial from Ripon cathedral, North Yorkshire, England: Differential diagnosis of a chest deformity. *International Journal of Osteoarchaeology, 13*(6), 358–368.

Hernandez, C. J., Loomis, D. A., Cotter, M. M., Schifle, A. L., Anderson, L. C., Elsmore, L., et al. (2009). Biomechanical allometry in hominoid thoracic vertebrae. *Journal of Human Evolution, 56*, 462e470.

Herrero-Beaumont, G., Roman-Blas, J. A., Castañeda, S., & Jimenez, S. A. (2009, October). Primary osteoarthritis no longer primary: Three subsets with distinct etiological, clinical, and therapeutic characteristics. In *Vol. 39. Seminars in arthritis and rheumatism* (pp. 71–80). WB Saunders. No. 2.

Hu, S. S., Tribus, C. B., Diab, M., & Ghanayem, A. L. (2008). Spondylolisthesis and spondylolysis. *Journal of Bone and Joint Surgery, 90*, 656.

Iwamoto, J., Takeda, T., & Wakano, K. (2004). Returning athletes with severe low back pain and spondylolysis to original sporting activities with conservative treatment. *Scandinavian Journal of Medicine & Science in Sports, 14*, 346.

Jellema, L. M., Latimer, B., & Walker, A. (1993). The rib cage. In A. Walker, & R. Leakey (Eds.), *The Nariokotome homo erectus skeleton* (p. 294e325). Berlin: Springer.

Jurmain, R. (1989). Skeletal evidence of trauma in African apes, with special reference to the Gombe chimpanzees. *Primates, 38*, 1.

Jurmain, R. D. (1999). *Stories from the skeleton. Behavioral reconstruction in human osteology.* Amsterdam: Gordon and Breach.

Jurmain, R. D. (2000). Degenerative joint disease in African great apes: An evolutionary perspective. *Journal of Human Evolution, 39*, 185.

Karlson, K. A. (1998). Rib stress fractures in elite rowers. *The American Journal of Sports Medicine, 26*(4), 516–519.

Keith, A. (1923). Hunterian lectures on Man's posture: Its evolution and disorders. Lecture II. The evolution of the orthograde spine. *The British Medical Journal, 1*, 409e502.

Klaus, H. D., Centurión, J., & Curo, M. (2010). Bioarchaeology of human sacrifice: Violence, identity and the evolution of ritual killing at Cerro Cerrillos, Peru. *Antiquity, 84*(326), 1102–1122.

Klaus, H. D., Spencer Larsen, C., & Tam, M. E. (2009). Economic intensification and degenerative joint disease: Life and labor on the postcontact north coast of Peru. *American Journal of Physical Anthropology, 139*(2), 204–221.

Latimer, B. (2005). The perils of being bipedal. *Annals of Biomedical Engineering, 33*, 3.

Latimer, B., & Ward, C. V. (1993). The thoracic and lumbar vertebrae. In A. Walker, & R. Leakey (Eds.), *The Nariokotome Homo erectus skeleton* (p. 266e293). Berlin: Springer.

Lord, M. J., Ha, K. I., & Song, K. S. (1996). Stress fractures of the ribs in golfers. *The American Journal of Sports Medicine, 24*(1), 118–122.

Lovell, N. (1990). *Patterns of injury and illness in the great apes: A skeletal analysis.* Washington, DC: Smithsonian Institution.

Lovell, N. C. (1994). Spinal arthritis and physical stress at bronze age Harappa. *American Journal of Physical Anthropology, 93*(2), 149–164.

Lovell, N. C. (1997). Trauma analysis in paleopathology. *American Journal of Physical Anthropology, 104*(S25), 139–170.

Lovell, N. C., & Dublenko, A. A. (1999). Further aspects of fur trade life depicted in the skeleton. *International Journal of Osteoarchaeology, 9*(4), 248–256.

Lowenstine, L. J., McManamon, R., & Terio, K. A. (2016). Comparative pathology of aging great apes: Bonobos, chimpanzees, gorillas, and orangutans. *Veterinary Pathology, 53*(2), 250–276.

Madison, C. A., Harter, R. A., Pickerill, M. L., & Housman, J. M. (2021). Rib stress injuries among female national collegiate athletic association rowers: A prospective epidemiological study. *International Journal of Athletic Therapy and Training, 1*(aop), 1–7.

Marks, M. K., Marden, K., & Mileusnic-Polchan, D. (2009). Forensic osteology of child abuse. In D. W. Steadman (Ed.), *Hard evidence: Case studies in forensic anthropology* (pp. 205–220). Upper Saddle River: Prentice-Hall.

Masharawi, Y. (2012). Lumbar shape characterization of the neural arch and vertebral body in spondylolysis: A comparative skeletal study. *Clinical Anatomy, 25*, 224.

Masharawi, Y. M., Alperovitch-Najenson, D., Steinberg, N., Dar, G., Peleg, S., Rothschild, B., et al. (2007). Lumbar facet orientation in spondylolysis: A skeletal study. *Spine, 32*(6), E176–E180.

Matos, V. (2009). Broken ribs: Paleopathological analysis of costal fractures in the human identified skeletal collection from the Museu Bocage, Lisbon, Portugal (late 19th to middle 20th centuries). *American Journal of Physical Anthropology, 140*(1), 25–38.

Mays, S. (2007). Spondylolysis in non-adult skeletons excavated from a medieval rural archaeological site in England. *International Journal of Osteoarchaeology, 17*, 504.

Merbs, C. (1996a). Spondylolysis and spondylolisthesis: A cost of being an erect biped or clever adaptation? *American Journal of Physical Anthropology, 101*, 201.

Merbs, C. F. (1996b). Spondylolysis of the sacrum in Alaskan and Canadian Inuit skeletons. *American Journal of Physical Anthropology, 101*(3), 357–367.

Meyer, C., Brandt, G., Haak, W., Ganslmeier, R. A., Meller, H., & Alt, K. W. (2009). The Eulau eulogy: Bioarchaeological interpretation of lethal violence in corded ware multiple burials from Saxony-Anhalt, Germany. *Journal of Anthropological Archaeology, 28*(4), 412–423.

Micheli, L. J., & Wood, R. (1995). Back pain in young athletes: Significant differences from adults in causes and patterns. *Archives of Pediatrics & Adolescent Medicine, 149*(1), 15–18.

Middleton, K., & Fish, D. E. (2009). Lumbar spondylosis: Clinical presentation and treatment approaches. *Current Reviews in Musculoskeletal Medicine, 2*(2), 94–104.

Milella, M., Caspari, G., Kapinus, Y., Sadykov, T., Blochin, J., Malyutina, A., et al. (2021). Troubles in Tuva: Patterns of perimortem trauma in a nomadic community from Southern Siberia (second to fourth c CE). *American Journal of Physical Anthropology, 174*(1), 3–19.

Moore, J., & Buckberry, J. (2016). The use of corsetry to treat Pott's disease of the spine from 19th century Wolverhampton, England. *International Journal of Paleopathology, 14*, 74–80.

Murphy, M. S., Gaither, C., Goycochea, E., Verano, J. W., & Cock, G. (2010). Violence and weapon-related trauma at Puruchuco-Huaquerones, Peru. *American Journal of Physical Anthropology, 142*(4), 636–649.

Pabst, M. A., Letofsky-Papst, I., Moser, M., Spindler, K., Bock, E., Wilhelm, P., et al. (2010). Different staining substances were used in decorative and therapeutic tattoos in a 1000-year-old Peruvian mummy. *Journal of Archaeological Science, 37*(12), 3256–3262.

Pickering, R. B. (1979). Hunter-gatherer/agriculturalist arthritic patterns: A preliminary investigation. *Henry Ford Hospital Medical Journal, 27*(1), 50–53.

Plomp, K. A., et al. (2022). *Palaeopathology and evolutionary medicine: An integrated approach.* Oxford, UK: Oxford University Press.

Plomp, K. A., Dobney, K., & Collard, M. (2020). Spondylolysis and spinal adaptations for bipedalism: The overshoot hypothesis. *Evolution, Medicine, and Public Health, 2020*(1), 35–44.

Plomp, K. A., Dobney, K., Weston, D. A., Strand Viðarsdóttir, U., & Collard, M. (2019). 3D shape analyses of extant primate and fossil hominin vertebrae support the ancestral shape hypothesis for intervertebral disc herniation. *BMC Evolutionary Biology, 19*(1), 226.

Plomp, K., Roberts, C., & Strand Vidarsdottir, U. (2015a). Does the correlation between schmorl's nodes and vertebral morphology extend into the lumbar spine? *American Journal of Physical Anthropology, 157*(3), 526–534.

Plomp, K. A., Viðarsdóttir, U. S., Weston, D. A., Dobney, K., & Collard, M. (2015b). The ancestral shape hypothesis: An evolutionary explanation for the occurrence of intervertebral disc herniation in humans. *BMC Evolutionary Biology, 15*(1), 1–10.

Pomeroy, E., Stock, J. T., Zakrzewski, S. R., & Lahr, M. M. (2010). A metric study of three types of artificial cranial modification from north-central Peru. *International Journal of Osteoarchaeology, 20*(3), 317–334.

Ponseti, I. V. (1991). History of orthopaedic surgery. *The Iowa Orthopaedic Journal, 11*, 59.

Redfern, R. C. (2017). Identifying and interpreting domestic violence in archaeological human remains: A critical review of the evidence. *International Journal of Osteoarchaeology, 27*(1), 13–34.

Rose, M. D. (1975). Functional proportions of primate lumbar vertebral bodies. *Journal of Human Evolution, 4,* 21.

Ross, A. H., & Juarez, C. A. (2016). Skeletal and radiological manifestations of child abuse: Implications for study in past populations. *Clinical Anatomy, 29*(7), 844–853.

Roussouly, P., Gollogly, S., Berthonnaud, E., Labelle, H., & Weidenbaum, M. (2005). Sagittal alignment of the spine and pelvis in the presence of L5-S1 isthmic lysis and low-grade spondylolisthesis. *Spine, 31,* 2484.

Russo, G. A. (2010). Presygapophyseal articular facet shape in the catarrhine thoracolumbar vertebral column. *American Journal of Physical Anthropology, 142,* 600e612.

Sanders, W. J., & Bodenbender, B. E. (1994). Morphometric analysis of lumbar vertebra UMP 67-28: Implications for spinal function and phylogeny of the Miocene Moroto hominoid. *Journal of Human Evolution, 26,* 203.

Sarzi-Puttini, P., Atzeni, F., Fumagalli, M., Capsoni, F., & Carrabba, M. (2005). Osteoarthritis of the spine. *Seminars in Arthritis and Rheumatism, 34*(6), 38.

Schultz, A. H. (1961). Vertebral column and thorax. In *Vol. 4. Primatologia.* Basel: Karger.

Selinsky, P. (2015). Celtic ritual activity at Gordion, Turkey: Evidence from mortuary contexts and skeletal analysis. *International Journal of Osteoarchaeology, 25*(2), 213–225.

Shapiro, L. (1991). *Functional morphology of the primate spine with special reference to the orthograde posture and bipedal locomotion.* Ph.D. Dissertation State University of New York at Stony Brook.

Shapiro, L. J. (1993). Evaluation of the "unique" aspects of human vertebral bodies and pedicles with consideration of *Australopithecus africanus. Journal of Human Evolution, 25,* 433.

Soler, T., & Calderón, C. (2000). The prevalence of spondylolysis in the Spanish elite athlete. *The American Journal of Sports Medicine, 28*(1), 57–62.

Spector, T. D., Cicuttini, F., Baker, J., Loughlin, J., & Hart, D. (1996). Genetic influences on osteoarthritis in women: A twin study. *BMJ, 312*(7036), 940–943.

Spector, T. D., & MacGregor, A. J. (2004). Risk factors for osteoarthritis: Genetics. *Osteoarthritis and Cartilage, 12,* 39–44.

Steadman, D. W. (2008). Warfare related trauma at Orendorf, a middle Mississippian site in west-central Illinois. *American Journal of Physical Anthropology, 136*(1), 51–64.

Stevenson, R. E., Hall, J. G., & Goodman, R. M. (1993). *Human malformations and related abnormalities.* Oxford: University Press.

Stirland, A. J., & Waldron, T. (1997). Evidence for activity related markers in the vertebrae of the crew of the Mary Rose. *Journal of Archaeological Science, 24*(4), 329–335.

Stone, P. K. (2009). A history of western medicine, labor, and birth. In *Childbirth across cultures* (pp. 41–53). Dordrecht: Springer.

Stone, P. K. (2012). Binding women: Ethnology, skeletal deformations, and violence against women. *International Journal of Paleopathology, 2*(2–3), 53–60.

Stone, P. K. (2020a). Bound to please: The shaping of female beauty, gender theory, structural violence, and bioarchaeological investigations. In *Purposeful pain* (pp. 39–62). Cham: Springer.

Stone, P. K. (2020b). Female beauty, bodies, binding, and the bioarchaeology of structural violence in the industrial era through the Lens of critical white feminism. In *The bioarchaeology of structural violence* (pp. 13–30). Cham: Springer.

Stone, P., & Walrath, D. (2006). The gendered skeleton: Anthropological interpretations of the bony pelvis. In R. Gowland, & C. Knüsel (Eds.), *The social archaeology of funerary remains.* Oxford: Oxbow Books.

Taylor, J. R. (1975). Growth of human intervertebral discs and vertebral bodies. *Journal of Anatomy, 120*(Pt 1), 49.

Tiesler, V., & Olivier, G. (2020). Open chests and broken hearts: Ritual sequences and meanings of human heart sacrifice in Mesoamerica. *Current Anthropology, 61*(2), 168–193.

Tomasto-Cagigao, E. (2009). Talking bones: Bioarchaeological analysis of individuals from Palpa. In M. Reindel, & G. A. Wagner (Eds.), *New technologies for archaeology: Multidisciplinary investigations in Palpa and Nasca, Peru* (pp. 141–158). New York: Springer.

Torres-Rouff, C. (2020). Cranial modification and the shapes of heads across the Andes. *International Journal of Paleopathology, 29,* 94–101.

Toyne, J. M. (2011). Interpretations of pre-hispanic ritual violence at Tucume, Peru, from cut mark analysis. *Latin American Antiquity, 22*(4), 505–523.

Turner, W. G. (1913). The treatment of tubercular spondylitis or Pott's disease. *Canadian Medical Association Journal, 3*(10), 852.

Turner, C. H. (2006). Bone strength: Current concepts. *Annals of the New York Academy of Sciences, 1068*(1), 429–446.

Ueyoshi, A., & Shima, Y. (1985). Studies on spinal braces. *International Orthopaedics, 9*(4), 255–258.

Walker, P. L., Collins Cook, D., & Lambert, M. (1997). Skeletal evidence for child abuse: A physical anthropological perspective. *Journal of Forensic Sciences, 42*(2), 196–207.

Ward, C. V. (1991). *The functional anatomy of the lower back and pelvis of the Miocene hominoid proconsul nyanzae from the Miocene of Mfangano Island, Kenya.* Ph.D. Dissertation Johns Hopkins University.

Ward, C. V., & Latimer, B. (2005). Human evolution and the development of spondylolysis. *Spine, 30,* 1808.

Ward, C. V., Latimer, B., Alander, D. H., Parker, J., Ronan, J. A., Holden, A. D., et al. (2007). Radiographic assessment of lumbar facet distance spacing and spondylolysis. *Spine, 32,* E85.

Ward, C. V., Mays, S. A., Child, S., & Latimer, B. (2010). Lumbar vertebral morphology and isthmic spondylolysis in a British medieval population. *American Journal of Physical Anthropology, 141,* 273.

Weiss, E., & Jurmain, R. (2007). Osteoarthritis revisited: A contemporary review of aetiology. *International Journal of Osteoarchaeology, 17*(5), 437–450.

Wescott, D. J., Brinsko, K., Faerman, M., Golda, S. D., Nichols, J., Spigelman, M., et al. (2010). A Fisk patent metallic burial case from Western Missouri: An interdisciplinary and comprehensive effort to reconstruct the history of an early settler of Lexington, Missouri. *Archaeological and Anthropological Sciences, 2*(4), 283–305.

Wheeler, S. M., Williams, L., Beauchesne, P., & Dupras, T. L. (2013). Shattered lives and broken childhoods: Evidence of physical child abuse in ancient Egypt. *International Journal of Paleopathology, 3,* 71–82.

Whitcome, K. K., Shapiro, L. J., & Lieberman, D. E. (2007). Fetal load and the evolution of lumbar lordosis in bipedal hominins. *Nature, 450,* 1075.

Williams, S. A., Prang, T. C., Meyer, M. R., Russo, G. A., & Shapiro, L. J. (2020). Reevaluating bipedalism in Danuvius. *Nature, 586*(7827), E1–E3.

Williams, S. A., & Russo, G. A. (2015). Evolution of the hominoid vertebral column: The long and short of it. *Evolutionary Anthropology, 24,* 15–32.

Zhang, H., Merrett, D. C., Jing, Z., Tang, J., He, Y., Yue, H., et al. (2017). Osteoarthritis, labour division, and occupational specialization of the late Shang China-insights from Yinxu (ca. 1250–1046 BC). *PLoS One, 12*(5), e0176329.

Zhao, Y., Guo, L., Xiao, Y., Niu, Y., Zhang, X., He, D., et al. (2020). Osteological characteristics of Chinese foot-binding in archaeological remains. *International Journal of Paleopathology, 28,* 48–58.

Zimmermann, E. A., & Ritchie, R. O. (2015). Bone as a structural material. *Advanced Healthcare Materials, 4*(9), 1287–1304.

C H A P T E R

8

Human behaviour and the pelvis

Sarah-Louise Decrausaz[a] and Natalie Laudicina[b]

[a]Department of Anthropology, University of Victoria, Victoria, BC, Canada, [b]Department of Biomedical Sciences, Grand Valley State University, Allendale, MI, United States

8.1 Introduction

The human pelvis is anatomically complex and frequently viewed as the site of several uniquely human skeletal features. These features are related to two of the central characteristics of anatomically modern humans (AMH), our method of locomotion and how we birth offspring, both of which strongly influenced the evolution of the human pelvis. This chapter outlines how researchers can interpret pelvic features to learn more about behaviour using examples drawn from bioarchaeology, palaeoanthropology, obstetric and gynaecological medicine, comparative anatomy and evolution, and public health. This chapter does not provide a completely comprehensive overview of the pelvis because perspectives on the pelvis will vary with discipline. For example, obstetric clinical literature may not focus on the same pelvic features as studies in sports medicine. The division of themes is presented for clarity of description and understanding. Thus the subsections in this chapter should not be seen as mutually exclusive concepts, but rather, understood as pieces of an integrated whole, many of which all intersect with one another.

The chapter opens with an overview of the growth and development of the pelvis, as well as some general principles on its variation. This is followed by a review of the differences and similarities between the human and nonhuman ape pelvis, the changing form of the pelvis in human ancestral species, and how these elements relate to behaviour. Next, we cover locomotion, gait, and movement, including bioarchaeological case studies demonstrating how pelvic evidence can provide information about locomotion. The chapter then continues with an examination of childbirth and the pelvis, as well as pelvic evidence of disease, and closes with a critical assessment of using skeletal markers on the pelvis to infer daily behaviours of those who have died in the past.

Behaviour in our Bones
https://doi.org/10.1016/B978-0-12-821383-4.00006-1

8.2 The human pelvis

The pelvis consists of four bones in adulthood (paired os coxae, the sacrum, and the coccyx) and can be divided into two portions: the 'false pelvis' is the upper portion of the pelvis, while the 'true pelvis' includes the pelvic, or birth, canal. The canal in the middle of the pelvis is divided into three planes—the inlet, the midplane, and the outlet (Fig. 8.1).

The pelvis begins developing during the fifth week of gestation and the cartilage model nears completion near the onset of the third uterine month (Adair, 1918; Young et al., 2019). The primary centres of ossification in the pelvis are located in the centre of the ilium, ischium, and the pubis. The ilium ossifies first during the beginning of the third month of gestation, while ossification of the ischium occurs during the fourth to fifth months of gestation (Verbruggen and Nowlan, 2017; Young et al., 2019). The pubis ossifies last in the fifth to sixth month of gestation, while the sacrum continues to ossify from the third month of gestation onwards (Verbruggen and Nowlan, 2017; Young et al., 2019). At birth, the primary centres of ossification are clearly visible via radiograph and overall growth of these centres is noticeably rapid until three years of age, followed by a reduction in growth speed (Verbruggen and Nowlan, 2017; Young et al., 2019). Fusion of the three centres of ossification at the acetabulum can occur as early as three years of age, though it more commonly takes place between five and eight years of age. These ossification centres fully fuse between 11.5 and 12.8 years old in females and 12.2 and 14.6 years of age in males (Grissom et al., 2018; Rainer et al., 2021; Young et al., 2019).

The dimensions of the human pelvis vary with biological sex (Brůžek et al., 2017; Genovés, 1959; Krogman and Isçan, 1986; Santos et al., 2019; Schultz, 1949; St Hoyme and Iscan, 1989) (Fig. 8.2), latitude (Auerbach et al., 2018; Kurki, 2013), and genetics and genetic pools (Betti et al., 2013; Grabowski et al., 2011; Young et al., 2019). They are also influenced by differential

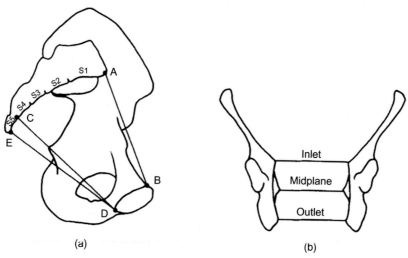

(a)   (b)

FIG. 8.1   Human pelvis depicting three major planes in the pelvic canal: inlet, midplane, outlet. *From Tague, R. G., & Lovejoy, C. O. (1986). The obstetric pelvis of AL 288-1 (Lucy). Journal of Human Evolution, 15(4), 237–255.*

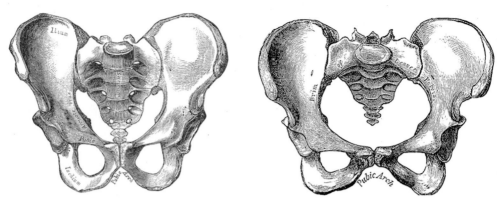

FIG. 8.2 Human male pelvis (left) and human female pelvis (right). *From Gray, H. (1918). Anatomy of the human body. Philadelphia: Lea & Febiger.*

migration in more recent evolutionary history (Betti and Manica, 2018), and growth trajectories during early and late development (Shirley et al., 2020). The canal and noncanal components of the pelvis (Figs 8.1 and 8.2) also diverge in their patterns of variability, with the pelvic canal being more variable in shape than the noncanal pelvis (Kurki and Decrausaz, 2016). In modern humans, the pelvis is commonly used to identify the biological sex of an adult individual, as the os coxae is one of the most sexually dimorphic bones (Brůžek et al., 2017; Genovés, 1959; Krogman and Işcan, 1986; Santos et al., 2019; St Hoyme and Işcan, 1989) (Fig. 8.2). Subadult pelves can be difficult to sex. This is due to many reasons including environmental and developmental factors that result in areas of the pelvis (e.g., sciatic notch) developing at different ages (Corron et al., 2021; Klales and Burns, 2017). This leads to decreased accuracy when estimating sex in juvenile vs adult pelves (Corron et al., 2021; Klales and Burns, 2017). Adult female pelves tend to have expanded true pelvic dimensions to serve as a reproductive passageway, while male pelvic canals tend to be more constricted with a larger overall pelvic girdle size (Buikstra and Ubelaker, 1994). However, modern human pelves exhibit a wide variety of sizes and shapes within and between the sexes. These differences are a result of sexual dimorphism (reproduction and stature), but also factors such as nutrition.

8.3 The human pelvis vs. the fossil and nonhuman primate pelvis

The hominin pelvis is easily distinguishable from those of other great ape species because of its overall shortened and broad structure (Fig. 8.3). These features, which include a wide sacrum and flaring iliac blades, allow hominins to engage with specific behaviours such as habitual bipedality and birthing large-brained infants. The reorganisation of the hominin pelvic structure arose early in the hominin lineage, over four million years ago (Johanson et al., 1982; Lovejoy et al., 2009; Robinson, 1972). Each hominin species shows differences in their pelvic morphology based on many factors, including behavioural activities. This section will cover some of the major differences seen in the pelvis along the fossil hominin lineage, concluding with anatomically modern humans.

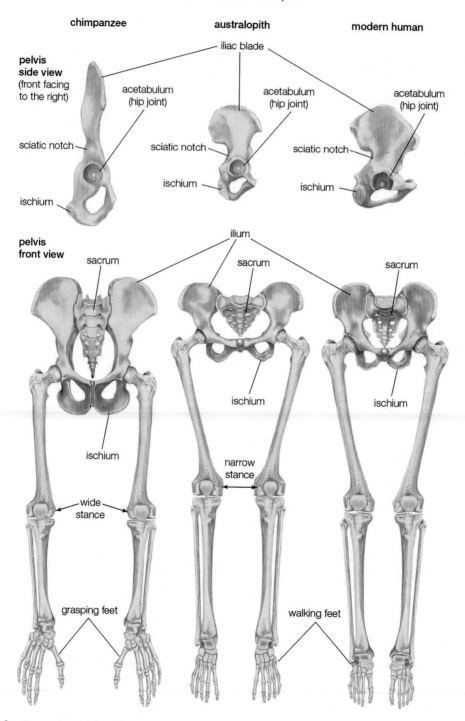

FIG. 8.3 Comparison of the pelvis and lower limbs of a chimpanzee (far left), an australopith (middle), and an anatomically modern human (right). *From Encyclopedia Britannica.*

Australopithecus is one of the earliest accepted hominin genera. Australopithecine female pelves are currently represented by four main specimens, representing three species—*Australopithecus afarensis* (A.L. 288–1), *Australopithecus africanus* (Sts 14, Sts 65), and *Australopithecus sediba* (MH2). Each pelvis is adapted to bipedality, with human-like shortened ilia and a broad sacrum. Unlike modern humans, however, Australopithecine pelves tend to be wider transversely and compressed anteroposteriorly, a pelvic shape known as platypelloid (Tague and Lovejoy, 1986). This extreme dorsoventral compression is especially seen in early Australopithecine pelves (A.L. 288–1, Sts 14, Sts 65). A later Australopithecine species, *Australopithecus sediba*, exhibits a more expanded anteroposterior pelvic shape, although not to the extent seen in modern humans (Kibii et al., 2011; Laudicina et al., 2019). This change from a platypelloid pelvic shape to a more anteroposteriorly expanded shape may reflect a transition from a more arboreal lifestyle in the earlier Australopithecines to more habitual bipedal locomotor pattern as the pelvic morphology adapts to accommodate shifting muscle attachments (see Movement Section 8.5 for more details) (Lewis et al., 2017; Lovejoy, 2005a; Robinson, 1972).

One hallmark of the hominin lineage is the increase in brain size. As brain size increases, so does the risk for obstruction (difficulty with the foetus descending through the birth canal even with strong uterine contractions) during birth. In modern humans, this results in a foetal rotation pattern unlike what is experienced in the other great apes (Fig. 8.4). The

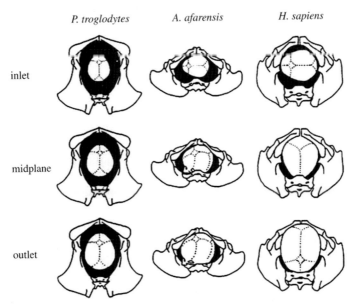

FIG. 8.4 "'Midwife's' (inferior or lithotomy position) view of the birth canal in a chimpanzee (*P. troglodytes*), *A. afarensis* (A.L. 288–1, 'Lucy'), and a modern human female. The position of the infant's head at the three major planes of the birth canal is shown. Note the nonrotational birth in apes in which the infant's head faces anteriorly throughout, the contrasting nonrotational pattern in the australopithecine in which the infant's head faces laterally throughout, and the rotational pattern in the human in which the infant's head enters the birth canal facing laterally, but then rotates at the midplane and exits the outlet facing posteriorly. Also note that in the two hominins, the infant's head exits the pelvis more anteriorly, under the ischiopubic rami" (Gruss and Schmitt, 2015: 3). *From Gruss, L. T., & Schmitt, D. (2015). The evolution of the human pelvis: Changing adaptations to bipedalism, obstetrics and thermoregulation. Philosophical Transactions of the Royal Society, B: Biological Sciences, 370(1663), 20140063.*

human birth canal changes shape: transversely wide at the cranial end, transversely constricted in the middle, and circular at the outlet (Rosenberg and Trevathan, 2002; Trevathan, 2015). The baby must turn not once, but twice to navigate these changing pelvic dimensions (Rosenberg and Trevathan, 2002). By contrast, a nonhuman ape neonate can pass relatively easily through the mother's birth canal without rotation (Rosenberg, 1992) (Fig. 8.4). Birthing larger-brained infants affects later hominin pelvic shape, but in early fossil hominins, the birth reconstructions are more influenced by the compressed pelvic morphology, as foetal head size was not much larger than a chimpanzee neonate (Berge and Goularas, 2010; Berge et al., 1984; Häusler and Schmid, 1995; Leutenegger, 1972, 1987; Rosenberg, 1992). Childbirth in *Australopithecus* has been described as ape-like with no necessary foetal rotations to a more human-like foetal rotational pattern (Berge et al., 1984; Berge and Goularas, 2010; Claxton et al., 2016; DeSilva et al., 2017; Kibii et al., 2011; Laudicina et al., 2019; Tague and Lovejoy, 1986). Therefore it can be assumed that the anteriorly compressed pelvic morphology in these early fossil hominins was primarily due to a shift to bipedality from arboreality rather than other behaviours.

The increasing stature and body size in the genus *Homo* results in an expanded pelvic size. A juvenile male specimen from Kenya, dubbed the Nariokotome Boy and dated to ~1.6 million years old, shows a human-like pelvis with tall and laterally flared ilia (Walker et al., 1993). The bi-iliac dimensions, adjusted for age, align with modern human populations and are thought to have been evolutionarily shaped by thermoregulation, specifically with heat dissipation in the warmer African climate (Walker et al., 1993). However, the early *Homo* story was complicated with the finding of a small, female pelvis from Ethiopia also assigned to *Homo erectus*. This pelvis, named the Gona pelvis, exhibits a wide, rounded birth canal thought to accommodate an encephalised (large-brained) neonate (Simpson et al., 2008). However, this pelvis also exhibits ilia that more resemble early Australopithecines compared to the ilia from the Nariokotome specimen (Simpson et al., 2008). These intraspecific variations may be a result of sex-specific differences in the pelvis, time differences, or simply a suggestion that morphological changes in the pelvis did not change in a linear fashion in our evolutionary history.

Neanderthals represent one of the most recent fossil hominin species where pelves are available. A wide female Neanderthal pelvis (Tabun C1) from Mount Carmel, Israel, exhibits a mediolaterally expanded pelvic inlet and midplane combined with an anteroposteriorly expanded outlet is also thought to support a neonate with a large cranium, approaching or exceeding modern human size (Ponce de León et al., 2008; Weaver and Hublin, 2009). A large male pelvis from Kebara shows a pelvis with externally rotated os coxae (Rak and Arensburg, 1987). The position of the hipbones results in extremely long superior pubic rami, exceeding what is seen in modern humans (Rak and Arensburg, 1987). Interestingly, the larger pubic ramus does not result in a larger-sized pelvic inlet compared to modern humans (Rak and Arensburg, 1987). However, the shape of the Tabun C1 pelvic canal is different compared to modern humans, exhibiting a transversely wide shape throughout the entirety of the canal (Ponce de León et al., 2008; Weaver and Hublin, 2009). In modern humans, the lower portion of the pelvic canal, the outlet, expands more anteroposteriorly to be relatively circular in shape, but the outlet in the Tabun C1 Neanderthal pelvis remains transversely wide (Ponce de León et al., 2008; Weaver and Hublin, 2009). However, it should be noted that the Tabun C1 pelvis is fragmentary, especially at this lower portion, which impacts reconstruction of the remains. The breadth of the Neanderthal pelvis may also be due, in part, to the overall increase in body breadth exhibited throughout the Neanderthal skeleton. In contrast to the thermoregulatory heat dissipation explanation for the above-mentioned *Homo erectus* specimen, Neanderthals are thought to be broader in body shape to help heat retention in colder climates (Holliday, 1997; Tilkens et al., 2007; Weaver, 2003).

The degree of variability in the hominin pelvic shape suggests that it has been influenced by many factors. This plasticity of the human pelvis can result in morphological variation in response to a wide array of factors, producing inter- and intraspecific size and shape patterns. The next section focuses on how the hominin pelvic shape alters the muscular attachments to the pelvis, and how this impacts how we as humans walk.

8.4 Childbirth

Reproduction is another major factor thought to shape the uniqueness of the human pelvis. This section opens with an outline of the process of childbirth to situate the reader with respect to how the shape and size of the pelvis influences human birth, as well as a brief discussion of the causes of death relating to childbirth in humans today. This information is provided to better contextualise the discussion of the human bony pelvis with respect to childbirth more broadly. This is followed by an outline of the evolutionary theories, including the obstetric dilemma hypothesis, on which forces influence human pelvic shape and size in light of childbirth, as well as specific examples showing how childbirth as a behaviour can be identified from the human skeleton or through bioarchaeological analyses.

8.4.1 The process of childbirth in humans

Humans have a rotational birth mechanism, which is necessary due to the tight fit between the long axis of the newborn's cranium and the planes of the maternal pelvic canal. Essentially, the newborn must negotiate the narrowest part of the birth canal (the midplane) as it is born. The newborn's head first rotates through the pelvic midplane and out of the canal, followed by the shoulders entering the pelvic canal, with the shoulders then beginning to rotate internally within the canal to align anterior-posteriorly with the outlet (Rosenberg, 1992; Rosenberg and Trevathan, 1995; Trevathan and Rosenberg, 2000).

Parturition (another term for childbirth) affects pelvic morphology through the expansion of female pelvic dimensions to accommodate a large newborn. That is, ligaments binding the pelvis together will relax to allow the pelvis to expand to allow the newborn to pass through the birth canal. In the last few weeks of pregnancy, the pelvic ligaments relax and become slack, thereby alleviating the tightness of the fit between the foetus and the bony pelvis (Marnach et al., 2003; Weinberg, 1954).

8.4.2 Childbirth as a cause of death today

Even with the medical support and intervention available today, childbirth remains a challenge for mother and baby. Childbirth and its associated complications account for 9% of maternal mortality globally (WHO, 2014). Such complications include, but are not limited to, eclampsia (a condition impacting a mother's blood pressure and potentially resulting in seizures), obstetric haemorrhage, and hypertensive disorders (Geller et al., 2018). In a survey of 417 studies across 115 countries, it was found that obstructed labour, a condition where even with the uterus contracting normally the newborn cannot exit the birth canal due to it being blocked, accounted for 69,000 deaths between 2003 and 2009 (Say et al., 2014). Every day in 2017, 810 women died from preventable causes related to pregnancy and childbirth (WHO, 2022).

8.4.3 The obstetric dilemma hypothesis

The scholarly focus on reproduction and locomotion to explain human pelvic evolution is predominantly framed by Washburn's (1960) suggestion that the tight fit between the maternal birth canal and the head of a newborn (neonate) is today, unique to humans. This is because the human pelvis must balance allowing the birth of a large-brained infant with the restraints imposed on pelvic morphology by bipedalism. Bipedality is a comparatively energetically efficient means of locomotion when compared to moving on four limbs (Lovejoy, 2005a,b), and from an evolutionary perspective, energetic economy allows for animals to gain access to water or food while also accessing mates and reproducing (Wall-Scheffler et al., 2020). The energetic benefits of bipedalism are characterised as one of the main drivers of this form of locomotion (Lovejoy, 2005a,b); however, the adaptations for bipedalism in the human pelvis have also impacted childbirth. For instance, a narrow pelvis can be considered more 'efficient' than a wide pelvis as friction is reduced at the hip joint, or the narrowness of the pelvis decreases the amount of muscle necessary to perform work at the hip joint (Wall-Scheffler et al., 2020). While a narrow pelvis may be effective for energetically efficient bipedalism, a wide pelvis provides a less constrictive passage for a newborn during birth. After the evolution of bipedalism, humans have also undergone strong selection for greater brain size (Gruss and Schmitt, 2015), resulting in conflicting pressures rendering childbirth biomechanically challenging and the birth of humans at a more altricial stage due to the mismatch between the size of the infant head and the maternal birth canal (Portmann, 1969; Washburn, 1960). Washburn (1960) termed the result of these conflicting pressures the 'obstetric dilemma' (OD).

Although the OD remained a fixture in characterising human birth as a necessary danger for some time, it has been re-examined in recent years (Dunsworth, 2018; Dunsworth et al., 2012; Haeusler et al., 2021; Warrener et al., 2015). For instance, using experimental models, researchers have demonstrated that there is no locomotor cost imposed by a wider pelvis, suggesting that variation seen in female pelvic shapes (including having a narrower birth canal) is caused by factors other than biomechanical necessity (Warrener et al., 2015). Other scholars propose that the OD varies in magnitude with different ecological settings and has not been fixed throughout human history (Wells et al., 2012). Wells et al. (2012) use the example of the advent of agriculture as a possible ecological setting that could have impacted the magnitude of the OD. They suggest that with the advent of agriculture, female growth and development, including that of the pelvis, were compromised by poorer dietary quality (Wells et al., 2012). Skeletal evidence shows that the transition to agriculture resulted in a decline in overall health in many populations (Larsen, 2006; Shuler et al., 2012), and Wells et al. (2012) suggest that the resulting delay in skeletal maturation led to compromised female pelvic capacity.

8.4.4 The human pelvis and bioarchaeological analyses

Bioarchaeologists have attempted to identify osteological evidence of childbirth on adult female skeletons; however, this is often made challenging by the poor preservation of the pelvis compared to other parts of the skeleton (Rosenberg and DeSilva, 2017). The pelvis is especially susceptible to damage to the pubis, due to the projection of the anterior portion of the pelvis—that is, when a human body is lying prone in a burial context, the pubis projects away from the centre of the body, taking a large part of the weight of soil or other matter. With time and the process of decomposition, the added weight of the soil often fractures the anterior portion of the pelvis, distorting the pelvis. The fragility of foetal remains contributes to their poor preservation too, making it doubly difficult to estimate if difficult childbirth was the cause of death when examining human remains.

Parturition is a complex interaction between bone and connective tissue (cartilage, tendon, ligament, and muscle). It has been suggested by some scholars that the muscular work of childbirth, that is, the interaction between bone and connective tissues, may leave evidence on the surface of bone (Angel, 1969; Bergfelder and Herrmann, 1980; Cox, 1989; Cox and Scott, 1992; Houghton, 1974; Putschar, 1976; Stewart, 1957; Ubelaker and De La Paz, 2012; Waltenberger et al., 2022). For example, the levator ani muscle group, which includes the pubococcygeus, iliococcygeus, and puborectalis, is the major muscle group acting during childbirth (Ashton-Miller and DeLancey, 2009). The iliococcygeus forms a relatively flat, almost horizontal shelf across the pelvic sidewalls, while the pubococcygeus (also known as the pubovaginalis muscle) originates at the pubis and attaches to the walls of the pelvic organs and the perineal body, and the puborectalis forms a type of sling around and posterior to the rectum (Ashton-Miller and DeLancey, 2009). These muscles tense the floor of the pelvis, support the organs of the pelvis, elevate and retract the anus, and flex the coccygeal joints in the pelvis (Martini et al., 2009). Voluntary contraction of the levator ani muscles at maximum strength (such as during childbirth) further compresses the distal part of the vagina, the mid-urethra and rectum against the pubic bone (Ashton-Miller and DeLancey, 2009). The maximum voluntary contraction of these muscles can further increase vaginal closure force by 46%, also significantly increasing intra-abdominal pressure (Ashton-Miller and DeLancey, 2009). These figures demonstrate the already significant contractile force of the levator ani muscles at rest; the further increase of contractile force output during the process of parturition has been suggested by scholars to leave surface changes to the bony pelvis.

There are a range of different surface changes to the bony pelvis that have been identified by scholars as possible evidence of childbirth, which have collectively been called 'parturition scarring' (Angel, 1969; Cox, 1989; Holt, 1978; Houghton, 1974; Stewart, 1957) and are thought to indicate vaginal birth (McArthur et al., 2016). For example, Stewart (1957) identified sclerotic growths at the margin of the pubic symphysis (Fig. 8.5) and pitting on the dorsal aspect of the pubis to be evidence of childbirth in a sample of Inuit skeletons of unknown origin. He argued that since these bone changes were only identified on female skeletons of childbearing age, that the most likely explanation was that they represented skeletal reactions to the stress of childbirth.

Similarly, Houghton (1974) examined the preauricular groove of the ilium in the skeletons of women from anatomical collections at the University of Otago, New Zealand (Fig. 8.6) and identified two different types of grooves—the first was present in both male

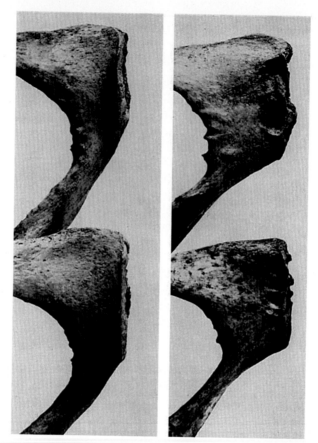

FIG. 8.5 Two male (left) and two female (right) pubic bones from an Inuit sample seen from behind. Note the pits undermining the dorsal edge of the symphyseal surface in the females (Stewart, 1957). *From Houghton, P. (1974). The relationship of the pre-auricular groove of the Ilium to pregnancy. American Journal of Physical Anthropology, 41(3), 381–389. Copyright © 1974 Wiley-Liss, Inc., A Wiley Company.*

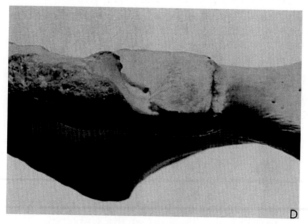

FIG. 8.6 'Groove of pregnancy', a form of preauricular groove on the left ilium. The inferior margin of the auricular surface is located on the right of the figure. *From Houghton, P. (1974). The relationship of the pre-auricular groove of the Ilium to pregnancy. American Journal of Physical Anthropology, 41(3), 381–389. Copyright © 1974 Wiley-Liss, Inc., A Wiley Company.*

and female pelves, the second only in females. Houghton suggested that the first groove type (the "groove of ligament") is caused by the pathological and physiological changes occurring at the site of attachment of the pelvic joint ligament (and not simply as a result of childbirth). Houghton proposed that the second groove type (the "groove of pregnancy") is caused by pregnancy. The sacroiliac joint is an important weight-bearing area that will undergo modifications to accommodate increased load during pregnancy, modification that is reflected by an active osteoclastic resorption of bone adjacent to ligamentous attachments (Houghton, 1974).

More recently, Igarashi et al. (2020) analysed a skeleton population of early 20th century Japanese women and did not find scarring on the pelves of nulliparous women (i.e., those who have not given birth), but did find that weak expressions of 'parturition scarring' indicated a lower number of pregnancies and birth events, and that more developed 'parturition scarring' indicated a higher number of pregnancies and birth events.

Unfortunately, however, the idea that bioarchaeologists can identify parous females in the archaeological record based on bony changes in the pelvis is not so clear cut. In particular, the findings of several other studies suggest that pelvic scarring does not accurately indicate that childbirth occurred (Holt, 1978; Snodgrass and Galloway, 2003; Suchey et al., 1979). Instead, other factors have been found to be associated with similar scarring on the dorsal pubis, including general age-related changes, urinary tract infections, lumbosacral anomalies, and obesity, as well as repeated minor trauma, surgery, general joint laxity and pelvic instability, variation in sciatic notch angle, and habitual postures, such as squatting (Ubelaker and De La Paz, 2012). Decrausaz (2014) investigated the presence and prevalence of 'parturition scarring' in male and female adults in two skeletal populations— the Maxwell Documented Skeletal Collection from University of New Mexico's Maxwell Museum of Anthropology in Albuquerque, New Mexico, and the 16th to 18th century individuals from Spitalfields, London. She found no evidence to support the assertion that there is an association between 'parturition scarring' and childbirth and indeed found parturition scarring present on the pelves of male individuals. Instead, Decrausaz (2014) found that dorsal pitting correlated weakly with bi-acetabular breadth, mediolateral pelvic inlet, anterior–posterior direction of the pelvic midplane, mediolateral pelvic outlet, and pubic length among females.

Some recent studies have instead suggested that 'parturition scarring' may be better used as an indicator of population childbirth status rather than individual childbirth experience, providing a counterpoint to studies that have demonstrated that pelvic scarring should not be used as a means of identifying childbirth on the human skeleton (Decrausaz, 2014; McFadden and Oxenham, 2018). Waltenberger et al. (2022) more recently examined pelvic scarring on the pelves of 150 women whose full-body scans were included in the New Mexico Decedent Image Database (NMDID), along with detailed information about their experiences of childbirth and pregnancy. They found that a size score of dorsal pitting helped estimate the mean childbirth status in groups of individuals, though was not precise enough to provide childbirth information for one person. Scarring on the surface of the pelvis cannot be used as an unequivocal indication of childbirth; however, it is possible that its presence may be impacted by childbirth or pregnancy. To this end, features that have been identified as 'parturition scarring' should be more accurately viewed as pelvic scarring and approached with caution as a means of identifying childbirth in the human skeleton.

8.5 Movement (locomotion and gait)

The human pelvis exhibits several evolutionary adaptations for habitual bipedality. Upright locomotion requires many bony adjustments, including extension at the hip joint and a vertically tilted pelvic girdle compared to what is seen in the other apes (Aiello and Dean, 2002). As humans walk, the pelvis undergoes movements along three primary planes: the sagittal, frontal, and transverse planes (Cappozzo et al., 2005; Lewis et al., 2017). Movement in the sagittal plane results in anterior/posterior pelvic tilt. As each foot is lifted in step, the pelvis lifts or drops along the frontal plane, and as the leg is swung, the pelvis rotates forward or backwards in the transverse plane (Cappozzo et al., 2005; Lewis et al., 2017; Warrener et al., 2015). All three motions are utilised during the human gait cycle as the pelvis must advance the body, adjust stride and step width, and maintain stability and balance. The human hip abductors create an increased moment arm to balance the centre of mass (Lewis et al., 2017; Lovejoy, 2005a; Robinson, 1972; Warrener et al., 2015). The conversion of these hip abductors to accommodate bipedal gait is reflected in the pelvic morphology of the human ilia: the iliac blades are shorter, broader, and possess larger anterior iliac spines compared to other apes, such as chimpanzee (Lewis et al., 2017; Lovejoy, 2005a; Robinson, 1972) (see Fig. 8.3). These anatomical changes are visible in some of the earliest hominin fossils, including the 3.8-million-year-old *Australopithecus afarensis* specimen, Lucy, from Ethiopia (Johanson et al., 1982; Lovejoy, 2005a; Robinson, 1972) (Fig. 8.3).

Understanding locomotor biomechanics of modern humans can enable the extrapolation of activities in the past. Pleistocene footprints from the Willandra Lakes region of Western New South Wales, Australia, are one such example (Webb et al., 2006). Through an examination of footprints dated to 19–23,000 years ago, researchers determined anthropometric measures of the individuals and calculated their approximate gait speed, enabling the researchers to distinguish which of the individuals were running and which were walking (Webb et al., 2006). These footprints showed that the group was comprised of people of different sexes and ages (Webb et al., 2006). The estimation of sex and age was based on the size of the footprints in comparison to modern Australian Aboriginal anthropomorphic datasets, with the larger footprints probably belonging to adult males and the smaller footprints belonging to females or children (Webb et al., 2006). Furthermore, sexual dimorphism in the pelvis results in different gait and stride patterns, which could be interpreted in the footprints. Specifically, compared to males, females tend to have more pelvic movement in the transverse plane when walking at a normal speed (forward/backward rotation) (Bruening et al., 2015; Chumanov et al., 2008; Lewis et al., 2017). The female pelvis tends to be mediolaterally broader and this allows for increased transverse movement, which is thought to compensate for a shorter stride width and conserves energy during locomotion (Wall-Scheffler and Myers, 2017). Differentiation between adults and children was determined by estimating the stature of the individuals, under the assumption that taller individuals would most likely be adults and shorter individuals would be adolescents or children (Webb et al., 2006). The mixed-sex and age analysis indicates collaborative activities, as opposed to a male-only hunting party. As there is a lake in close proximity to the footprints, the researchers suggest that the group was engaging in activities such as fishing or gathering shellfish (Webb et al., 2006). Through our understanding of how the pelvis influences how we walk, archaeologists can extrapolate lifestyles and behaviours from these fossil footprints. However, it is important to note that confounding factors such as gait speed and stature can affect these variables and conclusions (Kerrigan et al., 2001; Lewis et al., 2017; Murray, 1970).

In the archaeological record, pathologies at the hip joint can also point to specific gait patterns that can be used to reconstruct the behaviour. One such pathology is acetabular dysplasia, where the acetabulum develops as too shallow and does not provide enough coverage for the femoral head (Lewis et al., 2017). Due to this size discrepancy between the hip joint elements, pelvic motions are affected and can often lead to osteoarthritis (a disease of the articular cartilage, which breaks down as the disease progresses at the hip joint) (Birrell and Haslam, 2009; Lewis et al., 2017; Lievense et al., 2004; Reijman et al., 2005).

Excavations of a mediaeval cemetery from the site of St. Mary Spital, located in Spitalfields Market, London, in the United Kingdom resulted in analyses of over 500 adults, with 1.7% of these individuals having hip joint abnormalities, including acetabular dysplasia (Mitchell and Redfern, 2011). Whether this rate is high or low compared to modern times is hard to determine due to the modern overdiagnosis of acetabular or hip dysplasia during in utero ultrasounds but normal morphology presentation at or after birth (Mitchell and Redfern, 2007, 2011). However, the pathology in this sample can allow for lifestyle reconstruction at this time period (Mitchell and Redfern, 2007, 2011). As acetabular dysplasia can impede locomotion, some of these individuals may have required extra assistance. Identification of conditions such as hip dysplasia allows researchers to investigate whether factors such as medical interventions and increased nutritional quality impact the prevalence of such diseases. In the case of hip joint diseases at this site, researchers found no change in the incidence rate compared to modern times (Mitchell and Redfern, 2007, 2011). Thus they suggested that modern "improvements" in factors such as obstetric care or diet do not impact the rate of these conditions (Mitchell and Redfern, 2007, 2011). It is clear that the aetiology of these hip pathologies is multifactorial and complex, with a strong genetic component (Blatt, 2015; Mitchell and Redfern, 2011), but some researchers suggest that certain behaviours may cause these hip joint conditions.

Cultural behaviours such as swaddling infants have been shown to increase the risk for acetabular dysplasia, which may point to lifestyle behaviours in certain areas and time periods (Blatt, 2015; Mitchell and Redfern, 2011). While swaddling is thought to increase infant comfort by creating a secure environment, others advocate that swaddling contributes to the malposition of the infant's hips and lower limbs, leading to conditions such as acetabular dysplasia. This is thought to be a result of the prolonged extension and adduction of the hips and lower limbs in swaddled infants (Blatt, 2015; Mitchell and Redfern, 2011). One analysis of pelves from the Late Prehistoric Buffalo site in West Virginia (United States) investigated the rate of various hip dysplasia (DDH) conditions (Blatt, 2015). Acetabular dysplasia was found in 0.8% of the pelves in this population, which is "the highest reported global prevalence of DDH now and in prehistory" (Blatt, 2015, p. 126). Indigenous populations have a variety of swaddling techniques, yet almost all involve a tight binding around the hips and lower limbs (Blatt, 2015; Houston and Buhr, 1988). Therefore it is thought that the rate of dysplasia from this sample could in part be a result of this cultural practice (Blatt, 2015).

Abnormal pelvic bone morphology can also indicate when individuals in the past experienced certain diseases and, indirectly, behaviour of care in a population. For example, cerebral palsy is a musculoskeletal disease that results in a breakdown of the neuromuscular system (Schrenk and Martin, 2017). As muscles are affected, so are the bones. In one

Bronze Age (2200 BCE to 300 CE) site, Tell Abraq in the United Arab Emirates, an individual (~18 years old at time of death) was found with pathologically related bone changes indicative of cerebral palsy (Schrenk and Martin, 2017). Multiple areas of the skeleton showed signs of bone changes, with pathological changes on the pelvis and lower limbs indicating that the individual experienced paralysis and impaired gait during adulthood (Schrenk and Martin, 2017). Specifically, limb asymmetry, with the left lower limb being shorter than the right, is common in individuals with cerebral palsy (Schrenk and Martin, 2017). This lower limb asymmetry produces different gait movements, as the muscles' direction of pull is altered. This results in individuals with cerebral palsy walking with a flexed hip and knee (Schrenk and Martin, 2017). In the Bronze Age individual's pelvis, indicators of this type of gait pattern are expressed in the pelvis and lower limb: the left os coxae is asymmetrical, with the left sciatic notch being much narrower than the right side, and the left foot exhibits a calcaneo-cavo valgus deformity which is indicative of cerebral palsy (Schrenk and Martin, 2017). The researchers also note abnormalities in the sacrum, but beyond noting that the sacral morphology is both "both concavely and convexly pronounced" (Schrenk and Martin, 2017, p. 51), they do not expand upon this morphology's significance to the disease. If this individual did have cerebral palsy, Schrenk and Martin (2017) argue that special care is evident to ensure this individual's survival, as cerebral palsy symptoms are present at birth and this individual lived to be around 18 years old. Namely, mobility assistance was probably required for this individual to survive, as well as extra care for other potential complications associated with cerebral palsy such as urinary incontinence, which affects around 25% of cerebral palsy cases (Schrenk and Martin, 2017). This individual's bones reveal more than just a diagnosis of a pathology—they indicate societal structure, at least at the household level.

The adaptation to bipedal locomotion necessitated a conversion of the hominin pelvic structure and function compared to the other great apes. The hip joint morphology, including abnormalities, can tell us a lot about an individual's gait, behaviours, and even cultural and societal beliefs as seen in individuals who were swaddled or had pathological conditions. In the next section, we shift our focus to a discussion of another major function of the pelvis: serving as a reproductive passageway.

8.6 Health, disease, and trauma

Aside from the potential for injury during childbirth, the pelvis can be impacted by joint diseases, metabolic diseases, infectious diseases, and trauma. Diseases or pathological conditions of the pelvis will often impact locomotion, childbirth, or pelvic soft tissues that support reproductive and digestive organs. Considering the pivotal role that pelvis morphology plays in both locomotion and reproduction, pathological changes can often result in impairment and acute or chronic pain. Identifying such changes in archaeological skeletons may help us understand how these health issues were treated and handled in the past, including how impairment may have impacted the daily lives of not only those inflicted, but also the people who cared for them. These impacts may mean that people living with such diseases or trauma may have benefited from community support for everyday living.

8.6.1 Joint diseases of the pelvis

Joint diseases of the pelvis may appear because of overuse or injury in everyday life or may be the result of developmental conditions. They may impact locomotory behaviour by changing gait patterns, as the resulting pain from these conditions often prompt the affected individual to alter their movements to avoid increased pain or restrict movement as a result of bones fusing together at specific joints, such as the sacrum and the lumbar spine.

One such condition is ankylosing spondylitis, an inflammatory condition that may, over time, cause bone fusion at the sacroiliac joint, joints of the spine, and costovertebral joints (Šlaus et al., 2012). Inflammatory conditions occur when the body's immune system mistakenly attacks the body's own cells and tissues as if they were an invading pathogen or infection. Individuals with ankylosing spondylitis are generally genetically predisposed to the condition due to the presence of the HLA allele B27 (Simone et al., 2018). Šlaus et al. (2012) found evidence of ankylosing spondylitis in the skeletons of four adults dated to between the 9th and 14th centuries in Croatia. The sacrum and os coxae of one of the skeletons examined in this study, that of a female over 55 years old, were fused at the sacroiliac joint. Sacroiliac joint fusion may result in lower back pain, though as the condition progresses, individuals may become significantly immobilised, particularly as bone fusion around the ribcage may act to restrict chest expansion. This would mean that some of the individuals examined by Šlaus et al. (2012) may have become disabled as their conditions continued and they required community support to perform everyday activities such as gathering and preparing food and water, as well as engaging in labour activities.

As discussed before, childbirth can cause stress on the human pelvis, and when these stresses act on a pelvis that already has pathological changes, this can exacerbate the condition. These pathological changes can then be used to not only tell us about what disease affected the pelvis, but also whether the inflicted individual has given birth. For example, osteitis pubis is also an inflammatory condition that affects the public symphysis and the surrounding soft tissue, characterised by pain in the pubic symphysis area which worsens with physical activity (Athanasiou et al., 2022). Today, it is particularly prevalent among athletes participating in sports such as rugby and soccer (Athanasiou et al., 2022), though it also appears in pregnant women or women who have given birth (Nasrallah et al., 2020). Pfeiffer (2011) found evidence of osteitis pubis on the pelvis of a small, middle-aged woman who lived approximately 2000 years ago on the South African Cape Coast. Pfeiffer (2011) suggested that this injury, along with pelvic asymmetry caused by a difference in sacral costal process (or ala) thickness on one side, may have been exacerbated by childbirth. In women today, the separation of the pubic symphysis during challenging childbirth has been reported to occur between 1 in 600 and 1 in 30,000 deliveries (Nitsche and Howell, 2011). The pathological evidence identified by Pfeiffer (2011) provides what could be the most reliable indicator of childbirth in the past, as well as indirect evidence that may suggest community care behaviour for the mother during and after birth.

Diseases in the pelvis can also indicate behaviour in a more indirect way. For example, Clark et al. (2020) diagnosed a young to middle-aged woman buried in Lake County, Ohio, in a site dated to between 1320 and 1650 C.E. as having Legg–Calvé–Perthes disease. This disease is a childhood condition where the blood supply to the femoral head is temporarily interrupted, and the bone begins to die. This can potentially change the shape of the femoral head, cause pain and stiffness, and may lead to osteoarthritis (Waldron, 2008). Upon identifying evidence of Legg–Calvé–Perthes in this young woman, Clark et al. (2020) suggested that she developed the disease in childhood as a result of strenuous hip flexion and extension associated with agricultural work, specifically digging, carrying hay, and twisting the upper and lower body. It was not uncommon for children and young adults to be involved in agricultural labour in the past, leading the authors to suggest that Legg–Calvé–Perthes disease could be understood to indicate that this woman transitioned into a new social role of engaging in agricultural work, potentially as a part of maturing into an adult in that community.

8.6.2 Metabolic diseases of the pelvis

Metabolic diseases are those that interfere with the regular metabolic functions of the skeleton. Rickets, primarily caused by a lack of vitamin D, is a childhood disease that can impact the pelvis. Vitamin D is found in some foods (including oily fish) but is principally formed within skin tissue when ultraviolet (UV) rays touching the skin initiate the production of vitamin D_3, which is inactive (Waldron, 2008). Vitamin D_3 then moves to the liver and kidney where it is converted to the active form of vitamin D known as calcitriol, which binds with calcium to mineralise bone (Waldron, 2008). A lack of vitamin D prevents the binding of the active metabolite with calcium, meaning that new bone formed during normal bone deposition is not adequately mineralised. This under-mineralised bone is softer than healthy bone and may bend under loading, such as bearing the weight of the body. This is a particular issue with the pelvis, as with weight-bearing through the trunk, into the pelvis and down through the legs, over time the softness of the femora via the hip joint may impact the shape of the pelvis. The adult form of rickets is known as osteomalacia, during which the pubic symphysis may undergo dramatic changes, such as beginning to project anteriorly and potentially fracturing (Brickley et al., 2005).

Consistent evidence suggests that the prevalence of vitamin D deficiency is highest today in Asia, the Middle East, and Africa, as well as among immigrants from these regions living in countries at higher latitudes (Roth et al., 2018). This may be due to inadequate dietary sources of vitamin D, a lack of vitamin D supplement use during childhood and lactation, or inadequate exposure to sunlight due to mobility issues or excessive clothing covering or use of sun-blocking cosmetic lotions (Roth et al., 2018). For instance, in 2020, a 22-month-old girl presented with rickets at a primary care clinic in Selangor, Malaysia (Sodri et al., 2021). This condition resulted from the combination of inadequate sun exposure due to COVID-19 pandemic's isolation requirements to reduce transmission, as well as a lack of dietary vitamin D since she had digestive issues with formula milk (Sodri et al., 2021). In England, it has been suggested that rickets prevalence continues because of the lack of UV sunlight for at least six months of the year, and because of the higher proportion of immigrants to England who are more vulnerable to vitamin D deficiency (those with darker skin, as darker skin produces far less vitamin D than lighter skin per unit UV light exposure) (Uday and Högler, 2018).

During the Industrial Revolution in countries like England, the prevalence of rickets and osteomalacia increased due to lack of exposure to sunlight. Multiple studies found evidence of rickets and osteomalacia impacting pelvic development in the skeletons of adults, including examples of pelvic canal restriction in women living in poor urban areas in the UK in the 18th and 19th centuries (Brickley et al., 2005, 2007, 2014). Labour practices and the ways in which children were involved in different types of work may also have added to the prevalence of rickets. Ellis (2010) examined the remains of 86 subadult individuals, ranging from newborns to 15-year-olds, buried between 1820 and 1846 at the Spring Street Presbyterian Church in New York City, for evidence of vitamin D deficiency. Ellis (2010) found evidence of vitamin D deficiency in over 34% of the 86 subadult individuals examined. She interpreted this high prevalence to indicate that the children in this population were likely from lower socioeconomic status groups that not only had poor dietary sources of vitamin D, but also likely had to work long hours inside factories and workshops, where they had a chronic lack of sunlight, and therefore, vitamin D. Her interpretation is supported by the historical records of the area, which state that during this period in New York City, there was a transition from home and workshop apprenticeships to a market-driven labour system, where children were frequently hired as members of the labour force (Ellis, 2010).

Rickets and a lack of vitamin D did not only affect children in low socioeconomic classes, and instead, its presence in children from higher socioeconomic groups can provide unique insight into childcare practices of more wealthy families in the past. For example, palaeopathological analyses of the remains of nine child members of Italy's famous Medici family (16th–17th centuries) showed evidence of rickets (Giuffra et al., 2010). Giuffra et al. (2010) suggested that this may have resulted from cultural practices of breastfeeding infants until the age of two. Vitamin D levels in maternal milk drop notably after three months postpartum, suggesting that dietary vitamin D levels would have been very low for these children (Giuffra et al., 2010). This may have been compounded by extended periods where the children were not exposed to sunlight in the winter months, as children who were not yet walking during this period would have been swaddled for warmth and protection (Giuffra et al., 2010). Giuffra et al. (2010) study highlights the value of understanding the socioeconomic and cultural contexts that may impact on the health of the skeleton and should be routinely considered in bioarchaeology.

8.6.3 Infectious diseases of the pelvis

Infectious diseases are caused by biological agents such as prions, viruses, or bacteria and may leave evidence of infection on the pelves of people who were inflicted with infections in the past. At times, dietary behaviour can drive how a person may contract an infectious disease, whether through the consumption of contaminated milk or meat products, or by living in close proximity with husbandry animals. Tuberculosis (TB) is a bacterial infection that is frequently identified skeletally, sometimes including the pelvis. TB is caused by the organisms of the *Mycobacterium tuberculosis* complex and can be spread via the respiratory route (*M. tuberculosis*) or contracted from animals (*Mycobacterium bovis*), specifically via consuming contaminated milk or meat (Burke, 2011). TB can spread from the lungs or gastrointestinal tract through the blood and lymphatic systems to bones, mainly causing destruction of the bony tissue. The most frequent site of destruction is the spine (Pott's disease); the hip and knee joints are less so, but any bone may be affected (Filipek and Roberts, 2018). Lewis (2011) found evidence of a possible TB infection in the pelvis of an 11-year-old individual from the Romano-British Poundbury Camp (1st to 3rd centuries CE). Sparacello et al. (2017) found lesions suggestive of TB in a five-year-old child, dated to the Middle Neolithic of Liguria (Italy), or 5740 ± 30 BP. Both examples provide evidence that not only were cattle likely kept in close proximity to the households of the afflicted, but the children were likely consuming raw milk from the cows, thus infecting them with the bacteria.

8.6.4 Trauma to the pelvis

Trauma due to accidents and conflict might also impact the pelvis during life. Evidence of trauma on the pelvis will vary depending on whether it occurred perimortem or postmortem. The comparative fragility of the pelvis with respect to preservation (Rosenberg and DeSilva, 2017) might also limit interpretation of trauma patterns, which influences the comparative scarcity of palaeopathological reports on pelvic injuries.

In modern clinical literature, the height of a fall has been found to correlate with the severity of pelvic injury (Nau et al., 2021). Unstable pelvic ring fractures, where fractures occur at more than two points and involve severe bleeding, are typically seen in high-energy trauma such as falls from heights (Balogh et al., 2007). Willett and Harrod (2017) present a case study on the skeleton of a woman from the site known as Aztec Ruins in northern New Mexico, occupied between circa C.E. 1100 and 1290, who survived a severe pelvic ring fracture. This woman's pelvic injuries are consistent with a fall, showing fractures at the ischiopubic ramus, the acetabulum, and the ischial spine. If the ischial spine was injured at the same time as the rest of the pelvic ring, it would have likely caused pelvic floor disruption and associated injuries to the bladder. Willett and Harrod (2017) propose that these injuries would have resulted in decreased mobility and that the individual would have required near constant care from community members, beginning with trauma management and transforming into daily accommodation for their impairment. Overall, the trauma that this individual experienced, survived, and continued to live with indicates that both the afflicted and their community worked to rehabilitate and reintegrate them back into group practices.

Worne (2017) also presents a case study on the impact of pelvic trauma to the life of an adult female of about 46 years of age from a late prehistoric agricultural community, Averbuch, in Tennessee, United States of America. This female skeleton was interred in a burial dated to C.E. 1390 and C.E. 1470 and showed evidence of multiple fractures of the pelvic ring (as well as other injuries throughout the skeleton), some showing healing, and others likely occurring closer to death. For example, the most recent injury involved a compression fracture of the left pubic symphyseal face, causing the dorsal half of the surface to be depressed laterally and angled posteriorly. There was minimal evidence of healing for this injury, and, clinically, this type of injury is associated with increased risk of damage to internal abdominal and pelvic organs, bleeding, and death. Worne (2017) outlines that the combination of injuries in the pelvis and throughout the skeleton may indicate that this woman required direct support for essential tasks, such as walking, gathering food and water, and cleaning her wounds in the period right after these injuries occurred. Worne (2017) also incorporated analyses of the built structures of the Averbuch site into her evaluation of the impact of injuries to this woman's life, suggesting that the many stairs found throughout the village complex would have required this woman to have assistance to move around the site, given her pelvic instability.

Pelvic trauma also arises from different forms of conflict, reflecting different ways in which interpersonal behaviour may leave evidence on the human skeleton. For example,

Wheeler et al. (2013) provide the earliest documented case of child abuse in the archaeological record, focusing on the skeletal remains of a two-year-old child from a Romano-Christian period cemetery in the Dakhleh Oasis, Egypt, dated to between C.E. 50 and C.E. 450. The pattern of fractures and differential stages of healing in this child's pelvic bones indicate multiple traumatic events. Wheeler et al. (2013) identified a compression fracture on the right and left ilium, with new bone formation anterior to the fracture. They also found new bone formation on the ramus of the right ischium and the acetabular surface of the right pubis, showing a fracture in the early stages of healing. In clinical investigations, fractures of the pubic rami can suggest child sexual abuse (Starling et al., 2002). Wheeler et al. (2013) also made use of isotopic and histological analyses to assess physiological responses to trauma. They found that new bone formation patterns on the humeri and ribs were indicative of a traumatic aetiology, while isotopic values suggested a period of nutritional depletion or trauma, and a period of recovery, consistent with child abuse.

8.7 Everyday behaviours

Evidence pointing to daily activities or the use of specific cultural materials, such as clothing or personal items, may also be found on the human pelvis. The evidence left on the skeleton from the use of clothing items, such as corsets, or activities, such as horse riding, may vary from individual to individual as a result of the length of time engaging in the activity or wearing the item of clothing.

8.7.1 Corset use and its impacts on the pelvis

One common cultural behaviour of the past was the wearing of restrictive corsets. Wearing tight, waist-altering corsets was a popular fashion trend in Europe between the 1500s and 1900s. This included multiple variations in style and fit for men and women, as well as children, where corsets were used to mould the body into idealised shapes and sizes (Gibson, 2020). It is challenging to compare the impacts of corset wearing in past populations with corset wearing today, as corsets today are not as commonly worn as they were during the 1500s and 1900s, are often constructed using different materials, and are not worn by growing children. Aiming to understand the physiological impacts of corset wearing, Na (2015) examined changes in heart rate, blood flow, perspiration, metabolism, and subjective pressure sensation for five healthy women in their 20s when exercising wearing three different types of 18th century European corsets, all with plastic boning. Each corset was reconstructed from historical corset patterns. The results of the study showed that the corset with the most boning in its construction incurred the highest clothing pressure at two points on the body, under the armpit and near the side waist point, and resulted in the highest changes of energy metabolism for the wearer. However, this same corset showed the least changes of heart rate among those tested.

The pressure applied to the ribcage, spine, abdominal area, and upper parts of the pelvis has been suggested to impact the skeleton. Researchers have proposed that corseting may act to flatten the pelvis from front to back, though these studies do not include any specific examples of how this presents skeletally (Loudon, 1997; Ortner and Putschar, 1981;

Roberts and Manchester, 2007). More recently, Gibson (2020) examined the skeletons of St. Bride's Lower Churchyard (a parish cemetery dated to the 18th century) for skeletal evidence of corset wearing. She suggests that corseting puts radial pressure on the ribcage, distorting the usual wide-oval shape (broader than it is deep) into a much more circular profile. She identified such profiles on a number of individuals from St. Bride's and suggested that the thoracic spinous processes are diverted from their regular angles of about 45 degrees (measured from the body of the vertebrae), to angles between 10 and 30 degrees as a result of corset wearing. Gibson (2020) does not outline any pelvic indicators of corset use. Corsets have also been examined through a critical feminist lens as examples of purposeful pain inflicted on women's bodies (Stone, 2020), usually a means of controlling women's behaviours, both physically and morally (Stone, 2012), and as part of the suite of features that are connected to biomedical control of women's bodies (see Martin, 2001).

8.7.2 Horse riding and its impacts on the pelvis

Another common behaviour that may impact pelvic features is horse riding. Horse riding engages muscles throughout the lower body and the trunk, and involves repetitive impacts radiating through the lumbar spine, ischial tuberosities, pelvis, and sacrum (Pugh and Bolin, 2004). There are relatively few bioarchaeological studies that examine the impact of horse riding on the skeleton, despite the prevalence of this behaviour in past populations across the globe. This is an area that would benefit from a multidisciplinary approach to expand studies on the human skeleton and behaviour, ideally bringing together bioarchaeological scholars with clinicians specialising in human anatomy, bone health, and human movement (kinesiology).

In bioarchaeology, this might present as 'horseback riding syndrome', which includes a combination of skeletal changes such as osteoarthritic changes to the spine, Schmorl's nodes, variation in femoral neck length, and evidence of muscular action at the site of the quadricep and femoris muscles (see Chapter 9 in this book and Berthon et al., 2019 for complete list of sources). The ideal pelvis position for a rider is retroversion, where the pelvis is tilted backwards to allow for more effective absorption of vertical stresses (Auvinet, 1999; Humbert, 2000). In this retroverted position, the shape of the acetabulum may change as a result of the pressure from the femoral head on the anterior and anterior–superior aspects of the acetabular space (Baillif-Ducros et al., 2012). Zejdlik et al. (2021) found osteophytic development on the vertebrae and hypertrophic muscle attachment areas on the pelvis and leg bones of two adult males dated to the 18th century, likely members of the Szekler people of Transylvania, Romania. The authors of this study suggested that these skeletal features, along with historical information pointing to these individuals being mounted cavalry as a result of their noble status, were consistent with regular horse riding.

Interpreting everyday behaviour from the pelvis requires a foundation in anatomical knowledge, but also an understanding of how the pelvis contributes to and is impacted by movement. Material culture, such as extended periods of corset wearing, may alter the shape of pelvis, though more extensive research is required to understand the long-term physiological impacts of shaping the body in this way. Despite the high prevalence of horse riding behaviour in past populations, there are comparatively few bioarchaeological examinations of how these sorts of behaviours impacted the pelvis, providing a platform for future collaborative work between clinicians and bioarchaeologists.

8.8 Conclusion

The human pelvis is complex, rendering its study from a multidisciplinary perspective both appropriate and challenging. The pelvis serves not only as a supporting structure for our locomotor and visceral capacity but can also serve as a reproductive passageway in females. This chapter has outlined the factors that shape the pelvis due to our evolutionary history, development and growth, health status, and activities such as childbirth. Throughout our evolutionary history, the pelvis has continually adapted to the multitude of influences that can impact its morphology. A slight change in the structure of the acetabulum, for example, can impact the gait of an individual. Nutrient deficiencies impact bone health and stability, and trauma to the pelvis can indicate societal behaviours. Future research on the human pelvis must include data from studies including, but not limited to, those in biomechanics, human biology, child growth, obstetric medicine, palaeoanthropology, and comparative anatomy. Analyses of these data should also be incorporated with an understanding of the socioeconomic, political, and cultural factors that impact pelvic morphology—while some of these factors may not be possible to examine with palaeoanthropological or archaeological skeletal remains, they are vital to understanding pelvic morphology in humans today. In this sense, the human pelvis represents knowledge areas that connect humans to their past, as well as to their futures.

References

Adair, F. L. (1918). The ossification centers of the fetal pelvis. *The American Journal of Obstetrics and Diseases of Women and Children (1869–1919)*, 78(2), 175.

Aiello, L., & Dean, C. (2002). *An Introduction to Human Evolutionary Anatomy*. New York: Elsevier Academic Press.

Angel, J. L. (1969). The bases of paleodemography. *American Journal of Physical Anthropology*, 30(3), 427–437.

Ashton-Miller, J. A., & DeLancey, J. O. (2009). On the biomechanics of vaginal birth and common sequelae. *Annual Review of Biomedical Engineering*, 11, 163–176.

Athanasiou, V., Ampariotou, A., Lianou, I., Sinos, G., Kouzelis, A., & Gliatis, J. (2022). Osteitis pubis in athletes: a literature review of current surgical treatment. *Cureus*, 14(3).

Auerbach, B. M., King, K. A., Campbell, R. M., Campbell, M. L., & Sylvester, A. D. (2018). Variation in obstetric dimensions of the human bony pelvis in relation to age-at-death and latitude. *American Journal of Physical Anthropology*, 167(3), 628–643.

Auvinet, B. (1999). Lombalgies et équitation. *Synoviale, rhumatologie sportive*, 83, 25–31.

Baillif-Ducros, C., Truc, M. C., Paresys, C., & Villotte, S. (2012). Approche méthodologique pour distinguer un ensemble lésionnel fiable de la pratique cavalière. Exemple du squelette de la tombe 11 du site de «La Tuilerie» à Saint-Dizier (Haute-Marne), VI e siècle. *Bulletins et Mémoires de la Société d'Anthropologie de Paris*, 24(1), 25–36.

Balogh, Z., King, K. L., Mackay, P., McDougall, D., Mackenzie, S., Evans, J. A., et al. (2007). The epidemiology of pelvic ring fractures: A population-based study. *Journal of Trauma and Acute Care Surgery*, 63(5), 1066–1073.

Berge, C., & Goularas, D. (2010). A new reconstruction of Sts 14 pelvis (*Australopithecus africanus*) from computed tomography and three-dimensional modeling techniques. *Journal of Human Evolution*, 58(3), 262–272.

Berge, C., Orban-Segebarth, R., & Schmid, P. (1984). Obstetrical interpretation of the australopithecine pelvic cavity. *Journal of Human Evolution*, 13(7), 573–587.

Bergfelder, T., & Herrmann, B. (1980). Estimating fertility on the basis of birth-traumatic changes in the pubic bone. *Journal of Human Evolution, 9*(8), 611–613.

Berthon, W., Tihanyi, B., Kis, L., Révész, L., Coqueugniot, H., Dutour, O., et al. (2019). Horse riding and the shape of the acetabulum: Insights from the bioarchaeological analysis of early Hungarian mounted archers (10th century). *International Journal of Osteoarchaeology, 29*(1), 117–126.

Betti, L., & Manica, A. (2018). Human variation in the shape of the birth canal is significant and geographically structured. *Proceedings of the Royal Society B, 285*(1889), 20181807.

Betti, L., von Cramon-Taubadel, N., Manica, A., & Lycett, S. J. (2013). Global geometric morphometric analyses of the human pelvis reveal substantial neutral population history effects, even across sexes. *PLoS One, 8*(2), e55909.

Birrell, S., & Haslam, R. (2009). The effect of military load carriage on 3-D lower limb kinematics and spatiotemporal parameters. *Ergonomics, 52*(10), 1298–1304. https://doi.org/10.1080/00140130903003115. In press.

Blatt, S. H. (2015). To swaddle, or not to swaddle? Paleoepidemiology of developmental dysplasia of the hip and the swaddling dilemma among the indigenous populations of North America. *American Journal of Human Biology, 27*(1), 116–128.

Brickley, M., Mays, S., & Ives, R. (2005). Skeletal manifestations of vitamin D deficiency osteomalacia in documented historical collections. *International Journal of Osteoarchaeology, 15*(6), 389–403.

Brickley, M., Mays, S., & Ives, R. (2007). An investigation of skeletal indicators of vitamin D deficiency in adults: Effective markers for interpreting past living conditions and pollution levels in 18th and 19th century Birmingham, England. *American Journal of Physical Anthropology, 132*(1), 67–79.

Brickley, M. B., Moffat, T., & Watamaniuk, L. (2014). Biocultural perspectives of vitamin D deficiency in the past. *Journal of Anthropological Archaeology, 36*, 48–59.

Bruening, D. A., Frimenko, R. E., Goodyear, C. D., Bowden, D. R., & Fullenkamp, A. M. (2015). Sex differences in whole body gait kinematics at preferred speeds. *Gait & Posture, 41*(2), 540–545.

Brůžek, J., Santos, F., Dutailly, B., Murail, P., & Cunha, E. (2017). Validation and reliability of the sex estimation of the human os coxae using freely available DSP2 software for bioarchaeology and forensic anthropology. *American Journal of Physical Anthropology, 164*(2), 440–449.

Buikstra, J., & Ubelaker, D. (1994). *Standards for Data Collection From Human Skeletal Remains. 44.* Arkansas Archaeological Survey Research Series.

Burke, S. D. (2011). Tuberculosis: Past and present. *Reviews in Anthropology, 40*(1), 27–52.

Cappozzo, A., Della Croce, U., Leardini, A., & Chiari, L. (2005). Human movement analysis using stereophotogrammetry: Part 1: Theoretical background. *Gait & Posture, 21*(2), 186–196.

Chumanov, E. S., Wall-Scheffler, C., & Heiderscheit, B. C. (2008). Gender differences in walking and running on level and inclined surfaces. *Clinical Biomechanics, 23*(10), 1260–1268.

Clark, M. A., Bargielski, R., & Reich, D. (2020). Adult paleopathology as an indicator of childhood social roles: A case study of Perthes disease in a native Ohio female. *International Journal of Osteoarchaeology, 30*(1), 24–32.

Claxton, A. G., Hammond, A. S., Romano, J., Oleinik, E., & DeSilva, J. M. (2016). Virtual reconstruction of the Australopithecus africanus pelvis Sts 65 with implications for obstetrics and locomotion. *Journal of Human Evolution, 99*, 10–24.

Corron, L. K., Santos, F., Adalian, P., Chaumoitre, K., Guyomarc'h, P., Marchal, F., et al. (2021). How low can we go? A skeletal maturity threshold for probabilistic visual sex estimation from immature human os coxae. *Forensic Science International, 325*, 110854.

Cox, M. J. (1989). *An evaluation of the significance of scars of parturition in the Christ Church Spitalfields sample Doctoral dissertation.* University of London.

Cox, M., & Scott, A. (1992). Evaluation of the obstetric significance of some pelvic characters in an 18th century British sample of known parity status. *American Journal of Physical Anthropology, 89*(4), 431–440.

Decrausaz, S. L. (2014). *A morphometric analysis of parturition scarring on the human pelvic bone [unpublished Master's thesis].* University of Victoria.

DeSilva, J. M., Laudicina, N. M., Rosenberg, K. R., & Trevathan, W. R. (2017). Neonatal shoulder width suggests a semirotational, oblique birth mechanism in Australopithecus afarensis. *The Anatomical Record, 300*(5), 890–899.

Dunsworth, H. M. (2018). There is no" obstetrical dilemma": towards a braver medicine with fewer childbirth interventions. *Perspectives in Biology and Medicine, 61*(2), 249–263.

Dunsworth, H. M., Warrener, A. G., Deacon, T., Ellison, P. T., & Pontzer, H. (2012). Metabolic hypothesis for human altriciality. *Proceedings of the National Academy of Sciences of the United States of America, 109*(38), 15212–15216.

Ellis, M. A. (2010). The children of spring street: Rickets in an early nineteenth-century congregation. *Northeast Historical Archaeology, 39*(1), 7.

Filipek, K. L., & Roberts, C. A. (2018). Bioarchaeology of infectious diseases. In *The International Encyclopedia of Biological Anthropology* (pp. 1–9). John Wiley & Sons, Inc.

Geller, S. E., Koch, A. R., Garland, C. E., MacDonald, E. J., Storey, F., & Lawton, B. (2018). A global view of severe maternal morbidity: Moving beyond maternal mortality. *Reproductive Health, 15*(1), 31–43.

Genovés, S. (1959). L'estimation des différences sexuelles dans l'os coxal: différences métriques et différences morphologiques. *Bulletins et Mémoires de la Société d'Anthropologie de Paris, 10*(1), 3–95.

Gibson, R. (2020). The corset as a garment: Is it a representative of who wore it? In *The corseted skeleton* (pp. 57–91). Cham: Palgrave Macmillan.

Giuffra, V., Giusiani, S., Fornaciari, A., Villari, N., Vitiello, A., & Fornaciari, G. (2010). Diffuse idiopathic skeletal hyperostosis in the Medici, Grand Dukes of Florence (XVI century). *European Spine Journal, 19*(2), 103–107. https://doi.org/10.1007/s00586-009-1125-3. In press.

Grabowski, M. W., Polk, J. D., & Roseman, C. C. (2011). Divergent patterns of integration and reduced constraint in the human hip and the origins of bipedalism. *Evolution: International Journal of Organic Evolution, 65*(5), 1336–1356.

Gray, H. (1918). *Anatomy of the human body.* Philadelphia: Lea & Febiger.

Grissom, L. E., Harty, M. P., Guo, G. W., & Kecskemethy, H. H. (2018). Maturation of pelvic ossification centers on computed tomography in normal children. *Pediatric Radiology, 48*(13), 1902–1914.

Gruss, L. T., & Schmitt, D. (2015). The evolution of the human pelvis: Changing adaptations to bipedalism, obstetrics and thermoregulation. *Philosophical Transactions of the Royal Society, B: Biological Sciences, 370*(1663), 20140063.

Haeusler, M., Grunstra, N. D., Martin, R. D., Krenn, V. A., Fornai, C., & Webb, N. M. (2021). The obstetrical dilemma hypothesis: there's life in the old dog yet. *Biological Reviews, 96*(5), 2031–2057.

Häusler, M., & Schmid, P. (1995). Comparison of the pelves of Sts 14 and AL288-1: Implications for birth and sexual dimorphism in australopithecines. *Journal of Human Evolution, 29*(4), 363–383.

Holliday, T. W. (1997). Body proportions in late Pleistocene Europe and modern human origins. *Journal of Human Evolution, 32*(5), 423–448.

Holt, A. C. (1978). A re-examination of parturition scars on the human female pelvis. *American Journal of Physical Anthropology, 49*(1), 91–94.

Houghton, P. (1974). The relationship of the pre auricular groove of the Ilium to pregnancy. *American Journal of Physical Anthropology, 41*(3), 381–389.

Houston, C. S., & Buhr, R. H. (1988). Swaddling of Indian infants in northern Saskatchewan. *Musk-ox, 36*, 5–14.

Humbert, C. (2000). L'équitation et ses conséquences sur le rachis lombaire du cavalier. In *à propos de 123 observations [unpublished doctoral dissertation]* UHP-Université Henri Poincaré.

Igarashi, Y., Shimizu, K., Mizutaka, S., & Kagawa, K. (2020). Pregnancy parturition scars in the preauricular area and the association with the total number of pregnancies and parturitions. *American Journal of Physical Anthropology, 171*(2), 260–274.

Johanson, D. C., Lovejoy, C. O., Kimbel, W. H., White, T. D., Ward, S. C., Bush, M. E., et al. (1982). Morphology of the Pliocene partial hominid skeleton (AL 28~1) from the Hadar formation, Ethiopia. *American Journal of Physical Anthropology, 57*, 403–451.

Kerrigan, D., Riley, P., Lelas, J., & Croce, U. (2001). Quantification of pelvic rotation as a determinant of gait. *Archives of Physical Medicine and Rehabilitation, 82*(2), 217–220. https://doi.org/10.1053/apmr.2001.18063. In press.

Kibii, J. M., Churchill, S. E., Schmid, P., Carlson, K. J., Reed, N. D., De Ruiter, D. J., et al. (2011). A partial pelvis of *Australopithecus sediba. Science, 333*(6048), 1407–1411.

Klales, A. R., & Burns, T. L. (2017). Adapting and applying the Phenice (1969) adult morphological sex estimation technique to subadults. *Journal of Forensic Sciences, 62*(3), 747–752.

Krogman, W. M., & Işcan, M. Y. (1986). *The human skeleton in forensic medicine.* Illinois: Charles C. Thomas.

Kurki, H. K. (2013). Skeletal variability in the pelvis and limb skeleton of humans: Does stabilizing selection limit female pelvic variation? *American Journal of Human Biology, 25*(6), 795–802.

Kurki, H. K., & Decrausaz, S. L. (2016). Shape variation in the human pelvis and limb skeleton: Implications for obstetric adaptation. *American Journal of Physical Anthropology, 159*(4), 630–638.

Larsen, C. S. (2006). The agricultural revolution as environmental catastrophe: Implications for health and lifestyle in the Holocene. *Quaternary International, 150*(1), 12–20.

Laudicina, N. M., Rodriguez, F., & DeSilva, J. M. (2019). Reconstructing birth in *Australopithecus sediba. PLoS One, 14*(9), e0221871.

Leutenegger, W. (1972). Newborn size and pelvic dimensions of *Australopithecus. Nature, 240*(5383), 568–569.

Leutenegger, W. (1987). Neonatal brain size and neurocranial dimensions in Pliocene hominids: Implications for obstetrics. *Journal of Human Evolution, 16*(3), 291–296.

Lewis, M. E. (2011). Tuberculosis in the non-adults from Romano-British Poundbury camp, Dorset, England. *International Journal of Paleopathology, 1*(1), 12–23.

Lewis, C. L., Laudicina, N. M., Khuu, A., & Loverro, K. L. (2017). The human pelvis: Variation in structure and function during gait. *The Anatomical Record, 300*(4), 633–642.

Lievense, A. M., Bierma-Zeinstra, S. M., Verhagen, A. P., Verhaar, J. A., & Koes, B. W. (2004). Influence of hip dysplasia on the development of osteoarthritis of the hip. *The Annals of the Rheumatic Diseases, 63*, 621–626.

Loudon, I. (1997). Childbirth. In I. Loudon (Ed.), *The Oxford illustrated history of western medicine* (p. 214). Oxford: Oxford University Press.

Lovejoy, C. O. (2005a). The natural history of human gait and posture: Part 1. Spine and pelvis. *Gait & Posture, 21*(1), 95–112.

Lovejoy, C. O. (2005b). The natural history of human gait and posture: Part 2. Hip and thigh. *Gait & Posture, 21*(1), 113–124.

Lovejoy, C. O., Suwa, G., Spurlock, L., Asfaw, B., & White, T. D. (2009). The pelvis and femur of Ardipithecus ramidus: The emergence of upright walking. *Science, 326*(5949), 71-71e6.

Marnach, M., Ramin, K., Ramsey, P., Song, S.-W., Stensland, J., & An, K.-N. (2003). Characterization of the relationship between joint laxity and maternal hormones in pregnancy. *Obstetrics & Gynecology, 101*(2), 331–335. https://doi.org/10.1016/S0029-7844(02)02447-X. In press.

Martin, E. (2001). *The woman in the body: A cultural analysis of reproduction.* Boston: Beacon Press.

Martini, F. H., Timmons, M. J., & Tallitsch, R. B. (2009). *Human anatomy* (6th ed.). San Francisco: Pearson.

McArthur, T. A., Meyer, I., Jackson, B., Pitt, M. J., & Larrison, M. C. (2016). Parturition pit: The bony imprint of vaginal birth. *Skeletal Radiology, 45*(9), 1263–1267.

McFadden, C., & Oxenham, M. F. (2018). Sex, parity, and scars: A meta-analytic review. *Journal of Forensic Sciences, 63*(1), 201–206.

Mitchell, P. D., & Redfern, R. C. (2007). The prevalence of dislocation in developmental dysplasia of the hip in Britain over the past thousand years. *Journal of Pediatric Orthopaedics, 27*(8), 890–892.

Mitchell, P. D., & Redfern, R. C. (2011). Brief communication: Developmental dysplasia of the hip in medieval London. *American Journal of Physical Anthropology, 144*(3), 479–484.

Murray, M. P. (1970). Walking patterns of normal women. *Archives of Physical Medicine and Rehabilitation, 51*, 637–650.

Na, Y. (2015). Clothing pressure and physiological responses according to boning type of non-stretchable corsets. *Fibers and Polymers, 16*, 471–478. https://doi.org/10.1007/s12221-015-0471-5.

Nasrallah, K., Jammal, M., Khoury, A., & Liebergall, M. (2020). Adult female patient with osteitis pubis and pelvic instability requiring surgery: A case report. *Trauma Case Reports, 30*, 100357.

Nau, C., Leiblein, M., Verboket, R. D., Hörauf, J. A., Sturm, R., Marzi, I., et al. (2021). Falls from great heights: Risk to sustain severe thoracic and pelvic injuries increases with height of the fall. *Journal of Clinical Medicine, 10*(11), 2307.

Nitsche, J. F., & Howell, T. (2011). Peripartum pubic symphysis separation: A case report and review of the literature. *Obstetrical & Gynecological Survey, 66*(3), 153–158.

Ortner, D. J., & Putschar, W. (1981). *Identification of pathological conditions in human skeletal remains.* Washington DC: Smithsonian Institution Press.

Pfeiffer, S. (2011). Pelvic stress injuries in a small-bodied forager. *International Journal of Osteoarchaeology, 21*(6), 694–703.

Ponce de León, M. S., Golovanova, L., Doronichev, V., Romanova, G., Akazawa, T., Kondo, O., et al. (2008). Neanderthal brain size at birth provides insights into the evolution of human life history. *Proceedings of the National Academy of Sciences of the United States of America, 105*(37), 13764–13768.

Portmann, A. (1969). *A zoologist looks at humankind.* Basel: Schwabe. In press.

Pugh, T. J., & Bolin, D. (2004). Overuse injuries in equestrian athletes. *Current Sports Medicine Reports, 3*(6), 297–303.

Putschar, W. G. (1976). The structure of the human symphysis pubis with special consideration of parturition and its sequelae. *American Journal of Physical Anthropology, 45*(3), 589–594.

Rainer, W., Shirley, M. B., Trousdale, R. T., & Shaughnessy, W. J. (2021). The open Triradiate cartilage: How Young is too Young for Total hip arthroplasty? *Journal of Pediatric Orthopaedics, 41*(9), e793–e799

Rak, Y., & Arensburg, B. (1987). Kebara 2 Neanderthal pelvis: First look at a complete inlet. *American Journal of Physical Anthropology, 73*(2), 227–231.

Reijman, M., Hazes, J. M. W., Pols, H. A. P., Koes, B. W., & Bierma-Zeinstra, S. M. A. (2005). Acetabular dysplasia predicts incident osteoarthritis of the hip: The Rotterdam study. *Arthritis and Rheumatism*, 52(3), 787–793.

Roberts, C. A., & Manchester, K. (2007). *The archaeology of disease*. New York: Cornell University Press.

Robinson, J. T. (1972). *Early hominid posture and locomotion*. Chicago: University of Chicago Press.

Rosenberg, K. R. (1992). The evolution of modern human childbirth. *American Journal of Physical Anthropology*, 35(S15), 89–124.

Rosenberg, K. R., & DeSilva, J. M. (2017). Evolution of the human pelvis. *The Anatomical Record*, 300(5), 789–797.

Rosenberg, K., & Trevathan, W. (1995). Bipedalism and human birth: The obstetrical dilemma revisited. *Evolutionary Anthropology: Issues, News, and Reviews*, 4(5), 161–168.

Rosenberg, K., & Trevathan, W. (2002). Birth, obstetrics and human evolution. *BJOG: An International Journal of Obstetrics & Gynaecology*, 109(11), 1199–1206.

Roth, D. E., Abrams, S. A., Aloia, J., Bergeron, G., Bourassa, M. W., Brown, K. H., et al. (2018). Global prevalence and disease burden of vitamin D deficiency: a roadmap for action in low-and middle-income countries. *Annals of the New York Academy of Sciences*, 1430(1), 44–79.

Santos, F., Guyomarc'h, P., Rmoutilova, R., & Bruzek, J. (2019). A method of sexing the human os coxae based on logistic regressions and Bruzek's nonmetric traits. *American Journal of Physical Anthropology*, 169(3), 435–447.

Say, L., Chou, D., Gemmill, A., Tunçalp, Ö., Moller, A. B., Daniels, J., et al. (2014). Global causes of maternal death: A WHO systematic analysis. *The Lancet Global Health*, 2(6), e323–e333.

Schrenk, A. A., & Martin, D. L. (2017). Applying the index of care to the case study of a bronze age teenager who lived with paralysis: Moving from speculation to strong inference. In *New Developments in the Bioarchaeology of Care* (pp. 47–64). Cham: Springer.

Schultz, A. H. (1949). Sex differences in the pelves of primates. *American Journal of Physical Anthropology*, 7(3), 401–424.

Shirley, M. K., Cole, T. J., Arthurs, O. J., Clark, C. A., & Wells, J. C. (2020). Developmental origins of variability in pelvic dimensions: Evidence from nulliparous south Asian women in the United Kingdom. *American Journal of Human Biology*, 32(2), e23340.

Shuler, K. A., Hodge, S. C., Danforth, M. E., Funkhouser, J. L., Stantis, C., Cook, D. N., et al. (2012). In the shadow of Moundville: A bioarchaeological view of the transition to agriculture in the Central Tombigbee valley of Alabama and Mississippi. *Journal of Anthropological Archaeology*, 31(1), 586–603.

Simone, D., Al Mossawi, M. H., & Bowness, P. (2018). Progress in our understanding of the pathogenesis of ankylosing spondylitis. *Rheumatology*, 57(suppl_6), vi4-vi9.

Simpson, S. W., Quade, J., Levin, N. E., Butler, R., Dupont-Nivet, G., Everett, M., et al. (2008). A female Homo erectus pelvis from Gona, Ethiopia. *Science*, 322(5904), 1089–1092.

Šlaus, M., Novak, M., & Čavka, M. (2012). Four cases of ankylosing spondylitis in medieval skeletal series from Croatia. *Rheumatology International*, 32(12), 3985–3992.

Snodgrass, J. J., & Galloway, A. (2003). Utility of dorsal pits and pubic tubercle height in parity assessment. *Journal of Forensic Sciences*, 48(6), 1226–1230.

Sodri, N. I., Mohamed-Yassin, M. S., Nor, N. S. M., & Ismail, I. A. (2021). Rickets due to severe vitamin D and calcium deficiency during the COVID-19 pandemic in Malaysia. *The American Journal of Case Reports*, 22, e934216-1.

Sparacello, V. S., Roberts, C. A., Kerudin, A., & Müller, R. (2017). A 6500-year-old middle Neolithic child from Pollera cave (Liguria, Italy) with probable multifocal osteoarticular tuberculosis. *International Journal of Paleopathology*, 17, 67–74.

St Hoyme, L. E., & Iscan, M. Y. (1989). Determination of sex and race: Accuracy and assumptions. In *Reconstruction of Life from the Skeleton* (pp. 53–93). Liss.

Starling, S. P., Heller, R. M., & Jenny, C. (2002). Pelvic fractures in infants as a sign of physical abuse. *Child Abuse & Neglect*, 26(5), 475–480.

Stewart, T. D. (1957). Distortion of the pubic symphyseal surface in females and its effect on age determination. *American Journal of Physical Anthropology*, 15(1), 9–18.

Stone, P. K. (2012). Binding women: Ethnology, skeletal deformations, and violence against women. *International Journal of Paleopathology*, 2(2–3), 53–60.

Stone, P. K. (2020). Bound to please: The shaping of female beauty, gender theory, structural violence, and bioarchaeological investigations. In *Purposeful Pain* (pp. 39–62). Cham: Springer.

Suchey, J. M., Wiseley, D. V., Green, R. F., & Noguchi, T. T. (1979). Analysis of dorsal pitting in the os pubis in an extensive sample of modern American females. *American Journal of Physical Anthropology*, 51(4), 517–539.

218

8. Human behaviour and the pelvis

Tague, R. G., & Lovejoy, C. O. (1986). The obstetric pelvis of AL 288-1 (Lucy). *Journal of Human Evolution, 15*(4), 237–255.

Tilkens, M. J., Wall-Scheffler, C., Weaver, T. D., & Steudel-Numbers, K. (2007). The effects of body proportions on thermoregulation: An experimental assessment of Allen's rule. *Journal of Human Evolution, 53*(3), 286–291.

Trevathan, W. (2015). Primate pelvic anatomy and implications for birth. *Philosophical Transactions of the Royal Society, B: Biological Sciences, 370*(1663), 20140065.

Trevathan, W., & Rosenberg, K. (2000). The shoulders follow the head: Postcranial constraints on human childbirth. *Journal of Human Evolution, 39*(6), 583–586.

Ubelaker, D. H., & De La Paz, J. S. (2012). Skeletal indicators of pregnancy and parturition: A historical review. *Journal of Forensic Sciences, 57*(4), 866–872.

Uday, S., & Högler, W. (2018). Prevention of rickets and osteomalacia in the UK: Political action overdue. *Archives of Disease in Childhood, 103*(9), 901–906.

Verbruggen, S. W., & Nowlan, N. C. (2017). Ontogeny of the human pelvis. *The Anatomical Record, 300*(4), 643–652.

Waldron, T. (2008). *Palaeopathology*. Cambridge: Cambridge University Press.

Walker, A., Leakey, R. E., & Leakey, R. (1993). *The Nariokotome Homo erectus skeleton*. Harvard University Press.

Wall-Scheffler, C. M., Kurki, H. K., & Auerbach, B. M. (2020). *The evolutionary biology of the human pelvis: An integrative approach*. Cambridge: Cambridge University Press.

Wall-Scheffler, C. M., & Myers, M. J. (2017). The biomechanical and energetic advantages of a mediolaterally wide pelvis in women. *The Anatomical Record, 300*(4), 764–775.

Waltenberger, L., Rebay-Salisbury, K., & Mitteroecker, P. (2022). Are parturition scars truly signs of birth? The estimation of parity in a well-documented modern sample. *International Journal of Osteoarchaeology Early view, 32*(3), 619–629.

Warrener, A. G., Lewton, K. L., Pontzer, H., & Lieberman, D. E. (2015). A wider pelvis does not increase locomotor cost in humans, with implications for the evolution of childbirth. *PLoS One, 10*(3), e0118903.

Washburn, S. L. (1960). Tools and human evolution. *Scientific American, 203*(3), 62–75.

Weaver, T. D. (2003). The shape of the Neandertal femur is primarily the consequence of a hyperpolar body form. *Proceedings of the National Academy of Sciences of the United States of America, 100*(12), 6926–6929.

Weaver, T. D., & Hublin, J. J. (2009). Neandertal birth canal shape and the evolution of human childbirth. *Proceedings of the National Academy of Sciences of the United States of America, 106*(20), 8151–8156.

Webb, S., Cupper, M. L., & Robins, R. (2006). Pleistocene human footprints from the Willandra Lakes, southeastern Australia. *Journal of Human Evolution, 50*(4), 405–413.

Weinberg, A. (1954). Radiological estimation of pelvic expansion. *Journal of the American Medical Association, 154*(10), 822–823.

Wells, J. C., DeSilva, J. M., & Stock, J. T. (2012). The obstetric dilemma: An ancient game of Russian roulette, or a variable dilemma sensitive to ecology? *American Journal of Physical Anthropology, 149*(S55), 40–71.

Wheeler, S. M., Williams, L., Beauchesne, P., & Dupras, T. L. (2013). Shattered lives and broken childhoods: Evidence of physical child abuse in ancient Egypt. *International Journal of Paleopathology, 3*(2), 71–82.

Willett, A. Y., & Harrod, R. P. (2017). Cared for or outcasts: A case for continuous care in the Precontact US southwest. In *New developments in the bioarchaeology of care* (pp. 65–84). Cham: Springer.

World Health Organization (WHO). (2014). *Saving mothers' lives [infographic]*. WHO. https://www.who.int/reproductivehealth/publications/monitoring/infographic-mmr-2015/en/.

World Health Organization (WHO). (2022, March 1). *Maternal mortality*. WHO. https://www.who.int/news-room/fact-sheets/detail/maternal-mortality.

Worne, H. (2017). Inferring disability and care provision in late prehistoric Tennessee. In *New developments in the bioarchaeology of care* (pp. 85–100). Cham: Springer.

Young, M., Selleri, L., & Capellini, T. D. (2019). Genetics of scapula and pelvis development: An evolutionary perspective. *Current Topics in Developmental Biology, 132*, 311–349.

Zejdlik, K., Nyárádi, Z., & Gonciar, A. (2021). Evidence of horsemanship in two Szekler noblemen from the baroque period. *International Journal of Osteoarchaeology, 31*(1), 66–76.

Horse riding and the lower limbs

William Berthon[a,b], Christèle Baillif-Ducros[c,d], Matthew Fuka[e], and Ksenija Djukic[f]

[a]Chair of Biological Anthropology Paul Broca, École Pratique des Hautes Études (EPHE), PSL University, Paris, France, [b]Department of Biological Anthropology, University of Szeged, Szeged, Hungary, [c]French National Institute for Preventive Archaeological Research (INRAP), Great East Region, Châlons-en-Champagne, France, [d]UMR 6273 CRAHAM (Michel de Boüard Centre, Centre for Archaeological and Historical Research in Ancient and Medieval Times), CNRS/University of Caen Normandy, Caen, France, [e]Cultural Resource Analysts, Inc., Richmond, VA, United States, [f]Centre of Bone Biology, Faculty of Medicine, University of Belgrade, Belgrade, Serbia

9.1 Introduction

The lower limbs are essential for the human body to move and participate in physical activities involving walking, running, climbing, jumping, carrying, lifting, stabilising, or resisting force. All these physical behaviours necessarily involve muscle chains that allow individuals to maintain balance, adapt to gravity, and coordinate their gestures. The practice of horse riding particularly solicits the muscles of the lower limbs. To maintain their position on their mount, riders mainly use the adductor muscles in their legs. By moving the thighs inwards, the rider can transmit impulses to the mount, while also remaining firmly in the saddle. Overuse of the muscles involved in riding, which can result from regular and intense practice, can cause pathological changes and morphological adaptations on the riders' skeleton.

The 'horse and rider' couple has profoundly changed ancient societies, having led to further conquest of space and further control of time. The study of this interaction in past societies involves various disciplines, such as archaeology, bioanthropology, archaeozoology, history, and even medicine, and must therefore be carried out with a multidisciplinary approach. The organisation of a two-part session about the "Horseman-Horse Couple Through Time and Space" by Bindé M., Bede I., and Gergely, Cs. at the 27th Annual Meeting of the European Association of Archaeologists (EAA) in September 2021 highlighted not only the broad interest in the subject but also the multidisciplinary and interdisciplinary approaches that can be undertaken to advance research on this matter.

This chapter brings together recent and relevant research on the influence of the equestrian practice on an anatomical region particularly solicited by this physical activity: the lower limbs. The introductory section considers the equine partner and its place in ancient societies since its domestication, while also discussing various approaches relevant to the study of horseback riding in the past. The second section focuses on the anatomy of riding, by presenting the main muscles of the lower limbs involved in this activity and their functions. The injuries faced by modern riders are also outlined and discussed. The third section presents an overview of relevant and recent research on the influence of equestrianism on the human skeleton, with a focus on attachment sites of the rider's muscles on a macro- and micromorphological scale, as well as on the investigation of joint changes, morphological variants and adaptations, and skeletal trauma. Lastly, the fourth section emphasises the methodological limitations of studying the impact of horse riding behaviours on the skeleton and reflects on the identification of reliable skeletal indicators of horse riding with the modest prospect of limiting major methodological pitfalls in future research.

9.1.1 Why study horse riding?

The horse is omnipresent in human life. From around 35,000 BP, it is already a privileged animal, as we can observe in the themes of prehistoric art, such as in the Chauvet-Pont-d'Arc Cave, in southern France (Clottes, 2003). Although contested, some of the oldest proposed archaeological traces of the domestication of an equid (the Przewalski's horse, a wild species) were discovered on the Eneolithic site of Botai (in the northern steppe of Kazakhstan) and are dated to 3,500 BCE (Taylor and Barrón-Ortiz, 2021). The study by Outram et al. (2009) highlighted indices of primary domestication, namely the transition from the wild horse to the domestic horse (e.g. changes in the morphology of horses, characteristic signs possibly left by the port of the bit on the premolars and discovery of mare's milk residues on pottery) (Outram et al., 2009, 2021).

Through its domestication and training, the horse became a precious aid to the activities of humans, allowing them to increase their movement speed tenfold and acquire a new mastery of time and space (Outram et al., 2009). That way, horse riding, among all activities, brought profound and lasting changes in human cultural evolution concerning major aspects such as trade, settlement, warfare, subsistence, social organisation, and political ideology (Anthony, 2007; Anthony and Brown, 1991). The use of horses for transportation also considerably contributed to the circulation of languages, Indo-European ones in particular, as well as cultures and diseases, among other things (Anthony, 2007; de Barros Damgaard et al., 2018; Donoghue et al., 2015; Gaunitz et al., 2018; Librado et al., 2016; Outram et al., 2009).

During protohistoric and historic periods, horse riding was a very common activity. Horses were not only used in battle, but also for agricultural work, transport, hunting, and even for leisure purposes. This need for horses in everyday life made them valuable and cherished animals (Baillif-Ducros, 2018). The horse is an animal that requires a lot of attention as well as know-how for its reproduction, breeding (feeding and care), training (breaking in the foal), and the activity (training) for which it is intended (plough horse, draught horse, or mount). Thus owning horses requires specific infrastructure, equipment, and an assigned workforce for care and maintenance (Hyland, 1999). For this reason, riding a horse was often a sign of social distinction during different eras, such as in protohistoric Western Europe or during the Middle Ages (Baillif-Ducros et al., 2012). Horse riding was regularly performed by particular subgroups in societies and was, therefore, an activity associated with the identities of those specific groups and bore symbolic meaning (Hyland, 1999; Tichnell, 2012). Furthermore, horse riding had a dominant role in various nomadic populations from the steppes, such as the Scythians, Sarmatians, Alans, Huns, Avars, Magyars, and the Mongols, both in warfare and daily life. Through their mastery of the horse and art of war, characterised by redoubtable raids, those populations left an indelible mark in human history (Lebedynsky, 2007). Finally, in many cultures (e.g. Scythians, Celts, Gauls, Avars, Magyars), the horse had strong spiritual importance and was involved in various rituals or funerary practices (Bede, 2012; Garofalo, 2004; Jennbert, 2003).

With those aspects in mind, it is clear that a better understanding of who exactly rode horses during their life can shed light on different aspects of past societies. This is especially the case for some ancient populations of nomadic pastoralists where the sparseness of archaeological remains and the lack of internal textual evidence, among other causes, may result in a partial or oversimplified understanding regarding different societal aspects (Tichnell, 2012). Considering the extreme importance of horse riding in their life, the analysis of this practice in such populations can provide insight into the societal structure, socioeconomic organisation, sexual division of labour, or, more generally, lifestyle and behaviours in past groups.

9.1.2 A multidisciplinary approach

Horse riding in past societies can be investigated through the study of horse and human skeletal remains, archaeological artefacts (bits, spurs, elements of saddle, and stirrups), and iconographic and textual sources. The practice of burying the dead with a horse or burying horses without a person is widespread in Eurasian populations throughout the ages, as in Scythians, Celts, Avars, or Magyars. Furthermore, in the necropoles from equestrian societies, such burials show considerable diversity in funerary practices regarding aspects such as the position of horse and human remains in the grave, the simultaneity of the deposits, the preparation and organisation of the bones, or the presence of equestrian equipment (Anthony, 2007; Bede, 2012; Bede and Detante, 2014; Berthon et al., 2019, 2021a; Carver, 1998; Cross, 2011; Jennbert, 2003; Liancheng, 2015; Róna-Tas, 1999). In various cultural examples, archaeological investigations have identified that horses served important roles in mortuary practices as socioeconomic status markers and ritual sacrifices (e.g. Bugarski, 2009; Dizdar et al., 2017; Mikić Antonić, 2012, 2016; Rolle, 1989). In the Avar population, for instance, horse sacrifice stems from the belief that a warrior needed a horse in the afterlife (Kovačević, 2014; Pohl, 2018) and the horse was buried at the same time as its owner. The burials of some riders may also reflect the importance of their horses for their identity, as in some Scythian royal tombs, where remains of horses have been recovered along with ornamental bridles indicative of higher statuses (Rolle, 1989).

Although the study of archaeological goods, funerary practices, and ethnographic sources can be very informative, especially in the case of some nomadic pastoralist populations, the examination of skeletal remains for horse riding-related bone changes represents the most direct source of information (Tichnell, 2012). Thus various studies have attempted to infer activities from the palaeopathological analysis of horse skeletal remains (e.g. Bindé et al., 2019; Bulatović et al., 2014; Levine et al., 2000; Marković et al., 2014; Pluskowski et al., 2010; Taylor et al., 2015). In these studies, the presence of bit wear, a bevelling of the anterior surface of the second premolar caused by grinding or chewing the bit, is often considered as a direct and unmistakable sign of riding (Anthony, 2007; Bendrey, 2007; Brown and Anthony, 1998; Greenfield et al., 2018; Taylor et al., 2015). With the aim of developing a better understanding of their status and the activities in which they were involved in past societies, scholars also recently focused on the entheseal changes developed on the skeletons of riding, draft, and pack equids (Bendrey, 2008; Bertin et al., 2016; Bindé et al., 2019, 2021; Taylor et al., 2015). These bone changes are promising to help identify the practice of riding using horse skeletal remains, but the existence of a direct link between specific skeletal changes and horse riding has not yet been unarguably demonstrated in human skeletal remains. The remainder of this chapter explores the potential human skeletal changes associated with horse riding to clarify the evidence and further enable researchers to identify riders in the archaeological record.

9.2 Modern riders and sports medicine

Today, there are parts of the world where the horse represents an important economic factor in land cultivation, transport, or hunting. However, riding is predominantly recognised in many other modern cultural contexts as a sport and entertainment, primarily reserved for those members of society who nurture riding as a tradition, and who can afford to practice it. In addition to being very profitable and popular, riding is a dangerous sport and riders, regardless of riding style, are at risk of numerous injuries. Although the techniques and style of riding, the equipment, and the morphology of the horses used nowadays by modern riders are not directly comparable with those used by past populations in different regions of the world, clinical sources can offer relevant complementary data (McGrath, 2015). It seems likely that some features are comparable for both modern and ancient riders, especially those related to the posture on the horse and the muscles needed to maintain it (Baillif-Ducros, 2018).

9.2.1 Anatomy of riding

The aim of any rider is to control their position while riding (Auvinet and Estrade, 1998; Baillif-Ducros, 2018). The rider adapts their posture according to the equipment used (e.g. the type of flat or hollow seat saddle, the presence or absence of stirrups), the discipline exercised (dressage, racing, or jumping), but also to the different paces of the horse (walk, trot, canter, gallop) (Baillif-Ducros, 2018). Therefore different styles of riding require different muscle actions (German Equestrian Federation, 2017; Glosten, 2014; Willson, 2002). Horse riding uses all of the body's muscle groups, but specifically requires very strong and athletic postural muscles. During riding, the body activates the muscles of the abdominal musculature, some back muscles, and those of the upper and lower limbs (Glosten, 2014). Whatever the rider's posture, the discipline, and the paces of the horse, three anatomical regions are especially stressed and traumatised by the equestrian practice: the lumbar spine, the hip joint, and the knee joint, each of them involving different groups of muscles (Auvinet, 1980; Humbert, 2000; Lemaire, 1985). In particular, the main muscles of the lower extremities used in riding are the gluteals, hamstrings (posterior femoral muscles), adductors, hip rotators, quadriceps muscle, and the calf muscles (Glosten, 2014; Willson, 2002) (Table 9.1, Fig. 9.1).

TABLE 9.1 Main muscles of the lower limbs involved in horse riding and their functions.

Muscle	Muscle functions (Putz and Pabst, 2001)
Iliopsoas: iliacus, psoas major	Lumbar vertebral column lateral flexion, extension (increases lumbar lordosis) Hip flexion, medial rotation (lateral rotation when gluteal muscles contract simultaneously)
Quadriceps femoris: vastus medialis, vastus intermedius, vastus lateralis, rectus femoris	Hip flexion (*rectus femoris* only) Knee extension
Gracilis	Hip adduction, flexion, lateral rotation Knee flexion, medial rotation
Pectineus	Hip adduction, flexion, lateral rotation
Adductor brevis	Hip adduction, flexion, lateral rotation
Adductor longus	Hip adduction, flexion, lateral rotation (most anterior fibres rotate medially)
Adductor magnus	Hip adduction, lateral rotation, flexion (anterior part), extension (posterior part)
Adductor minimus	Hip adduction, lateral rotation
Obturator externus	Hip lateral rotation, adduction (when in flexion)
Piriformis	Hip lateral rotation, extension, adduction
Gluteus maximus	Cranial part: Hip extension, lateral rotation, abduction Caudal part: Hip extension, lateral rotation, adduction Knee extension (via iliotibial tract)
Gluteus medius	Ventral part: Hip abduction, flexion, medial rotation Dorsal part: Hip abduction, extension, lateral rotation
Gluteus minimus	Ventral part: Hip abduction, flexion, medial rotation Dorsal part: Hip abduction, extension, lateral rotation
Biceps femoris	Hip extension, adduction, lateral rotation Knee flexion, lateral rotation
Semitendinosus	Hip extension, adduction, lateral rotation Knee flexion, medial rotation
Semimembranosus	Hip extension, adduction, lateral rotation Knee flexion, medial rotation
Triceps surae: soleus, gastrocnemius	Knee flexion (*gastrocnemius* only) Ankle plantar flexion Talotarsal supination

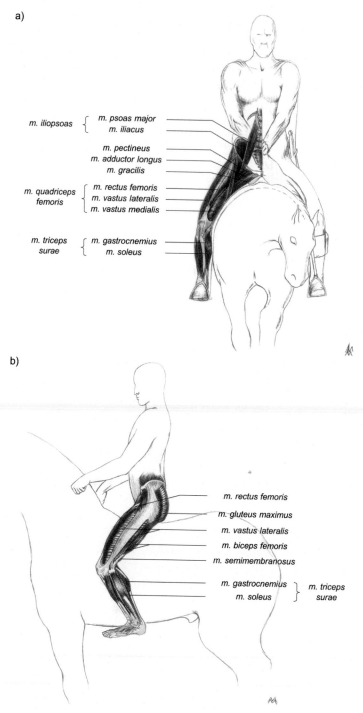

a)

m. iliopsoas { m. psoas major
m. iliacus

m. pectineus
m. adductor longus
m. gracilis

m. quadriceps
femoris { m. rectus femoris
m. vastus lateralis
m. vastus medialis

m. triceps
surae { m. gastrocnemius
m. soleus

b)

m. rectus femoris
m. gluteus maximus
m. vastus lateralis
m. biceps femoris
m. semimembranosus

m. gastrocnemius } m. triceps
m. soleus } surae

FIG. 9.1 Main muscles of the lower limbs used by horse riders: (A) anterior view, and (B) lateral view (drawing by Milutin Micic). *Modified from Djukic, K., Miladinovic-Radmilovic, N., Draskovic, M., & Djuric, M. (2018). Morphological appearance of muscle attachment sites on lower limbs: Horse riders versus agricultural population. International Journal of Osteoarchaeology, 28(6), 656–668. https://doi.org/10.1002/oa.2680, reproduced with permissions from John Wiley and Sons, licence 4898190287327.*

The gluteal muscles (*gluteus maximus*, *gluteus medius*, and *gluteus minimus*) are involved in hip extension, abduction, and medial and lateral rotation (Putz and Pabst, 2001). Riders use the gluteal muscles especially in the jumping position and in the rising trot (Willson, 2002). With the adductors (*adductor brevis*, *adductor longus*, *adductor magnus*, *adductor minimus*, *pectineus*, *gracilis*, and *obturator externus*), mainly involved in hip adduction, but also, for instance, in hip flexion and rotation, they allow the rider to stay on their moving mount (Glosten, 2014; Putz and Pabst, 2001). The adductor muscles serve the function of keeping the legs pressed against the horse while riding, which is not a common action otherwise in daily life (Willson, 2013). This group of muscles, in conjunction with the hip rotators, such as the *piriformis* and *obturator externus*, controls the position of the whole leg against the saddle and the horse's side. Riders tend to overuse these muscles and often incorrectly grip the saddle in modern riding techniques (Glosten, 2014; Willson, 2002).

The good seat on the saddle requires a position in retroversion (i.e. tilted backward) of the rider's pelvis (Auvinet, 1999). This position in an 'arched kidney' causes the coxofemoral joint to be put forward and the adductor muscles to be tensioned at the level of their pubic insertions, in particular for the *gracilis* and long adductor muscles (Baillif-Ducros, 2018). The abdominal muscles are contracted as well as the posterior muscles of the hip and the thigh (*gluteus maximus* and hamstring muscles) (German Equestrian Federation, 2017; Glosten, 2014; Willson, 2002). The hamstrings (*semimembranosus*, *semitendinosus*, and *biceps femoris*) are involved in knee flexion and hip extension (Putz and Pabst, 2001). By stabilising the heel and lower leg against the horse's side, they are used in combination with the abdominal muscles to keep the seat deep in the saddle (Glosten, 2014; Willson, 2002).

The *quadriceps femoris* muscle (a group that includes the *rectus femoris*, *vastus lateralis*, *vastus intermedius*, and *vastus medialis*) is involved in knee extension, as well as hip flexion via the *rectus femoris* (Putz and Pabst, 2001). There is little or no activity in the quadriceps during standing (Standring, 2005). A rider using stirrups rests their feet on them, helping to support roughly 16% of their bodyweight at the trotting gait (Lang, 1995). Although the impact of the use of stirrups on the rider may vary according to the style and equipment, the heels are generally pushed down by putting more weight onto the stirrups. In short stirrups, the quadriceps work hard to hold the body weight and control the position of the seat in and out of the saddle (Willson, 2002).

The calf muscle, or *triceps surae*, is made up of two muscles, the *gastrocnemius* and *soleus*. They are the primary plantar flexors of the ankle and are also involved in talotarsal supination and knee flexion (Putz and Pabst, 2001). In riders, the *gastrocnemius*, in particular, needs to be long, and thus regularly stretched, as the heel is pushed down below the level of the toes (German Equestrian Federation, 2017; Glosten, 2014; Willson, 2002). In addition, the *iliopsoas* muscle (*iliacus* and *psoas major*), which notably acts as the strongest flexor of the hip, plays a role in stabilisation and shock absorption as it contracts to bend the trunk and pelvis forwards against resistance (Glosten, 2014; Putz and Pabst, 2001; Standring, 2005).

9.2.2 Injuries of modern riders

Horse riding is one of the most dangerous sporting activities according to EuroSafe statistics (EuroSafe, 2014), with a higher risk of injury than automobile racing, motorcycle riding, or skiing, and at least a similar risk as rugby (Ball et al., 2007). Due to the speed, height, weight, and strength of the horse, horse riding puts the rider at risk for different types of injuries, ranging from minor soft tissue lacerations to skeletal fractures, and paralysis; it is also sometimes considered to be the sport with the highest mortality (Ball et al., 2007). Sports medicine studies provide data concerning the nature and location of injuries in modern riders. The main injuries are related to the fall of the rider, with or without their mount. The average jockey weighs less than 55 kg, yet they must control a thoroughbred horse 10 times their weight that may act unpredictably, whether at rest or in full gallop. Falls during races can result in various types of injuries (Balendra et al., 2008; Ryan et al., 2020). For example, an analysis of European flat racing suggests that 40%–59% of all falls result in head injuries (including concussions, fractures, and haematomas) (Rueda et al., 2010). Other injuries can be caused by the animal itself, while handling or taking care of the horse on foot (kicking, biting, or stepping on the handler) (Balendra et al., 2008; Ball et al., 2007; Bixby-Hammett and Brooks, 1990; Havlik, 2010; Ryan et al., 2020).

The most frequently injured anatomical areas are the upper limbs, skull, spine, lower limbs, thorax, and the pelvis, although the order of the most affected regions varies between studies (e.g. Altgärde et al., 2014; Ball et al., 2007; Bixby-Hammett and Brooks, 1990; Campagne, 2011; Havlik, 2010; Lacourt, 2012; Loder, 2008; Zeier, 2006). Moreover, modern riders under the age of 21 years show a high rate of head injuries that can lead to death (Bixby-Hammett, 1992). Altgärde et al. (2014) found, in particular, that head and neck injuries were more common among children (under the age of 19 years) than adults (19 years of age or older), reflecting the higher frequency of falling accidents in younger riders. These results might be interpreted as potentially related to experience level, individuals with less experience being more likely to suffer falls with severe injuries (Balendra et al., 2007; Mayberry et al., 2007; Rueda et al., 2010).

The severity and frequency of injuries during active horse riding is not the only health risk for riders (Ryan et al., 2020). Mackinnon et al. (2019) report that the most prevalent health outcome of retired jockeys is osteoarthritis, occurring at a rate of 7.5 times higher than in the reference population. In order to improve health safety and prevention of injuries, modern medical practices recommend that the sports medicine practitioners caring for jockeys develop a working knowledge of nutrition, bone health, and mental health of the athlete, as well as the management of on-field trauma (Ryan et al., 2020).

9.3 Biological anthropology of horse riding

Mainly from the 1990s onwards, and then with a new dynamic in recent years, many biological anthropology studies have developed and tested methods of identifying osteological changes of the human skeleton that are suggestive of habitual horse riding. While methodologically varying from one another, these studies focus on the analysis of the same types of skeletal changes. The following sections cover recent and relevant research to describe how macro- and micromorphological analyses of entheseal changes, the analysis of joint changes, the identification of morphological variants and adaptations, and the study of skeletal trauma have been used to investigate evidence of habitual horse riding.

A summary of key bioarchaeological studies on horse riding-related skeletal changes that are mentioned in this section is provided in Table 9.2.

TABLE 9.2 Summary table of key bioarchaeological studies investigating the influence of horse riding on the human skeleton.

References	Samples	Evidence for horse riding practice	Methods and objectives	Key results (skeletal changes related to horse riding)
Miller and Reinhard (1991), Miller (1992), Reinhard et al. (1994)	Late 18th–early 19th c. Omaha and Ponca cemeteries from northeastern Nebraska ($n = 32$ comprising 18 considered as riders and 14 as non-riders)	Historic and ethnographic sources: riding practice by Native American populations since the reintroduction of the horse on the American continent by Europeans	Macromorphological analysis of a suite of skeletal changes considered to be related to the riding practice beforehand	- Superior elongation of the acetabulum - Extension of the articular surface of the femoral head onto the neck - Entheseal changes at the insertion sites of the *gluteus medius* and *minimus*, *adductor magnus* and *brevis*, *vastus lateralis*, and *gastrocnemius* muscles on the coxal bone and femur - Acute trauma ('fall' type fractures, dislocations) - Schmorl's nodes and spondylolysis - Degenerative changes at the vertebrae, knees, hips, feet, and elbows, as well as in sternal ends of ribs and clavicles - Osteoarthritis of the first metatarsal (put in relation with the use of toe stirrups)
Pálfi (1992), Pálfi and Dutour (1996), Pálfi (1997)	Hungarian Conquest period (10th c. CE) cemetery from eastern Hungary ($n = 263$); Comparison group from a Late Antiquity (4th c. CE) cemetery from southeastern France ($n = 91$)	Historical and archaeological sources: Hungarian Conquest period populations were semi-nomadic societies with armies of mounted archers and great importance of the horse; Grave goods related to the riding practice (equipment or horse bones) associated with some individuals	General palaeopathological analysis with a special focus on activity-related skeletal changes	A suite of skeletal changes forming a 'Horseback-riding syndrome': - Superior extension of the acetabulum - Extension of the femoral head onto the neck - Entheseal changes at the insertions of the *gluteus medius* and *maximus*, *adductor magnus* and *brevis*, *biceps femoris*, *semimembranosus* and *semitendinosus*, *quadriceps femoris*, *pectineus*, and *gastrocnemius* muscles on the coxal bone and femur; In addition, this 'syndrome' was associated with a higher frequency of fractures and sprains, spinal and extraspinal degenerative changes, and spondylolysis
Courtaud and Rajev (1998)	Iron Age necropolises from southwestern Siberia ($n = 38$)	Historical and archaeological sources: populations of horse riders as attested by the art, texts, and funerary practices; Grave goods (harness elements) related to the riding practice associated with some individuals	Palaeopathological analysis with the objective of investigating the effect of horse riding on the skeleton	- Ovalization of the acetabulum connected with arthrosis and often with osteochondrosis - Entheseal changes at the medial lip of the *linea aspera* on the femur (adductor muscles)

TABLE 9.2 Summary table of key bioarchaeological studies investigating the influence of horse riding on the human skeleton—cont'd

References	Samples	Evidence for horse riding practice	Methods and objectives	Key results (skeletal changes related to horse riding)
Erickson et al. (2000)	Native American Arikara sites from South Dakota: - Late 17th–early 18th c. CE site (37 acetabula from presumed non-riders) - Early 19th c. CE site (24 acetabula from presumed riders)	Historical sources: little to no use of horses during the period of occupation of the oldest site, and extensive use of horses regarding the most recent site	Fourier analysis of the shape of the acetabulum to identify changes associated with the riding practice	Expanded anterior-superior acetabular rims in the group of presumed riders
Tichnell (2012)	Series of known non-riders and riders from different cultures: - Pre- (n = 25) versus Post-Contact (n = 67) Arikara populations representing bareback riding style - Post-medieval British country-dwelling (n = 52) versus city-dwelling (n = 51) populations representing British saddle-riding style - Xiongnu (n = 54) and Mongol (n = 32) populations both representing Mongolian riding style	Historical sources: - Pre-Contact Arikara before the reintroduction of horses on the American continent by Europeans versus Post-contact Arikara known as great horse riders - Post-medieval British country dwellers known for use of horses versus city dwellers consisting of poor Londoners unlikely to ride horses; Historical, ethnographic, and archaeological sources: horse riding of great importance in Xiongnu (4th c. BCE–4th c. CE) and Mongol (13th–14th c. CE) nomadic societies	Macromorphological analysis of entheseal changes to test the hypotheses that they can be used to identify horse riders and that different riding styles develop different suites of entheseal changes	Significant correlations with horse riding for entheseal changes at the gluteal tuberosity (gluteus maximus) and adductor tubercle (adductor magnus) on the femur in some cases; However, the results did not allow identifying any universal suite of riding-related skeletal changes and suggest that culturally specific suites of features are needed to identify riders in different societies
Đukić (2016), Djukic et al. (2018)	6th–7th c. CE Avar necropolises from Serbia (n = 48); Comparison group from a medieval (11th–16th c. CE) agricultural population (n = 34) from Serbia	Historical and archaeological sources: Avars known to be societies of skilled riders where horses were of great importance; Comparison group mainly engaged in agrarian affairs and craft activities and no evidence of horse riding practice	Macromorphological analysis of entheseal changes of the lower limb to investigate the influence of habitual activities on the morphology of the muscle insertion sites through a comparison between a horse riding and an agricultural population	Entheseal changes at the ischial tuberosity on the coxal bone and at the linea aspera and adductor tubercle on the femur; The adductors being the most specific muscles for riding, entheseal changes at the adductor tubercle (adductor magnus) are considered as the most reliable indicators for riding

Reference	Sample	Sources	Aim / Indicators	Results / Markers identified
Fuka (2018)	Early and Middle Bronze Age (n = 11), Late Bronze and Early Iron Age (n = 16), and Xiongnu (n = 26) nomadic pastoralist populations from Mongolia	Archaeological sources (horse remains and riding-related grave goods) and zooarchaeological sources (morphological alterations on horse skeletal remains) suggesting that horse riding emerged during the Late Bronze and Early Iron Age period	Exploratory macromorphological analysis of entheseal changes to investigate their link with horse riding	Entheseal changes related to composite muscle groups forming the shoulder, elbow, and hip joints; A left side dominance for the upper limb is indicative of left-handed riding, which is the style of present-day Mongolian riders
Baillif-Ducros (2018)	Merovingian (5th–7th c. CE) necropolises from northeastern Gaul (n = 217), including at least one individual per site with horse riding-related grave goods (total of 8 individuals with goods)	Historical and archaeological sources: horses and horse riding of great importance in Merovingian societies; Horse riding equipment (bits, spurs, elements of saddle, and stirrups) associated with the individuals in the grave	Selection of potential bone indicators for the distinction of riders in past societies based on anthropological studies and sports medicine literature: - Entheseal changes at the pubis (*adductor longus*) and ischiopubic ramus (*gracilis*) - Poirier's facet at the femoral head-neck junction - Joint changes of the hip and knee - Scheuermann's disease (wedged vertebrae, Schmorl's nodes) - Acute trauma - Osteometric analysis of the shape of the acetabulum	Equestrian stress markers identified: - Poirier's facet at the femoral head-neck junction - Ovalization of the acetabulum - Joint changes of the hip - Osteoarthritis on the upper outline of the patellar surface (in case of use of stirrups)
Berthon (2019), Berthon et al. (2019, 2021a)	Hungarian Conquest period (10th c. CE) cemetery from eastern Hungary (n = 67, comprising 17 with horse riding-related grave goods); Comparison group of known non-riders from the modern documented collection of Lisbon (n = 47 of known occupation)	Historical and archaeological sources: Hungarian Conquest period populations were semi nomadic societies with armies of mounted archers and great importance of the horse; Grave goods related to the riding practice (equipment or horse bones) directly associated with the individuals	Systematic anthropological and palaeopathological analyses for the identification of reliable horse riding-related skeletal changes: - Entheseal changes - Joint changes - Morphological variants - Fractures - Schmorl's nodes - Spondylolysis - Indices of shape and robusticity of lower limb bones	Skeletal changes that can contribute to identify, statistically, the presence of riders in a population: - Entheseal changes at the adductor tubercle (*adductor magnus*) of the femur and the calcaneal tuberosity (*triceps surae, plantaris*), in particular - Poirier's facet at the femoral head-neck junction - Vertical ovalization of the acetabulum - Schmorl's nodes at the thoracolumbar transition - Clavicle fractures (especially at midshaft)

Note: We included in this table methodological studies aiming to investigate the influence of the riding practice on the human skeleton based on evidence of this activity and key studies that ... Studies reporting observations made on skeletal remains of a single or a handful of individuals were not included

9.3.1 Entheseal changes

Entheses serve as muscle attachment sites that act as stress relievers for the muscles, using the structural support of the skeleton as an anchor for muscle-induced stress, which in turn can leave signs on the bone where these tendons attach, allowing one to get insight into the activity of the bone (Benjamin et al., 2002; Jurmain et al., 2012). Entheseal changes refer to physical alterations to the entheses, which may result from muscle-induced stress, among other causes (Benjamin et al., 2002). The analysis of entheseal changes has become an important tool for the bioarchaeologist in researching evidence of horse riding in the archaeological record (Djukic et al., 2018). The following section reviews how previous studies have developed methods and attempted to better understand habitual horse riding and its effects on the lower limbs through the analysis of entheseal changes.

9.3.1.1 Macromorphological analysis of entheseal changes

Macromorphological study of entheseal changes involves the analysis of specific muscles or composite muscle groups. The muscles related to the movements of the hips and knees are the most important to study in regard to horse riding (see Section 9.2.1). The primary functions of the hip muscle group are flexion and stabilisation of the hip joint as well as external and internal hip rotation (Putz and Pabst, 2001; Stone and Stone, 2003). These mechanical functions are necessary as riders use their hips to maintain balance while riding (de Cocq et al., 2013; Taylor and Tuvshinjargal, 2018). According to studies on modern horse riding and repetitive riding, the shock absorption due to the force when the horse makes contact with the ground causes mechanical stress on the *os coxae* and vertebrae of the rider (Biau et al., 2013; de Cocq et al., 2013; Nicol et al., 2014; Pugh and Bolin, 2004; Tsirikos et al., 2001; Zeng et al., 2017). In archaeological contexts, various pathological signs on the vertebrae and the main bones of the lower limbs were considered to be related to the mechanical pressure on the body during horse riding (e.g. Khudaverdyan et al., 2016; Molleson, 2007; Sandness and Reinhard, 1992; Wescott, 2008).

A possible link between modifications of the entheses and the practice of horse riding was discussed by scholars since as early as the 1980s (Angel et al., 1987; Blondiaux, 1989; Pap, 1985; Snow and Fitzpatrick, 1989). However, the work of Miller and collaborators on historic Native American Omaha and Ponca skeletal remains (Miller, 1992; Miller and Reinhard, 1991; Reinhard et al., 1994) and the palaeopathological investigations by Pálfi and Dutour on a Hungarian population of the 10th century CE (Pálfi, 1992, 1997; Pálfi and Dutour, 1996) played a particularly influential role on archaeological equestrian research during the next decades. Recently, some studies focusing on the influence of horse riding on the skeleton, particularly on the lower limbs, included a more precise and systematic recording of macromorphological entheseal changes, allowing for statistical analyses and more reliable interpretations.

Tichnell (2012) performed multiple comparisons between pre- versus post-contact Arikara Indigenous populations from South Dakota, as well as between post-medieval British city- versus country-dwelling populations. The author observed significant correlations with horse riding for two entheses, the gluteal tuberosity and the adductor tubercle, and also observed that the greater and lesser trochanters had no predictive value for identifying horse riders. The results suggested, however, that the entheses that can be used to identify horse riders are culturally specific, depending notably on the riding style (Tichnell, 2012).

Later, Djukic and colleagues (Djukic et al., 2018; Đukić, 2016) compared an equestrian population (Avars) to a medieval agricultural population, both from Serbia, using entheseal changes to determine the effects horse riding has on the lower limbs. The macromorphological analysis of 10 entheses of the lower limbs, including both fibrous and fibrocartilaginous entheses, was performed on a sample of a total of 82 adult individuals, relying on previous scoring systems (Alves Cardoso and Henderson, 2010; Villotte, 2006, 2012). Compared to the agricultural population, the Avar population displayed more robust entheseal changes concerning especially the *iliopsoas* muscle in younger adult males. Overall, the equestrian population showed higher entheseal change scores and significantly more pronounced entheseal changes at the ischial tuberosity on the coxal bone, as well as the *linea aspera* and adductor tubercle on the femur, in particular. The authors conclude that the entheseal changes at the attachment sites of the adductor muscles proved to be the most reliable indicator of the prevalence of horse riding as these muscles are responsible for the action of keeping the legs together, which is not common in everyday physical activity other than in horse riding (Djukic et al., 2018; Đukić, 2016).

Fuka (2018) investigated whether entheseal changes could be used to determine evidence of horse riding among Bronze and Iron Age nomadic pastoralist populations from Mongolia. The macromorphological analysis of entheseal changes focused on fibrocartilaginous entheses as their scoring was based on the 'Coimbra method' (Henderson et al., 2016). Composite muscle groups that form the major joints of the upper and lower limbs were selected for analysis. The results of the study suggested a positive association between certain composite muscle groups and horse riding. Specifically, the muscles from the hips, left shoulder, and left elbow in Late Bronze and Early Iron Age males statistically had higher entheseal change scores than other muscle groups. Fuka comes to the conclusion that the particular entheseal change pattern observed in Late Bronze and Early Iron Age males in the study likely is due to habitual horse riding as it is supported by previous ethnoarchaeological and zooarchaeological studies in Mongolia (Fuka, 2018, p. 79). In addition, Fuka (2018) suggests that the left skew observed among Late Bronze and Early Iron Age males' upper limbs in his study is indicative of left-handed riding, which is the riding style of present-day Mongolian riders (Taylor and Tuvshinjargal, 2018). This style of riding is suspected of being used during the Late Bronze and Early Iron Age as zooarchaeological studies demonstrate that horses from Mongolian sites from that period show signs of asymmetrical bone remodelling that disproportionally affect the left side of the skull (Taylor et al., 2015; Taylor and Tuvshinjargal, 2018). Likewise, the high scoring in the hips of Late Bronze and Early Iron Age males is taken to indicate the practice of horse riding. The author admits, however, that the same activity pattern fails to appear in the Xiongnu population, in a period where horse riding was widespread in Mongolia (Brosseder, 2009; Di Cosmo, 1994, 2011; Erdélyi, 2000; Honeychurch, 2014; Johannesson, 2011). Elite socioeconomic status, the emergence of stirrups, and a higher proportion of young adults in the study are all suggested as reasons for the lack of evidence of horse riding among the Xiongnu population (Fuka, 2018).

Lastly, a more recent study focused on a population from the Hungarian Conquest period (10th century CE) by comparing individuals with and without horse riding deposit in their grave between them as well as with a population of non-riders (Berthon, 2019; Berthon et al., 2021b). Various entheseal changes were selected and scored on the main bones of the lower limbs, based on previous anthropological studies and taking into consideration the muscle functions involved during the practice of horse riding. When restricting the analysis to young and mature adults only, which limited the influence of ageing on the development of entheseal changes, significant differences were observed between groups, suggesting that the entheseal changes at the adductor tubercle of the distal femur (Fig. 9.2) and the calcaneal tuberosity are the most reliable indicators for the practice of horse riding, combined with other types of changes of the spine and the hip joint, in particular (Berthon, 2019).

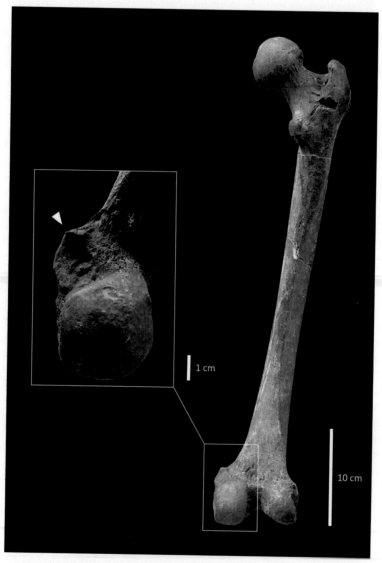

FIG. 9.2   Entheseal changes at the adductor tubercle on the femur (posterior view), insertion site of *m. adductor magnus*, considered to be a promising skeletal marker for the practice of horse riding according to several recent studies. *Image courtesy of William Berthon.*

9.3.1.2 *Micromorphological analysis of entheseal changes*

It is a widely accepted opinion that morphological changes of muscle attachment sites follow the overall bone adaptation to different mechanical stress. Consequently, macromorphological changes of the bone on the muscle attachment site should correlate with microarchitectural changes in the underlying bone (Djukic et al., 2015). Analysis of bone microstructure via histology or microcomputed tomography (micro-CT) may offer further evidence regarding the structural consequence of muscle loading, as persistent or repeated mechanical loading may drive bone microarchitectural adaptation, as proposed by the 'bone functional adaptation law' (Đurić et al., 2012; Heřt, 1994; Mulder et al., 2020; Roux, 1881; Ruff et al., 2006; Skedros and Baucom, 2007; Wolff, 1892). The microstructural evaluation of entheses may provide knowledge of the biomechanical relevance of entheseal gross morphology, further clarifying the purpose of entheseal morphology in the interpretation of physical activities of past populations (Berthon et al., 2015, 2016b; Djukic et al., 2015). Some studies have, however, suggested that surface morphology at entheses may not be responsive, in a straightforward manner, to the local strain environment (Rabey et al., 2015; Schlecht, 2012; Turcotte et al., 2022; Wallace et al., 2017; Zumwalt, 2006). Consequently, it was not unexpected that Djukic et al. (2015) found little to suggest that there was a link between trabecular or cortical bone microarchitecture and entheseal surface morphology using micro-CT. Nonetheless, histological analyses of cadaveric human hand entheses do confirm a connection between bone elevation and biomechanical load, further suggesting that the internal architecture of the underlying bone at the entheses has potential to assist with the understanding of habitual loading (Karakostis et al., 2019). In addition, experimental evidence suggests that the gross morphology of entheses is a less reliable proxy than internal bone structure for making inferences regarding past human behaviour (Rabey et al., 2015; Wallace et al., 2017).

The microstructural analysis of entheses has been concerned with more general activity levels rather than specific physical activities, such as, for instance, horse riding (Djukic et al., 2015; Djukić et al., 2020). However, the analysis of the microarchitecture of underlying bones at attachment sites that are specific to particular activities may allow a better understanding of the biomechanical relevance of entheseal gross morphology, further clarifying the contribution of entheseal changes in the interpretation of human behaviours (Djukic et al., 2015). Although the use of microstructural analysis in the reconstruction of some specific activities is not applicable at present, several important questions have arisen. The following addresses questions of how the micro-CT analysis of entheses, especially those of the lower limbs, can contribute to a better perception of attachment sites and, therefore, to a more accurate interpretation of the macromorphological appearance of entheses.

If we accept that entheseal internal morphology is related to a functional character and, consequently, that it represents an important aspect for behavioural reconstruction, then the primary tasks should be to establish the most appropriate scoring system for entheseal changes, based on the microarchitectural picture, thus forming the basis for the interpretation of physical activities in ancient populations.

Focusing on examining the possible successive nature of the three-stage scale of entheseal macroscopic changes, Djukic et al. (2015) used microcomputed tomography and 3D methods to compare those scores with the microarchitectural features at four lower limb insertions (femur and tibia). They noted that, among all microarchitectural parameters of cortical and trabecular bone, only the cortical thickness was significantly different between entheseal change stages (Djukic et al., 2015). Hence, it seems the connection between bone microstructural design and macroscopic scores is not altogether straightforward. To better understand the relationship between the micro and macro aspects of entheses, it is essential to evaluate the origin of different macroscopic entheseal forms. From the macromorphological perspective, it seems that two main forms occur at entheses: those characterised by predominant signs of bony resorption, and those with pronounced signs of bony formation (Djukić et al., 2020). Analysis of the *m. gluteus maximus*, conducted using microcomputed tomography, offers a possible explanation of the origin of the aforementioned macroscopic entheseal forms. From the initial flat surfaces at the point of the enthesis, two different microarchitectural patterns are evident (Fig. 9.3). The first is characterised by the existence of a clearly expressed cortical surface prominence with increased trabecularisation of the underlying bone. In the second pattern, a surface defect exists (in the form of bony resorption) but it is not accompanied by trabecularisation of the underlying bone (Djukić et al., 2020). It could be suggested that these variations in the development of entheseal changes result from variations in the load magnitude, a different loading pattern, or changes in the angle of the tendon/bone during the motions (Djukić et al., 2020).

Studying entheseal changes is but one method that can be used by researchers to determine evidence of horse riding within a population and the effects riding has on the human skeleton. Methods of scoring and analysing entheseal changes are still tied to prospective studies, and reviews of these studies have revealed a lack of uniformity in their data collection and results. A holistic approach that integrates other lines of skeletal evidence is necessary for researchers to properly study horse riding. The study of entheseal changes in the lower limbs must often take other types of skeletal changes into account in order for the analyses to provide comprehensive and more conclusive results.

9.3.2 Joint changes

Osteoarthritis represents one of the main research areas in the reconstruction of activities, partly because it has been used widely and relatively early as a marker of activity (Angel, 1982; Jurmain, 1977; Merbs, 1983; Ortner, 1968). The bone changes observed at the synovial joints include marginal osteophytes and contour deformation, as well as porosity, bone formation, and eburnation on the joint surface (Buikstra and Ubelaker, 1994; Jurmain et al., 2012; Waldron, 2009). These alterations appear following the breakdown of the articular cartilage of the joint and the inflammation in the synovial membrane. The discussions on the link between osteoarthritis and activities are based on the fact that it has been considered as resulting from repetitive mechanical loading and that repetitive tasks could, therefore, lead to severe osteoarthritic features on specific joints (Weiss and Jurmain, 2007). Nowadays, it is acknowledged from clinical research that the aetiology of osteoarthritis is multifactorial (Jurmain et al., 2012; Larsen, 1997; Ortner, 2003; Resnick, 2002; Waldron, 2009; Weiss and Jurmain, 2007).

In horse riders, the main joints of the lower limbs, i.e. the hip, the knee, and the ankle, are particularly stressed in relation to their posture and the movements that are involved in riding (Glosten, 2014). Therefore joint changes at those locations are very commonly mentioned in bioarchaeology and palaeopathology as being related to the intense and regular practice

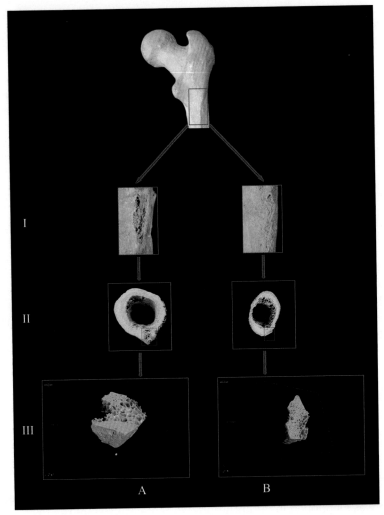

FIG. 9.3   Two possible microarchitectural patterns for entheseal changes: (A) trabecularisation of the cortical bone under the lacuna, and (B) regular cortical bone morphology under the lacuna (not to scale). I—Macroscopic appearance. II—Cross-section made at the most prominent point of enthesis of *m. gluteus maximus*. III—Volume rendering of acquired micro-CT tomograms. *From Djukić, K., Milovanović, P., Milenković, P., & Djurić, M. (2020). A microarchitectural assessment of the gluteal tuberosity suggests two possible patterns in entheseal changes. American Journal of Physical Anthropology, 172(2), 291–299. https://doi.org/10.1002/ajpa.24038, reproduced with permissions from John Wiley and Sons, licence 4898181436644.*

of horse riding during life (e.g. Bradtmiller, 1983; Courtaud and Rajev, 1998; Fornaciari et al., 2007, 2014; Miller and Reinhard, 1991; Pálfi, 1992; Sarry et al., 2016; Üstündağ and Deveci, 2011). Systematic methodological research on this aspect is, however, rather uncommon. Baillif-Ducros (2018) systematically scored, in Merovingian populations, the presence of 'osteophytic hip' when there was the combined presence of perifoveal osteophytes and osteophytosis on the margin of the femoral head (Baillif-Ducros, 2018). The results suggested that this criterion was not specific to riders but that it could be a good indicator of equestrian behaviours when observed in young adults and associated with other changes in relation to the repeated sitting position (see Section 9.3.3). In addition, joint changes at the femoro-patellar joint were specifically observed in riding populations and assumed to be related to the use of stirrups (Baillif-Ducros, 2018; Baillif-Ducros and McGlynn, 2013). An analysis of joint changes in a Hungarian Conquest period population with riding deposit in their graves did not, however, allow to make conclusions regarding a link between joint changes and the practice of horse riding, although this might be related to the numerous methodological limiting factors that must be considered when studying joint changes for the reconstruction of activities (Berthon, 2019).

9.3.3 Morphological variants and adaptations

The seated position repeated over time by the rider on their mount implies a flexion of the coxofemoral joints. The accentuated position in retroversion of the rider's pelvis induces a pressure applied by the femoral heads on the anterosuperior rim of the acetabulum (Auvinet and Estrade, 1998; Baillif-Ducros, 2018). This position of the femora may result in a vertical elongation (or ovalization) of the acetabulum and can also be the cause of a femoroacetabular impingement (Baillif-Ducros, 2018). This femoroacetabular impingement is due to repeated abnormal contact between the femoral head-neck junction area and the acetabular rim (Villotte and Knüsel, 2009). This contact is the result of either a bone abnormality in the pelvis (pincer type) or in the neck of the femur (cam type) (Ganz et al., 2003). The pincer type results from an excessive acetabular coverage over the femoral head, while the cam type is caused by a non-spherical portion of the femoral head (Bonin et al., 2008; Ganz et al., 2003; Villotte and Knüsel, 2009). The final consequence of this deformation would be the extension of the articular surface of the femoral head onto the neck, which is more known under the name 'Poirier's facet' (Finnegan, 1978). The presence of this osseous non-metric variation of the proximal femur would be expected with frequent squatting or hyperflexion of the hip (Verna et al., 2014). All of these features have been observed and used by researchers to interpret an individual as a habitual rider in archaeological contexts.

Regarding the changes of the shape of the acetabulum, their relation with a possible practice of horse riding was mentioned in the anthropological literature since the 1990s (e.g. Courtaud and Rajev, 1998; Miller and Reinhard, 1991; Pálfi, 1992). The results of a visual examination greatly depend on the experience of the observer, thereby limiting the reliability and reproducibility of the observations. Therefore different methodologies have been experimented to identify precise, reliable, and repeatable criteria to describe the acetabular shape changes, starting with the study by Erickson et al. (2000), who performed a Fourier analysis of the shape of the acetabulum on adult males from two Indigenous Arikara populations (South Dakota), respectively, presumed riders and non-riders. They observed significant differences (at a set level of $\alpha = 0.1$) between both groups, with expanded anterosuperior acetabular rims in the presumed rider group (Erickson et al., 2000). Although this methodology seems particularly relevant to capture the changes of the shape of the entire acetabular rim, it is however limited by the fact that several portions of the acetabular rim are frequently not preserved in ancient materials. This research has paved the way for further investigations following different methods with the aim of analysing the entire shape of the acetabulum (e.g. Eng et al., 2012; Garofalo, 2004) but those failed to identify a correlation between changes of the rim shape and the practice of horse riding, although this could be due to sampling and methodological limitations.

Inspired by the research by Erickson and colleagues, Baillif-Ducros et al. (2012) calculated a vertical/horizontal diameter index, via direct measurements of the acetabulum, to analyse the anterosuperior elongation of the acetabular rim on the remains of one probable Merovingian horse rider found in France. The coxal bones of the individual revealed very elongated acetabula compared to a reference collection. Following the same approach, Sarry et al. (2016) did not observe particularly elongated acetabula in a La Tène period mass burial (France) involving eight horses and eight human individuals, but bone preservation was a limiting factor. This same methodology was then used in a more extensive way on Merovingian series from northeastern Gaul, including individuals with riding-related grave goods (Baillif-Ducros, 2018). Elongated acetabula were observed in a notable part of these populations (with a predominance in males), including in three out of eight individuals with riding deposits in their graves. Later, a systematic study following the same approach, i.e. the calculation of a vertical/horizontal diameter index, was performed in a population from the Hungarian Conquest period (10th century CE), known to be composed of mounted archers. The statistical analyses revealed a vertical ovalization of the acetabulum (Fig. 9.4), significantly more important in the individuals with a riding-related deposit in their graves compared to the group without a deposit and a comparison group of presumed non-riders, which confirmed that this skeletal change can be, indeed, included among a set of skeletal indicators for the practice of horse riding (Berthon, 2019; Berthon et al., 2019). This methodology has been recently applied by Bühler and Kirchengast (2022) on an Avar population from an Austrian cemetery (ca. 625–800 CE). They observed significantly higher indices of ovalization of the acetabulum in males than in females as well as differences between the phases of the cemetery for females. The authors

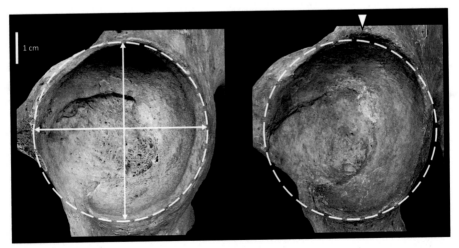

FIG. 9.4 Vertical ovalization of the acetabulum (right, with a *white arrow* showing the anterosuperior rim elongation) compared to a rather circular acetabulum (left), and measurements of the vertical and horizontal diameters of the acetabulum for the calculation of the index of ovalization of the acetabulum (left). The *discontinuous lines* represent exact circles for reference. See Baillif-Ducros et al. (2012) and Berthon et al. (2019) for the definition of the measurements. *Images courtesy of William Berthon.*

interpreted these findings in terms of sex and chronological differences in horse riding practice and lifestyle in this early medieval population in which horse riding is assumed to have been a habitual activity.

On another note, the variations of the anterior aspect of the femoral head-neck junction (i.e. Poirier's facet, the anteroiliac plaque, and Allen's fossa) are the most frequently cited skeletal changes assumed to be related to the practice of horse riding in the anthropological literature (e.g. Anđelinović et al., 2015; Bartsiokas et al., 2015; Belcastro et al., 2001; Blänkle, 1992; Bühler and Kirchengast, 2022; Dutour and Buzhilova, 2014; Fornaciari et al., 2014; Garofalo, 2004; Khudaverdyan et al., 2016; Miller and Reinhard, 1991; Molleson and Blondiaux, 1994; Molleson and Hodgson, 1993; Novotny et al., 2014; Pálfi, 1992; Panzarino and Sublimi Saponetti, 2017; Pulcini, 2014; Üstündağ and Deveci, 2011; Willey, 1997). These include, however, various changes that are often not differentiated, leading to confusion regarding their presumed link with the practice of horse riding. In particular, a distinction between three separate changes at the femoral head-neck junction, which are Poirier's facet (the most commonly cited one), the anteroiliac plaque, and Allen's fossa, has not always been made (Capasso et al., 1999; Finnegan, 1978; Kostick, 1963; Verna et al., 2014). In view of this, a more precise definition of those morphological variants was proposed by Radi et al. (2013). Thus, while Poirier's facet is smooth, on the same plane, and in continuity with the articular surface of the femoral head, the anteroiliac plaque corresponds to a rough imprint located on the anterior margin of the femoral neck close to the head. It is a bony overgrowth that may be on the same plane as the femoral head, lower than the femoral neck plane, or on an intermediate level, and may be delimitated by a bony rim. It can be present in addition to Poirier's facet and Allen's fossa, which is a circumscribed area on the anterior portion of the femoral neck, close to the border of the head, characterised by a cortical erosion with exposition of the trabecular bone or an area with clustered pores on the cortical surface (Radi et al., 2013).

Among all studies that mentioned a possible link between Poirier's facet and horse riding, rare were those relying on a systematic methodological investigation or on a sample of confirmed riders. Two recent studies, focusing on Merovingian (Baillif-Ducros, 2018) and Hungarian Conquest period (Berthon, 2019) populations, performed a comparison between individuals with and without riding deposits in their grave. They demonstrated that Poirier's facet (Fig. 9.5) is a reliable indicator for the practice of horse riding, as a consequence of the femoroacetabular impingement related to the repeated sitting position of the rider (see earlier). Recently, Poirier's facet was notably used by Bühler and Kirchengast (2022) as an indicator to investigate sex differences in habitual horse riding in an early medieval population of nomadic Avars from Austria. Along with other results, a higher prevalence in Poirier's facets observed in males compared to females led the authors to conclude that 'horse riding was a typical male activity' in this population, although not restricted to male individuals.

9.3.4 Skeletal trauma

Although traumatic lesions can be considered as anecdotical, reflecting accidental wounds or interpersonal injuries, they can provide insights on the lifestyles and behaviours in past

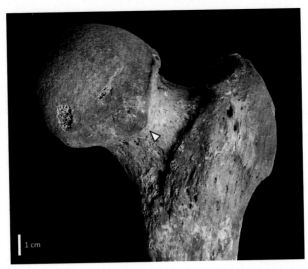

1 cm

FIG. 9.5 Poirier's facet (*white arrow*) on the femoral head and neck junction. *Image courtesy of William Berthon.*

populations, completed by the analysis of other types of skeletal changes (see Sections 9.3.1–9.3.3) (Buzon and Richman, 2007; de la Cova, 2012; Dutour, 2000; Henderson, 2009; Jurmain, 1999; Larsen, 1997; Paine et al., 2007; Pálfi, 1992; Tung, 2007). In the anthropological literature, horse riding was previously mentioned as a possible explanation for some traumatic lesions, mostly including fractures and *myositis ossificans* (e.g. Dutour and Buzhilova, 2014; Karstens et al., 2018; Langlois and Gallien, 2006; Pálfi, 1992; Willey, 1997). They concern all parts of the skeleton, from the skull, spine, trunk, and upper and lower limbs, and a fall from a horse is one of the main possible explanations that is given by scholars. The link between trauma and horse riding is, however, highly hypothetical, as most of studies lack direct archaeological evidence for the practice of horse riding, or they concern only a few individuals or a single one (e.g. Khudaverdyan et al., 2016; Wentz and de Grummond, 2009). In the end, only scarce studies include a systematic and statistical investigation of the frequency and location of traumatic lesions in populations with direct evidence of horse riding practice.

In a sample of Merovingian populations from France (Baillif-Ducros, 2018), most of the acute trauma recorded involved the bones of the forearm. However, individuals with these lesions did not have any equestrian equipment in their grave and horse riding was not specifically presented as an explanation for those trauma. Later on, a systematic analysis of acute fractures was performed on a population from the Hungarian Conquest period (Berthon, 2019; Berthon et al., 2021a). A comparison between individuals with and without riding-related deposit in their grave, and a modern population of non-riders, revealed that the main differences observed between groups concerned the bones of the upper limb, with higher frequencies in the Hungarian archaeological groups, and a statistically significant difference in clavicle fractures, in particular (Berthon et al., 2021a). Even though it remains impossible to assess with certainty the connection between the trauma and a specific activity, these findings are consistent with the data from sports medicine that describe the fall from the horse as the main cause of injuries in riders and the frequent fractures of the upper limb bones and the clavicle, specifically, as a consequence of this type of accident (Balendra et al., 2007; Bixby-Hammett and Brooks, 1990; Havlik, 2010; Khan et al., 2009; Loder, 2008). The type of clavicle fractures could especially be a further indicator as clinical data reveal that midshaft clavicle fractures are often related to a fall from a height, such as in equestrian sports (Berthon et al., 2021a; Khan et al., 2009; Kihlström et al., 2017).

9.4 The identification of reliable horse riding-related skeletal changes

As presented in this chapter, the influence of horse riding on the human skeleton has interested many bioarchaeologists and palaeopathologists for several decades. The identification of this determinant activity in past populations can, indeed, allow understanding better the organisation of these societies, regarding aspects such as mobility, economics, hierarchy, division of tasks, warfare, as well as cultural, spiritual, and funerary practices (Anthony, 2007). Despite the large number of anthropological studies referring to horse riding-related skeletal changes in the past few decades (see Section 9.3), the reliability of the link between those changes of various types and the activity remained debated due to multiple methodological pitfalls.

9.4.1 Methodological limitations and reflections

A significant issue in the research field of activity reconstruction is the lack of specificity of the activity-related skeletal changes, which is closely related to their multifactorial aetiologies (Dutour, 1992). Those bone changes are, indeed, not only related to mechanical stress but also to other internal and external factors. Age, sex, body weight, body size, ancestry, diet, genetic factors, or pathological conditions can, for instance, influence the expression of entheseal changes, osteoarthritis, or bone geometry (e.g. Crubézy et al., 2002; Dutour, 1992; Jurmain et al., 2012; Knüsel, 2000; Milella et al., 2012; Niinimäki, 2011; Niinimäki and Baiges Sotos, 2013; Rogers and Waldron, 1995; Ruff and Larsen, 2014; Schrader, 2019; Weiss et al., 2012; Weiss and Jurmain, 2007; Wilczak, 1998). These systemic factors can expose an individual to risk of developing changes, while other factors can also have a local influence, such as musculoskeletal anomalies, joint alignment, or injury, which can, for instance, lead to the development of osteoarthritis of a joint (Dutour, 1992; Schrader, 2019). Scholars who aim to investigate the influence of a specific activity, such as horse riding, on the human skeleton should therefore acknowledge those limiting factors in their research when selecting the individuals in their samples. Microstructural investigations may also contribute to addressing the question of the multifactorial aetiology of the changes (see Section 9.3.1.2).

The selection of a pertinent skeletal series is also one of the most determinant conditions for the reconstruction of activities in past populations. Unfortunately, there is no historical or modern reference collection of skeletons of horse riders, therefore the selection of a pertinent sample represents a major difficulty. Although the existence of historical or ethnological sources that suggest that a population of interest used to practice horse riding is particularly valuable, the activity must be attested by strong evidence, such as the presence of horse

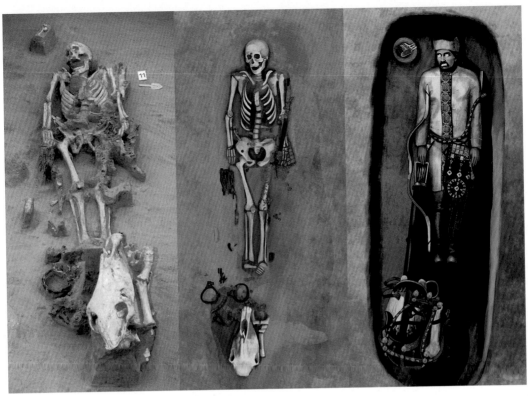

FIG. 9.6 Grave with horse riding equipment and horse bones from the Hungarian Conquest period cemetery of Karos-Eperjesszög (Karos III/11): picture of the grave (left; Révész, 1999); and reconstructions (middle and right) made by Pazirik Informatikai Kft. for the exhibition "Elit alakulat" of the Herman Ottó Museum (Miskolc, Hungary). *Reproduced with permissions from László Révész and Pazirik Informatikai Kft., https://pazirik.hu/. From Révész, L. (1999). Emlékezzetek utatok kezdetére… Régészeti kalandozások a magyar honfoglalás és államalapítás korában. Timp.*

remains and riding-related grave goods, in order to assure the reliability of its association with skeletal changes (Fig. 9.6).

On a similar note, only the analysis of comparative samples of known performers and non-performers (e.g. riders and non-riders) can allow distinguishing skeletal markers that are specific to an activity as the most significant differences observed between them should reflect the consequences of this activity (Tichnell, 2012). The identification of reliable skeletal changes associated with horse riding should, therefore, include the analysis of comparative samples of known riders and non-riders, ideally from the same group or similar populations (Garofalo, 2004; Tichnell, 2012). Such comparisons are however only exceptionally provided in the literature, which may partly be explained by the difficulty that it represents to find adequate series. In fact, the only certain cases of non-riding populations in the archaeological record are prehistoric ones who lived before the domestication of the horse and later populations from territories where the horse was absent. This is, for example, the case concerning pre-contact Indigenous peoples, who lived before its reintroduction on the American continent by Europeans during the contact period (Tichnell, 2012).

Despite this difficulty, some scholars made a valuable effort to include comparisons in their studies, based on lifestyle differences of populations, such as nomadic pastoralists versus agricultural populations from Mongolia and China (Eng, 2007), pre- versus post-contact Arikara populations and post-medieval British city- versus country-dwelling populations (Tichnell, 2012), or also a medieval Avar population versus a medieval agricultural population from Serbia (Djukic et al., 2018). Another approach consists in using modern documented collections, with recorded information such as the sex, age, cause of death, or occupation available for each individual, and where the practice of horse riding in a regular and intense way can be excluded (Berthon, 2019; Berthon et al., 2019, 2021a; Eng et al., 2012; Garofalo, 2004). Finally, the comparison, within the same population, of individuals who are presumed riders on the basis of their grave goods (horse bones, riding equipment) with the individuals who lack such direct evidence is another relevant option. In that case, the possible differences in observed skeletal changes may reflect the differences in grave deposits between the two subgroups, and thus, hypothetically reflect variations in the practice of horse riding (Baillif-Ducros, 2018; Baillif-Ducros et al., 2012; Belcastro and Facchini, 2001; Belcastro et al., 2007; Berthon, 2019; Berthon et al., 2016a, 2019, 2021a,b). This approach has also been adopted in the case of other physical activities in ancient populations, such as archery (Ryan et al., 2018; Thomas, 2014; Tihanyi et al., 2015, 2020). It must, however, be kept in mind that the presence of deposit related to an activity in the grave represents a solid argument but no indisputable evidence for the practice of riding during life, and, similarly, that the absence of riding-related deposit in the grave does not necessarily mean either that an individual was not practising this activity.

In addition, another major challenge that considerably limits the analyses is the rather limited sample sizes, which is particularly common concerning archaeological series. Studies relying on very few or single individuals do not allow observation of a repetition of skeletal changes, or the use of statistical analyses to test the significance of their occurrence, which consequently decreases the reliability of the interpretations. Furthermore, even though there are problems of transposition between the modern sports and the activities performed in past populations, clinical data from sports medicine can still provide useful indicators of the possible lesions and changes resulting from an activity and give more credibility to, if not confirm, the association between a skeletal change and an activity (Dutour, 1992, 2000). In this regard, only a few studies refer to sports medicine literature to support the riding hypothesis (e.g. Baillif-Ducros, 2018; Baillif-Ducros et al., 2012; Berthon, 2019; Berthon et al., 2019, 2021a; Blondiaux, 1994; Đukić, 2016; Pálfi and Dutour, 1996).

Lastly, another question that must be addressed for the identification of reliable skeletal changes related to the practice of horse riding is the existence of various riding styles and equipment through cultures and eras. Ancient and modern European equestrian treaties reveal that there is no universal riding posture but that postures are imposed on riders by the societies to which they belong (Baillif-Ducros, 2018). In his treaty *On Horsemanship*, the ancient Greek author Xenophon advocates a strict position for the rider on their horse (Xenophon, 1897):

But now, supposing the rider fairly seated, whether bareback or on a saddle-cloth, a good seat is not that of a man seated on a chair, but rather the pose of a man standing upright with his legs apart. In this way he will be able to hold on to the horse more firmly by his thighs; and this erect attitude will enable him to hurl a javelin or to strike a blow from horseback, if occasion calls, with more vigorous effect (pp. 53–54).

In the 18th century, the modern rider must adopt 'la belle posture' recommended by La Guérinière (de) (2000): "Immediately after placing the bridle hand, one should sit right in the middle of the saddle, with waist and buttocks forward, so as not to be seated near the back of the saddle tree" (p. 84). Therefore a Greek rider of the classical era is not seated in the saddle in the same way as a French rider of the modern times, and this example expands to many other societies and periods.

Likewise, the equestrian equipment varies through historical times. The rider interacts directly with the animal through their equipment, the spur, and that of their equine partner, the saddle and the stirrup(s). The equipment used in riding differs culturally; generally speaking, traditional Greek saddles were *ephippium* (carpet, cloth, or blanket), while Roman and Germanic saddles were flat, Celtic saddles were with horns, and Merovingian saddles had hollow seats like the Hunnic ones (Baillif-Ducros, 2018). The stirrups, for their part, are an invention of the nomadic populations from the steppes of Central Asia or Siberia. The oldest ones, made of iron, were discovered in Xiongnu tombs and date from the 3rd to 1st centuries BCE (Tseveendorj, 1996). It is less than a millennium later that the stirrups reached the West, where they were previously unknown throughout Antiquity. At the end of the 6th century CE, stirrups appeared among grave goods of Merovingian burials (Baillif-Ducros, 2018). This multiplicity of postures and equipment used in the different regions of the world and over the centuries necessarily implies variability of changes on the rider's skeleton as different postures may indeed involve different movements and muscular action (Baillif-Ducros, 2018, 2021; Djukic et al., 2018; Tichnell, 2012).

Only a few studies mention the potential impact of the riding style on the development of skeletal changes (Baillif-Ducros, 2018; Eng, 2016; Eng et al., 2012; Fuka, 2018; Tichnell, 2012). In particular, Tichnell (2012) includes this question in her study, testing the hypothesis that different riding styles use different muscles by comparing bareback, British, and Mongolian styles. Her results did not allow identifying any universal set and suggest, therefore, that "culturally-specific suites of features are needed to identify activity performers in different societies" (Tichnell, 2012, p. 137). Likewise, the impact of stirrup use on the rider's knee has been demonstrated by analyses conducted on the skeletons of a Mongolian rider-archer (McGlynn et al., 2012) and an Avar rider (Baillif-Ducros and McGlynn, 2013) which showed the presence of osteoarthritis on the upper outline of the patellar surface. The use of stirrups by these two young male individuals (20–29 years) is evidenced by their inclusion as a grave good within each burial, both of which are dated to the 7th century CE.

Future comparative analyses including different nomadic and semi-nomadic populations should allow identifying the possible existence of specific or common features according to the riding style. More detailed analyses of the upper body (entheseal and joint changes of the upper limbs, as well as vertebral lesions, in particular) may also contribute not only to give a more complete picture of the influence of horse riding on the skeleton but also to give insights into the riding style question. Some valuable information might as well be provided by observations made on present-day nomadic populations who still follow a traditional lifestyle (such as in Mongolia), or on individuals who, nowadays, perform reconstructions of historical equestrian techniques used in various cultures, thus following an approach that can be considered as experimental ethnoarchaeology.

9.4.2 Perspectives

The numerous methodological limitations that were mentioned before result in a blurred vision regarding which skeletal changes should be considered as reliable and repeatable indicators of horse riding practice. On this basis, we suggest that a study to identify reliable skeletal changes related to the practice of horse riding should, in the best circumstances:

- rely on a pertinent population with a well-documented archaeological, historical, or ethnological context, ideally with direct archaeological evidence (such as grave goods) that confirm the practice of horse riding for a part of the population at least;
- have an adequate sample size of individuals from this population, preferably young and mature adults, with a sex distinction, in order to allow a repetition of the observations and comparisons, and to avoid methodological biases;
- use a precise methodology with a systematic recording of a large scale of qualitative and quantitative features for statistical analyses;
- cautiously deal with the data in case of pathological conditions that can affect the analysis of some activity-related changes (i.e. exclude individuals with metabolic or inflammatory disorders such as the diffuse idiopathic skeletal hyperostosis or spondyloarthropathies for the entheseal changes, or with trauma for the entheseal and joint changes, and metrics);
- discuss the validity of the observed skeletal changes in light of functional morphology (i.e. by referring to the riding posture characteristics) and sports medicine data;
- include a comparison sample of known or presumed non-riders;
- acknowledge that the identified riding-related skeletal changes may be specific to a culture, depending on the style and equipment used.

Following these criteria may lead to the identification of skeletal changes reliably associated with horse riding. Thereafter, the presence of several skeletal changes together can contribute to identifying, statistically, the practice of horse riding in a group or population of interest, and, cautiously, to regard particular individuals as potential riders. Considering the importance of horse riding in the history of human cultural evolution, this research area can significantly contribute to improving our understanding of the lifestyles and organisation of numerous past populations.

References

Altgärde, J., Redéen, S., Hilding, N., & Drott, P. (2014). Horse-related trauma in children and adults during a two year period. *Scandinavian Journal of Trauma, Resuscitation and Emergency Medicine, 22*(1), 40. https://doi.org/10.1186/s13049-014-0040-8.

Alves Cardoso, F., & Henderson, C. Y. (2010). Enthesopathy formation in the humerus: Data from known age-at-death and known occupation skeletal collections. *American Journal of Physical Anthropology, 141*(4), 550–560. https://doi.org/10.1002/ajpa.21171.

Anđelinović, Š., Anterić, I., Škorić, E., & Bašić, Ž. (2015). Skeleton changes induced by horse riding on medieval skeletal remains from Croatia. *The International Journal of the History of Sport, 32*(5), 708–721. https://doi.org/10.1080/09523367.2015.1038251.

Angel, J. L. (1982). Osteoarthritis and occupation (ancient and modern). In V. V. Novotny (Ed.), *Second anthropological congress of ales Hrdlicka* (pp. 443–446). Prague—Humpolec, Czech Republic: Universitas Carolina Pragensis.

Angel, J. L., Kelley, J. O., Parrington, M., & Pinter, S. (1987). Life stresses of the free black community as represented by the first African Baptist Church, Philadelphia, 1823-1841. *American Journal of Physical Anthropology, 74*(2), 213–229. https://doi.org/10.1002/ajpa.1330740209.

Anthony, D. W. (2007). *The horse, the wheel, and language: How bronze-age riders from the Eurasian steppes shaped the modern world*. Princeton, NJ: Princeton University Press.

Anthony, D. W., & Brown, D. R. (1991). The origins of horseback riding. *Antiquity, 65*(246), 22–38. https://doi.org/10.1017/S0003598X00079278.

Auvinet, B. (1980). La hanche du cavalier. *Médecine Du Sport, 54*(5), 281–285.

Auvinet, B. (1999). Lombalgies et équitation. *Synoviale, 83*, 25–31.

Auvinet, B., & Estrade, M. (1998). *La santé du cavalier : conseils pratiques pour une équitation sans risque*. Paris, France: Chiron.

Baillif-Ducros, C. (2018). *La pratique de la monte à cheval au haut Moyen Âge (fin Ve-VIIe siècle) dans le nord-est de la Gaule. État des connaissances archéologiques, recherche méthodologique sur le syndrome du cavalieret application d'un nouveau protocole d'étude aux populations mérovingiennes [doctoral dissertation]*. Caen, France: Université de Caen Normandie.

Baillif-Ducros, C. (2021). The Merovingian rider and his horse: Impact of equestrian equipment on the rider's posture and skeleton [conference presentation]. In *27th annual meeting of the european association of archaeologists (EAA), 6–11 September 2021*.

Baillif-Ducros, C., & McGlynn, G. (2013). Stirrups and archaeological populations: Bio-anthropological considerations for determining their use based on the skeletons of two steppe riders. *Bulletin de La Société Suisse d'Anthropologie, 19*(2), 43–44.

Baillif-Ducros, C., Truc, M.-C., Paresys, C., & Villotte, S. (2012). Approche méthodologique pour distinguer un ensemble lésionnel fiable de la pratique cavalière. Exemple du squelette de la tombe 11 du site de « La Tuilerie » à Saint-Dizier (Haute-Marne), VIe siècle. *Bulletins et Mémoires de La Société d'Anthropologie de Paris, 24*(1–2), 25–36. https://doi.org/10.1007/s13219-011-0049-8.

Balendra, G., Turner, M., & McCrory, P. (2008). Career-ending injuries to professional jockeys in British horse racing (1991–2005). *British Journal of Sports Medicine, 42*(1), 22–24. https://doi.org/10.1136/bjsm.2007.038950.

Balendra, G., Turner, M., McCrory, P., & Halley, W. (2007). Injuries in amateur horse racing (point to point racing) in Great Britain and Ireland during 1993–2006. *British Journal of Sports Medicine, 41*(3), 162–166. http://doi.org/10.1136/bjsm.2006.033894.

Ball, C. G., Ball, J. E., Kirkpatrick, A. W., & Mulloy, R. H. (2007). Equestrian injuries: Incidence, injury patterns, and risk factors for 10 years of major traumatic injuries. *The American Journal of Surgery, 193*(5), 636–640. https://doi.org/10.1016/j.amjsurg.2007.01.016.

Bartsiokas, A., Arsuaga, J.-L., Santos, E., Algaba, M., & Gómez-Olivencia, A. (2015). The lameness of king Philip II and Royal Tomb I at Vergina, Macedonia. *Proceedings of the National Academy of Sciences of the United States of America, 112*(32), 9844–9848. https://doi.org/10.1073/pnas.1510906112.

Bede, I. (2012). The status of horses in late Avar-period society in the Carpathian Basin. In R. Annaert, K. De Groote, Y. Hollevoet, F. Theuws, D. Tys, & L. Verslype (Eds.), *The very beginning of Europe? Cultural and social dimensions of early-medieval migration and colonisation (5th–8th century). Archaeology in contemporary Europe (conference Brussels – May 17–19, 2011). Relicta Monografieën 7* (pp. 41–59). Brussels, Belgium: Flanders Heritage Agency.

Bede, I., & Detante, M. (2014). *Rencontre autour de l'animal en contexte funéraire. IVe Rencontre du Groupe d'anthropologie et d'archéologie funéraire, Saint-Germain-en-Laye, 30 et 31 mars 2012*. Saint-Germain-en-Laye, France: Groupe d'anthropologie et d'archéologie funéraire.

Belcastro, M. G., & Facchini, F. (2001). Anthropological and cultural features of a skeletal sample of horsemen from the medieval necropolis of Vicenne-Campochiaro (Molise, Italy). *Collegium Antropologicum, 25*(2), 387–401.

Belcastro, M. G., Facchini, F., Neri, R., & Mariotti, V. (2001). Skeletal markers of activity in the early middle ages necropolis of Vicenne-Campochiaro (Molise, Italy). *Journal of Paleopathology, 13*(3), 9–20.

Belcastro, M. G., Rastelli, E., Mariotti, V., Consiglio, C., Facchini, F., & Bonfiglioli, B. (2007). Continuity or discontinuity of the life-style in Central Italy during the Roman imperial age-early middle ages transition: Diet, health, and behavior. *American Journal of Physical Anthropology, 132*(3), 381–394. https://doi.org/10.1002/ajpa.20530.

Bendrey, R. (2007). New methods for the identification of evidence for bitting on horse remains from archaeological sites. *Journal of Archaeological Science, 34*(7), 1036–1050. https://doi.org/10.1016/j.jas.2006.09.010.

Bendrey, R. (2008). An analysis of factors affecting the development of an equid cranial enthesopathy. *Veterinarija Ir Zootechnika, 41*(63), 25–31.

Benjamin, M., Kumai, T., Milz, S., Boszczyk, B. M., Boszczyk, A. A., & Ralphs, J. R. (2002). The skeletal attachment of tendons—tendon 'entheses.'. *Comparative Biochemistry and Physiology Part A: Molecular & Integrative Physiology, 133*(4), 931–945. https://doi.org/10.1016/S1095-6433(02)00138-1.

Berthon, W. (2019). *Bioarchaeological analysis of the mounted archers from the Hungarian conquest period (10th century): Horse riding and activity-related skeletal changes [doctoral dissertation]*. Paris, France – Szeged, Hungary: EPHE, PSL University—University of Szeged.

Berthon, W., Rittemard, C., Tihanyi, B., Pálfi, G., Coqueugniot, H., & Dutour, O. (2015). Three-dimensional microarchitecture of entheseal changes: Preliminary study of human radial tuberosity. *Acta Biologica Szegediensis, 59*(1), 79–90.

Berthon, W., Tihanyi, B., Révész, L., Coqueugniot, H., Pálfi, G., & Dutour, O. (2016a). A contribution to the definition of "horse riding syndrome": The mounted archers from the Hungarian conquest (Xth century AD) [conference poster presentation]. In *21st European meeting of the paleopathology association, Moscow, Russia, 15-19 August 2016*.

Berthon, W., Tihanyi, B., Kis, L., Révész, L., Coqueugniot, H., Dutour, O., et al. (2019). Horse riding and the shape of the acetabulum: Insights from the bioarchaeological analysis of early Hungarian mounted archers (10th century). *International Journal of Osteoarchaeology, 29*(1), 117–126. https://doi.org/10.1002/oa.2723.

Berthon, W., Tihanyi, B., Váradi, O. A., Coqueugniot, H., Dutour, O., & Pálfi, G. (2021a). Riding for a fall: Bone fractures among mounted archers from the Hungarian conquest period (10th century CE). *International Journal of Osteoarchaeology, 31*(5), 926–940. https://doi.org/10.1002/OA.3010.

Berthon, W., Tihanyi, B., Révész, L., Coqueugniot, H., Dutour, O., & Pálfi, G. (2021b). Were they riders? Identification of reliable horse riding-related skeletal changes from the bioarchaeological analysis of a Hungarian Conquest period population [conference presentation]. In *27th annual meeting of the European Association of Archaeologists (EAA), 6–11 September 2021*.

Berthon, W., Tihanyi, B., Pálfi, G., Dutour, O., & Coqueugniot, H. (2016b). Can micro-CT and 3D imaging allow differentiating the main aetiologies of entheseal changes? In S. S. Gál (Ed.), *The talking dead. New results of the central and eastern European Osteoarchaeology. Proceedings of the first conference of the Török Aurél anthropological association from Târgu-Mureş, 13–15 November 2015* (pp. 29–43). Cluj-Napoca, Romania: MEGA Publishing House.

Bertin, L., Saladie, P., Lignereux, Y., Moigne, A.-M., & Boulbes, N. (2016). Les équidés du Clos d'Ugnac (Aude, France) : force de travail, ressource carnée et source de matière première, au Moyen Âge. *Ethnozootechnie, 101*, 53–56.

Biau, S., Gilbert, C. H., Gouz, J., Roquet, C. H., Fabis, J., & Leporcq, B. (2013). Preliminary study of rider back biomechanics. *Computer Methods in Biomechanics and Biomedical Engineering, 16*(sup1), 48–49. https://doi.org/10.1080/10255842.2013.815845.

Bindé, M., Cochard, D., & Knüsel, C. J. (2019). Exploring life patterns using entheseal changes in equids: Application of a new method on unworked specimens. *International Journal of Osteoarchaeology, 29*(6), 947–960. https://doi.org/10.1002/oa.2809.

Bindé, M., Cochard, D., & Knüsel, C. J. (2021). Mounted or not mounted? Identification of riding by using entheseal changes in horses associated with the burial of Childeric I [conference presentation]. In *27th annual meeting of the European Association of Archaeologists (EAA), 6–11 September 2021*.

Bixby-Hammett, D. M. (1992). Pediatric equestrian injuries. *Pediatrics, 89*(6), 1173–1176.

Bixby-Hammett, D. M., & Brooks, W. H. (1990). Common injuries in horseback riding. A review. *Sports Medicine, 9*(1), 36–47.

Blänkle, P. H. (1992). *Skelette erzählen. Abhandlungen des Offenbacher Vereins für Naturkunde, Band 8. Begleitheft zur gleichnamigen Ausstellung im Stadtmuseum, Offenbach am Main vom 26. Mai bis 25. Oktober 1992*. Offenbach, Germany: Offenbacher Verein für Naturkunde.

Blondiaux, J. (1989). *Essai d'anthropologie physique et de paléopathologie des populations du nord de la Gaule au haut Moyen Age [doctoral dissertation]*. Lille, France: Université de Lille III.

Blondiaux, J. (1994). À propos de la dame d'Hochfelden et de la pratique cavalière : discussion autour des sites fonctionnels fémoraux. In L. Buchet (Ed.), *Actes des 6e Journées Anthropologiques (Valbonne, 9–11 juin 1992). Dossier de Documentation Archéologique n° 17* (pp. 97–109). Paris, France: CNRS Éditions.

Bonin, N., Tanji, P., Cohn, J., Moyere, F., Ferret, J.-M., & Dejour, D. (2008). Relation entre conflit fémoro-acétabulaire et coxarthrose ? Place de l'arthroscopie dans les lésions pré-arthrosiques de la hanche. In *Journées Lyonnaises de Chirurgie de la Hanche* (pp. 103–109). France: Lyon.

Bradtmiller, B. (1983). The effect of horseback riding on Arikara arthritis patterns [conference presentation]. In *82nd annual meeting of the American Anthropological Association, Chicago, IL, 16–20 November 1983*.

Brosseder, U. (2009). Xiongnu terrace tombs and their interpretation as elite burials. In J. Bemmann, H. Parzinger, E. Pohl, & D. Tseveendorzh (Eds.), *Current archaeological research in Mongolia. Papers from the first international conference on "archaeological research in Mongolia" held in Ulaanbaatar, august 19th-23rd, 2007* (pp. 247–280). Bonn, Germany: Vor- und Frühgeschichtliche Archäologie, Rheinische Friedrich-Wilhelms-Universität Bonn.

Brown, D., & Anthony, D. (1998). Bit Wear, horseback riding and the Botai site in Kazakstan. *Journal of Archaeological Science, 25*(4), 331–347. https://doi.org/10.1006/jasc.1997.0242.

Bugarski, I. (2009). *Nekropole iz doba antike i ranog srednjeg veka na lokalitetu Čik*. Belgrade–Bečej, Serbia: Arheološki institut, Gradski muzej Bečej.

Bühler, B., & Kirchengast, S. (2022). Horse riding as a habitual activity among the early medieval Avar population of the cemetery of Csokorgasse (Vienna): Sex and chronological differences. *International Journal of Osteoarchaeology.* https://doi.org/10.1002/oa.3107.

Buikstra, J. E., & Ubelaker, D. H. (Eds.). (1994). *Standards for data collection from human skeletal remains. Proceedings of a seminar at the Field Museum of natural history, organized by Jonathan Haas.* Fayetteville, AR: Arkansas Archaeological Survey.

Bulatović, J., Bulatović, A., & Marković, N. (2014). Paleopathological changes in an early iron age horse skeleton from the Central Balkans (Serbia). *International Journal of Paleopathology, 7,* 76–82. https://doi.org/10.1016/J.IJPP.2014.07.001.

Buzon, M. R., & Richman, R. (2007). Traumatic injuries and imperialism: The effects of Egyptian colonial strategies at Tombos in upper Nubia. *American Journal of Physical Anthropology, 133*(2), 783–791. https://doi.org/10.1002/ajpa.20585.

Campagne, G. (2011). *Accidents d'équitation, ampleur du risque et degré d'urgence : nouveau regard médical sur un sport en plein développement, à partir d'une étude menée au service d'Accueil et de Traitement des Urgences-SMUR du Centre hospitalier de Boulogne-sur-Mer [MD thesis].* Lille, France: Université Lille 2 Droit et Santé.

Capasso, L., Kennedy, K. A. R., & Wilczak, C. A. (1999). *Atlas of occupational markers on human remains.* Teramo, Italy: Edigrafital.

Carver, M. (1998). *Sutton Hoo: Burial ground of kings?.* London, UK: British Museum Press.

Clottes, J. (2003). *Chauvet cave: The art of earliest times.* Salt Lake City, UT: University of Utah Press.

Courtaud, P., & Rajev, D. (1998). Osteomorphological features of nomadic riders: Some examples from Iron age populations located in southwestern Siberia. In M. Pearce, & M. Tosi (Eds.), *Papers from the EAA third annual meeting at Ravenna, 1997. BAR international series 717, Vol. 1, pre- and protohistory* (pp. 110–113). Oxford, UK: Archaeopress.

Cross, P. J. (2011). Horse burial in first millennium AD Britain: Issues of interpretation. *European Journal of Archaeology, 14*(1–2), 190–209. https://doi.org/10.1179/146195711798369409.

Crubézy, E., Goulet, J., Bruzek, J., Jelinek, J., Rouge, D., & Ludes, B. (2002). Epidemiology of osteoarthritis and enthesopathies in a European population dating back 7700 years. *Joint, Bone, Spine, 69*(6), 580–588.

de la Cova, C. (2012). Patterns of trauma and violence in 19th-century-born African American and Euro-American females. *International Journal of Paleopathology, 2*(2), 61–68. https://doi.org/10.1016/j.ijpp.2012.09.009.

de Barros Damgaard, P., Martiniano, R., Kamm, J., Moreno-Mayar, J. V., Kroonen, G., Peyrot, M., et al. (2018). The first horse herders and the impact of early bronze age steppe expansions into Asia. *Science, 360*(6396), eaar7711. https://doi.org/10.1126/science.aar7711.

de Cocq, P., Muller, M., Clayton, H. M., & van Leeuwen, J. L. (2013). Modelling biomechanical requirements of a rider for different horse-riding techniques at trot. *The Journal of Experimental Biology, 216*(10), 1850–1861. https://doi.org/10.1242/jeb.070938.

Di Cosmo, N. (1994). Ancient inner Asian nomads: Their economic basis and its significance in Chinese history. *The Journal of Asian Studies, 53*(4), 1092–1126. https://doi.org/10.2307/2059235.

Di Cosmo, N. (2011). Ethnogenesis, coevolution and political morphology of the earliest steppe empire: The Xiongnu question revisited. In U. Brosseder, & B. K. Miller (Eds.), *Xiongnu archaeology: Multidisciplinary perspectives of the first steppe empire in inner Asia* (pp. 35–48). Bonn, Germany: Vor- und Frühgeschichtliche Archäologie, Rheinische Friedrich-Wilhelms-Universität Bonn.

Dizdar, M., Rapan Papesa, A., & Rimpf, A. (2017). Rezultati zaštitnih istraživanja kasnoavarodobnoga groblja Šarengrad—Klopare. *Annales Instituti Archaeologici, XIII*(1), 9–18.

Djukić, K., Milovanović, P., Milenković, P., & Djurić, M. (2020). A microarchitectural assessment of the gluteal tuberosity suggests two possible patterns in entheseal changes. *American Journal of Physical Anthropology, 172*(2), 291–299. https://doi.org/10.1002/ajpa.24038.

Djukic, K., Milovanovic, P., Hahn, M., Busse, B., Amling, M., & Djuric, M. (2015). Bone microarchitecture at muscle attachment sites: The relationship between macroscopic scores of entheses and their cortical and trabecular microstructural design. *American Journal of Physical Anthropology, 157*(1), 81–93. https://doi.org/10.1002/ajpa.22691.

Djukic, K., Miladinovic-Radmilovic, N., Draskovic, M., & Djuric, M. (2018). Morphological appearance of muscle at-tachment sites on lower limbs: Horse riders versus agricultural population. *International Journal of Osteoarchaeology*, *28*(6), 656–668. https://doi.org/10.1002/oa.2680.

Donoghue, H. D., Taylor, G. M., Marcsik, A., Molnár, E., Pálfi, G., Pap, I., et al. (2015). A migration-driven model for the historical spread of leprosy in medieval Eastern and Central Europe. *Infection, Genetics and Evolution, 31*, 250–256. https://doi.org/10.1016/j.meegid.2015.02.001.

Đukić, K. (2016). *Bone macromorphology at muscle attachment sites: Its relationship with the microarchitecture of the un-derlying bone and possible implications for the reconstruction of habitual physical activities of past populations [doctoral dissertation].* Belgrade, Serbia: University of Belgrade.

Đurić, M., Milovanović, P., Đonić, D., Minić, A., & Hahn, M. (2012). Morphological characteristics of the developing proximal femur: A biomechanical perspective. *Srpski Arhiv za Celokupno Lekarstvo, 140*(11–12), 738–745. https://doi.org/10.2298/sarh1212738d.

Dutour, O. (1992). Activités physiques et squelette humain : le difficile passage de l'actuel au fossile. *Bulletins et Mémoires de La Société d'Anthropologie de Paris, 4*(3–4), 233–241. https://doi.org/10.3406/bmsap.1992.2319.

Dutour, O. (2000). Chasse et activités physiques dans la Préhistoire : les marqueurs osseux d'activités chez l'homme fossile. *Anthropologie et Préhistoire, 111*, 156–165.

Dutour, O., & Buzhilova, A. (2014). Palaeopathological study of Napoleonic mass graves discovered in Russia. In C. J. Knüsel, & M. Smith (Eds.), *The Routledge handbook of the bioarchaeology of human conflict* (pp. 511–524). London, UK – New York, NY: Routledge.

Eng, J. T. (2007). *Nomadic pastoralists and the Chinese empire: A bioarchaeological study of China's northern frontier [doctoral dissertation].* Santa Barbara, CA: University of California.

Eng, J. T. (2016). A bioarchaeological study of osteoarthritis among populations of northern China and Mongolia during the bronze age to Iron age transition to nomadic pastoralism. *Quaternary International, 405*(part B), 172–185. https://doi.org/10.1016/j.quaint.2015.07.072.

Eng, J. T., Baker, A., Tang, P., Thompson, S., & Gomez, J. M. (2012). Morphometric analysis of acetabular rim shape among ancient Mongolian pastoralists [conference poster presentation]. In *19th annual meeting of the Midwest Bioarchaeology and Forensic Anthropology Association, Carbondale, IL, October 2012*.

Erdélyi, I. (2000). *Archaeological expeditions in Mongolia*. Budapest, Hungary: Mundus Hungarian University Press.

Erickson, J. D., Lee, D. V., & Bertram, J. E. A. (2000). Fourier analysis of acetabular shape in native American Arikara populations before and after acquisition of horses. *American Journal of Physical Anthropology, 113*(4), 473–480. https://doi.org/10.1002/1096-8644(200012)113:4<473::AID-AJPA3>3.0.CO;2-5.

EuroSafe. (2014). *Injuries in the European Union. Summary of injury statistics for the years 2010–2012.* Amsterdam, The Netherlands: EuroSafe. Retrieved from https://www.eurosafe.eu.com/uploads/inline-files/IDB_Report_2014_final%202010-2012.pdf.

Finnegan, M. (1978). Non-metric variation of the infracranial skeleton. *Journal of Anatomy, 125*(1), 23–37.

Fornaciari, G., Vitiello, A., Giusiani, S., Giuffra, V., Fornaciari, A., & Villari, N. (2007). The Medici project: First an-thropological and paleopathological results of the exploration of the Medici tombs in Florence. *Medicina nei Secoli, 19*(2), 521–543.

Fornaciari, G., Bartolozzi, P., Bartolozzi, C., Rossi, B., Menchi, I., & Piccioli, A. (2014). A great enigma of the Italian renaissance: Paleopathological study on the death of Giovanni dalle Bande Nere (1498-1526) and historical rel-evance of a leg amputation. *BMC Musculoskeletal Disorders, 15*, 301. https://doi.org/10.1186/1471-2474-15-301.

Fuka, M. R. (2018). *Activity markers and horse riding in Mongolia: Entheseal changes among bronze and Iron age human skeletal remains [master thesis].* West Lafayette, IN: Purdue University.

Ganz, R., Parvizi, J., Beck, M., Leunig, M., Nötzli, H., & Siebenrock, K. A. (2003). Femoroacetabular impinge-ment: A cause for osteoarthritis of the hip. *Clinical Orthopaedics and Related Research, 417*, 112–120. https://doi.org/10.1097/01.blo.0000096804.78689.c2.

Garofalo, E. (2004). *The osteologic markers of horseback riding: An examination of two medieval English populations [master thesis].* Bradford, UK: University of Bradford.

Gaunitz, C., Fages, A., Hanghøj, K., Albrechtsen, A., Khan, N., Schubert, M., et al. (2018). Ancient genomes revisit the ancestry of domestic and Przewalski's horses. *Science, 360*(6384), 111–114. https://doi.org/10.1126/science.aao3297.

German Equestrian Federation. (2017). *The principles of riding. Basic training for horse and rider.* Shrewsbury, UK: Kenilworth Press.

Glosten, B. (2014). *The riding doctor: A prescription for healthy, balanced, and beautiful riding, now and for years to come.* North Pomfret, VT: Trafalgar Square Books.

Greenfield, H. J., Shai, I., Greenfield, T. L., Arnold, E. R., Brown, A., Eliyahu, A., et al. (2018). Earliest evidence for equid bit wear in the ancient near east: The "ass" from early bronze age tell eṣ-Ṣâfi/Gath, Israel. *PLoS One, 13*(5). https://doi.org/10.1371/journal.pone.0196335, e0196335.

Havlik, H. S. (2010). Equestrian sport-related injuries: A review of current literature. *Current Sports Medicine Reports, 9*(5), 299–302. https://doi.org/10.1249/JSR.0b013e3181f32056.

Heřt, J. (1994). A new attempt at the interpretation of the functional architecture of the cancellous bone. *Journal of Biomechanics, 27*(2), 239–242. https://doi.org/10.1016/0021-9290(94)90214-3.

Henderson, C. Y. (2009). *Musculo-skeletal stress markers in bioarchaeology: Indicators of activity levels or human variation? A re-analysis and interpretation [doctoral dissertation].* Durham, UK: University of Durham.

Henderson, C. Y., Mariotti, V., Pany-Kucera, D., Villotte, S., & Wilczak, C. A. (2016). The new 'Coimbra method': A biologically appropriate method for recording specific features of fibrocartilaginous Entheseal changes. *International Journal of Osteoarchaeology, 26*(5), 925–932. https://doi.org/10.1002/oa.2477.

Honeychurch, W. (2014). Alternative complexities: The archaeology of pastoral nomadic states. *Journal of Archaeological Research, 22*(4), 277–326. https://doi.org/10.1007/s10814-014-9073-9.

Humbert, C. (2000). *L'équitation et ses conséquences sur le rachis lombaire du cavalier : à propos de 123 observations [doctoral dissertation].* Nancy, France: Université Henri Poincaré Nancy I.

Hyland, A. (1999). *The horse in the middle ages.* Stroud, UK: Sutton Publishing.

Jennbert, K. (2003). Animal graves: Dog, horse and bear. *Current Swedish Archaeology, 11,* 139–152.

Johannesson, E. G. (2011). Grave matters: Reconstructing a Xiongnu identity from mortuary stone monuments. In U. Brosseder, & B. K. Miller (Eds.), *Xiongnu archaeology: Multidisciplinary perspectives of the first steppe empire in inner Asia* (pp. 201–212). Bonn, Germany: Vor- und Frühgeschichtliche Archäologie, Rheinische Friedrich-Wilhelms-Universität Bonn.

Jurmain, R. D. (1977). Stress and the etiology of osteoarthritis. *American Journal of Physical Anthropology, 46*(2), 353–365. https://doi.org/10.1002/ajpa.1330460214.

Jurmain, R. D. (1999). *Stories from the skeleton. Behavioral reconstruction in human osteology.* Amsterdam, The Netherlands: Gordon and Breach.

Jurmain, R. D., Alves Cardoso, F., Henderson, C. Y., & Villotte, S. (2012). Bioarchaeology's holy grail: The reconstruction of activity. In A. L. Grauer (Ed.), *A companion to paleopathology* (pp. 531–552). Chichester, UK: Wiley-Blackwell.

Karakostis, F. A., Jeffery, N., & Harvati, K. (2019). Experimental proof that multivariate patterns among muscle attachments (entheses) can reflect repetitive muscle use. *Scientific Reports, 9*(1), 16577. https://doi.org/10.1038/s41598-019-53021-8.

Karstens, S., Littleton, J., Frohlich, B., Amgaluntugs, T., Pearlstein, K., & Hunt, D. (2018). A palaeopathological analysis of skeletal remains from bronze age Mongolia. *Homo, 69*(6), 324–334. https://doi.org/10.1016/j.jchb.2018.11.002.

Khan, L. A. K., Bradnock, T. J., Scott, C., & Robinson, C. M. (2009). Fractures of the clavicle. *The Journal of Bone and Joint Surgery, 91*(2), 447–460. https://doi.org/10.2106/jbjs.h.00034.

Khudaverdyan, A., Khachatryan, H., & Eganyan, L. (2016). Multiple trauma in a horse rider from the late Iron age cemetery at Shirakavan, Armenia. *Bioarchaeology of the Near East, 10,* 47–68.

Kihlström, C., Möller, M., Lönn, K., & Wolf, O. (2017). Clavicle fractures: Epidemiology, classification and treatment of 2 422 fractures in the Swedish fracture register; an observational study. *BMC Musculoskeletal Disorders, 18,* 82. https://doi.org/10.1186/s12891-017-1444-1.

Knüsel, C. J. (2000). Bone adaptation and its relationship to physical activity in the past. In M. Cox, & S. Mays (Eds.), *Human osteology in archaeology and forensic science* (pp. 381–402). London, UK: Greenwich Medical Media Ltd.

Kostick, E. L. (1963). Facets and imprints on the upper and lower extremities of femora from a Western Nigerian population. *Journal of Anatomy, 97*(Pt 3), 393–402.

Kovačević, J. (2014). *Avarski kaganat.* Sremska Mitrovica, Serbia: Blago Sirmijuma.

La Guérinière (de), F. R. (2000). *École de cavalerie, contenant la connoissance, l'instruction et la conservation du cheval, avec figures en taille douce.* Paris, France: Belin (Original work published 1733).

Lacourt, G. (2012). *Mise en place d'un programme de prévention de l'accidentologie liée à la pratique équestre à partir d'une étude épidémiologique sur le département de l'Aube [MD thesis].* Reims, France: Université de Reims Champagne-Ardenne.

Lang, N. (1995). *Pathologie du rachis chez le cavalier [MD thesis].* Créteil, France: Université de Créteil.

Langlois, J.-Y., & Gallien, V. (2006). L'église de Notre-Dame-de-Bondeville et sa population (VIIe-IXe siècles, Seine-Maritime). In L. Buchet, C. Dauphin, & I. Séguy (Eds.), *La paléodémographie. Mémoire d'os, mémoire d'hommes. Actes des 8e journées d'anthropologie de Valbonne* (pp. 249–257). Antibes, France: Éditions APDCA.

Larsen, C. S. (1997). *Bioarchaeology: Interpreting behavior from the human skeleton*. Cambridge, UK: Cambridge University Press.

Lebedynsky, I. (2007). *Les nomades : les peuples nomades de la steppe, des origines aux invasions mongoles (IXe siècle av. J.-C. - XIIIe siècle apr. J.-C.)*. Paris, France: Errance.

Lemaire, M. (1985). Le genou du cavalier. In *Médecine et Sports Équestres (Ve Congrès des Pays Francophones)* (pp. 214–220). Saumur, France: Groupe d'Étude "Médecine des Sports Équestres.".

Levine, M. A., Bailey, G., Whitwell, K. E., & Jeffcott, L. B. (2000). Palaeopathology and horse domestication: The case of some Iron age horses horn the Altai Mountains, Siberia. In G. Bailey, R. Charles, & N. Winder (Eds.), *Human ecodynamics. Symposia of the Association for Environmental Archaeology 19* (pp. 123–133). Oxford, UK: Oxbow Books.

Liancheng, L. (2015). Chariot and horse burials in ancient China. *Antiquity, 67*(257), 824–838. https://doi.org/10.1017/S0003598X0006381X.

Librado, P., Fages, A., Gaunitz, C., Leonardi, M., Wagner, S., Khan, N., et al. (2016). The evolutionary origin and genetic makeup of domestic horses. *Genetics, 204*(2), 423–434. https://doi.org/10.1534/genetics.116.194860.

Loder, R. T. (2008). The demographics of equestrian-related injuries in the United States: Injury patterns, orthopedic specific injuries, and avenues for injury prevention. *Journal of Trauma and Acute Care Surgery, 65*(2), 447–460. https://doi.org/10.1097/TA.0b013e31817dac43.

Mackinnon, A. L., Jackson, K., Kuznik, K., Turner, A., Hill, J., Davies, M. A. M., et al. (2019). Increased risk of musculoskeletal disorders and mental health problems in retired professional jockeys: A cross-sectional study. *International Journal of Sports Medicine, 40*(11), 732–738. https://doi.org/10.1055/a-0902-8601.

Marković, N., Stevanović, O., Nešić, V., Marinković, D., Krstić, N., Nedeljković, D., et al. (2014). Palaeopathological study of cattle and horse bone remains of the ancient Roman city of Sirmium (Pannonia/Serbia). *Revue de Médecine Vétérinaire, 165*(3–4), 77–88.

Mayberry, J. C., Pearson, T. E., Wiger, K. J., Diggs, B. S., & Mullins, R. J. (2007). Equestrian injury prevention efforts need more attention to novice riders. *Journal of Trauma - Injury, Infection and Critical Care, 62*(3), 735–739. https://doi.org/10.1097/TA.0B013E318031B5D4.

McGlynn, G., Immler, F., & Zapf, S. (2012). Anthropologische Untersuchung. In J. Bemmann (Ed.), *Steppenkrieger, Reiternomaden des 7.-14. Jahrhunderts aus der Mongolei* (pp. 228–235). Landschaftsverbandes Rheinland/LVR-LandesMuseum Bonn: Darmstadt, Germany.

McGrath, M. S. (2015). *A review of the archaeological and sports medicine literature to determine the biomechanical markers of equestrian activity*. University of New Brunswick. Academia.edu. Retrieved from http://www.academia.edu/12094909.

Merbs, C. F. (1983). Patterns of activity-induced pathology in a Canadian Inuit population. In *National Museum of man mercury series, archaeological survey of Canada no. 119*. Ottawa, Canada: National Museums of Canada.

Mikić Antonić, B. (2012). *Nekropola iz perioda avarske dominacije - lokalitet Pionirska ulica u Bečeju*. Bečej, Serbia: Gradski muzej Bečej.

Mikić Antonić, B. (2016). *Tragom konjanika ratnika—200 godina vladavine Avara*. Bečej, Serbia: Gradski muzej Bečej.

Milella, M., Belcastro, M. G., Zollikofer, C. P. E., & Mariotti, V. (2012). The effect of age, sex, and physical activity on entheseal morphology in a contemporary Italian skeletal collection. *American Journal of Physical Anthropology, 148*(3), 379–388. https://doi.org/10.1002/ajpa.22060.

Miller, E. (1992). The effect of horseback riding on the human skeleton [conference poster presentation]. In *19th Annual meeting of the Paleopathology Association, Las Vegas, NV, 31 March-1 April 1992*.

Miller, E., & Reinhard, K. J. (1991). The effect of horseback riding on the Omaha and Ponca: Paleopathological indications of European contact [conference presentation]. In *Plains anthropology conference, Lawrence, KS, 13–16 November 1991*.

Molleson, T. (2007). A method for the study of activity related skeletal morphologies. *Bioarchaeology of the Near East, 1*, 5–33.

Molleson, T., & Blondiaux, J. (1994). Riders' bones from Kish, Iraq. *Cambridge Archaeological Journal, 4*(2), 312–316. https://doi.org/10.1017/S095977430000113X.

Molleson, T., & Hodgson, D. (1993). A cart driver from Ur. *Archaeozoologia, VI*(11), 93–106.

Mulder, B., Stock, J. T., Saers, J. P. P., Inskip, S. A., Cessford, C., & Robb, J. E. (2020). Intrapopulation variation in lower limb trabecular architecture. *American Journal of Physical Anthropology, 173*(1), 112–129. https://doi.org/10.1002/ajpa.24058.

Nicol, G., Arnold, G. P., Wang, W., & Abboud, R. J. (2014). Dynamic pressure effect on horse and horse rider during riding. *Sports Engineering*, 17(3), 143–150. https://doi.org/10.1007/s12283-014-0149-z.

Niinimäki, S. (2011). What do muscle marker ruggedness scores actually tell us? *International Journal of Osteoarchaeology*, 21(3), 292–299. https://doi.org/10.1002/oa.1134.

Niinimäki, S., & Baiges Sotos, L. (2013). The relationship between intensity of physical activity and entheseal changes on the lower limb. *International Journal of Osteoarchaeology*, 23(2), 221–228. https://doi.org/10.1002/oa.2295.

Novotny, F., Spannagl-Steiner, M., & Teschler-Nicola, M. (2014). Anthropologische Analyse der menschlichen Skelette aus Gobelsburg, Niederösterreich. *Archaeologia Austriaca*, 97–98, 141–153.

Ortner, D. J. (1968). Description and classification of degenerative bone changes in the distal joint surfaces of the humerus. *American Journal of Physical Anthropology*, 28(2), 139–155. https://doi.org/10.1002/ajpa.1330280212.

Ortner, D. J. (Ed.). (2003). *Identification of pathological conditions in human skeletal remains* (2nd ed.). San Diego, CA: Academic Press.

Outram, A. K., Stear, N. A., Bendrey, R., Olsen, S., Kasparov, A., Zaibert, V., et al. (2009). The earliest horse harnessing and milking. *Science*, 323(5919), 1332–1335. https://doi.org/10.1126/science.1168594.

Outram, A. K., Bendrey, R., Evershed, R. P., Orlando, L., & Zaibert, V. F. (2021). Rebuttal of Taylor and Barrón-Ortiz 2021 rethinking the evidence for early horse domestication at Botai. *Zenodo*. https://doi.org/10.5281/zenodo.5142604.

Paine, R. R., Mancinelli, D., Ruggieri, M., & Coppa, A. (2007). Cranial trauma in iron age Samnite agriculturists, Alfedena, Italy: Implications for biocultural and economic stress. *American Journal of Physical Anthropology*, 132(1), 48–58. https://doi.org/10.1002/ajpa.20461.

Pálfi, G. (1992). Traces des activités sur les squelettes des anciens Hongrois. *Bulletins et Mémoires de La Société d'Anthropologie de Paris*, 4(3–4), 209–231. https://doi.org/10.3406/bmsap.1992.2318.

Pálfi, G. (1997). Maladies dans l'Antiquité et au Moyen-Âge. Paléopathologie comparée des anciens Gallo-Romains et Hongrois. *Bulletins et Mémoires de La Société d'Anthropologie de Paris*, 9(1–2), 1–205. https://doi.org/10.3406/bmsap.1997.2472.

Pálfi, G., & Dutour, O. (1996). Activity-induced skeletal markers in historical anthropological material. *International Journal of Anthropology*, 11(1), 41–55. https://doi.org/10.1007/BF02442202.

Panzarino, G. A., & Sublimi Saponetti, S. (2017). Chi era e come viveva: profilo bio-antropologico dell'uomo sepolto nella tomba 6. In G. Perrino, & S. S. Saponetti (Eds.), *Una finestra sulla storia. Un cavaliere a Castiglione tra angioini e aragonesi* (pp. 75–85). Conversano, Italy: Società di Storia Patria per la Puglia, "Sezione Sudest Barese.".

Pap, I. (1985). A Dabas (Gyón)-paphegyi XI. századi embertani széria (the anthropological series of Dabas (Gyón)-paphegy from the 11th century). In N. Ikvai (Ed.), *Régészeti tanulmányok Pest megyéből. Studia Comitatensia 17* (pp. 387–407). Szentendre, Hungary: Museums of Pest County.

Pluskowski, A., Seetah, K., & Maltby, M. (2010). Potential osteoarchaeological evidence for riding and the military use of horses at Malbork Castle, Poland. *International Journal of Osteoarchaeology*, 20(3), 335–343. https://doi.org/10.1002/oa.1048.

Pohl, W. (2018). Narratives of origin and migration in early medieval Europe: Problems of interpretation. *The Medieval History Journal*, 21(2), 192–221. https://doi.org/10.1177/0971945818775460.

Polet, C., & Villotte, S. (2021). Osteobiography of a 19th century Belgian infantry captain and its contribution to the knowledge of skeletal markers of horse riding [conference presentation]. In *27th annual meeting of the European Association of Archaeologists (EAA), 6–11 September 2021*.

Pugh, T. J., & Bolin, D. (2004). Overuse injuries in equestrian athletes. *Current Sports Medicine Reports*, 3(6), 297–303. https://doi.org/10.1007/s11932-996-0003-6.

Pulcini, M. L. (2014). *La necropoli di Olmo di Nogara (Verona). Studio paleobiologico dei resti umani per la ricostruzione dell'organizzazione di una comunità dell'Età del bronzo padana* [doctoral dissertation]. Padova, Italy: Università di Padova.

Putz, R., & Pabst, R. (Eds.). (2001) (Vols 1 & 2). *Sobotta atlas of human anatomy* (13th ed.). Philadelphia, PA: Lippincott Williams & Wilkins.

Rabey, K. N., Green, D. J., Taylor, A. B., Begun, D. R., Richmond, B. G., & McFarlin, S. C. (2015). Locomotor activity influences muscle architecture and bone growth but not muscle attachment site morphology. *Journal of Human Evolution*, 78, 91–102. https://doi.org/10.1016/j.jhevol.2014.10.010.

Radi, N., Mariotti, V., Riga, A., Zampetti, S., Villa, C., & Belcastro, M. G. (2013). Variation of the anterior aspect of the femoral head-neck junction in a modern human identified skeletal collection. *American Journal of Physical Anthropology*, 152(2), 261–272. https://doi.org/10.1002/ajpa.22354.

Reinhard, K. J., Tiezen, L., Sandness, K. L., Beiningen, L. M., & Miller, E. (1994). Trade, contact, and female health in Northeast Nebraska. In C. S. Larsen, & G. R. Milner (Eds.), *In the wake of contact: Biological responses to conquest* (pp. 63–74). New York, NY: Wiley-Liss.

Resnick, D. (2002). *Diagnosis of bone and joint disorders* (4th ed.). Philadelphia, PA: Saunders.

Révész, L. (1999). *Emlékezzetek utatok kezdetére.. Régészeti kalandozások a magyar honfoglalás és államalapítás korában.* Budapest, Hungary: Timp.

Rogers, J., & Waldron, T. (1995). *A field guide to joint disease in archaeology.* Chichester, UK: John Wiley & Sons.

Rolle, R. (1989). *The world of the Scythians.* Berkeley – Los Angeles, CA: University of California Press.

Róna-Tas, A. (1999). *Hungarians and Europe in the early middle ages: An introduction to early Hungarian history.* Budapest, Hungary: Central European University Press.

Roux, W. (1881). *Der Kampf der Teile im Organismus.* Leipzig, Germany: Wilhelm Engelmann.

Rueda, M. A., Halley, W. L., & Gilchrist, M. D. (2010). Fall and injury incidence rates of jockeys while racing in Ireland, France and Britain. *Injury, 41*(5), 533–539. https://doi.org/10.1016/j.injury.2009.05.009.

Ruff, C. B., & Larsen, C. S. (2014). Long bone structural analyses and the reconstruction of past mobility: A historical review. In K. J. Carlson, & D. Marchi (Eds.), *Reconstructing mobility: Environmental, behavioral, and morphological determinants* (pp. 13–29). New York, NY: Springer.

Ruff, C. B., Holt, B. M., & Trinkaus, E. (2006). Who's afraid of the big bad Wolff?: "Wolff's law" and bone functional adaptation. *American Journal of Physical Anthropology, 129*(4), 484–498. https://doi.org/10.1002/ajpa.20371.

Ryan, K. D., Brodine, J., Pothast, J., & McGoldrick, A. (2020). Medicine in the sport of horse racing. *Current Sports Medicine Reports, 19*(9), 373–379. https://doi.org/10.1249/JSR.0000000000000750.

Ryan, J., Desideri, J., & Besse, M. (2018). Bell Beaker archers: Warriors or an ideology? *Journal of Neolithic Archaeology, 20*(S4), 97–122. https://doi.org/10.12766/jna.2018S.6.

Sandness, K. L., & Reinhard, K. J. (1992). Vertebral pathology in prehistoric and historic skeletons from northeastern Nebraska. *Plains Anthropologist, 37*(141), 299–309. https://doi.org/10.2307/25669124.

Sarry, F., Courtaud, P., & Cabezuelo, U. (2016). La sépulture multiple laténienne du site de Gondole (Le Cendre, Puy-de-Dôme). *Bulletins et Mémoires de la Société d'anthropologie de Paris, 28*(1–2), 72–83. https://doi.org/10.1007/s13219-016-0151-z.

Schlecht, S. H. (2012). Understanding entheses: Bridging the gap between clinical and anthropological perspectives. *The Anatomical Record, 295*(8), 1239–1251. https://doi.org/10.1002/ar.22516.

Schrader, S. (2019). Activity, diet and social practice. In *Addressing everyday life in human skeletal remains.* Cham, Switzerland: Springer International Publishing.

Skedros, J. G., & Baucom, S. L. (2007). Mathematical analysis of trabecular 'trajectories' in apparent trajectorial structures: The unfortunate historical emphasis on the human proximal femur. *Journal of Theoretical Biology, 244*(1), 15–45. https://doi.org/10.1016/j.jtbi.2006.06.029.

Snow, C. C., & Fitzpatrick, J. (1989). Human osteological remains from the Battle of the little Bighorn. In D. D. Scott, R. A. Fox, M. A. Connor, & D. Harmon (Eds.), *Archaeological perspectives on the Battle of the little Bighorn* (pp. 243–282). Norman, OK: University of Oklahoma Press.

Standring, S. (Ed.). (2005). *Gray's anatomy: The anatomical basis of clinical practice* (39th ed.). Edingurgh, UK – New-York, NY: Elsevier Churchill Livingstone.

Stone, R. J., & Stone, J. A. (2003). *Atlas of skeletal muscles* (4th ed.). Boston, MA: McGraw-Hill.

Taylor, W. T. T., & Barrón-Ortiz, C. I. (2021). Rethinking the evidence for early horse domestication at Botai. *Scientific Reports, 11*, 7440. https://doi.org/10.1038/s41598-021-86832-9.

Taylor, W., & Tuvshinjargal, T. (2018). Horseback riding, asymmetry, and changes to the equine skull: Evidence for mounted riding in Mongolia's late bronze age. In L. Bartosiewicz, & E. Gál (Eds.), *Care or neglect? Evidence of animal disease in archaeology. Proceedings of the 6th meeting of the animal Palaeopathology working Group of the International Council for Archaeozoology (ICAZ), Budapest, Hungary, 2016* (pp. 134–154). Oxford, UK: Oxbow Books.

Taylor, W., Bayarsaikhan, J., & Tuvshinjargal, T. (2015). Equine cranial morphology and the identification of riding and chariotry in late Bronze Age Mongolia. *Antiquity, 89*(346), 854–871. https://doi.org/10.15184/aqy.2015.76.

Thomas, A. (2014). Bioarchaeology of the middle Neolithic: Evidence for archery among early European farmers. *American Journal of Physical Anthropology, 154*(2), 279–290. https://doi.org/10.1002/ajpa.22504.

Tichnell, T. A. (2012). *Invisible horsewomen: Horse riding and social dynamics on the steppe [doctoral dissertation].* East Lansing, MI: Michigan State University.

Tihanyi, B., Bereczki, Z., Molnár, E., Berthon, W., Révész, L., Dutour, O., et al. (2015). Investigation of Hungarian conquest period (10th c. AD) archery on the basis of activity-induced stress markers on the skeleton—Preliminary results. *Acta Biologica Szegediensis, 59*(1), 65–77.

Tihanyi, B., Berthon, W., Kis, L., Váradi, O. A., Dutour, O., Révész, L., et al. (2020). "Brothers in arms": Activity-related skeletal changes observed on the humerus of individuals buried with and without weapons from the 10th-century CE Carpathian Basin. *International Journal of Osteoarchaeology, 30*(6), 798–810. https://doi.org/10.1002/oa.2910.

Tseveendorj, D. (1996). Horse and Mongols. In G. E. Afanas'ev (Ed.), *The prehistory of Asia and Oceania. Colloquia of the XIII international congress of prehistoric and protohistoric sciences* (pp. 89–93). Forli, Italy: A.B.A.C.O. Edizioni.

Tsirikos, A., Papagelopoulos, P. J., Giannakopoulos, P. N., Boscainos, P. J., Zoubos, A. B., Kasseta, M., et al. (2001). Degenerative spondyloarthropathy of the cervical and lumbar spine in jockeys. *Orthopedics, 24*(6), 561–564.

Tung, T. A. (2007). Trauma and violence in the Wari empire of the Peruvian Andes: Warfare, raids, and ritual fights. *American Journal of Physical Anthropology, 133*(3), 941–956. https://doi.org/10.1002/ajpa.20565.

Turcotte, C. M., Rabey, K. N., Green, D. J., & McFarlin, S. C. (2022). Muscle attachment sites and behavioral reconstruction: An experimental test of muscle-bone structural response to habitual activity. *American Journal of Biological Anthropology, 177*(1), 63–82. https://doi.org/10.1002/ajpa.24410.

Üstündağ, H., & Deveci, A. (2011). A possible case of Scheuermann's disease from Akarçay Höyük, Birecik (Şanlıurfa, Turkey). *International Journal of Osteoarchaeology, 21*(2), 187–196. https://doi.org/10.1002/oa.1120.

Verna, E., Piercecchi-Marti, M. D., Chaumoitre, K., Panuel, M., & Adalian, P. (2014). Mise au point sur les caractères discrets du membre inférieur: définition, épidémiologie, étiologies. *Bulletins et Mémoires de la Société d anthropologie de Paris, 26*(1), 52–66. https://doi.org/10.1007/s13219-013-0090-x.

Villotte, S. (2006). Connaissances médicales actuelles, cotation des enthésopathies : nouvelle méthode. *Bulletins et Mémoires de La Société d'Anthropologie de Paris, 18*(1–2), 65–85.

Villotte, S. (2012). *Practical protocol for scoring the appearance of some fibrocartilaginous entheses on the human skeleton.* Retrieved from https://www.academia.edu/1427191/Practical_protocol_for_scoring_the_appearance_of_some_fibrocartilaginous_entheses_on_the_human_skeleton.

Villotte, S., & Knüsel, C. J. (2009). Some remarks about femoroacetabular impingement and osseous non-metric variations of the proximal femur. *Bulletins et Mémoires de La Société d'Anthropologie de Paris, 21*(1–2), 95–98.

Waldron, T. (2009). *Palaeopathology.* Cambridge, UK – New-York, NY: Cambridge University Press.

Wallace, I. J., Winchester, J. M., Su, A., Boyer, D. M., & Konow, N. (2017). Physical activity alters limb bone structure but not entheseal morphology. *Journal of Human Evolution, 107*, 14–18. https://doi.org/10.1016/j.jhevol.2017.02.001.

Weiss, E., & Jurmain, R. D. (2007). Osteoarthritis revisited: A contemporary review of aetiology. *International Journal of Osteoarchaeology, 17*(5), 437–450. https://doi.org/10.1002/oa.889.

Weiss, E., Corona, L., & Schultz, B. (2012). Sex differences in musculoskeletal stress markers: Problems with activity pattern reconstructions. *International Journal of Osteoarchaeology, 22*(1), 70–80. https://doi.org/10.1002/oa.1183.

Wentz, R. K., & de Grummond, N. T. (2009). Life on horseback: Palaeopathology of two Scythian skeletons from Alexandropol, Ukraine. *International Journal of Osteoarchaeology, 19*(1), 107–115. https://doi.org/10.1002/oa.964.

Wescott, D. J. (2008). Biomechanical analysis of humeral and femoral structural variation in the Great Plains. *Plains Anthropologist, 53*(207), 333–355. http://www.jstor.org/stable/25671004.

Wilczak, C. A. (1998). Consideration of sexual dimorphism, age, and asymmetry in quantitative measurements of muscle insertion sites. *International Journal of Osteoarchaeology, 8*(5), 311–325. https://doi.org/10.1002/(SICI)1099-1212(1998090)8:5<311::AID-OA443>3.0.CO;2-E.

Willey, P. S. (1997). *Osteological analysis of human skeletons excavated from the Custer National Cemetery.* Lincoln, NE: U.S. Department of the Interior, Midwest Archeological Center, National Park Service.

Willson, A. (2002). *Applied posture riding: Training the abdominal and posture muscles for riding.* Willson Annette.

Willson, A. (2013). *The riding muscles and their function in riding.* Retrieved from http://fliphtml5.com/arwv/huaq/basic.

Wolff, J. (1892). *Das Gesetz der Transformation der Knochen.* Berlin, Germany: A. Hirschwald.

Xenophon. (1897). On Horsemanship (Trans.). In H. G. Dakyns (Ed.), *The works of Xenophon, vol. III, part II* (pp. 37–69). London, UK: Macmillan and Co. (Original work published ca. 362–350 BCE).

Zeier, G. (2006). *Epidémiologie des traumatismes sportifs de l'enfant et de l'adolescent [MD thesis].* Lausanne, Switzerland: Université de Lausanne.

Zeng, X., Trask, C., & Kociolek, A. M. (2017). Whole-body vibration exposure of occupational horseback riding in agriculture: A ranching example. *American Journal of Industrial Medicine, 60*(2), 215–220. https://doi.org/10.1002/ajim.22683.

Zumwalt, A. (2006). The effect of endurance exercise on the morphology of muscle attachment sites. *Journal of Experimental Biology, 209*(3), 444–454. https://doi.org/10.1242/jeb.02028.

10

Locomotion and the foot and ankle

Kimberleigh A. Tommy[a] and Meir M. Barak[b]

[a]Human Variation and Identification Research Unit, School of Anatomical Sciences, University of the Witwatersrand, Johannesburg, South Africa, [b]Department of Veterinary Biomedical Sciences, College of Veterinary Medicine, Long Island University, Brookville, NY, United States

10.1 Introduction

The ankle (tibiotalar) joint constitutes the convergence between the lower leg (tibia and fibula) and the foot. The foot and ankle form a complex system consisting of 28 bones and 33 joints acting together to fulfil several purposes, including supporting the body weight, providing balance, absorbing shock, and transferring ground reaction forces. These skeletal elements of the foot are the first to make contact with the ground as we navigate the world on two legs. The morphology and internal bony structure of the foot and ankle support the idea of a relationship between form and function and, as such, the bones of the foot and ankle are of particular interest and importance when it comes to understanding locomotion in living humans and the evolution of bipedalism throughout the hominin lineage.

Bone is a vital and dynamic tissue that constantly changes and adjusts in response to internal and external stimuli via the processes of bone modelling and remodelling (Barak, 2020). The response of bone to external mechanical stimuli (i.e. loads) is known as 'bone functional adaptation' (Table 10.1), and it postulates that bone will respond to repeated dynamic mechanical stimuli or the lack of them (i.e. disuse) by adjusting its shape (i.e. cortical bone modelling) and its tissue structure (i.e. trabecular bone modelling and cortical bone remodelling). In line with this prediction, bone shape and structure should hold a functional signal of how the bone was loaded. This concept, if supported by experimental evidence, could help us move beyond the challenges that come with analyses of fossil hominin morphology, which is often fragmentary, and the long-standing palaeoanthropological debates founded on differing functional interpretations of external morphology, as they relate to the evolution of bipedalism. This chapter, which is divided into two main sections, will communicate a comprehensive overview on the relationship between foot and ankle bone structure, and their function in locomotion. The first section discusses experimental support for bone functional adaptation in the lower leg, ankle, and foot. Based on this support, the second section focuses on bone functional adaptation signals derived from extinct hominins fossil remains

TABLE 10.1   List of terms and abbreviations in this chapter and their meaning.

Term/abbreviation	Description
Ankle	The joint between the tibia/fibula and the talus, where the foot and the leg meet
Bone functional adaptation	The concept of bone function adaptation postulates that bone tissue responds to mechanical stimuli by adjusting (modelling) its architecture accordingly. This response is present in both cortical and trabecular bone tissue, although trabecular bone demonstrates a higher plasticity due to a larger surface area to volume ratio. Thus it predicts that the newly adjusted trabecular structure is mechanically in-line with the habitual in vivo loading direction. This concept, if found to be accurate, will help us to predict the locomotion behaviour of extinct hominins
Bone volume fraction (BV/TV)	The proportion (%) of volume occupied by bone out of the total volume measured
Foot	The distal and terminal part of the leg, below the ankle joint, which bears weight and allows locomotion

and Holocene humans. Collectively, these data extend our understanding of the origin and evolution of human bipedalism.

10.2 Experimental studies that support bone functional adaptation in the ankle and foot

The following section discusses experimental work that supports the principle of bone functional adaptation (Table 10.1) and that demonstrates a relationship between physiological loading and adjustments of bone structure. In order to concentrate on work closely related to humans, we will present only studies done on animals from the class Mammalia (i.e. excluding studies on birds and reptiles).

To start with, we need to address several aspects that could potentially influence how, and to what degree, bone functional adaptation will take place. These aspects include (1) the type of load—dynamic vs. static (Berman et al., 2015; Fritton et al., 2005; Gross et al., 2002; Turner et al., 1991); (2) age—young vs. adult (Brodt and Silva, 2010; Chen et al., 1994; Holguin et al., 2014; Lieberman et al., 2001, 2003; Lynch et al., 2011; Main et al., 2014); (3) the individual underlining genetic 'blueprint'—i.e. nature vs. nurture (Akhter et al., 1998; Holguin et al., 2013; Judex et al., 2002; Kodama et al., 1999, 2000; Wallace et al., 2007); and (4) the distinction between the effect of bone functional adaptation on cortical vs. trabecular bone (Grimston et al., 2012; Kelly et al., 2016; Parfitt, 2002; Yang et al., 2019).

10.2.1 The lower leg—Tibia

This section focuses on the leg, the part between the knee and the foot, also known in humans as the crus, which consists of two bones: the tibia and fibula. Yet, since the fibula is non-weight bearing in humans and practically in all quadrupedal mammals (except when it is fused to the tibia), the literature and our discussion focus on the tibia.

10.2.1.1 Tibial response to loading—Cortical modelling

As cortical bone is the main load-bearing bony component of the tibia, it is not surprising that most research focused on cortical bone modelling in response to load when testing tibial bone functional adaptation (Table 10.1). These experiments, which mostly used mice and rats as the animal model, applied in vivo whole-bone loading of the tibia using various contraptions, either axially between the knee and the ankle (Berman et al., 2015; De Souza et al., 2005; Weatherholt et al., 2013), or in four-point and cantilever bending with the medial surface in compression and the lateral surface in tension (Akhter et al., 1998; Gross et al., 2002; Weatherholt et al., 2013). Worth noting is a collection of related studies, which investigated tibial cortical bone functional adaptation in mice of various ages (Grimston et al., 2012; Holguin et al., 2014; Lynch et al., 2011; Main et al., 2014; Yang et al., 2019). These studies found that tibial cortical bone increased in volume, cross-sectional area, and cortical thickness in response to cyclic loading of peak forces ranging between 7 and 14 N. While cortical bone formation was most evident in young growing mice, these studies found that adult mice (12 months of age) and even ageing mice (22 months of age) demonstrated a noticeable, yet reduced, tibial cortical bone formation (Brodt and Silva, 2010; Holguin et al., 2014; Silva et al., 2012). Other experiments applied treadmill exercise to rats (Chen et al., 1994) and sheep (Lieberman et al., 2001, 2003, 2004; Wallace et al., 2014), or investigated the effect of physical activity in young boys aged 9–11 years (Macdonald et al., 2009). The overall finding of all these studies is quite uniform—increased loading of the tibia triggered a cortical bone modelling response that increased cortical bone mass, cortical bone area, and cortical bone thickness. Consequently, exercised subjects tend to demonstrate an adopted tibial structure that is better in resisting bending and torsion, predicted from the ratio between maximal and minimal second moments of area (I_{max}/I_{min}) and the polar moment of area (J), respectively (Shaw and Stock, 2013).

While these types of experiments and their conclusions may seem to be straightforward, it is difficult to pinpoint exactly what component of the loading signal is the main trigger to the modelling outcome. Although many of the studies demonstrate that tibial cortical modelling is triggered in response to load in a dose-dependent manner (i.e. load magnitude) (Berman et al., 2015; De Souza et al., 2005), other studies point to the duration of loading (i.e. length of the experiment) (Fritton et al., 2005; Iwamoto et al., 1999), or even the number of load cycles per session (Turner et al., 1991; Yang et al., 2017) as the leading factor to cortical bone modelling and tibial bone functional adaptation (Table 10.1). Yet the complexity of tibial cortical bone functional adaptation does not stop here; several studies have pointed out that while tibial cortical bone functional adaptation indeed takes place, it does not always correspond to the expected bone surfaces (Demes et al., 2001; Lieberman et al., 2004; Wallace et al., 2014). Lieberman et al. (2004) found that the tibia's maximal second moments of area (I_{max}) in sheep that were running on a treadmill did not correspond with the plane in which the bone was bending, as should be by the prediction of bone functional adaptation. Similarly, Wallace et al. (2014) demonstrated that while treadmill running induced peak strains in the anterior (tensile) and posterior (compressive) cortices of sheep tibia, the most significant amount of bone deposition occurred at the medial cortex. In other words, while tibial cortical bone deposition occurs in response to increased loading, this does not always correlate with the actual bone surfaces that endure the highest stresses and strains. Consequently, studying only tibial cortical bone shape, cross-sectional area, and the resistance to bending and torsion (I_{max}/I_{min} and J, respectively) in fossilised bones may not be able to tell us accurately their locomotor behaviour and level and type of physical activity. This viewpoint was supported in a study where tibial rigidity measurements could not differentiate between adult long-distance runners and adult non-runners even though runners started their practice during adolescence and have been competing for about 10 years (Shaw and Stock, 2013).

10.2.1.2 Tibial response to loading—Trabecular modelling

Like cortical bone, studies of trabecular bone in the proximal and distal tibia have repeatedly demonstrated responsiveness to mechanical stimuli marked by a gain in trabecular bone mass (Iwamoto et al., 1999; Swift et al., 2010). De Souza et al. (2005) and others (e.g. Berman et al., 2015; Holguin et al., 2013; Iwamoto et al., 1999; Lynch et al., 2011; Main et al., 2014; Yang et al., 2017) have applied non-invasive in vivo axial loads (7–13 N), for periods ranging from 1 week and up to 6 weeks to mice tibiae of various ages (8–26 weeks of age). While some variance existed in their experimental outcomes (lower loads and shorter periods tended to induce lower modelling responses and older mice usually demonstrated reduced bone modelling), their results consistently revealed that trabecular bone volume fraction (BV/TV) (Table 10.1) in the proximal tibia significantly increased primarily due to an increase in trabecular thickness (Tb.Th), rather than an increase in trabecular number (Tb.N). This finding is a strong indication that trabecular bone response to an increase in loading by deposition of new bone on existing trabeculae, rather than the creation of new trabeculae.

When a constant mode of loading persists (e.g. habitual locomotion behaviour), it can be said that the bone is loaded, on average, in a predicted direction. Trabecular bone surfaces that tend to experience under these loads, on average, low stresses and strains will be removed (resorbed), and surfaces that tend to experience, on average, high stresses and strains will promote bone deposition. In the case of habitual locomotion, this phenomenon of tibial bone functional adaptation (Table 10.1) will start as soon as the young animal or the child will begin locomoting (Raichlen et al., 2015; Tanck et al., 2001) and thus this further eliminates the issue of decreased bone adaptation in adults.

Tanck et al. (2001) studied the architecture and mechanical parameters of trabecular bone from the proximal tibia of pigs from young to adult. The authors noticed that trabecular bone functional adaptation (Table 10.1) to habitual locomotion had two distinct phases. First, at a young age (6 weeks), trabecular BV/TV (Table 10.1) increases in response to physiological loading (i.e. locomotion). Next, at 23 weeks, trabecular bone starts to adjust its structure to better accommodate to the direction of loading (for similar results in mice, see De Souza et al., 2005). This is indicated by an increase in trabecular degree of anisotropy. The time lag between the increase in trabecular BV/TV and degree of anisotropy suggests that the initial response to increased loads is to add bone mass and later refine that bone into a more mechanically efficient, anisotropic structure. Furthermore, the angle between the stiffest direction of the trabecular tissue (calculated from finite element analysis) and the loading direction during locomotion decreased with age (i.e. became aligned as trabecular architecture adopts to loading), demonstrating that trabecular bone functional adaptation took place in response to habitual locomotion.

Raichlen et al. (2015) observed a similar behaviour in trabecular bone from the distal tibia of children aged 1–8 years old and were able to link structural changes to locomotor kinematic changes in the ankle joint during ontogeny. While young children demonstrated a lower degree of anisotropy and a large variation in values between individuals, manifesting the variation in loading due to kinematic instability as they refine their locomotion skills, older children (6–8 years of age) demonstrate a higher degree of anisotropy and decreased variance in values between individuals. Together, these two studies demonstrate how habitual locomotion affects trabecular bone structure and how numerical data derived from trabecular architecture (namely degree of anisotropy and principal trabecular orientation) hold a functional signal.

In a sequence of two studies, Barak et al. (2011, 2013b) advanced the understanding of trabecular bone functional adaptation (Table 10.1) by demonstrating a close correlation between principal trabecular orientation and locomotor behaviour. In the first study, Barak et al. (2011) used young sheep to test whether changing their hock joint loading orientation would cause a corresponding adjustment in trabecular orientation in the distal tibia. One group of exercising sheep trotted on a level treadmill and the other group trotted on an inclined treadmill (7°), while a third group of sedentary sheep served as a control. In order to maintain a constant change in loading orientation, when the inclined group was not trotting on the treadmill, they were fitted with platform shoes on their front legs that reproduced the same hock joint angle measured during incline trotting. The results demonstrated that both exercising groups had a significant increase in their distal tibia trabecular BV/TV (Table 10.1). More importantly, the results revealed a close correlation between the change in hock joint angle and the principal trabecular orientation in the distal tibia. These findings demonstrated for the first time that trabecular bone dynamically realigns itself in relation to changes in peak loading direction during locomotion.

In a following study, Barak et al. (2013b) tested the relationship between ankle joint angle at the point of peak ground reaction force during the gait cycle, and principal trabecular orientation in the distal tibia of chimpanzees (locomoting quadrupedally) and humans (walking bipedally). Consistent with the previous findings in sheep, the difference in ankle joint angle corresponded closely to the difference in principal trabecular orientation, which was found to differ significantly between the two species (Fig. 10.1). These findings were then used to deduce from the principal trabecular orientation of fossilised distal tibiae assigned to *Australopithecus africanus* that this species was locomoting similar to modern humans—with a relatively extended posture (extended hips and knees, see Section 10.3.1.4).

10.2.1.3 Tibial response to loading—Summary

Literature supports the principle of tibial bone functional adaptation (Table 10.1) provided that the loading is dynamic, above a certain threshold (magnitude or duration) and the response is local (i.e. the directly loaded bone(s)). In addition, tibial bone functional adaptation is more pronounced in growing animals and children but is still evident in adults. Finally, trabecular bone architecture and principal trabecular orientation appear to hold a more accurate signal to loading direction than cortical bone. The logic behind this may be that tibial cortical bone is the key mechanical component of the bone and as such it must be robust and resist loading in various orientations. In contrast, tibial trabecular bone serves to distribute and dissipate stresses that were transmitted from the knee and ankle joints, and as such, it is more sensitive to the habitual loading direction during locomotion.

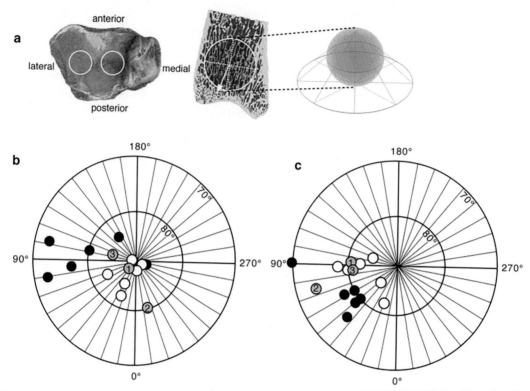

FIG. 10.1 Measurements of 3D-PTO in the distal tibia. (A) Schematic showing location of the lateral and medial VOIs (volume of interest) in the distal tibia and how the 3D spheres were visualised in 2D using an equal-angle stereoplot. A stereoplot is a 2D map which is created by projecting points from a surface of a sphere to a tangential plane. (B) The stereoplot projections of the lateral VOI. (C) The stereoplot projections of the medial VOI. *Filled circles*, chimpanzees; *open circles*, humans; *grey circle 1*, StW 358; *grey circle 2*, StW 389; *grey circle 3*, StW 567. Angles 0°, 90°, 180°, and 270° correspond to the anatomical directions: posterior, lateral, anterior, and medial, respectively. *From Barak, M. M., Lieberman, D. E., Raichlen, D., Pontzer, H., Warrener, A. G., & Hublin, J.-J. (2013). Trabecular evidence for a human-like gait in Australopithecus africanus. PLOS ONE, 8(11), e77687. https://doi.org/10.1371/journal.pone.0077687.*

10.2.2 The foot

The anatomy of the human foot (Table 10.1) is unique among primates (and mammals in general) in features that have been interpreted as relating to our unique form of locomotion, bipedalism. Some of these features include a large calcaneus (heel bone); short phalanges; and an adducted, non-opposable hallux. Humans exhibit plantigrade locomotion whereby the toes and metatarsals are flat on the ground while walking, using the foot as a stiff lever, propelling the body forward with each step. The foot is the initial point of contact between the body and the ground, dissipating forces from the ground into the leg and pelvis. Animal experimental studies on the foot (or pes) have been conducted primarily on the calcaneus (or astragalus) in mammals. This section explores animal experimental models as well as additional support for bone modelling in the foot and has been subdivided into the hindfoot (heel), midfoot, and forefoot.

10.2.2.1 The hindfoot

The hindfoot is the posterior-most region of the foot, consisting of the talus and calcaneus bones. The talus together with the tibia and fibula forms the talocrural or ankle joint (Table 10.1), and the talus together with the calcaneus forms the subtalar joint.

Animal experimental models testing the response of cortical and trabecular bone of the calcaneus to different loading conditions have been conducted primarily on artiodactyls, specifically deer and sheep (Lanyon, 1974; Sinclair et al., 2013; Skedros and Baucom, 2007; Skedros et al., 2004, 2012, 2019; Skerry and Lanyon, 1995). Artiodactyls are digitigrade (i.e. walk on their digits) and as such, the calcaneus does not contact the ground during locomotion. The calcaneus is interpreted as having a relatively simple mechanical role during locomotion, classified as a 'low-complexity load category' (Skedros et al., 2012). For these reasons, the wild and domesticated artiodactyl calcaneus models are ideal for studying trabecular and cortical bone adaptations for habitual bending. There are also data that support the idea that artiodactyl calcanei are useful in comparative studies because they have trabecular and cortical modelling activities and dynamics that are more similar to humans (Skerry and Lanyon, 1995; Turner et al., 2000) when compared to other test animals such as mice and rats (Sinclair et al., 2013; Skedros and Baucom, 2007; Skedros et al., 2004, 2012, 2019).

In initial in vivo strain gauge studies on sheep, Lanyon (1974) tested the relationship between strain and bone modelling by placing rosette strain gauges on the lateral surface at midshaft of the calcaneus. The sheep were made to walk and trot on a treadmill for several consecutive days before the limbs were radiographed. Lanyon (1974) observed that the underlying trabeculae in the calcaneus were arranged in two distinct tracts (i.e. dorsal and plantar, respectively) that intersected to form an arch and concluded that the tracts and their intersection coincided with the direction of the principal compressive (i.e. trabeculae in the dorsal tract) and tensile strains (i.e. trabeculae in the plantar tract), providing support to the idea of a relationship between principal strain direction and trabecular modelling/organisation. A subsequent study by Skerry and Lanyon (1995) examined the response of trabecular bone to different loading conditions at the hock joint (i.e. the joint between the tarsals and the tibia in quadrupedal digitigrade/unguligrade animals) using external fixators placed on the left calcaneus in order to alter joint function and loading. The hock joint of the control group was not impeded by a fixator, enabling normal joint movement and function; a second group was fitted with fixators from the tibia to the metatarsus which inhibited joint loading; and a third group had fixators attached to the hock joints that were removed for 20 min per day. The sheep were then made to walk on a treadmill over 12 weeks and bone mineral content was measured across all groups. In the control group that was unimpeded, there was no significant reduction in bone mineral content; however, within the two test groups where normal loading was inhibited, even for a short time, the bone mineral content reduced by up to 22%. This change in bone mineral content was interpreted by the authors as indicating remodelling changes that coincided with changes in loading. Extensive studies of cortical bone in sheep and deer calcanei have also shown that differences in dorsal versus plantar cortical, thickness, mineral content, and many histomorphological characteristics are correlated with differences in strain mode (e.g. tension, compression, and shear) and/or magnitude (Sinclair et al., 2013; Skedros and Baucom, 2007; Skedros et al., 2004, 2012, 2019).

Yet another study further investigated the relationship between trabecular structure and principal strains and their orientations in the calcaneus of potoroos, an Australian marsupial

(Biewener et al., 1996). The authors analysed the response of trabecular structure in the calcaneus to a loss of mechanical function, and therefore stimuli, through an Achilles tenotomy. Twelve potoroos were dived into two groups: a control group where the animals were trained over a six-day period to run on a motorised treadmill with a rosette strain gauge surgically attached to the lateral aspect of the left calcaneus to record functional strains imposed on the bone through in vivo locomotor activity and the second group, where animals had 8–10 mm of the right Achilles tendon surgically removed to isolate the calcaneus from functional load bearing. Trabecular structure in the second group where the Achilles tendon had been compromised demonstrated significantly reduced bone volume fraction (BV/TV, see Table 10.1), trabecular thickness, and number in comparison to the control group, although no significant differences were detected in degree of anisotropy between the two groups. This study's (Biewener et al., 1996) result supported previous observations on sheep calcanei that the underlying trabeculae are oriented with the principal strain direction in the cortical surface.

The talus transmits forces encountered from the foot to the leg, making it ideal for studying the relationship between biomechanics and morphology. Although animal experimental models are ideal for studying bone response to loading in a controlled fashion, other methods of strain modelling such as finite element analysis have been used as an effective tool in successfully quantifying and validating mechanical properties and functional significance of trabecular bone (Kabel et al., 1999; Ulrich et al., 1999). Finite element micro-modelling of the talus in healthy adult humans demonstrated that the trabecular network contributed towards a stiffer but lightweight calcaneus and that the trabecular network is key in dissipating loads applied to the cortical surface (Parr et al., 2013). Recent studies involving trabecular mapping of the talus in non-human great apes using sliding semilandmarks across the articular surface and multiple volumes of interest (VOIs) immediately deep to the cortical shell demonstrated that trabecular parameters (i.e. degree of anisotropy and trabecular separation) are not homogeneously distributed within the tali of different species, possibly in response to variation in locomotor-related loading (Sylvester and Terhune, 2017).

10.2.2.2 The midfoot

The midfoot is composed of the cuboid, cuneiform bones, and the navicular (i.e. central tarsal bone). Bone modelling in response to loading in the navicular, in particular, has been explored using naturally occurring animal experimental models. Fatigue fractures of the navicular are common among human athletes and dancers and are known to occur naturally in animal athletes such as dogs and horses as well (Bennell et al., 2016; Boudrieau et al., 1984; Devas, 1961; Khan et al., 1994; Muir et al., 1999). These fracture types are hypothesised to be caused by an accumulation of bone microdamage with repetitive (i.e. cyclic) loading (Muir et al., 1999).

A study conducted by Johnson et al. (2000) on the navicular in greyhounds demonstrated changes in bone structure that correlate with asymmetrical loading imposed by regular racing on a circular track. The navicular in a group of racing greyhounds was compared to the navicular of a group of retired greyhounds (i.e. not racing for more than 4 months), using a combination of low-resolution Computed Tomography (CT) scans (i.e. to measure bone mineral density) and histomorphometry (i.e. to determine percentage bone area). The left and right naviculars were compared between (i.e. racing vs retired) and within groups (i.e. left vs right navicular in racing dogs and retired dogs, respectively). Within each group (i.e. racing and retired, respectively) the bone mineral density of the right navicular was significantly greater than the left ($p < 0.001$). Between groups, the bone mineral density in the racing group was significantly greater than the retired group in both the left and right navicular ($p < 0.005$). Through analyses of the histologic sections, the authors concluded that the bone in the dorsal region of the right navicular of racing greyhounds had undergone site-specific adaptive modelling which resulted in significantly thickened trabeculae that were more compact compared to the left side of the same individuals ($p < 0.001$) as well as the same site in the non-racing group ($p < 0.001$). No significant differences in percentage bone area were detected between the left and right naviculars of individuals in the retired group. The authors attributed the differences observed between the left and right navicular within the same individual (most notably in racing greyhounds) to the asymmetric loading of the limbs as greyhounds run at a high speed across a relatively short distance in a counterclockwise direction with the right limb on the outside. This study also demonstrated that bone in retired greyhounds had remodelled in response to a lack of asymmetrical cyclic loading, losing the site-specific adaptations observed in the navicular (particularly the right) of racing dogs. Additional studies have also found changes in cross-sectional geometry of the cuboidal bones of racehorses that have undergone cyclic loading from training and racing (Nunamaker et al., 1989; Young et al., 1991).

10.2.2.3 The forefoot

The forefoot is the most distal region of the foot which is composed of the 5 metatarsals and 14 phalanges. Griffin et al. (2010) analysed trabecular structure across three regions of interest in the second metatarsal heads of different great ape species (*Homo*, *Pan*, *Pongo*, and *Gorilla*) using high-resolution X-ray computed tomography. They found that although BV/TV (Table 10.1) did not differ significantly between humans and non-human great apes, human metatarsal heads demonstrated significantly more anisotropic trabecular architecture, most notably in the dorsal regions when compared to the same regions in non-human great apes. The authors suggested that this increased anisotropy was reflective of the habitual use of the forefoot in humans for propulsion during gait as modern humans dorsiflex the joint between their metatarsals and phalanges (i.e. metatarsophalangeal joint) to form a stiff propulsive lever during the end of stance phase.

10.2.2.4 The foot—Summary

Studies exploring the various skeletal elements of the foot through multiple methods have demonstrated bone functional adaptation (Table 10.1) in the foot that, to some extent, reflects locomotor-related loading and strain. These studies are primarily focused on the hindfoot (as it relates to the enlarged human heel) and the forefoot (as it relates to a loss of manipulation ability and instead an increase in strength to act as a lever for forward propulsion during bipedal walking). The experimental support for bone functional adaptation in the bones of the foot is of particular importance in interpreting locomotor-related loading and inferring postural and locomotor behaviour in fossil hominins that exhibit variable morphology.

10.3 Development in archaeological and anthropological research

Exploration of the concept of bone functional adaptation has extended beyond experimental work using animal models in controlled environments and has been applied within the fields of palaeoanthropology and archaeology. This section will focus on fossils of the foot and ankle belonging to various hominin species, from the very earliest proposed bipeds to the appearance of anatomically modern humans. Fossils attributed to each species and the findings of various studies focusing on morphology and bone functional adaptation are highlighted, which further our understanding into the evolution of bipedalism and locomotor variability among hominins. This section also focuses on modern humans and the applications to the archaeological record which provide further information on the drivers of intraspecific variation within a single species (i.e. what affects bone functional adaptations in obligate bipedal walkers).

10.3.1 The evolution of bipedalism as evidenced by the foot and ankle

Homo sapiens is the only extant primate that is an obligate biped (Table 10.1). This evolved trait is attributed to an extensive adaptation of the postcranial skeleton, particularly the lower part of the leg and the foot. These distal skeletal elements are directly affected by bipedal locomotion and therefore they are the main components under strong selective pressure. In this section, we focus on how information of bone functional adaptation (Table 10.1) derived from extinct hominins fossil remains can extend our understanding of the origin and evolution of bipedalism.

10.3.1.1 *Evidence of locomotion—Ardipithecus ramidus*

Information relating to locomotion in *Ardipithecus ramidus* (4.4 mya, Aramis and Gona, Ethiopia) is derived mostly from an incomplete foot and a few isolated metatarsal bones (Lovejoy et al., 2009; White et al., 2009). The talus was found to be ape-like, and its shape implies that *A. ramidus* was capable of extensive inversion and dorsiflexion of its ankle, indicative of climbing and potentially an arboreal lifestyle. Of special importance is the medial cuneiform bone that, while crushed, shows that the hallux was prehensile and functioned for grasping. Other midtarsal bones demonstrated intermediate features between humans and non-human hominoids. While the shape of the lateral metatarsal bones and their articulations with the midtarsal bones indicate that the lateral midfoot of *A. ramidus* was stiff and thus adapted to some form of terrestrial bipedal locomotion, other aspects of the medial midfoot were more ape-like, indicating that *A. ramidus* was adapted to both arboreal and terrestrial bipedal locomotion. The proximal and intermediate phalanges also demonstrate a mix of ape-like (bones are long and curved) and human-like (bases are dorsal canted) characteristics. In conclusion, current known remains of *Ardipithecus ramidus* point to a mixed terrestrial/arboreal locomotion of this species.

10.3.1.2 *Evidence of locomotion—Australopithecus anamensis*

The only meaningful data related to locomotion for *A. anamensis* are inferred from the proximal and distal thirds of a single tibia (4.12 mya, Kanapoi, Kenya). Multiple studies concluded based on various parameters that the tibia demonstrates clear signs of being adapted to accommodate habitual bipedal locomotion. Both Leakey et al. (1995) and Ward et al. (2001) commented on the vertical orientation of the shaft relative to the talar facet, which suggests a human-like valgus knee, and the shape of the proximal articular surface and the concave condyles of roughly equal area as indicators of *A. anamensis* habitual bipedality. Similarly, based on a multivariate discriminant function analysis, Zipfel et al. (2011) found that the

A. anamensis tibial shape clusters with humans and not with non-human hominoids. Finally, DeSilva (2009) established that *A. anamensis* distal tibia articular surface is square shaped and perpendicularly oriented relative to the long axis of the bone, both indicating a more human-like morphology. DeSilva (2009) concluded that these features do not functionally correlate with vertical climbing done by chimpanzees and thus, if *A. anamensis* was engaged in arboreal climbing, it was distinct from chimpanzees. These data combined indicate that *A. anamensis* was at least a habitual biped, spending a significant amount of time in some form of terrestrial bipedal locomotion.

10.3.1.3 Evidence of locomotion—Australopithecus afarensis

Several tibiae attributed to *A. afarensis* (c.3.2 mya, Hadar, Ethiopia) were all found to be distinct from non-human hominoids and similar to humans by the vertical orientation of their shaft relative to the talar facet and the talocrural joint axis (i.e. ankle, Table 10.1), both indicative of a human-like valgus knee (DeSilva, 2009; Zipfel et al., 2011). Furthermore, similar to humans, the sagittal orientation of *A. afarensis* distal tibia articular facet, termed the tibial arch angle, is tilted anteriorly (DeSilva and Throckmorton, 2010), indicating the presence of a longitudinal plantar arch—a structure that functions in storing elastic energy and stiffening the foot during bipedal locomotion. A different approach compared the tibia proximal and distal joints sizes of *A. afarensis* to humans and non-human hominoids (Jungers, 1988). Humans have enlarged hindlimb joints relative to other extant primates to accommodate for larger loads due to a reliance on hindlimbs for locomotion; *A. afarensis* was found to exhibit a modest increase in hindlimb joint size, although not on par with modern humans. Yet, contrary to the tibia, *A. afarensis* fossilised fibulae reveal many more similarities to non-human primates, such as an oblique proximal margin of the talar facet. While these adaptations have no clear functional significance, according to some studies they may signify adaptations for arboreal locomotion (Stern and Susman, 1983).

The talus, calcaneus, and navicular bones of *A. afarensis* were studied for their external structure and interarticular surfaces: talus and tibia/fibula (talocrural joint), talus and calcaneus (subtalar joint), and talus and navicular bone (talonavicular joint). Using multivariate discriminant function analysis, Prang (2016) found that *A. afarensis* is human-like in talar and calcaneal articular facet orientations and thus probably had a longitudinal arch and was committed to terrestrial locomotion (but see Stern and Susman (1983) and Stern (2000) for a different view). Similarly, DeSilva (2009) showed that the *A. afarensis* talar angle (i.e. the axis of rotation of the ankle and the superior plane of the ankle) was human-like. They concluded that *A. afarensis* possessed an everted foot and that it was incapable of the same range of ankle dorsiflexion and foot inversion as chimpanzees during vertical climbing. Latimer and Lovejoy (1990) and Zipfel et al. (2011) both emphasised the similarity between *A. afarensis* and modern human calcanei, expressed as a discrete lateral process and a flattened posterior talocalcaneal joint. They suggested that these adaptations are related to the restricted range of joint mobility and the force applied to the calcaneus during the heel strike phase of human-like bipedal locomotion. In addition, using multivariate discriminant function analysis, Prang (2016) stated that, contrary to what was previously suggested (Sarmiento and Marcus, 2000), the two navicular bones attributed to *A. afarensis* were most similar to modern humans. He suggested that the relatively large navicular tuberosity that was previously attributed to the lack of longitudinal arch in *A. afarensis* was reflecting developed leg musculature associated with ankle plantarflexion.

Finally, several studies focused on *A. afarensis* forefoot bones, specifically a complete set of five metatarsal heads and five proximal and two middle phalanges, all from a single individual, as well as several isolated metatarsals and phalanges. Stern and Susman (1983) noted a mix of ape-like and human-like features in the foot bones; these included long slender and curved proximal phalanges that suggest arboreal locomotion but also a dorsal rim at the basal articular surface of these phalanges, which is indicative of a bipedal behaviour. A similar human-like anatomy of the base of a proximal phalanx (3.5 mya, Lake Turkana, Kenya) was described by Ward et al. (1999). Latimer and Lovejoy (1990) and Ward et al. (2011) noted that corresponding to the dorsal rim at the base of the proximal phalanges, the MT heads are domed dorsally, which would enable human-like dorsiflexion angles during the push-off phase of bipedal locomotion. Furthermore, Ward et al. (2011) reported that the fourth metatarsal bone exhibits rotation of the head relative to the base, indicative of a transverse plantar arch, as well as plantarly and distally inclined proximal and distal articular surfaces, respectively, indicative of a longitudinal arch. These characteristics, together with a flat metatarsal base that implies the absence of a chimpanzee-like midtarsal break, show that *A. afarensis* foot was human-like and adapted for terrestrial bipedalism. Yet, based on multivariate discriminant function analysis, Fernández et al. (2016) determined that the metatarsal heads are intermediate in their morphology between humans and chimpanzees and thus suggests a less efficient push-off kinematics than bipedal humans. Nevertheless, Latimer and Lovejoy (1990) have explained the presence of these and other primitive ape-like features as phylogenetic inertia that were retained by stabilising selection from a committed arboreal ancestor. Alternatively, the recent discovery of a juvenile *A. afarensis* foot (3.3 mya, Dikika, Ethiopia) with evidence of a grasping implies that juvenile *A. afarensis* were more arboreal than the adults and that some of these immature features may be retained in mature individuals, giving their skeleton some ape-like anatomical characteristics (DeSilva et al., 2018).

In conclusion, current information strongly suggests that *A. afarensis* was considerably involved in terrestrial bipedal locomotion, yet this bipedality was probably somewhat different than the way modern bipedal human locomote.

10.3.1.4 *Evidence of locomotion—Australopithecus africanus*

Our information regarding *A. africanus* is mostly derived from bones uncovered at Sterkfontein (StW), South Africa (2.0–2.8 mya). Several distal tibiae and one proximal tibia were analysed for their external shape and joint orientation as well as for their internal trabecular BV/TV (Table 10.1), structure, and orientation. Berger and Tobias (1996) addressed the *A. africanus* tibia as ape-like based on several external morphologies of the proximal tibia, of which the most notable is the convex lateral condyles at the knee joint that suggested a more mobile knee than humans. However, they have also noted the lack of distinct localisation of the tibial tuberosity which is more human-like. Other studies (DeSilva, 2009; DeSilva and Throckmorton, 2010; Zipfel et al., 2011) looked at the distal tibiae and noted, similar to *A. anamensis* and *A. afarensis*, that they are more human-like, having a square-shaped articular surface that is perpendicularly oriented relative to the long axis of the bone and tilted anteriorly in the sagittal plane. These characteristics suggest that *A. africanus* had a human-like valgus knee, was not able to dorsiflex and invert its foot like chimpanzees, and possessed a longitudinal arch. Chirchir et al. (2015) measured trabecular

BV/TV (Table 10.1) at various bones, including the distal tibia, of non-human primates, extinct hominins, and modern humans. The authors reported that the distal tibia of extinct hominins, and specifically *A. africanus*, had significantly higher trabecular BV/TV than modern humans, which is indicative of the high activity level of these early hominins compared to recent sedentary humans. Barak (2013a) also reported that *A. africanus* had significantly higher trabecular BV/TV than modern humans and other trabecular properties values that fall between those of chimpanzees and humans. More interestingly, they have found that the principal trabecular orientation of *A. africanus* distal tibia was human-like and not ape-like (Fig. 10.1), indicating that they have loaded their ankles in a relatively extended posture like modern humans.

Five tali assigned to *A. africanus* give us insights into the bones' external structure, ankle joint angle, and inner trabecular structure. Contrary to *A. afarensis*, Prang (2016) found that *A. africanus* talonavicular joint angle is ape-like, which implies a low or absent longitudinal arch. However, DeSilva (2009) has noted that *A. africanus* talar angle is within human range and different from chimpanzees and gorillas. Investigating tali trabecular BV/TV, architecture, and orientation, DeSilva and Devlin (2012) found that *A. africanus* tali are human-like in most respects, but that trabecular parameters largely overlapped between human and non-human primates. Hence, they have suggested that trabecular architecture in the primate talus is conserved and does not reflect the locomotion behaviour of different species. Su and Carlson (2017) also found large variations in tali trabecular bone parameters but concluded that the trabecular structure did demonstrate characteristics of terrestrial bipedal locomotion. Similarly, Zeininger et al. (2016) suggested from the calcaneus trabecular architecture that *A. africanus* experienced more variable loading than humans but less than other non-human primates. Finally, Dowdeswell et al. (2017) and Zipfel et al. (2009) studied the fifth metatarsal bone. Both studies noted a mix of ape-like (e.g. cortical bone thickness) and human-like (e.g. second moment of area) features.

Summarising the earlier information, it is difficult to clearly infer the way *A. africanus* was locomoting. This is attributed to the probability that not all these bones are really *A. africanus* bones (DeSilva et al., 2019). Still, it is probably safe to say that while *A. africanus* was capable of bipedal locomotion, they had an arboreal component to their locomotor behaviour. This fact is interesting as it seems that overall, *A. africanus* lower leg and foot bones are less adapted to human-like bipedal locomotion than its predecessor *A. afarensis*.

10.3.1.5 Evidence of locomotion—Australopithecus sediba

There are two partial skeletons attributed to *A. sediba*—MH1 and MH2 from Malapa, South Africa (1.98 mya) (Berger et al., 2010). While this species is, evolutionary speaking, younger than the previous discussed Australopithecines, it does not show any clear progress towards modern human-like bipedal locomotion but a mosaic of adaptations towards terrestrial bipedalism and arboreality (Rein et al., 2017). Both distal tibiae were found to cluster with other human and fossil *Homo* tibiae and demonstrated an orthogonal ankle joint in relation to the long axis of the tibia. However, both tibiae have also extremely robust ape-like medial malleoli indicating that *A. sediba* may have frequently loaded its foot in inversion such as during climbing (Zipfel et al., 2011). Alternatively, DeSilva et al. (2013) hypothesised that *A. sediba* may have locomoted bipedally differently than modern humans—inverting its foot during the swing phase and landing on the lateral aspect of the foot. Thus, as the foot rolled and the weight was transferred medially into the toe-off phase, high shear stresses were introduced on the medial malleolus, which explains its robustness. In addition, Chirchir (2019) noted that trabecular BV/TV (Table 10.1) in the *A. sediba* distal tibia is similar to non-human primates and other australopithecines, and significantly higher than humans, reflecting the higher activity levels in these early hominins.

The talus has a human-like trochlea, which suggests that the ankle was not habitually loaded in dorsiflexion, yet several analyses positioned it and its associated gracile ape-like calcaneus intermediate between apes and humans (DeSilva et al., 2018; Zipfel et al., 2011). The orientations of the calcaneus and talus articular facets implied a lack of longitudinal arch (Prang, 2015), yet others interpreted the slightly plantarly angle of the calcaneocuboid joint as an emerging longitudinal arch in this species (Zipfel et al., 2011). The fourth metatarsal also reveals a mosaic of adaptations, key of them is its dorsoplantarly convexity, which implies an ape-like midtarsal break (Zipfel et al., 2011).

10.3.1.6 Evidence of locomotion—Paranthropus

Due to spatial and temporal overlap with *Homo* fossil remains (e.g. Sterkfontein, Member 5 East infill), it is difficult to confidently attribute many of these bones to the genus *Paranthropus*. As a result, the potentially interspecies variability of the fossil remains complicates our ability to infer the locomotor behaviour of *P. robustus* and *P. boisei*. Nevertheless, much of the data in the literature support the prediction that *P. robustus* and *P. boisei* were committed terrestrial bipeds with a human-like foot that revealed a minimal longitudinal arch, a stable lateral column, and a medial weight transfer through the ankle towards an adducted hallux.

DeSilva (2009) reported that while a distal tibia (1.89 mya, Turkana Basin, Kenya) and four tali (1.7–1.9 mya, Turkana Basin, Kenya and Olduvai Gorge, Tanzania) revealed some ape-like adaptations, they also demonstrated human-like features that would have prevented *P. robustus* and *P. boisei* ankle to dorsiflex and invert like modern chimpanzees during vertical climbing. Similarly, using multivariate discriminant function analysis, Zipfel et al. (2011) found the shape of *P. boisei* tibia to cluster with human tibiae. Conversely, looking at tali articular facets, Prang (2015) reported talonavicular joint angles that were outside the human range, indicating that *Paranthropus* lacked, or at best had a minimal, human-like longitudinal arch. In line with these findings, the trabecular bone of the tali was found to be generally ape-like, yet both tali demonstrated highly oriented trabeculae in the anteromedial portion of the talar body, a feature consistent with human-like medial weight transfer through the ankle towards the hallux (Su and Carlson, 2017; Su et al., 2013).

Moving distally to the forefoot, metatarsal bones attributed to *P. robustus* and *P. boisei* (1.0–2.0 mya, Swartkrans, South Africa and Omo, Ethiopia) demonstrate different degrees of ape-like and human-like characteristics, implying terrestrial bipedal locomotion but a low medial longitudinal arch and a reduced role of the big toe in propelling the body forward during the toe-off phase (DeSilva et al., 2019; Dowdeswell et al., 2017; Zipfel et al., 2009). Investigating the internal trabecular structure of first and fifth metatarsal heads, Chirchir et al. (2015) noted that, similar to other early hominins, *P. robustus* had comparable trabecular BV/TV (Table 10.1) values to non-human primates, which was attributed to the high activity level of these early hominins compared to recent more sedentary humans. In a second paper looking at the same first metatarsal bones, Komza and Skinner

(2019) noted distinct trabecular architecture patterns and low degree of anisotropy that suggested an increase in first metatarsal mobility and a more variable bipedal locomotion behaviour in these hominins.

Maybe the most promising fossil remains, attributed to *P. boisei*, are an almost complete foot—OH 8, from Olduvai, East Africa (although some attribute it to a juvenile *Homo habilis*). The foot is comprised of the three tarsal bones and all five metatarsal bones (except for their distal heads). Multiple studies found the various bones and their association with each other to be mostly human-like with a medial cuneiform that suggests a non-opposable big toe (adducted hallux) and an interlocking calcaneocuboid joint that infer a stable lateral column and the presence of a longitudinal arch (DeSilva et al., 2019; Prang, 2016; Sarmiento and Marcus, 2000). Yet, other characteristics like a large talar head and a short first metatarsal may indicate a somewhat different bipedal gait kinematics compared to modern humans.

10.3.1.7 Evidence of locomotion—Early Homo and Neanderthals

While it is accepted that all hominins in the genus *Homo* were committed bipeds, minor differences still existed between the leg and foot anatomy of Pleistocene *Homo* species and modern *H. sapiens*. As of now, the oldest foot bone attributed to *Homo* is a talus from Omo, Ethiopia, dated to about 2.2 mya (Gebo and Schwartz, 2006). The talus is very human-like and thus there is no question that this individual was committed to bipedal locomotion. However, it does show evidence for a low talar declination angle, a feature that is associated with a low longitudinal arch and flat feet, suggesting that these Pleistocene *Homo* species were applying bipedal locomotion somewhat differently than modern humans.

The earliest evidence of early humans migrating out of Africa, a key event in our evolutionary timeline, is fossil remains attributed to *H. erectus* found at Dmanisi, Georgia, and dated to 1.8 mya (Pontzer et al., 2010). The lower leg and foot fossil material include a tibia and a talus; a medial cuneiform; and a first, third, fourth, and fifth metatarsal bones, all of them predominantly human-like. The ankle joint is human-like, and the foot reveals developed transverse and longitudinal arches with an adducted big toe, all indicating that the Dmanisi lower leg and foot were functioning like in modern humans. Pontzer et al. (2010) suggest, based on stone tools at the site, that the Dmanisi were active hunters or scavengers, linking these activities to improved bipedal locomotion in early *Homo*. Yet, retention of some primitive features, such as the shape of the first metatarsal head and the robusticity of the metatarsal bone shafts, suggests that the Dmanisi foot was still loaded somewhat different than modern humans, indicating that locomotor evolution continued through the early Pleistocene (Dowdeswell et al., 2017; Pontzer et al., 2010). This continued evolution of bipedal locomotion also finds support from a partial *H. erectus* foot (Koobi Fora, Kenya), dated to 1.53 mya, which demonstrates a shortening of the intermediate phalanges, a feature that was associated with a reduced metabolic cost during running (Rolian et al., 2009).

The earliest foot bones attributed to *H. sapiens* are from Omo, Ethiopia, dated 195,000 years ago, just before the onset of the Late Pleistocene (Pearson et al., 2008), which is more or less the time when modern human bipedal locomotion was adopted by *H. sapiens*. Nevertheless, changes in our lifestyle (e.g. footwear), terrain (e.g. flatland vs. mountains), and activity levels (i.e. becoming more sedentary) continue to have effect in adapting our lower leg and foot bones. These parameters will be discussed in the next section.

10.3.1.8 Evidence of locomotion—Summary

Mirroring the experimental data, most of the information derived from hominins fossil remains is related to external morphology and cortical bone parameters rather than trabecular bone. Yet, the seeming easiness of obtaining external morphology data from individual bones is offset by the complex overall 3D structure of the entire lower leg and foot, which creates an intricate biomechanical apparatus that is hard to analyse and interpret. Adding to this difficulty is the limited number of isolated fossilised hominin lower leg and foot bones and the even scarcer partial feet. Nevertheless, in the last decade, with the decreasing costs of micro-CT machines and their increased resolution, together with improved segmentation techniques, we see a significant increase in bone functional adaptation information collected from the internal bony structure of fossilised bones. The combination of cortical and trabecular structural data opens a window to the unknown locomotion behaviour of our extinct hominin ancestors and helps us learn how we have evolved to become the sole extant obligate primate biped species (using two legs as the predominant way of locomoting).

10.3.2 Holocene human activity and the effects on bone modelling/remodelling

Analysing populations from the past provides a unique opportunity to assess known risk factors under different environmental conditions and shed light on modern conditions in a way that might not be possible from studying modern populations in isolation (Wallace et al., 2017). For most of the evolutionary history of anatomically modern humans, the dominant subsistence strategy was unshod/minimally shod (i.e. barefoot or with basic footwear) foraging before the introduction of pastoralism/agriculture and subsequent later technological advances, including the introduction of increasingly complex footwear. This section will discuss how changes in activity levels, substrates, and footwear have impacted the morphology of the leg, ankle, and foot.

10.3.2.1 Effects of activity levels/subsistence strategies on bone modelling/remodelling in Holocene humans

When compared to other living non-human primates, extinct hominins, and archaeological foraging populations, modern humans exhibit relatively gracile postcranial skeletons (Chirchir, 2019; Chirchir et al., 2015; Holt, 2003; Marchi, 2008; Marchi et al., 2011; Nowlan et al., 2011; Ruff, 1999; Ruff et al., 2015 1998; 1993; Stock, 2006). This recent trend in gracilisation has been attributed to a declining mechanical loading of the postcranium associated with shifts in subsistence strategies in modern humans (i.e. Holocene *H. sapiens*), most notably from highly active foragers towards increasing sedentism (e.g. Chirchir, 2019; Chirchir et al., 2015; Holt, 2003; Marchi et al., 2011; Ruff, 2007; Ruff and Hayes, 1983; Ruff et al., 1984, 1993, Stock, 2006). Populations that are presumed to engage in long-distance walking and running are often hypothesised to have stronger lower limb bones (i.e. femur and tibia) in response to repetitive loading for prolonged periods of time.

Postcranial robusticity in lower limb bones of members of the genus *Homo* was initially explored through comparisons of femoral diaphyseal and articular properties and concluded that there was a general decline in diaphyseal robusticity from Pleistocene *Homo* throughout human evolution until modern Holocene humans (Ruff et al., 1993). Further studies utilising limb diaphyseal cross-sectional geometric of the femur and tibia have supported the hypothesis that limb bone relative robusticity, as well as shape change, can be related to changes in subsistence strategies (Chirchir et al., 2015; Holt, 2003; Marchi, 2008; Ogilvie and Hilton, 2011; Shaw and Ryan, 2012; Sládek et al., 2006; Stock and Pfeiffer, 2001). Diaphyseal shape change is most characterised by a greater bone deposition in the anteroposterior plane resulting in oval shaped cross-sections in more active populations, while an increasingly circular cross-section is associated with increased sedentism (Bridges, 1989; Larsen, 1995; Ruff, 1987; Ruff et al., 1984; Sládek et al., 2006).

Changes in trabecular structure throughout the lower limb have also been found to correspond with changes in subsistence strategies although these studies have concluded that this process of gracilisation is not systemic or uniform throughout human evolution (Chirchir et al., 2015; Ryan and Shaw, 2015; Saers et al., 2016). Activity levels appear to positively correlate with retaining bone mass (Iwamoto et al., 1999; Mori et al., 2003), for example, mobile foraging populations show significantly greater bone density (as measured by BV/TV, Table 10.1) and thicker trabeculae in the proximal femur (and throughout the lower limb in general) than more sedentary agricultural populations, associated with a pattern of decreasing terrestrial mobility (Chirchir et al., 2015; Ruff et al., 1984, 2015; Ryan and Shaw, 2015; Saers et al., 2016). Trabecular parameters such as thickness and number also appear to be affected by activity levels as evidenced by a recent study of trabecular structure throughout the lower limb of three populations representing a gradient of mobility from highly terrestrial mobile, intermediate mobility, and highly sedentary demonstrated more robust trabecular structure in highly mobile groups (i.e. higher BV/TV composed of relatively fewer thicker trabecular struts) (Saers et al., 2016).

10.3.2.2 *Effects of substrate/terrain on bone modelling/remodelling in Holocene humans*

Previous studies have suggested that variation in terrain could possibly influence bone modelling, and in turn, biomechanical reconstructions of mobility patterns in archaeological populations (Carlson et al., 2007; Ruff, 1999). Some studies have focused on populations that are known to have occupied mountainous regions in order to understand the possible influence of rugged terrain on diaphyseal morphology in the lower limb; these studies concluded that long-distance travel over rugged terrain increases overall lower limb mechanical loading evidenced by higher relative femoral and tibial robusticity (Marchi, 2008; Ruff, 1999; Ruff et al., 2006; Sparacello and Marchi, 2008). Similar to changes in bone shape and diaphyseal geometry observed in highly active populations, those exposed to rugged terrain also demonstrate increased bending strength and more anteroposterior oriented cross-sectional shapes (i.e. buttressing across the sagittal plane) (Larsen, 1995; Ruff, 1999). Increased medio-lateral forces due to stabilisation on an uneven substrate have also been suggested; however, the effect is not as pronounced as anteroposterior cross-section (Holt and Whittey, 2019; Sparacello and Marchi, 2008). In general, the strongest impact of terrain as demonstrated by lower limb cross-sectional properties is observed in highly active foragers (i.e. hunter-gatherers) and even more sedentary groups (i.e. agriculturists) from hilly/mountainous regions, with less effect in modern industrial and post-industrial groups (Holt and Whittey, 2019).

10.3.2.3 Effects of footwear on bone modelling/remodelling in Holocene humans

Previous studies on the effect of footwear on the morphology of the foot have focused mainly on clinical aspects as it relates to forefoot pain or metatarsalgia (Capener, 1965; Mafart, 2007); however, research has expanded to include archaeological populations. The foot–ground interaction has historically involved a barefoot interacting with a natural substrate; it is not until recently that humans became habitually shod although indigenous footwear probably remained very minimal (D'AoÛt et al., 2009; Kuttruff et al., 1998; Trinkaus, 2005; Trinkaus and Shang, 2008; Zipfel and Berger, 2007). The archaeological record dates footwear to the middle Upper Palaeolithic (Gravettian, c.25,000 years ago) in parts of Europe (Trinkaus, 2005). Subsequently, modern footwear has become increasingly more complex to include features to improve heel and arch support among other features (Lieberman et al., 2010; Trinkaus, 2005; Willems et al., 2017). The use of footwear has been found to restrict the range of motion of the foot specifically during the push-off phase and can therefore affect the biomechanics of the foot and ankle, likely because of prolonged constriction and accommodation (Morio et al., 2009; Sim-Fook and Hodgson, 1958). It has also been suggested that habitual use of footwear can cause pathological changes in the foot (e.g. hallux valgus) (Fong Yan et al., 2013; Zipfel and Berger, 2007).

Some studies have further suggested that the habitually unshod foot is healthier than a habitually shod foot as evidenced by lower frequencies of osteological modification, most notably in the forefoot (Mafart, 2007; Zipfel and Berger, 2007). Studies on trabecular bone of the various skeletal elements comprising the foot have primarily focused on an evolutionary context to further understand when modern human-like foot morphology and gait emerged (Griffin et al., 2010; Su et al., 2013; Zeininger et al., 2016); however, these methods could potentially expand to include intraspecific studies on various human populations that are habitually shod or unshod.

10.3.2.4 Factors affecting bone modelling/remodelling in Holocene humans—Summary

Understanding variation among living and archaeological human populations allows us a time frame to explore and understand how changes in lifestyle affect bone modelling. Activity levels appear to be the primary influence on bone functional adaptation across modern *H. sapiens* populations, coinciding with the transition in foraging strategies. However, more research into the effects of substrate and footwear is needed in order to understand possible confounding variables.

10.4 Conclusion

The aim of the first section of this chapter was to bring experimental support for bone functional adaptation in the lower leg, ankle, and foot. Many studies have demonstrated active bone functional adaptation in response to physiological dynamic loading, especially in relation to trabecular bone architecture and principal trabecular orientation. This experimental support is key for interpreting postural and locomotor behaviour in fossil hominins that exhibit variable morphology, which is the focus of the second section of the chapter. In the second section of this chapter, we have demonstrated that a study of both external (cortical) and internal (trabecular) bone structures is imperative for the interpretation of bone functional adaptation signals derived from extinct hominins fossil remains.

While our analysis indicates that the locomotor behaviour of different species and even human populations can be inferred from their external and internal bone morphology, our current body of knowledge concerning early hominin locomotion behaviour suffers from several key limitations. First, despite the recent considerable growth of our hominin fossil record, it is still relatively small, especially when we just consider the number of leg and foot bones, and even more so when we account for the ones that articulate with each other and can be confidently attributed to a single individual. Accordingly, the locomotor behaviour interpretations of entire species are based on a small and inadequate number of bones from just one or a few individuals. Seeing that each foot consists of 26 bones, interpreting the foot mechanical functioning from just one or a few bones is restricting our ability to accurately depict function from structure. Second, and directly related to the previous point, due to the small number of bones we currently have per hominin species, we may mistake intraspecies normal variability as an indication for diverse bipedal locomotion behaviour. Third, due to spatial and temporal overlap of early hominin fossil remains, it is difficult in some cases to confidently attribute these bones to a specific species. Finally, we should not forget that some of these mixed bone morphologies may simply be primitive traits that were retained due to stabilising selection, pleiotropic effect (i.e. one gene that influences two or more seemingly unrelated phenotypic traits), or ontogeny (e.g. more arboreal behaviour in subadults and terrestrial bipedal locomotion in adults). We can only hope that ongoing field research will yield additional hominin bones that will help us overcome these limitations.

Nevertheless, our current body of knowledge supports the concept that terrestrial bipedal locomotion evolved in hominins through the Pleistocene and that hominin species had a relatively large diversity of bipedal locomotion behaviours. A combined approach of studying fossil bones' external morphology and their articulation with adjacent bones, together with their internal trabecular architecture and principal trabecular orientation (when possible), would bring us closer to solve many of these unanswered questions of extinct hominin locomotion and the evolution of human bipedalism.

References

Akhter, M. P., Cullen, D. M., Pedersen, E. A., Kimmel, D. B., & Recker, R. R. (1998). Bone response to in vivo mechanical loading in two breeds of mice. *Calcified Tissue International, 63*(5), 442–449. https://doi.org/10.1007/s002239900554.

Barak, M. M. (2020). Bone modeling or bone remodeling: That is the question. *American Journal of Physical Anthropology, 172*(2), 153–155. https://doi.org/10.1002/ajpa.23966.

Barak, M. M., Lieberman, D. E., & Hublin, J.-J. (2011). A Wolff in sheep's clothing: Trabecular bone adaptation in response to changes in joint loading orientation. *Bone, 49*(6), 1141–1151. https://doi.org/10.1016/j.bone.2011.08.020.

Barak, M. M., Lieberman, D. E., & Hublin, J.-J. (2013a). Of mice, rats and men: Trabecular bone architecture in mammals scales to body mass with negative allometry. *Journal of Structural Biology, 183*(2), 123–131. https://doi.org/10.1016/j.jsb.2013.04.009 (Special Issue in Recognition of Dr. Steve Weiner's Scientific Accomplishments).

Barak, M. M., Lieberman, D. E., Raichlen, D., Pontzer, H., Warrener, A. G., & Hublin, J.-J. (2013b). Trabecular evidence for a human-like gait in Australopithecus africanus. *PLoS One, 8*(11). https://doi.org/10.1371/journal.pone.0077687, e77687.

Bennell, K. L., Malcolm, S. A., Thomas, S. A., Wark, J. D., & Brukner, P. D. (2016). The incidence and distribution of stress fractures in competitive track and field athletes: A twelve-month prospective study. *The American Journal of Sports Medicine.* https://doi.org/10.1177/036354659602400217.

Berger, L. R., de Ruiter, D. J., Churchill, S. E., Schmid, P., Carlson, K. J., Dirks, P. H. G. M., et al. (2010). Australopithecus sediba: A new species of homo-like Australopith from South Africa. *Science, 328*(5975), 195–204. https://doi.org/10.1126/science.1184944.

Berger, L. R., & Tobias, P. V. (1996). A chimpanzee-like tibia from Sterkfontein, South Africa and its implications for the interpretation of bipedalism in Australopithecus africanus. *Journal of Human Evolution, 30*(4), 343–348. https://doi.org/10.1006/jhev.1996.0027.

Berman, A. G., Clauser, C. A., Wunderlin, C., Hammond, M. A., & Wallace, J. M. (2015). Structural and mechanical improvements to bone are strain dependent with axial compression of the tibia in female C57BL/6 mice. *PLoS One, 10*(6). https://doi.org/10.1371/journal.pone.0130504, e0130504.

Biewener, A. A., Fazzalari, N. L., Konieczynski, D. D., & Baudinette, R. V. (1996). Adaptive changes in trabecular architecture in relation to functional strain patterns and disuse. *Bone, 19*(1), 1–8. https://doi.org/10.1016/8756-3282(96)00116-0.

Boudrieau, R. J., Dee, J. F., & Dee, L. G. (1984). Central tarsal bone fractures in the racing greyhound: A review of 114 cases. *Journal of the American Veterinary Medical Association, 184*(12), 1486–1491.

Bridges, P. S. (1989). Changes in activities with the shift to agriculture in the Southeastern United States. *Current Anthropology, 30*(3), 385–394. https://doi.org/10.1086/203756.

Brodt, M. D., & Silva, M. J. (2010). Aged mice have enhanced endocortical response and normal periosteal response compared with young-adult mice following 1 week of axial tibial compression. *Journal of Bone and Mineral Research, 25*(9), 2006–2015. https://doi.org/10.1002/jbmr.96.

Capener, N. (1965). Hallux Valgus, allied deformities of the forefoot and Metatarsalgia. *British Medical Journal, 2*(5472), 1231–1232. https://www.ncbi.nlm.nih.gov/pmc/articles/PMC1846557/.

Carlson, K. J., Grine, F. E., & Pearson, O. M. (2007). Robusticity and sexual dimorphism in the postcranium of modern hunter-gatherers from Australia. *American Journal of Physical Anthropology, 134*(1), 9–23. https://doi.org/10.1002/ajpa.20617.

Chen, M. M., Yeh, J. K., Aloia, J. F., Tierney, J. M., & Sprintz, S. (1994). Effect of treadmill exercise on tibial cortical bone in aged female rats: A histomorphometry and dual energy X-ray absorptiometry study. *Bone, 15*(3), 313–319. https://doi.org/10.1016/8756-3282(94)90294-1.

Chirchir, H. (2019). Trabecular bone fraction variation in modern humans, fossil hominins and other primates. *The Anatomical Record, 302*(2), 288–305. https://doi.org/10.1002/ar.23967.

Chirchir, H., Kivell, T. L., Ruff, C. B., Hublin, J.-J., Carlson, K. J., Zipfel, B., et al. (2015). Recent origin of low trabecular bone density in modern humans. *Proceedings of the National Academy of Sciences, 112*(2), 366. https://doi.org/10.1073/pnas.1411696112.

D'AoÛt, K., Pataky, T. C., Clercq, D. D., & Aerts, P. (2009). The effects of habitual footwear use: Foot shape and function in native barefoot walkers. *Footwear Science, 1*(2), 81–94. https://doi.org/10.1080/19424280903386411.

De Souza, R. L., Matsuura, M., Eckstein, F., Rawlinson, S. C. F., Lanyon, L. E., & Pitsillides, A. A. (2005). Non-invasive axial loading of mouse tibiae increases cortical bone formation and modifies trabecular organization: A new model to study cortical and cancellous compartments in a single loaded element. *Bone, 37*(6), 810–818. https://doi.org/10.1016/j.bone.2005.07.022.

Demes, B., Qin, Y.-X., Stern, J. T., Larson, S. G., & Rubin, C. T. (2001). Patterns of strain in the macaque tibia during functional activity. *American Journal of Physical Anthropology, 116*(4), 257–265. https://doi.org/10.1002/ajpa.1122.

DeSilva, J. M. (2009). Functional morphology of the ankle and the likelihood of climbing in early hominins. *Proceedings of the National Academy of Sciences, 106*(16), 6567–6572. https://doi.org/10.1073/pnas.0900270106.

DeSilva, J. M. (2009). Functional morphology of the ankle and the likelihood of climbing in early hominins. *PNAS, 106*, 6567–6572.

DeSilva, J. M., & Devlin, M. J. (2012). A comparative study of the trabecular bony architecture of the talus in humans, non-human primates, and Australopithecus. *Journal of Human Evolution, 63*(3), 536–551. https://doi.org/10.1016/j.jhevol.2012.06.006.

DeSilva, J. M., Gill, C. M., Prang, T. C., Bredella, M. A., & Alemseged, Z. (2018). A nearly complete foot from Dikika, Ethiopia and its implications for the ontogeny and function of Australopithecus afarensis. *Science Advances, 4*(7). https://doi.org/10.1126/sciadv.aar7723, eaar7723.

DeSilva, J. M., Holt, K. G., Churchill, S. E., Carlson, K. J., Walker, C. S., Zipfel, B., et al. (2013). The lower limb and mechanics of walking in Australopithecus sediba. *Science, 340*(6129). https://doi.org/10.1126/science.1232999.

DeSilva, J., McNutt, E., Benoit, J., & Zipfel, B. (2019). One small step: A review of Plio-Pleistocene hominin foot evolution. *American Journal of Physical Anthropology, 168*(S67), 63–140. https://doi.org/10.1002/ajpa.23750.

DeSilva, J. M., & Throckmorton, Z. J. (2010). Lucy's flat feet: The relationship between the ankle and rearfoot arching in early hominins. *PLoS One, 5*(12). https://doi.org/10.1371/journal.pone.0014432, e14432.

Devas, M. B. (1961). Compression stress fractures in man and the greyhound. *The Journal of Bone and Joint Surgery. British Volume, 43*-B(3), 540–551. https://doi.org/10.1302/0301-620X.43B3.540.

Dowdeswell, M. R., Jashashvili, T., Patel, B. A., Lebrun, R., Susman, R. L., Lordkipanidze, D., et al. (2017). Adaptation to bipedal gait and fifth metatarsal structural properties in Australopithecus, Paranthropus, and Homo. *Comptes Rendus Palevol, 16*(5), 585–599. https://doi.org/10.1016/j.crpv.2016.10.003.

Fernández, P. J., Holowka, N. B., Demes, B., & Jungers, W. L. (2016). Form and function of the human and chimpanzee forefoot: Implications for early hominin bipedalism. *Scientific Reports, 6*(1), 30532. https://doi.org/10.1038/srep30532.

Fong Yan, A., Sinclair, P. J., Hiller, C., Wegener, C., & Smith, R. M. (2013). Impact attenuation during weight bearing activities in barefoot vs. shod conditions: A systematic review. *Gait & Posture, 38*(2), 175–186. https://doi.org/10.1016/j.gaitpost.2012.11.017.

Fritton, J. C., Myers, E. R., Wright, T. M., & van der Meulen, M. C. H. (2005). Loading induces site-specific increases in mineral content assessed by microcomputed tomography of the mouse tibia. *Bone, 36*(6), 1030–1038. https://doi.org/10.1016/j.bone.2005.02.013.

Gebo, D. L., & Schwartz, G. T. (2006). Foot bones from Omo: Implications for hominid evolution. *American Journal of Physical Anthropology, 129*(4), 499–511. https://doi.org/10.1002/ajpa.20320.

Griffin, N. L., D'Août, K., Ryan, T. M., Richmond, B. G., Ketcham, R. A., & Postnov, A. (2010). Comparative forefoot trabecular bone architecture in extant hominids. *Journal of Human Evolution, 59*(2), 202–213. https://doi.org/10.1016/j.jhevol.2010.06.006.

Grimston, S. K., Watkins, M. P., Brodt, M. D., Silva, M. J., & Civitelli, R. (2012). Enhanced periosteal and endocortical responses to axial tibial compression loading in conditional connexin43 deficient mice. *PLoS One, 7*(9). https://doi.org/10.1371/journal.pone.0044222, e44222.

Gross, T. S., Srinivasan, S., Liu, C. C., Clemens, T. L., & Bain, S. D. (2002). Noninvasive loading of the murine tibia: An in vivo model for the study of mechanotransduction. *Journal of Bone and Mineral Research, 17*(3), 493–501. https://doi.org/10.1359/jbmr.2002.17.3.493.

Holguin, N., Brodt, M. D., Sanchez, M. E., Kotiya, A. A., & Silva, M. J. (2013). Adaptation of tibial structure and strength to axial compression depends on loading history in both C57BL/6 and BALB/c mice. *Calcified Tissue International, 93*(3), 211–221. https://doi.org/10.1007/s00223-013-9744-4.

Holguin, N., Brodt, M. D., Sanchez, M. E., & Silva, M. J. (2014). Aging diminishes lamellar and woven bone formation induced by tibial compression in adult C57BL/6. *Bone, 65*, 83–91. https://doi.org/10.1016/j.bone.2014.05.006.

Holt, B. M. (2003). Mobility in upper paleolithic and mesolithic Europe: Evidence from the lower limb. *American Journal of Physical Anthropology, 122*(3), 200–215. https://doi.org/10.1002/ajpa.10256.

Holt, B., & Whittey, E. (2019). The impact of terrain on lower limb bone structure. *American Journal of Physical Anthropology, 168*(4), 729–743. https://doi.org/10.1002/ajpa.23790.

Iwamoto, J., Yeh, J. K., & Aloia, J. F. (1999). Differential effect of treadmill exercise on three cancellous bone sites in the young growing rat. *Bone, 24*(3), 163–169. https://doi.org/10.1016/S8756-3282(98)00189-6.

Johnson, K. A., Muir, P., Nicoll, R. G., & Roush, J. K. (2000). Asymmetric adaptive modeling of central tarsal bones in racing greyhounds. *Bone, 27*(2), 257–263. https://doi.org/10.1016/S8756-3282(00)00313-6.

Judex, S., Donahue, L.-R., & Rubin, C. (2002). Genetic predisposition to low bone mass is paralleled by an enhanced sensitivity to signals anabolic to the skeleton. *The FASEB Journal, 16*(10), 1280–1282. https://doi.org/10.1096/fj.01-0913fje.

Jungers, W. L. (1988). Relative joint size and hominoid locomotor adaptations with implications for the evolution of hominid bipedalism. *Journal of Human Evolution, 17*(1), 247–265. https://doi.org/10.1016/0047-2484(88)90056-5.

Kabel, J., Odgaard, A., van Rietbergen, B., & Huiskes, R. (1999). Connectivity and the elastic properties of cancellous bone. *Bone, 24*(2), 115–120. https://doi.org/10.1016/S8756-3282(98)00164-1.

Kelly, N. H., Schimenti, J. C., Ross, F. P., & van der Meulen, M. C. H. (2016). Transcriptional profiling of cortical versus cancellous bone from mechanically-loaded murine tibiae reveals differential gene expression. *Bone, 86*, 22–29. https://doi.org/10.1016/j.bone.2016.02.007.

Khan, K. M., Brukner, P. D., Kearney, C., Fuller, P. J., Bradshaw, C. J., & Kiss, Z. S. (1994). Tarsal navicular stress fracture in athletes. *Sports Medicine, 17*(1), 65–76. https://doi.org/10.2165/00007256-199417010-00006.

Kodama, Y., Dimai, H. P., Wergedal, J., Sheng, M., Malpe, R., Kutilek, S., et al. (1999). Cortical tibial bone volume in two strains of mice: Effects of sciatic neurectomy and genetic regulation of bone response to mechanical loading. *Bone, 25*(2), 183–190. https://doi.org/10.1016/S8756-3282(99)00155-6.

Kodama, Y., Umemura, Y., Nagasawa, S., Beamer, W. G., Donahue, L. R., Rosen, C. R., et al. (2000). Exercise and mechanical loading increase periosteal bone formation and whole bone strength in C57BL/6J mice but not in C3H/ Hej mice. *Calcified Tissue International*, 66(4), 298–306. https://doi.org/10.1007/s002230010060.

Komza, K., & Skinner, M. M. (2019). First metatarsal trabecular bone structure in extant hominoids and Swartkrans hominins. *Journal of Human Evolution*, 131, 1–21. https://doi.org/10.1016/j.jhevol.2019.03.003.

Kuttruff, J. T., DeHart, S. G., & O'Brien, M. J. (1998). 7500 Years of prehistoric footwear from Arnold Research Cave, Missouri. *Science*, 281(5373), 72–75. https://doi.org/10.1126/science.281.5373.72.

Lanyon, L. E. (1974). Experimental support for the trajectorial theory of bone structure. *The Journal of Bone and Joint Surgery. British Volume*, 56-B(1), 160–166. https://doi.org/10.1302/0301-620X.56B1.160.

Larsen, C. S. (1995). Biological changes in human populations with agriculture. *Annual Review of Anthropology*, 24(1), 185–213. https://doi.org/10.1146/annurev.an.24.100195.001153.

Latimer, B., & Lovejoy, C. O. (1990). Metatarsophalangeal joints of Australopithecus afarensis. *American Journal of Physical Anthropology*, 83(1), 13–23. https://doi.org/10.1002/ajpa.1330830103.

Leakey, M. G., Feibel, C. S., McDougall, I., & Walker, A. (1995). New four-million-year-old hominid species from Kanapoi and Allia Bay, Kenya. *Nature*, 376(6541), 565–571. https://doi.org/10.1038/376565a0.

Lieberman, D. E., Devlin, M. J., & Pearson, O. M. (2001). Articular area responses to mechanical loading: Effects of exercise, age, and skeletal location. *American Journal of Physical Anthropology*, 116(4), 266–277. https://doi.org/10.1002/ajpa.1123.

Lieberman, D. E., Pearson, O. M., Polk, J. D., Demes, B., & Crompton, A. W. (2003). Optimization of bone growth and remodeling in response to loading in tapered mammalian limbs. *Journal of Experimental Biology*, 206(18), 3125–3138. https://doi.org/10.1242/jeb.00514.

Lieberman, D. E., Polk, J. D., & Demes, B. (2004). Predicting long bone loading from cross-sectional geometry. *American Journal of Physical Anthropology*, 123(2), 156–171. https://doi.org/10.1002/ajpa.10316.

Lieberman, D. E., Venkadesan, M., Werbel, W. A., Daoud, A. I., D'Andrea, S., Davis, I. S., et al. (2010). Foot strike patterns and collision forces in habitually barefoot versus shod runners. *Nature*, 463(7280), 531–535. https://doi.org/10.1038/nature08723.

Lovejoy, C. O., Latimer, B., Suwa, G., Asfaw, B., & White, T. D. (2009). Combining prehension and propulsion: The foot of Ardipithecus ramidus. *Science*, 326(5949), 72–72e8. https://doi.org/10.1126/science.1175832.

Lynch, M. E., Main, R. P., Xu, Q., Schmicker, T. L., Schaffler, M. B., Wright, T. M., et al. (2011). Tibial compression is anabolic in the adult mouse skeleton despite reduced responsiveness with aging. *Bone*, 49(3), 439–446. https://doi.org/10.1016/j.bone.2011.05.017.

Macdonald, H. M., Cooper, D. M. L., & McKay, H. A. (2009). Anterior–posterior bending strength at the tibial shaft increases with physical activity in boys: Evidence for non-uniform geometric adaptation. *Osteoporosis International*, 20(1), 61–70. https://doi.org/10.1007/s00198-008-0636-9.

Mafart, B. (2007). Hallux valgus in a historical French population: Paleopathological study of 605 first metatarsal bones. *Joint, Bone, Spine*, 74(2), 166–170. https://doi.org/10.1016/j.jbspin.2006.03.011.

Main, R. P., Lynch, M. E., & van der Meulen, M. C. H. (2014). Load-induced changes in bone stiffness and cancellous and cortical bone mass following tibial compression diminish with age in female mice. *Journal of Experimental Biology*, 217(10), 1775–1783. https://doi.org/10.1242/jeb.085522.

Marchi, D. (2008). Relationships between lower limb cross-sectional geometry and mobility: The case of a Neolithic sample from Italy. *American Journal of Physical Anthropology*, 137(2), 188–200. https://doi.org/10.1002/ajpa.20855.

Marchi, D., Sparacello, V., & Shaw, C. (2011). Mobility and lower limb robusticity of a pastoralist neolithic population from North-Western Italy. In *Human bioarchaeology of the transition to agriculture* (pp. 317–346). John Wiley & Sons, Ltd. https://onlinelibrary.wiley.com/doi/abs/10.1002/9780470670170.ch13.

Mori, T., Okimoto, N., Sakai, A., Okazaki, Y., Nakura, N., Notomi, T., et al. (2003). Climbing exercise increases bone mass and trabecular bone turnover through transient regulation of marrow osteogenic and osteoclastogenic potentials in mice. *Journal of Bone and Mineral Research*, 18(11), 2002–2009. https://doi.org/10.1359/jbmr.2003.18.11.2002.

Morio, C., Lake, M. J., Gueguen, N., Rao, G., & Baly, L. (2009). The influence of footwear on foot motion during walking and running. *Journal of Biomechanics*, 42(13), 2081–2088. https://doi.org/10.1016/j.jbiomech.2009.06.015.

Muir, P., Johnson, K. A., & Ruaux-Mason, C. P. (1999). In vivo matrix microdamage in a naturally occurring canine fatigue fracture. *Bone*, 25(5), 571–576. https://doi.org/10.1016/s8756-3282(99)00205-7.

Nowlan, N. C., Jepsen, K. J., & Morgan, E. F. (2011). Smaller, weaker, and less stiff bones evolve from changes in subsistence strategy. *Osteoporosis International*, 22(6), 1967–1980. https://doi.org/10.1007/s00198-010-1390-3.

Nunamaker, D. M., Butterweck, D. M., & Provost, M. T. (1989). Some geometric properties of the third metacarpal bone: A comparison between the thoroughbred and standardbred racehorse. *Journal of Biomechanics*, 22(2), 129–134. https://doi.org/10.1016/0021-9290(89)90035-3.

Ogilvie, M. D., & Hilton, C. E. (2011). Cross-sectional geometry in the humeri of foragers and farmers from the prehispanic American southwest: Exploring patterns in the sexual division of labor. *American Journal of Physical Anthropology*, 144(1), 11–21. https://doi.org/10.1002/ajpa.21362.

Parfitt, A. M. (2002). Misconceptions (2): Turnover is always higher in cancellous than in cortical bone. *Bone*, 30(6), 807–809. https://doi.org/10.1016/S8756-3282(02)00735-4.

Parr, W. C. H., Chamoli, U., Jones, A., Walsh, W. R., & Wroe, S. (2013). Finite element micro-modelling of a human ankle bone reveals the importance of the trabecular network to mechanical performance: New methods for the generation and comparison of 3D models. *Journal of Biomechanics*, 46(1), 200–205. https://doi.org/10.1016/j.jbiomech.2012.11.011.

Pearson, O. M., Royer, D. F., Grine, F. E., & Fleagle, J. G. (2008). A description of the Omo I postcranial skeleton, including newly discovered fossils. *Journal of Human Evolution*, 55(3), 421–437. https://doi.org/10.1016/j.jhevol.2008.05.018.

Pontzer, H., Rolian, C., Rightmire, G. P., Jashashvili, T., Ponce de León, M. S., Lordkipanidze, D., et al. (2010). Locomotor anatomy and biomechanics of the Dmanisi hominins. *Journal of Human Evolution*, 58(6), 492–504. https://doi.org/10.1016/j.jhevol.2010.03.006.

Prang, T. C. (2015). Rearfoot posture of Australopithecus sediba and the evolution of the hominin longitudinal arch. *Scientific Reports*, 5(1), 17677. https://doi.org/10.1038/srep17677.

Prang, T. C. (2016). Reevaluating the functional implications of Australopithecus afarensis navicular morphology. *Journal of Human Evolution*, 97, 73–85. https://doi.org/10.1016/j.jhevol.2016.05.008.

Raichlen, D. A., Gordon, A. D., Foster, A. D., Webber, J. T., Sukhdeo, S. M., Scott, R. S., et al. (2015). An ontogenetic framework linking locomotion and trabecular bone architecture with applications for reconstructing hominin life history. *Journal of Human Evolution*, 81, 1–12. https://doi.org/10.1016/j.jhevol.2015.01.003.

Rein, T. R., Harrison, T., Carlson, K. J., & Harvati, K. (2017). Adaptation to suspensory locomotion in Australopithecus sediba. *Journal of Human Evolution*, 104, 1–12.

Rolian, C., Lieberman, D. E., Hamill, J., Scott, J. W., & Werbel, W. (2009). Walking, running and the evolution of short toes in humans. *Journal of Experimental Biology*, 212(5), 713–721. https://doi.org/10.1242/jeb.019885

Ruff, C. (1987). Sexual dimorphism in human lower limb bone structure: Relationship to subsistence strategy and sexual division of labor. *Journal of Human Evolution*, 16(5), 391–416. https://doi.org/10.1016/0047-2484(87)90069-8.

Ruff, C. B. (1999). Skeletal structure and behavioral patterns of prehistoric Great Basin populations. In *Prehistoric lifeways in the great basin wetlands: Bioarchaeological reconstruction and interpretation*. https://ci.nii.ac.jp/naid/10018120432/.

Ruff, C. B. (2007). Biomechanical analyses of archaeological human skeletons. In *Biological anthropology of the human skeleton* (pp. 183–206). John Wiley & Sons, Ltd. https://onlinelibrary.wiley.com/doi/abs/10.1002/9780470245842.ch6.

Ruff, C. B., & Hayes, W. C. (1983). Cross-sectional geometry of Pecos Pueblo femora and tibiae—A biomechanical investigation: II. Sex, age, and side differences. *American Journal of Physical Anthropology*, 60(3), 383–400. https://doi.org/10.1002/ajpa.1330600309.

Ruff, C. B., Holt, B., Niskanen, M., Sladek, V., Berner, M., Garofalo, E., et al. (2015). Gradual decline in mobility with the adoption of food production in Europe. *Proceedings of the National Academy of Sciences*, 112(23), 7147–7152. https://doi.org/10.1073/pnas.1502932112.

Ruff, C. B., Holt, B. M., Sládek, V., Berner, M., Murphy, W. A., Zur Nedden, D., et al. (2006). Body size, body proportions, and mobility in the Tyrolean "Iceman". *Journal of Human Evolution*, 51(1), 91–101. https://doi.org/10.1016/j.jhevol.2006.02.001.

Ruff, C. B., Larsen, C. S., & Hayes, W. C. (1984). Structural changes in the femur with the transition to agriculture on the Georgia coast. *American Journal of Physical Anthropology*, 64(2), 125–136. https://doi.org/10.1002/ajpa.1330640205.

Ruff, C., Strasser, E., Fleagle, J. G., Rosenberger, A. L., & McHenry, H. M. (1998). Evolution of the hominid hip. In *Primate locomotion: Recent advances* (pp. 449–469). Springer US. https://doi.org/10.1007/978-1-4899-0092-0_23.

Ruff, C. B., Trinkaus, E., Walker, A., & Larsen, C. S. (1993). Postcranial robusticity in Homo. I: Temporal trends and mechanical interpretation. *American Journal of Physical Anthropology*, 91(1), 21–53. https://doi.org/10.1002/ajpa.1330910103.

Ryan, T. M., & Shaw, C. N. (2015). Gracility of the modern *Homo sapiens* skeleton is the result of decreased biomechanical loading. *Proceedings of the National Academy of Sciences*, 112(2), 372–377. https://doi.org/10.1073/pnas.1418646112.

Saers, J. P. P., Cazorla-Bak, Y., Shaw, C. N., Stock, J. T., & Ryan, T. M. (2016). Trabecular bone structural varia-tion throughout the human lower limb. *Journal of Human Evolution, 97*, 97–108. https://doi.org/10.1016/j.jhevol.2016.05.012.

Sarmiento, E. E., & Marcus, L. F. (2000). The os navicular of humans, great apes, OH 8, Hadar, and Oreopithecus: Function, phylogeny, and multivariate analysis. *American Museum Novitates*, (3288). http://digitallibrary.amnh.org/handle/2246/2980.

Shaw, C. N., & Ryan, T. M. (2012). Does skeletal anatomy reflect adaptation to locomotor patterns? Cortical and trabecular architecture in human and nonhuman anthropoids. *American Journal of Physical Anthropology, 147*(2), 187–200. https://doi.org/10.1002/ajpa.21635.

Shaw, C. N., & Stock, J. T. (2013). Extreme mobility in the Late Pleistocene? Comparing limb biomechanics among fossil Homo, varsity athletes and Holocene foragers. *Journal of Human Evolution, 64*(4), 242–249. https://doi.org/10.1016/j.jhevol.2013.01.004.

Silva, M. J., Brodt, M. D., Lynch, M. A., Stephens, A. L., Wood, D. J., & Civitelli, R. (2012). Tibial loading increases osteogenic gene expression and cortical bone volume in mature and middle-aged mice. *PLoS One, 7*(4). https://doi.org/10.1371/journal.pone.0034980.

Sim-Fook, L., & Hodgson, A. (1958). A comparison of foot forms among the non-shoe and shoe-wearing Chinese population. *Journal of Bone and Joint Surgery, 40*(5), 1058–1062. http://insights.ovid.com.

Sinclair, K. D., Farnsworth, R. W., Pham, T. X., Knight, A. N., Bloebaum, R. D., & Skedros, J. G. (2013). The artio-dactyl calcaneus as a potential 'control bone' cautions against simple interpretations of trabecular bone adap-tation in the anthropoid femoral neck. *Journal of Human Evolution, 64*(5), 366–379. https://doi.org/10.1016/j.jhevol.2013.01.003.

Skedros, J. G., & Baucom, S. L. (2007). Mathematical analysis of trabecular 'trajectories' in apparent trajectorial struc-tures: The unfortunate historical emphasis on the human proximal femur. *Journal of Theoretical Biology, 244*(1), 15–45. https://doi.org/10.1016/j.jtbi.2006.06.029.

Skedros, J. G., Hunt, K. J., & Bloebaum, R. D. (2004). Relationships of loading history and structural and material characteristics of bone: Development of the mule deer calcaneus. *Journal of Morphology, 259*(3), 281–307. https://doi.org/10.1002/jmor.10167.

Skedros, J. G., Knight, A. N., Farnsworth, R. W., & Bloebaum, R. D. (2012). Do regional modifications in tissue min-eral content and microscopic mineralization heterogeneity adapt trabecular bone tracts for habitual bending? Analysis in the context of trabecular architecture of deer calcanei. *Journal of Anatomy, 220*(3), 242–255. https://doi.org/10.1111/j.1469-7580.2011.01470.x.

Skedros, J. G., Su, S. C., Knight, A. N., Bloebaum, R. D., & Bachus, K. N. (2019). Advancing the deer calcaneus model for bone adaptation studies: Ex vivo strains obtained after transecting the tension members suggest an unrecog-nized important role for shear strains. *Journal of Anatomy, 234*(1), 66–82. https://doi.org/10.1111/joa.12905.

Skerry, T. M., & Lanyon, L. E. (1995). Interruption of disuse by short duration walking exercise does not prevent bone loss in the sheep calcaneus. *Bone, 16*(2), 269–274. https://doi.org/10.1016/8756-3282(94)00039-3.

Sládek, V., Berner, M., & Sailer, R. (2006). Mobility in central European late Eneolithic and early bronze age: Femoral cross-sectional geometry. *American Journal of Physical Anthropology, 130*(3), 320–332. https://doi.org/10.1002/ajpa.20372.

Sparacello, V., & Marchi, D. (2008). Mobility and subsistence economy: A diachronic comparison between two groups settled in the same geographical area (Liguria, Italy). *American Journal of Physical Anthropology, 136*(4), 485–495. https://doi.org/10.1002/ajpa.20832.

Stern, J. T. (2000). Climbing to the top: A personal memoir of Australopithecus afarensis. *Evolutionary Anthropology: Issues, News, and Reviews, 9*(3), 113–133. https://doi.org/10.1002/1520-6505(2000)9:3<113::AID-EVAN2>3.0.CO;2-W.

Stern, J. T., & Susman, R. L. (1983). The locomotor anatomy of Australopithecus afarensis. *American Journal of Physical Anthropology, 60*(3), 279–317. https://doi.org/10.1002/ajpa.1330600302.

Stock, J. T. (2006). Hunter-gatherer postcranial robusticity relative to patterns of mobility, climatic adaptation, and selection for tissue economy. *American Journal of Physical Anthropology, 131*(2), 194–204. https://doi.org/10.1002/ajpa.20398.

Stock, J., & Pfeiffer, S. (2001). Linking structural variability in long bone diaphyses to habitual behaviors: Foragers from the southern African Later Stone Age and the Andaman Islands. *American Journal of Physical Anthropology, 115*(4), 337–348. https://doi.org/10.1002/ajpa.1090.

Su, A., & Carlson, K. J. (2017). Comparative analysis of trabecular bone structure and orientation in South African hominin tali. *Journal of Human Evolution, 106*, 1–18. https://doi.org/10.1016/j.jhevol.2016.12.006.

Su, A., Wallace, I. J., & Nakatsukasa, M. (2013). Trabecular bone anisotropy and orientation in an Early Pleistocene hominin talus from East Turkana, Kenya. *Journal of Human Evolution, 64*(6), 667–677. https://doi.org/10.1016/j.jhevol.2013.03.003.

Swift, J. M., Gasier, H. G., Swift, S. N., Wiggs, M. P., Hogan, H. A., Fluckey, J. D., et al. (2010). Increased training loads do not magnify cancellous bone gains with rodent jump resistance exercise. *Journal of Applied Physiology, 109*(6), 1600–1607. https://doi.org/10.1152/japplphysiol.00596.2010.

Sylvester, A. D., & Terhune, C. E. (2017). Trabecular mapping: Leveraging geometric morphometrics for analyses of trabecular structure. *American Journal of Physical Anthropology, 163*(3), 553–569. https://doi.org/10.1002/ajpa.23231.

Tanck, E., Homminga, J., van Lenthe, G. H., & Huiskes, R. (2001). Increase in bone volume fraction precedes architectural adaptation in growing bone. *Bone, 28*(6), 650–654. https://doi.org/10.1016/S8756-3282(01)00464-1.

Trinkaus, E. (2005). Anatomical evidence for the antiquity of human footwear use. *Journal of Archaeological Science, 32*(10), 1515–1526. https://doi.org/10.1016/j.jas.2005.04.006.

Trinkaus, E., & Shang, H. (2008). Anatomical evidence for the antiquity of human footwear: Tianyuan and Sunghir. *Journal of Archaeological Science, 35*(7), 1928–1933. https://doi.org/10.1016/j.jas.2007.12.002.

Turner, C. H., Akhter, M. P., Raab, D. M., Kimmel, D. B., & Recker, R. R. (1991). A noninvasive, in vivo model for studying strain adaptive bone modeling. *Bone, 12*(2), 73–79. https://doi.org/10.1016/8756-3282(91)90003-2.

Turner, C. H., Hsieh, Y.-F., Müller, R., Bouxsein, M. L., Baylink, D. J., Rosen, C. J., et al. (2000). Genetic regulation of cortical and trabecular bone strength and microstructure in inbred strains of mice. *Journal of Bone and Mineral Research, 15*(6), 1126–1131. https://doi.org/10.1359/jbmr.2000.15.6.1126.

Ulrich, D., van Rietbergen, B., Laib, A., & Rüegsegger, P. (1999). The ability of three-dimensional structural indices to reflect mechanical aspects of trabecular bone. *Bone, 25*(1), 55–60. https://doi.org/10.1016/S8756-3282(99)00098-8.

Wallace, I. J., Demes, B., Mongle, C., Pearson, O. M., Polk, J. D., & Lieberman, D. E. (2014). Exercise-induced bone formation is poorly linked to local strain magnitude in the sheep tibia. *PLoS One, 9*(6). https://doi.org/10.1371/journal.pone.0099108.

Wallace, I. J., Worthington, S., Felson, D. T., Jurmain, R. D., Wren, K. T., Maijanen, H., et al. (2017). Knee osteoarthritis has doubled in prevalence since the mid-20th century. *Proceedings of the National Academy of Sciences, 114*(35), 9332–9336. https://doi.org/10.1073/pnas.1703856114.

Wallace, J. M., Rajachar, R. M., Allen, M. R., Bloomfield, S. A., Robey, P. G., Young, M. F., et al. (2007). Exercise-induced changes in the cortical bone of growing mice are bone- and gender-specific. *Bone, 40*, 1120–1127.

Ward, C. V., Kimbel, W. H., & Johanson, D. C. (2011). Complete fourth metatarsal and arches in the foot of Australopithecus afarensis. *Science, 331*, 750–753. https://doi.org/10.1126/science.1201463.

Ward, C. V., Leakey, M. G., Brown, B., Brown, F., Harris, J., & Walker, A. (1999). South Turkwel: A new Pliocene hominid site in Kenya. *Journal of Human Evolution, 36*(1), 69–95. https://doi.org/10.1006/jhev.1998.0262.

Ward, C. V., Leakey, M. G., & Walker, A. (2001). Morphology of Australopithecus anamensis from Kanapoi and Allia Bay, Kenya. *Journal of Human Evolution, 41*(4), 255–368. https://doi.org/10.1006/jhev.2001.0507.

Weatherholt, A. M., Fuchs, R. K., & Warden, S. J. (2013). Cortical and trabecular bone adaptation to incremental load magnitudes using the mouse tibial axial compression loading model. *Bone, 52*(1), 372–379. https://doi.org/10.1016/j.bone.2012.10.026.

White, T. D., Asfaw, B., Beyene, Y., Haile-Selassie, Y., Lovejoy, C. O., Suwa, G., et al. (2009). Ardipithecus ramidus and the paleobiology of early hominids. *Science, 326*(5949), 64–86. https://doi.org/10.1126/science.1175802.

Willems, C., Stassijns, G., Cornelis, W., & D'Août, K. (2017). Biomechanical implications of walking with indigenous footwear. *American Journal of Physical Anthropology, 162*(4), 782–793. https://doi.org/10.1002/ajpa.23169.

Yang, H., Embry, R. E., & Main, R. P. (2017). Effects of loading duration and short rest insertion on cancellous and cortical bone adaptation in the mouse tibia. *PLoS One, 12*(1). https://doi.org/10.1371/journal.pone.0169519, e0169519.

Yang, H., Xu, X., Bullock, W., & Main, R. P. (2019). Adaptive changes in micromechanical environments of cancellous and cortical bone in response to in vivo loading and disuse. *Journal of Biomechanics, 89*, 85–94. https://doi.org/10.1016/j.jbiomech.2019.04.021.

Young, D., Richardson, D., Markel, M., & Nunamaker, D. (1991). Mechanical and morphometric analysis of the third carpal bone of thoroughbreds. *American Journal of Veterinary Research, 52*(3), 402–409. http://europepmc.org/abstract/MED/2035913.

Zeininger, A., Patel, B. A., Zipfel, B., & Carlson, K. J. (2016). Trabecular architecture in the StW 352 fossil hominin calcaneus. *Journal of Human Evolution, 97*, 145–158. https://doi.org/10.1016/j.jhevol.2016.05.009.

Zipfel, B., & Berger, L. R. (2007). Shod versus unshod: The emergence of forefoot pathology in modern humans? *The Foot, 17*(4), 205–213. https://doi.org/10.1016/j.foot.2007.06.002.

Zipfel, B., DeSilva, J. M., & Kidd, R. S. (2009). Earliest complete hominin fifth metatarsal—Implications for the evolution of the lateral column of the foot. *American Journal of Physical Anthropology, 140*(3), 532–545. https://doi.org/10.1002/ajpa.21103.

Zipfel, B., DeSilva, J. M., Kidd, R. S., Carlson, K. J., Churchill, S. E., & Berger, L. R. (2011). The foot and ankle of *Australopithecus sediba*. *Science, 333*(6048), 1417–1420. https://doi.org/10.1126/science.1202703.

Injury, disease, and recovery: Skeletal adaptations to immobility and impairment

Rebecca J. Gilmour[a], Liina Mansukoski[b], and Sarah Schrader[c]

[a]Department of Sociology & Anthropology, Mount Royal University, Calgary, AB, Canada,
[b]Department of Health Sciences, University of York, Heslington, York, United Kingdom,
[c]Faculty of Archaeology, Leiden University, Leiden, Netherlands

11.1 Introduction

The human experience of disease, illness, or injury, along with other socially mandated changes in the mechanical environment (e.g. space flight, bed rest), can lead to reduced loading forces on bone (Giangregorio and McCartney, 2006; Kulkarni et al., 1998; Lang et al., 2004; Leblanc et al., 1990; Morbeck et al., 1991). As discussed throughout this book, bone responds and adapts to mechanical stimuli in predictable ways that can be understood and interpreted by applying mechanical engineering principles to biological tissues (see Chapter 2 for more detailed information). Just as bone can be deposited and shaped in response to increased mechanical loading stimuli, it can also be resorbed in the absence of, or due to, decreased strain (i.e. mechanical unloading). Disuse and immobility leads to mechanical unloading, which first weakens and atrophies muscle tissues. Consequently, muscular contractions and gravitational and ground forces acting on the bone are also diminished. Much like how increased stresses on bone can lead to apposition and shape changes, reduced deformation forces acting on the skeletal tissue will also trigger bone remodelling. This leads to bone resorption and potentially to the development of osteopenia and/or osteoporosis, which, in turn, can increase the risk of future fracture (Alexandre and Vico, 2011; Bloomfield, 2010; Burr, 1997; Osipov et al., 2018). Circumstances during disuse can therefore result in both localised and systemic atrophic changes to bone density, thickness, shape, and histological structure.

Behaviour in our Bones
https://doi.org/10.1016/B978-0-12-821383-4.00002-4

In bioarchaeology, there is a great need to recognise the true and diverse range of human functional abilities to accurately interpret individual lived experiences and group lifeways in the past. In order to understand the breadth of human behaviours, activities, and evolutionary histories, researchers must consider the full range of variation in all individuals, healthy and with pathology alike (Gilmour and Plomp, 2022). To this end, this chapter explores how bone adapts and changes in response to pathologically altered loading environments, highlighting how various methods are employed in palaeopathological studies to make interpretations about behaviour and recovery after illness and injury.

11.2 Progression of disuse bone loss

Many studies of how bone adapts to unloaded environments involve animal subjects (e.g. rats, dogs, bears, sheep, non-human primates; see Jee and Ma, 1999; McGee et al., 2008; Thomas et al., 1996; Uhthoff and Jaworski, 1978; Weinreb et al., 1989; Young et al., 1986). Some influential research in this area was conducted on beagle limbs by Uhthoff and Jaworski (1978) and Jaworski et al. (1980) who proposed three main stages in the process of bone adaptation to immobilisation: an initial rapid bone loss phase (Phase I: Rapid Loss) followed by a longer and slower phase of loss (Phase II: Slow and Prolonged Loss), the final stage is an inactive stage where bone becomes stable, but the lost bone persists (Phase III: Steady State) (Fig. 11.1). The progression of disuse-related bone loss through these general phases has been documented in humans related to causes such as spinal cord injury, bed rest, stroke, and time spent in microgravity (Frotzler et al., 2008; Garland et al., 1992; Jørgensen and Jacobsen, 2001; Lau and Guo, 2011; Uebelhart et al., 1995; Vico et al., 2000). This section explains these three stages of bone loss and outlines the important and distinguishing features that are recognisable in skeletal material.

11.2.1 Phase I: Rapid loss

With the onset of unloading, bone immediately begins to adapt to the new mechanical environment and disuse. Skeletal changes during this acute period can progress rapidly for approximately six weeks (Lau and Guo, 2011; Schäfer et al., 2012; Uhthoff and Jaworski, 1978). Trabecular bone is highly sensitive to mechanical loading and is often the first to respond to disuse, and structural and microstructural trabecular bone changes are evident after approximately two weeks (Coupaud et al., 2015; Jaworski et al., 1980; Jiang et al., 2006; Lau and Guo, 2011; Schäfer et al., 2012). As the trabecular bone is resorbed, the struts start to thin and eventually disconnect from each other completely, resulting in bone that is less resistant to applied strains (Guo and Kim, 2002).

Cortical bone is typically described as less involved in this initial stage of rapid bone loss. However, research by Rittweger et al. reports that in the first five weeks of immobilisation, the tibial subendocortical layer is "readily transformed into trabecular bone" (Rittweger et al., 2009, p. 618). In this study, the amount of bone lost at this transitional cortical–trabecular zone differed within a single tibia, with the greatest bone loss occurring in regions with the most endocortical surface area, such as in the epiphyses. This finding is

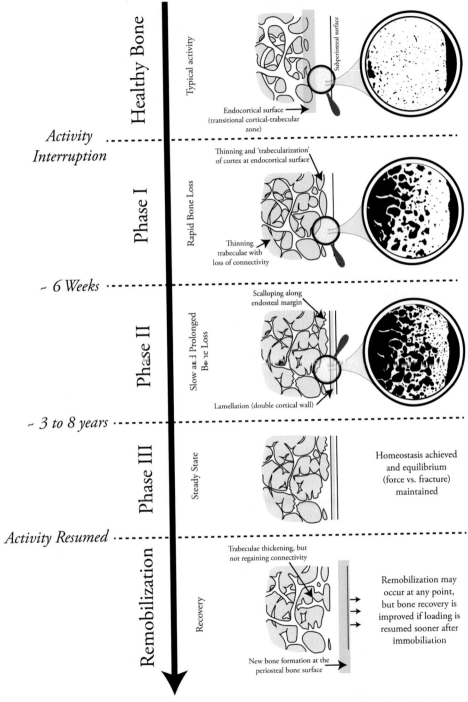

FIG. 11.1 Stages of bone loss. Progression and stages of bone loss from healthy bone, through periods of immobilisation and unloading, to a state of potential remobilisation and recovery. Time ranges are approximate and vary according to a number of factors. *Image courtesy of Rebecca J. Gilmour.*

supported by others, such as Schäfer et al. (2012), who studied hands 30 days after a finger amputation. They reported significantly decreased metacarpal cortical thicknesses and bone mineral densities (BMD), a measure that reflects the mass of mineral in a volume of bone and is indicative of bone quantity, quality, and strength (Tabensky et al., 1996). Complementing these findings, Kazakia et al. (2014) report that in addition to the expected trabecular bone changes, cortical bone pore spaces also increase significantly during a six-week non-weight-bearing period. Using high-resolution 3D methods, these studies affirm that while trabecular bone changes are rapid and marked during the acute unloading period, cortical bone microstructure is also affected during this period of immobilisation. There are still unanswered questions regarding the acute and rapid loss of bone in the first six weeks of immobilisation, but it is clear that a reduction and reorganisation of existing trabecular structures, as well as trabecularisation and increasing porosity of cortical bone, characterise this initial stage of mechanical unloading.

11.2.2 Phase II: Slow and prolonged loss

The next phase of disuse is characterised by slower bone loss that more clearly involves cortical bone changes. While trabecular changes continue through this second stage, cortical bone changes become more readily observable, with some studies reporting radiographically identifiable alterations as early as eight weeks into immobilisation (Arnstein, 1972). Specifically, these radiographically observable changes consist of uniform osteoporosis, spotty osteoporosis, subchondral/metaphyseal radiolucent bands, and lamellation or scalloping at cortical margins (Jones, 1969). These changes are related to the continuing increase in cortical bone porosity and decrease in BMD (Lang et al., 2004; Minaire, 1989; Schlecht, 2012). The trabecularisation of subendocortical bone reported by Rittweger et al. (2009) likely continues into this phase, as cortical bone is noted to thin primarily at the endosteal surface in instances of disuse bone loss (Eser et al., 2004; Lang et al., 2004). This said, the study by Schäfer et al. (2012) suggested that, in addition to endosteal bone loss, the periosteal surface was also reduced following amputation. Presumably this can be attributed to absent or markedly reduced voluntary muscle forces needed to signal periosteal bone maintenance. The duration of this second stage varies and lasts until the bone reaches the next and final phase of disuse loss, referred to by Minaire (1989) as the 'Inactive' Phase, and by others as a stable or steady state (Garland et al., 1992).

11.2.3 Phase III: Steady state

If immobilisation and disuse continue, bone loss eventually slows and stabilises. This is the final phase of bone loss. The onset timing of this steady state varies; Uhthoff and Jaworski's (1978) study of beagles suggested that bone loss began to stabilise at approximately 40 weeks (~ 10 months) after immobility, whereas an early synthesis by Minaire (1989) suggested the steady state begins after five to six months (~ 20–24 weeks). Garland et al.'s (1992) study of knees in spinal cord injury patients found that an equilibrium and steady state was established 16 months post-injury; this was characterised as the point when these paralysed patients reached a BMD that did not differ significantly from the BMD of chronic patients who had been paralysed for 10 years. Others report that bone mineral content of spinal cord injury patients reached a steady state in the proximal tibia within two years; however, bone loss in the femur did not reach equilibrium within the four years they were studied (Biering-Sørensen et al., 1990). More recently, it has been suggested that this steady state is reached three to eight years after spinal cord injury, depending on the bone parameter considered (Eser et al., 2004; Frotzler et al., 2008). Once this steady state is reached, changes in femoral and tibial bone properties are minor, with most tibiae showing no detectable bone changes at all (Frotzler et al., 2008). Based on the results of these more recent studies, it seems most likely that this steady state of equilibrium is not reached until at least a year, and likely longer, following the onset of disuse.

Clearly, the timeline in which bone loss reaches homeostasis varies based on factors such as the skeletal element and location considered, the methods used, and the parameter measured (Frotzler et al., 2008). Many suggest that time is the greatest predictor of the amount of bone loss following immobilisation and unloading; the longer an individual remains unloaded, the more bone that will be lost before equilibrium is reached (Sievänen, 2010). However, bone loss does eventually cease when each element/location reaches the minimum level necessary to support function without failing (Minaire, 1989; Skerry, 2008). The link between homeostasis and fracture resistance is supported by evidence from spinal cord injury patients whose BMD reached homeostasis close to their estimated fracture threshold of $1.0\,\text{g/cm}^2$ (Garland et al., 1992).

The magnitude of disuse bone loss will vary between individuals based on factors such as diet, hormones, and activity, as well as the condition implicated in the immobilisation, and the specific element and location of the injury (Sievänen, 2010). While it is not possible to state an exact amount of bone loss expected following immobilisation and disuse, some studies of paralysis following stroke and spinal cord injury report BMD decreases of up to 25% in the affected elements (Jørgensen and Jacobsen, 2001). BMD reductions between 7% and 50% have also been recorded in the bone segments and elements distal to tibial fracture sites (Eyres and Kanis, 1995). Bone loss can also be concentrated in different locations within a single element. For example, the greatest losses of tibial bone mass are reported in the epiphyses of spinal cord injury patients, where bone mass was reduced by up to 50% in epiphyses, compared to only 30% in diaphyses (Rittweger et al., 2009). Coupaud et al. (2015) also found that femoral and tibial cortical bone BMD decreased by 2%–3% per year and trabecular bone by 15%–20% following a spinal cord injury. Importantly, these authors note that although the percent change in cortical bone may appear relatively small, cortical bone comprises approximately 80% of the peripheral skeleton's volume (Coupaud et al., 2015). This means that even small cortical changes have the potential to greatly impact the skeleton's composition and fragility, especially when they are coupled with the higher magnitude trabecular bone loss.

11.2.4 Distribution of bone loss

The distribution of bone loss throughout the skeleton varies based on the condition that caused the mechanical unloading. In instances of bed rest or paralysis, bone loss can be systemic and diffuse, affecting all elements except the skull, which is not greatly affected by changes to gravitational or muscular loading forces (Garland et al., 1992). However, even in these circumstances, the most significant amount of bone loss frequently occurs in the lower limb, specifically at the metaphyseal and epiphyseal regions of the distal femur and proximal tibia (Jiang et al., 2006). This can be explained by the large reduction in forces associated with weight bearing and gravity, as well as the concentration of trabeculae in these regions. As long as some upper body mobility is maintained, arm bones will routinely demonstrate little to no significant bone loss following spinal cord injury or even as a result of space travel (Giangregorio and Blimkie, 2002). In cases of fracture or stroke, it would be appropriate to expect some degree of asymmetry to develop between affected and unaffected limbs. Initially, following an injury, systemic bone loss may initially occur due to a period of bed rest. However, much greater amounts of bone will eventually be lost from isolated disuse of the affected limb (Osipov et al., 2018). In these instances of disuse hypotrophy associated with fractures, the area affected tends to be isolated to the injured limb and, more specifically, usually only affects the fracture site and the bone distal to the fracture site (Eyres and Kanis, 1995; Quek and Peh, 2002).

11.3 Asymmetry

In scenarios where immobilisation only impacts one limb, asymmetries can develop in the skeleton and the surrounding soft tissue. Various types of asymmetries exist that can be identified through the analysis of skeletal remains. Fluctuating asymmetry is related to underlying stresses endured during development, whereas directional asymmetry is better associated with forces from the mechanical environment (Lazenby, 2002). Lower limb bones, such as tibiae, typically have symmetrical antimeres (opposing sides) due to loading forces that are evenly applied to the legs during ambulation (walking) (Auerbach and Ruff, 2006). If lower limb asymmetry is present, it can sometimes be explained by uneven habitual activity that applies unequal loading forces (e.g. differences in the kicking versus support legs of American football players) (Hart et al., 2016; Weatherholt and Warden, 2016). Evaluation of asymmetry related to laterality (e.g. handedness and activity) and pathological conditions (e.g. trauma, amputation, and poliomyelitis) can provide anthropologists with insight into aspects of human behaviour, activity, and health. This section introduces these various types of human asymmetries, examines their aetiologies, and discusses the development of specific identifiable characteristics in bone such as hypertrophy or atrophy of tissue.

Fluctuating asymmetry has no apparent measure of heritability and is defined as small, random deviations from bilateral symmetry (Møller and Swaddle, 1997; Palmer and Strobeck, 1986; Zaidi, 2011). This type of asymmetry is thought to be associated with instabilities caused by environmental, physical, and nutritional pressures, essentially arising due to an individual's inability to adjust to stress during development (Barrett et al., 2012; Møller and Swaddle, 1997; Zaidi, 2011). Anthropological studies of fluctuating asymmetry often consider dental evidence, such as asymmetric discrepancies in tooth size (see Barrett et al., 2012), but fluctuating asymmetry has also been investigated using human long bone lengths, palm and finger ridges, and ear lobe lengths (Palmer and Strobeck, 1986). The randomised nature of fluctuating asymmetry means that the asymmetric variations tend to be normally distributed around a mean of zero (Møller and Swaddle, 1997; Palmer and Strobeck, 1986).

This suggests that at the population level, any asymmetric discrepancies that fluctuating asymmetry causes will not greatly impact assessments of the magnitude of directional asymmetry associated with bone functional adaptation (Møller and Swaddle, 1997; Palmer and Strobeck, 1986; Zaidi, 2011).

Directional asymmetry is primarily associated with skeletal adaptations to applied stresses and strains from function and activity (Palmer and Strobeck, 1986; Zaidi, 2011). Varying patterns of loading strains can result in different adaptations between right and left sides, allowing directional asymmetry to manifest either through hypertrophy (enlargement) or by atrophy (wasting) of bone tissue (Auerbach and Ruff, 2006). In anthropological research, investigations of directional asymmetry contribute to the understanding of human behaviour and activity, including laterality and mobility (for more details, see Auerbach and Ruff, 2006; Biewener and Bertram, 1993; Henderson et al., 2018; Marchi et al., 2006; Milella et al., 2015; Reichel et al., 1990; Rhodes and Knusel, 2005; Shaw, 2011; Stirland, 1993). Side-biassed mechanical adaptations associated with increased magnitudes and frequencies of mechanical strains are more prominent in the upper limbs than the lower limbs (Auerbach and Ruff, 2006; Milella et al., 2012; Ruff and Jones, 1981). The mechanical strains acting on the lower limb are more uniform and primarily involve equal stresses associated with locomotion and balance (Acosta et al., 2017; Auerbach and Ruff, 2006; Campanacho and Santos, 2013; Shaw, 2011). Consequently, the lower degrees of asymmetry apparent in legs may help recognise any asymmetry related to pathological causes. In comparison, healthy upper limbs can be highly lateralised and typically demonstrate some degree of expected asymmetry. This means that asymmetries associated with upper limb pathological conditions may be obscured by the typical asymmetries associated with handedness, for example. Therefore asymmetries related to pathological conditions may be more readily recognised in the lower than the upper limbs.

Limb directional asymmetry can also develop consequent to pathological conditions, such as infectious disease (e.g. poliomyelitis; Morbeck et al., 1991; Thompson, 2012) and trauma (e.g. fractures and amputations; Eyres and Kanis, 1995; Kulkarni et al., 1998; Lazenby and Pfeiffer, 1993; Sevastikoglou et al., 1969; Sherk et al., 2008). Injury and some cases of paralysis (e.g. stroke and poliomyelitis) commonly occur unilaterally, meaning that only one side is likely to experience immobilisation and unloading resulting in bone loss. The inability to use a limb may lead to accentuated compensatory activity by the opposing limb and/or other supporting tissues, resulting in the possible hypertrophy of the compensating side (Becker et al., 1995; Segal et al., 2014; Wang et al., 2010). Through the combination of bone atrophy consequent to immobilisation, and hypertrophy through coping and compensatory response, limbs of affected individuals may develop larger than expected amounts of directional/bilateral asymmetry.

Measurements of skeletal directional asymmetry preserved in bone lengths, widths, thicknesses, and circumferences are characterised and compared using various equations. Asymmetry differences can be standardised using percentages to identify side bias and facilitate comparisons of directional asymmetry among individuals (Eq. 11.1) (Auerbach and Ruff, 2006). As the directional asymmetry equation indicates right/left side biases, another calculation for per cent absolute asymmetries (%AA) is used to evaluate the overall magnitude of asymmetry independent of sided influences (Eq. 11.2).

Equation 11.1 Percentage of directional asymmetry (%DA)

This method preserves any side bias present and will express elements with larger left values as a negative number on one side of zero and elements with larger right values as positives on the other side (Auerbach and Ruff, 2006).

$$\%DA = \frac{(\text{right} - \text{left})}{\text{average of left and right}} \times 100 \tag{11.1}$$

Equation 11.2 Percentage of absolute asymmetry (%AA)

This equation allows for the comparison of overall asymmetry present without the influence of side (Auerbach and Ruff, 2006).

$$\%AA = \frac{(\text{maximum} - \text{minimum})}{\text{average of maximum and } \textit{minimum}} \times 100 \tag{11.2}$$

These calculations are simple ratios based on side differences compared to total averages, and so they do not always fully account for other complex factors that might impact results and interpretations. It is typically good practice to correct cross-sectional measures using measures of body size, such as body mass and bone length (Ruff, 2008). By correcting the cross-sectional measures by these attributes, more meaningful comparisons among individuals of the same and different groups can be conducted to ensure the results reflect bone shape and robusticity and not just the overall size of the person.

11.4 Recovery following disuse

There is dispute about the potential for bone to recover following a period of disuse. Some studies indicate full recovery, while others report permanent skeletal tissue losses. Within these reports, two factors seem to greatly influence an individual's ability to recover lost bone: their age and the length of time that they were immobilised. Younger individuals recover bone more easily than older individuals (Ceroni et al., 2013). This age-related pattern of bone response to reloading following a period of disuse has been documented in rats; the older subjects had an impaired ability to recover lost bone and often retained long-term deficits in bone mass (Cunningham et al., 2018). These life course differences in recovery can be explained by age-related changes in bone cell activity. Specifically, younger individuals have relatively balanced osteoblast and osteoclast activity, as bone is removed, the bone is replaced. As individuals age, the bone cell activity becomes increasingly imbalanced, resulting in reduced osteoblast and increased osteoclast activity (Chan and Duque, 2002; Lau and Guo, 2011). Numerous studies on age-related bone loss and osteoporosis discuss these cellular factors in great depth, principles that can also be applied to recovery from unloading or disuse osteoporosis (Eastell et al., 2016; Farr and Khosla, 2019; Titorenko, 2018).

In addition to age, one of the greatest factors influencing bone recovery is the total duration of disuse; longer periods of unloading are associated with less favourable recovery outcomes (Sievänen, 2010). Early research by Jaworski and Uhthoff (1986) and Minaire (1989) suggested that bone recovered best if an individual was able to resume function within the active stages of bone loss, that is within approximately the first six months (Phases I and II). More recent research supports this assertion, reporting that individuals who resume activity after a relatively short period of unloading recover from disuse bone atrophy, with bone properties returning to near-original levels. For example, Sibonga et al. (2007) found that 50% of the bone that astronauts lost during unloading periods of up to approximately four months could be restored within an average period of nine months. Jørgensen and Jacobsen (2001) describe a similar outcome in their study of the humeri of stroke patients, while also reporting on the importance of functional ability in recovery. They found that greater functional abilities following the stroke helped mitigate bone losses on the paretic side; functional outcomes predicted BMD changes and the individuals with the least impairment also exhibited the least bone loss (Jørgensen and Jacobsen, 2001). This demonstrates that in addition to age and duration of disuse, a person's potential for, and preservation of, function is also influential on their ability to recover lost bone.

It is possible to recover some bone following an injury or immobilisation event, but complete and total bone restoration following atrophy is not guaranteed. Even if some activity is eventually resumed, many studies report that some levels of disuse bone loss may persist for months and years after the immobilisation event (Schäfer et al., 2012; Takata and Yasui, 2001). This persistent bone loss is especially true for individuals who experience spinal cord injury or another type of trauma, stroke, or illness that involves sustained paralysis and/or immobilisation. The lack of mechanical forces acting on the bone inhibits the strains needed to signal the remodelling responses of bone cells (mechanostat theory, see Chapter 2). If activity is resumed following disuse, it does not always directly equate with skeletal response and recovery. Some researchers suggest that sometimes, even if normal function is resumed, bone loss may persist indefinitely (Eyres and Kanis, 1995). Although some bone may be regained following the resumption of activity, it is possible that in these cases, the bone may never return to its original levels or properties. Longitudinal studies exist that report bone equilibrium and loss in the years following chronic paralysis, but similar studies that quantify bone recovery in the years following immobilisation are less forthcoming and therefore represent an important area for future research.

When bone recovers following an unloading event, clear differences in recovery pattern according to region and the measured bone property emerge. In a study of tibiae after 90-day bed rest, Rittweger and Felsenberg (2009) reported that diaphyseal bone mineral content had fully recovered 12 months after bed rest ended, but that significant, albeit small, amounts of loss persisted in epiphyses following reloading. These different patterns in regional recovery likely reflect the predominant type of bone present in each area (trabecular or cortical). Epiphyses have proportionally more trabecular bone than diaphyses, meaning it is likely that the persistent loss reported in the epiphyseal region reflects unrecovered trabecular bone. This finding emphasises the important differences in how trabecular and cortical bone respond to resumed loading. While it is possible to recover cortical bone thickness, once trabecular connections are lost, they are typically not regenerated (Ozcivici and Judex, 2014). However, this does not mean that trabecular bone is permanently compromised. If loading is resumed after immobilisation, trabecular bone can restore its strength and compensate for the lost connectivity by increasing the thicknesses of the remaining trabeculae (Ozcivici and Judex, 2014). This concept is reported in a study on remobilisation following disuse in dogs (Li et al., 2003). In this work, the authors found that although trabecular bone mass recovered following disuse (due to thickening trabeculae), the architectural changes (lost connectivity) were not reversible. As such, it is possible that persistent epiphyseal bone loss, such as that noted by Rittweger and Felsenberg (2009), represents these important differences in cortical and trabecular bone recovery following atrophy.

In addition to differences in trabecular and cortical bone recovery, there are also differences evident in cortical bone geometric response to reloading. Ageing and osteoporosis literature has noted that cortical bone is primarily lost from the endosteal surface but, as a mechanism to protect against fracture, bone is then laid down on the periosteal surface, albeit at a slower rate (Lazenby, 1990). This process of bone recovery from immobilisation is clearly indicated in a study of non-human primates that outlines compensatory periosteal apposition as a part of recovery (Young et al., 1986). Additionally, in their study of bone recovery following space travel, Lang et al. (2006) report that recovery of bone mineral content following re-ambulation was explained by increases in bone volume and cross-sectional area. In this study, periosteal apposition was implicated as a compensatory response following bone loss that had primarily occurred at the endosteal margin. Essentially, this pattern of cortical bone resorption and apposition involves expansion of the medullary canal during active bone loss, followed by expansion of the outer bone width via subperiosteal apposition once remobilisation occurs. This demonstrates that bone mineral density and content measures may initially indicate a full recovery, but that the bone dimensions may have measurably changed (Lang et al., 2006). As such, it is important to also consider bone geometric properties in studies of disuse and reloading. Furthermore, different causes for immobilisation may yield differing bone responses. As new bone formation is, in part, a response to mechanically applied forces and voluntary muscle forces, periosteal apposition will likely only occur in individuals with some retained functional ability in the affected region. Individuals experiencing chronic paralysis and/or immobilisation will typically not experience the forces necessary to initiate periosteal remodelling responses. This is evidenced in studies of spinal cord injury and amputation patients who frequently exhibit resorption at *both* the periosteal and endosteal bone surfaces (Schäfer et al., 2012).

11.5 Bioarchaeological studies

Investigations into the functional experiences and recovery from past pathological conditions are not new. Some biological anthropologists have combined functional morphology and biomechanical analyses to understand physical deficits in the past (see Gilmour and Plomp, 2022 for a systematic review of related literature on shape analyses in palaeopathology). Some of these studies have even uncovered bone loss and altered cross-sectional properties suggestive of sustained unloading consequent to pathology (e.g. Sparacello et al., 2016). This section reviews the bioarchaeological evidence for bone loss secondary to pathology, specifically highlighting systemic bone loss associated with bed rest and frailty, and investigating evidence for impairment following asymmetric conditions such as injury. Additionally, as not all pathological conditions are actually debilitating, this section also discusses the evidence for recovery and adaptation in archaeological collections.

11.5.1 Bed rest and frailty: Diffuse/systemic bone loss

Analyses of skeletal properties in people after prolonged periods of inactivity following an illness have been used to help assess how disuse manifests in bone. These natural experiments are valuable in estimating the extent that a given length and severity of immobility can be expected to result in measurable changes in bone. The literature mostly consists of modern medical studies, but some work has focused on historical and archaeological samples known to have endured conditions such as spinal cord injury or communicable diseases that may have resulted in reduced mobility and diffuse bone loss due to extended periods of bed rest. The following section will briefly summarise evidence of systemic bone loss in individuals who are thought or known to have experienced long periods of relative immobility, or lack of bone and muscle function, as well as the measures and methods that were used to quantify bone loss and, in the case of bioarchaeological studies, to determine the condition responsible for the skeletal changes.

In bioarchaeological populations, evidence of paralysis and related severe impairment is scarce, but existing case studies describe changes to bone similar to those expected in the clinical literature. So far, one of the earliest known individuals with paralysis is an adult male from the Neolithic Man Bac site in Vietnam, who is thought to have had congenital fusion of the spine in childhood (Oxenham et al., 2009). The individual, who lived sometime before 3500 BP, had significantly reduced diaphyseal diameters in all preserved long bones compared to other adults from the same site. Multiple bones in his upper limbs showed signs of osteoarthritis, indicating that some upper body movement had been possible. The spinal fusion is thought to have occurred before adolescence, as post-adolescence illness would more likely have resulted in endosteal absorption and enlarged medullary cavity, instead of the observed smaller diaphyseal diameter (Oxenham et al., 2009). Whatever the precise cause of his impairment, the authors believe that this individual lived approximately 10 years with his condition and received significant care and assistance from members of his community (Tilley and Oxenham, 2011).

Commonly, the timing or precise nature of the illness or injury cannot be determined in the bioarchaeological context and differential diagnosis remains the only option. In paralysed individuals, potential disorders can include cerebral palsy, poliomyelitis, spinal cord injury, and encephalitis (Novak et al., 2014). To gather further information to confirm a diagnosis, there has been some effort to supplement bioarchaeological investigations of paralysis with biogeochemical evidence. For instance, stable isotope analysis has been used to investigate the life course of a Bronze Age female with paraplegia from the Tomb of Tell Abraq in the United Arab Emirates (Schrenk et al., 2016). Before this investigation, both cerebral palsy and paralytic poliomyelitis were considered possible diagnoses due to the individual's gracile femora, tibiae, and fibula combined with notable limb asymmetry. However, as the isotope analysis revealed the individual was a recent migrant to the area, it was thought unlikely that she could have travelled with a severe case of cerebral palsy, even if assisted. Further, being new to the area, she may have experienced increased immunological risk to develop paralytic poliomyelitis, which is more likely to manifest in the paralytic form, especially when contracted later in life (Schrenk et al., 2016).

Once a likely condition or diagnosis has been established, bioarchaeologists often try to construct a timeline of events that may have resulted in the presumed reduction in physical activity and subsequent bone loss. Macroscopic analysis of entheseal changes is one such method used by bioarchaeologists to describe activity changes (Schrader and Buzon, 2017). For instance, the remains of an adult male from New Mexico dating to the Late Period of Indigenous occupation at the Gran Quivira site (1550–1672 CE) suggest that the individual had juvenile chronic arthritis at multiple skeletal locations that began in childhood and lasted into adulthood (Hawkey, 1998). Juvenile chronic arthritis is a debilitating condition that begins in childhood and is indicated by various skeletal expressions including, for example, premature fusion of the epiphyses, deformity of mandibular condyles, and symmetrical involvement of more than four joints (for a full description, please see Hawkey, 1998). The author concludes that the entheseal changes identified on the bones of this individual contrast sharply with that of other group members and broadly support joint mobility estimates indicating that by young adulthood, the individual had lost mobility in their knees and ankles. Specifically, neither leg had significant entheseal changes other than those related to the flexion contracture action, which is interpreted to indicate that the lower extremities were unlikely to have been utilised in the later stages of the individual's life (Hawkey, 1998).

Besides diseases and conditions that result in total or partial paralysis or loss of movement, there is some evidence that certain communicable diseases can contribute to skeletal frailty even in the absence of significant or severe skeletal pathologies. Notably, tuberculosis has been linked with reduced bone strength and skeletal frailty in bioarchaeological studies. In Italy, the remains of an adolescent Neolithic individual from Liguria showing signs of osteoarticular tuberculosis also had extremely gracile postcranial skeletal elements compared to a sample of individuals without skeletal manifestations of tuberculosis from the same site (Sparacello et al., 2016). The authors suggest that this gracility may relate to periods of compromised periosteal apposition during key periods of bone growth that could be the result of a chronic tuberculosis infection. Similar findings were later reported in a documented skeletal collection of adult male individuals from 19th to early 20th century Finland (Mansukoski and Sparacello, 2018). Individuals whose cause of death was recorded as tuberculosis by the coroner or medical examiner had, when standardised for body size, smaller femoral and humeral total cross-sectional areas compared to individuals with a non-tuberculosis cause of death. As with the study from Liguria, it was not possible to ascertain what mechanism(s) contributed to the relative skeletal frailty of the individuals with tuberculosis, but the authors concluded the findings may reflect biological 'frailty' in terms of susceptibility to infection, reduced childhood activity due to chronic infection, or vitamin D deficiency, which is very common in northern latitudes. Vitamin D deficiency could have potentially influenced both subperiosteal development during adolescence, and later, susceptibility of contracting and dying of tuberculosis (Mansukoski and Sparacello, 2018).

11.5.2 Asymmetric impairment

In addition to looking for instances where an individual's health has led to prolonged bed rest, immobilisation, and consequent systemic bone loss, bioarchaeologists have also investigated activity indicators to identify functional differences arising from pathological conditions that asymmetrically affect the body. A variety of pathological conditions can result in asymmetric use of the limbs, including, but not limited to, stroke, injury, some infectious diseases, and congenital conditions. Bioarchaeologists have considered the functional implications of these conditions in their research, some of which are described as follows.

In living humans, cerebrovascular accidents (e.g. stroke) are one of the most common causes of impairments that include gait asymmetries and paresis (Duncan, 1994; Sánchez et al., 2021). In the bioarchaeological literature, stroke is frequently presented as a potential diagnosis in skeletons that exhibit asymmetry (Ciesielska and Stark, 2020; Novak et al., 2014), but we were unable to identify any published research that explicitly interprets disuse as a result of cerebrovascular accident in archaeological material. Cerebrovascular accidents (stroke) assuredly were present in the past; however, it is possible that these cases are not being recognised archaeologically as younger individuals may regain some function after a stroke, mitigating skeletal changes, while older individuals may not survive long enough to enact readily observable bone changes. In some of the archaeological cases that present stroke as a differential diagnosis, the authors argue that the asymmetry they observe could also be explained by cerebral palsy. While cases of cerebral palsy are rare in the archaeological literature (see Novak et al., 2014 for a review), some individuals who may have had hemiplegic cerebral palsy are interpreted based on the skeletal evidence, which includes morphological and length asymmetries present in both the upper and lower limbs (Novak et al., 2014; Tesorieri, 2016).

Poliomyelitis is another common differential diagnosis presented in complex and often non-specific cases of asymmetry. Like cerebral palsy, reported cases of poliomyelitis are relatively rare in the archaeological literature, but it is probable that this condition was present in the past, and therefore should have affected some individuals (Berner et al., 2021). Poliomyelitis is diverse in its presentation, but one of the most common manifestations involves paralysis of the lower limb and/or foot (Sharrard, 1955). Novak et al. (2014) describe the first archaeological case of possible triplegic cerebral palsy or poliomyelitis in a medieval Croatian female. In this case, bone atrophy and shortening, features indicative of limb paralysis from childhood, are evident in the morphology, diameters, circumferences, and lengths of elements in both legs and the right arm. Novak et al.'s (2014) observations of atrophy and shortening are similar to those of most bioarchaeological studies involving poliomyelitis, which rely on the observation of limb lengths, morphological differences, entheseal changes, and external cortical bone dimensions to identify disuse-related changes (Berner et al., 2021; Castells-Navaro et al., 2017; Ciesielska and Stark, 2020; Thompson, 2012). More rarely, histological and trabecular bone density analyses are included in bioarchaeological investigations of individuals with possible poliomyelitis (Berner et al., 2021; Kozłowski and Piontek, 2000). In some cases, discrepancies of leg lengths with no evidence for atrophy are used to indicate the onset of poliomyelitis in childhood with subsequent recovered function (Thompson, 2012). Similarly, Castells-Navaro et al. (2017) interpret muscle insertions on the smaller leg to suggest the individual maintained some function, which is possible due to "spontaneous recovery of muscle activity, characteristic in poliomyelitis" (p. 42). So far, the bioarchaeological evidence available suggests that functional indicators of activity associated with paralytic poliomyelitis vary individually, with some exhibiting continued paralysis and bone loss, while others retain or regain their function.

Instances of asymmetric use are much more clearly interpreted following skeletal trauma. Trauma with potential functional consequences has been observed in fossil hominin species, such as the *Australopithecus* from Sterkfontein known as "Little Foot" (StW 573) (Heile et al., 2018). In this case, the authors identify asymmetry in the curvature of the forearm as explained by a fracture sustained in adolescence, which, if true, may have affected supination and pronation at the hand (Heaton et al., 2019; Heile et al., 2018). Post-traumatic asymmetries in cross-sectional shape, size, and bone length have also been identified in Neanderthal upper limbs and are discussed and contextualised by Trinkaus et al. (1994) (e.g. Shanidar 1). Spikins et al. (2018) have extended these interpretations of Neanderthal functional consequences using a bioarchaeology of care lens. Through this perspective, dysfunction experienced by individuals, such as Shanidar 1 with an atrophied right arm and other potentially impairing conditions, could indicate some level of social care among Neanderthals similar to modern humans (Spikins et al., 2018).

Archaeological examples of reported dysfunction following injury increase over time, likely due to the increased prevalence of human remains from more recent time periods. For example, cases of upper limb asymmetry likely associated with injuries have been reported in Palaeolithic individuals by authors, including Churchill and Formicola (1997), Hershkovitz et al. (1993), and Chevalier (2019) (see Section 11.6 for further elaboration on Hershkovitz et al., 1993). Bone loss and atrophied muscle attachments attributed to disuse and post-traumatic dysfunction are described in several more modern individuals, such as the 19–20th century cases of lower limb injuries presented by Mariotti et al. (2013) and Neri and Lancellotti (2004).

Many of these studies successfully bring together a unique combination of methods from biomechanical, to morphological, histological, and even chemical, to thoroughly investigate pathologically affected activity. For example, Miszkiewicz et al. (2020) examined an individual from the Philippines with a thoroughly remodelled, ankylosed hip joint using simple bone morphometry and morphology, along with bone histology and synchrotron-sourced Fourier transform infrared microspectroscopy (sFTIRM), the latter of which was used to characterise the chemical composition of the bone (carbonate and phosphate). This study found asymmetries in the area and circularity of Haversian canals, the density of vessels, and the carbonate:phosphate ratios between the left and right femora. Miszkiewicz et al.'s (2020) results indicate that the right femur was denser than the left, indicating compromised bone quality likely due to disuse-related loss. This study is an important contribution that demonstrates a unique use of a technique (sFTIRM) that is not typically used in bioarchaeology. Specifically, these results show how the application of a broad range of methods can greatly inform studies of pathology and disuse that may involve bone remodelling changes that occur close to death.

Other studies also incorporate macro- and microscopic analyses to yield detailed information about individual adaptive choices following injury. Chevalier (2019) used CT scans of the upper limb in an Upper Palaeolithic female to quantify cross-sectional geometric properties and trabecular bone fraction as it related to a fracture in the left radius and possible trauma to the right humerus. The integration of both types of bone (trabecular and cortical) in this analysis not only allowed an interpretation of functional loss from fractures to the right humerus and left radius but also revealed that the individual likely switched dominant arms from right to left following the injuries. What is particularly remarkable about this study is the insight into the functional impacts of these traumatic lesions. Namely, the less severe looking right humeral trauma was evidently more impairing for the individual, resulting in the switch from right to left hand dominance, despite the presence of a left-radial fracture. This incredible amount of insight into this Upper Palaeolithic female's functional life was made possible through the integration of multiple lines of evidence, a practice that should be encouraged in future studies of palaeopathology and function.

Not all investigations of traumatic-related disuse are as clear cut as those involving skeletal fracture; not all injuries affect the skeleton, and some soft tissue injuries may irreparably harm nerves important for function without ever touching the bone. For example, Lieverse et al. (2008) describe a case of complete brachial palsy that occurred prior to skeletal maturity, resulting in marked bilateral asymmetries of the entire upper limb from the clavicle to the hand phalanges. In this case, the authors identified no clear skeletal pathology to explain the paralysis or paresis of the individual's right arm. This is typically a barrier to making an argument for injury-related dysfunction in skeletal remains; however, in this instance the severity and pattern of changes suggested a soft tissue-related cause. They found that not only were the bones of the affected side smaller in length, midshaft diameters, and joint dimensions, but they also had a "complete lack of observable or palpable muscle attachment sites [that] set them well outside the normal range for this or any other population" (Lieverse et al., 2008, p. 232). Based on their observations, the authors argue that this individual had complete disuse or severely limited use of the affected arm.

11.5.3 Amputation, adaptation, and assistive devices

Disuse may also occur following an amputation, which can result from either a severe traumatic injury or an intentional surgery. Skeletal changes following amputation have been outlined in clinical contexts in Section 11.2 and have also been documented in bioarchaeological studies (see a review of bioarchaeological evidence for amputation by Mays, 1996). In the last two decades, other bioarchaeological studies have noted amputations of limbs in varying temporal and cultural contexts (e.g. Van Cant, 2018; Więckowski, 2016). For example, asymmetry in lower limb cortical thicknesses and cross-sectional properties are described by Więckowski (2016) in an individual with an amputated foot. In this case, it is unclear if the subperiosteal surface dimensions are smaller on the amputated side, but X-Rays show thinning at the endosteal envelope. Bone atrophy is also documented by Van Cant (2018), who described a European late medieval period individual with an amputation at the mid-femur. In this case, the amputated limb clearly exhibits smaller subperiosteal diameters than the contralateral pair. These findings agree with the previously described clinical reports in that bone is primarily lost from the endosteal surface, and following prolonged disuse and absence of voluntary muscle forces (as would be the case with amputation), bone on the subperiosteal surface can also be significantly impacted.

In addition to the evidence for amputation, Van Cant (2018) also described hypertrophy of the right humeral entheses. They interpret this increased robusticity as indicative of underarm crutch use. Following the amputation or impairment of a limb/joint, humans may adapt to their altered bodies and functional abilities by adopting mobility aids, such as prosthetic devices and crutches. A number of cases of adaptive techniques, ranging from compensation in other body parts to adoption of functional aids, have been described in the bioarchaeological literature and are discussed as follows.

It is important to remember that not all instances of amputations can or should be associated with prosthetic use. For example, Mays (1996) describes a medieval British individual with an arm amputation who, based on scant evidence for upper-limb prostheses in this temporal and cultural context, seems unlikely to have used one. However, in other archaeological contexts, prosthetic upper and lower limbs were more common, and some bioarchaeological studies document artefactual and osteological evidence supporting the use of various external mobility aids. A history of artificial limbs is reviewed by Thurston (2007) who highlights evidence for artificial limbs found across varied geographical and cultural contexts, including ancient Egypt, Greece, and Rome. More recent bioarchaeological studies have provided further examples of prosthetic limbs and their impact on the skeleton in the archaeological record. For example, a paper by Micarelli et al. (2018) reports a 6–8th century CE individual from Italy with an amputated right forearm and a closely associated buckle and knife blade that are suggested to be a weapon-like prosthesis cinched to the forearm. Using 3D methods, bilateral asymmetric bone changes were quantified in the individual's humeri and considerable cortical bone loss was noted in the amputated limb with the prosthesis, indicative of diminished loading.

Evidence for lower limb prostheses has also been recovered from archaeological contexts. Binder et al. (2016) report an adult male from Austria (6th century CE) with an amputated left foot (above the ankle). This individual was buried along with their prosthetic replacement limb. Another case by Li et al. (2013) describes an older male from Western China (300–200 BCE) who used a prosthetic apparatus, which was buried with the individual and recovered from the grave upon excavation. The authors believe that this apparatus assisted with mobility following ankylosis of their knee, likely due to tuberculosis. This particular case is different as the individual did not undergo limb amputation, but rather strapped a peg leg-like lower limb prosthetic to their thigh. In both examples, the researchers provide good descriptive accounts of bone atrophy in the lower leg based on radiographic and CT evidence, specifically as it influences the trabecular bone. Neither of these studies incorporated cross-sectional measures to quantify the shape asymmetry that also may have manifested following the onset of disuse. This type of quantification may be outside the scope of these studies and also is limited in application due to taphonomic alterations and skeletal preservation but poses a point for future consideration in this type of research. In both these cases, the individuals are thought to have been functionally resilient and resumed or continued movement. Pronounced muscle attachments observed by Li et al. (2013) and degenerative joint changes in the left shoulder that may be associated with the use of a crutch (in addition to the prosthetic leg) recorded by Binder et al. (2016) provide further evidence that these individuals likely maintained their mobility using these prosthetic devices.

Examples of other types of medical devices and mobility aids have been identified archaeologically. Knüsel et al. (1995) describe, for example, two copper-alloy plates thought to be part of a right-sided knee splint recovered from a medieval burial. In this burial, the right arm is uniquely positioned relative to the torso, which the authors say might be explained by the inclusion of a crutch that has since decomposed. To our knowledge, traditional arm-propelled crutches have not yet been tangibly recovered in association with skeletal remains, but we do know that past peoples used crutches based on various artistic and literary depictions (Epstein, 1972; Hernigou, 2014; Loebl and Nunn, 1997) (Fig. 11.2). Bioarchaeologists can carefully extrapolate what skeletal changes might be expected and

FIG. 11.2 Crutches have been depicted as mobility aids in a wide variety of art from different cultures, geographic regions, and time periods; a selection of these depictions are indicated in this figure. (A) Stele of Roma the Doorkeeper, 18th Dynasty, New Kingdom, 1403–1365 BCE; (B) Man using hand trestles, marginal drawing in *Topographia Hiberniae, Expugnatio Hibernica, Itinerarium Kambriae; De purgatorio sancti Patrici, Excerpts from the Chronicle of Eusebius of Caesarea,* c.1196–1223 CE; (C) Figure on crutches representing old age in *Roman de la Rose,* Netherlands, c.1490–1500 CE; (D) Man on crutches in John Arderne's *Liber Medicinarium,* England, early 15th century CE; (E) The Immortal Li Tieguai Crossing the River on a Sword, China, 15–16th century CE. *Images courtesy of Ny Carlsberg Glyptotek, Copenhagen (A); The British Library, (Royal 13 B VIII f. 30v) (B); The British Library, (Harley 4425 f. 10v) (C), The British Library, (Sloane 56 f. 86v) (D); Shi Ke/Freer Gallery of Art, Smithsonian Institution, Washington, DC: Gift of Charles Lang Freer, F1916.37 (E).*

associated with altered loads from ambulatory aids. As such, crutch use has been interpreted in various archaeological contexts based on skeletal evidence, such as scapular stress fractures, accentuated upper limb muscle attachments, joint degeneration, and bone hypertrophy (Belcastro and Mariotti, 2000; Darton, 2010; Knüsel and Goeggel, 1993; Knüsel et al., 1995). Studies by Knüsel et al. (1992, 1995) and Knüsel and Goeggel (1993) considered the cross-sectional attributes of the shoulders, arms, wrists, and hands of individuals with trauma, trauma and leprosy, and a slipped femoral epiphysis. These studies all reported asymmetry in cross-sectional dimensions as well as hypertrophied muscle attachments, suggesting that the upper bodies of these individuals were used to help distribute their weight while walking. Other studies have also used accentuated and occasionally asymmetric musculoskeletal markers as evidence of crutch use (Belcastro and Mariotti, 2000). Because upper limb asymmetry can be induced from a variety of habitual loading activities, a combination of changes typically indicative of crutches should be used to infer the habitual use of this type of mobility aid.

One case study by Lazenby and Pfeiffer (1993) is particularly relevant to this discussion due to its application of a wide variety of methods to interpret habitual function after amputation and as a result of prosthetic limb use. This study describes a historically identified man from 19th century Ontario, Canada, who had a left below-knee amputation some time prior to his death and regularly wore a homemade wooden prosthetic secured to his waist and knee. Lazenby and Pfeiffer (1993) take a thorough and innovative approach to this case, examining the material properties of the wooden prosthetic, as well as the macro and microstructural properties of the paired femora, including enthesophyte development, cross- sectional geometry, and intracortical remodelling (measured histologically). Their results show an accentuated development of specific enthesophytes around the hip, which were interpreted to mean that there was "an increased and protracted functional obligation for hip flexion following amputation and probable impairment of other thigh flexors that cross the knee" (Lazenby and Pfeiffer, 1993, p. 21). Furthermore, cross-sectional geometric analyses from CT scans also found that bone was maintained in the proximal end of the amputated left femur, likely due to continued loading facilitated by a prosthesis that did not shield this region from stress. However, attributes expected in clinical cases of disuse atrophy, specifically endosteal bone loss, as well as a very high rate of intracortical bone turnover, were identified at, and distal to, the femoral midshaft, which were regions protected by the prosthetic's rigidity (Lazenby and Pfeiffer, 1993). The combinations of methods used in this study help elucidate the mobility and adaptation of this individual and demonstrate the importance of incorporating multiple approaches in studies of bone adaptation.

11.5.4 Adaptation: Recovery after pathology

Although today it is common to use assistive devices to aid our mobility following an injury or illness, this is not necessarily required, and many individuals are able to function and be mobile without the use of external supports. Therefore, while mobility aids may have existed in the past, it is also possible that some people simply continued moving using other personal and physical coping strategies.

Bone functional adaptation is exemplified in various biological archaeological case studies (e.g. Holt et al., 2002; Neri and Lancellotti, 2004). Lovell (2016) presents a case study of an older Roman individual with a healed hip fracture and a shortened leg length. She also describes degeneration in the lower limb joints and pathological changes to the bones of the foot on the same side as the fracture. The combination of these changes with the lack of other skeletal evidence for crutch use suggests that this individual maintained their mobility by walking on tiptoe (Lovell, 2016). Another similar study by Holt et al. (2002) describes an individual with an ankle fracture and residual, healed deformity. In this case, Holt et al. (2002) evaluate lower limb cross-sectional asymmetry and conclude that, although a high level of asymmetry was present, this individual did not exhibit disuse hypotrophy. This indicates that they "continued to walk for an extended period of time, albeit in a mechanically altered manner" (p. 405). In both cases, the individual's mobility was altered following their injuries. However, both these cases also serve as an important reminder that not all individuals with injuries or illnesses may require assistance or support, and that some may adapt alternative ways to respond to changes in their body's morphology and physiology.

These studies are not anomalous in their support of a general return to, or maintenance of, altered function following a pathological condition. In addition, a growing body of research investigates physical function as it relates to congenital and juvenile onset conditions. While some studies identify marked levels of bone loss that they associated with large-scale impairment (and care) (e.g. Tilley and Oxenham, 2011), many other studies report archaeological individuals who did not appear to be functionally impaired by their various conditions. Cowgill et al. (2012) describe a juvenile from the Upper Palaeolithic with anteriorly bowed femora and other indicators of non-specific stress. Using comparative cross-sectional analyses, the authors found that the tibiae and humeri were within the range of normal activity for a Late Pleistocene juvenile. In the humeri, a large amount of asymmetry was identified, but both sides were within the normal range, leading the authors to suggest the larger, right side was used for "a variety of heavy duty manipulative activities" (Cowgill et al., 2012, p. 182). Like Cowgill et al. (2012), Trinkaus et al. (2001) discuss an Upper Palaeolithic individual with developmental differences. This individual's cross-sectional properties, albeit asymmetric, were not inconsistent with the typical level of activity at that time and in this region. In both cases, although the individuals exhibited congenital skeletal differences, there was no biomechanical indication that they were anything other than active and participating contributors to their society.

11.6 Discussion and ways forward

Bioarchaeology provides a unique perspective on injury and care in the past. Bolstered by clinical research and in some cases historical or archaeological resources, we can gain better insight into not only the injury itself, but also life after the injury and the road to recovery. Building upon previous research, this section will identify some potential areas of future research as well as consider limitations to current approaches. We acknowledge that there is a large body of literature on the bioarchaeology of care that considers the experiences and provision of care for individuals with various impairing conditions (see, for example, Tilley, 2015; Tilley and Oxenham, 2011; Tilley and Schrenk, 2017). Summarising this work is beyond the scope of this chapter, but this lens does provide an important perspective on community and healing in the past and the methods and approaches we have outlined could assist in interpretation of care.

In general, we believe a multimethod approach is particularly useful when examining impairment in the past because we are often examining a single individual and thus, obtaining as much data as possible from each studied individual is vital. The Lazenby and Pfeiffer (1993) case study (introduced in Section 11.5.3) is a good model for how bioarchaeologists can incorporate multiple lines of evidence to construct a detailed argument that enables interpretation of the variation in adaptive responses present, even within a single element. Through consideration of cross-sectional geometry and the endosteal margin, histological methods to examine remodelling, observation of enthesophyte asymmetries, and assessment of the material properties of the prosthetic device, these authors were able to comprehensively argue for atrophy following amputation and provide skeletal evidence to support the individual's maintenance of mobility using the prosthetic device. The lesson that all bioarchaeologists interested in studies of disuse, immobilisation, and impairment should take from this is the importance of integrating as many lines of evidence as possible to strengthen our understanding of human experiences following injury or illness.

Another important takeaway is that impairment is not something that can immediately be inferred based on the perceived 'severity' of the pathological condition. Just because the skeletal alterations may initially appear drastic, it does not necessarily mean that the individual was consequently impaired. Gilmour et al. (2019) work on Roman lower limb fractures and functional disuse illustrates this. In this study, fracture malunion is measured and cortical bone asymmetries associated with injuries are quantified. The study found that none of the individuals with fractures had cortical bone asymmetries outside the normal ranges for the assemblages studied. Even those fractures with the greatest amounts of malunion, which typically are clinically associated with poor functional outcomes, did not exhibit cortical changes to indicate that these Romans experienced altered mobility after their fracture healed. These findings were interpreted to suggest that these individuals likely continued or regained function following their injuries (Gilmour et al., 2019). Unfortunately, it is not possible to know how the individuals felt about these injuries, or qualify their experiences of other symptoms, such as pain, which they very well may have continued to experience. However, we can infer that even if pain were involved, the most 'severe' looking injuries did not actually act to impede function in a way that resulted in cortical bone loss or shape changes.

As bioarchaeologists interested in quantifying activity and investigating impairment and immobilisation, it might be tempting to see marked skeletal changes and speculate on the individual's suffering. However, the findings reported by researchers such as Gilmour et al. (2019) and Lovell (2016) should serve to remind us that it is vital to remember that even the most 'severe' and marked skeletal changes cannot be immediately equated with impairment. Instead, various methods for analysing activity, like those outlined in this book, should be integrated in palaeopathological analyses to more accurately represent evidence for the individual's activity as it is preserved in the bone (see also the systematic review of quantitative shape analyses in palaeopathology by Gilmour and Plomp, 2022). This approach should involve both evaluating the asymmetry present within each individual, but also comparing the individual(s) in question and their skeletal indicators of activity to the larger assemblage to identify if they fall within the group's normal ranges. This approach also applies to evaluating diffuse and systemic bone loss. When systemic bone loss is suspected, it is important to compare the individuals with others in the same skeletal collection or archaeological site. Unless the individual has bone properties that lay outside a group's normal ranges, perceived fragility, frailty, or bone phenotype could simply reflect a small or gracile individual. In all instances, only when there is reason to believe that the individual's skeletal properties are significantly different from the group's normal ranges (e.g. via outlier analyses), should functional impairment be interpreted as a possible consequence of the pathological condition.

To illustrate this point and emphasise the importance of contextualising these activity-related changes in relation to other members in an individual's group, we can examine a Palaeolithic individual (Ohalo 2) described first by Hershkovitz et al. (1993) and more recently by Trinkaus (2018). This individual, a male in his mid-30s, was initially interpreted by Hershkovitz et al. (1993) to have asymmetric humeral diaphyses with reduced muscle attachments due to a possible brachial plexus injury and abnormal growths present in their thorax. Based on these observations, the individual was thought to have experienced significant dysfunction that would have affected his life and community activities as a fisher-hunter-gatherer. While the identification of a possible soft tissue injury would have been remarkable from dry-bone remains, this individual was reanalysed by Trinkaus (2018), who compared the asymmetry and cross-sectional measures of this individual against others from the Upper Palaeolithic. Through this comparison, it was illuminated that Ohalo 2's asymmetries and cross-sectional attributes, while gracile, all fell within the range of normal for other Upper Palaeolithic humans. The fact that they fall within this range is an important observation, and one that reminds us that we must not over-interpret evidence and should always compare our possible cases of bone loss to the group range and average. Ultimately, this individual is identified as a case of possible limited function, but as a result of changes in the thorax and ribs, and not because of the upper limb changes. This case stands to remind us that reduced muscle attachments and gracile limbs in a single individual alone are insufficient for a final interpretation of impairment.

Of the case studies that discuss the functional consequences of trauma, many focus on descriptions of morphological changes, including those to the outer (periosteal) part of the cortical bone, as well as reduced entheseal changes and total bone lengths. These are valuable observations and should be included in studies of asymmetry to help interpret functional consequences of pathological conditions. However, fewer palaeopathological studies incorporate biomechanical analyses that also consider trabecular bone or the endosteal margin of cortical bone. In adults, bone changes predominantly occur on the endosteal bone surface, meaning that it is not possible to identify bone hypo- or hypertrophy based on a bone's outer appearance (see Section 11.2). There are some exceptions, for example, an archaeological case presented by Holt et al. (2002) uses biomechanical methods that evaluate the endosteal bone margin to determine the amount of cortical bone present, and the previously discussed case by Chevalier (2019) incorporates quantifications of trabecular bone. However, studies of adult impairment that neglect the endosteal margin or trabecular bone (where most changes will occur in adults) may be missing most instances of disuse-related bone loss. This may explain, in part, why functional consequences have not been discussed more routinely in earlier analyses of injury and disease. The relatively few studies of long-term trauma repercussions may suggest that long-term dysfunction after a fracture is either not being recognised in archaeological collections or that it is relatively rare. Either way, both are interesting findings and suggest that in bioarchaeology, we should more regularly employ methods capable of recognising changes associated with immobilisation to better understand trauma treatment and recovery.

Finally, it is important to acknowledge limitations to bioarchaeological studies of impairment. Some pathological conditions will affect bone in such a way that cross-sectional properties cannot be reliably measured, at least not in the traditional ways presented in this chapter. In cases where pathological conditions are localised, it is possible to investigate bone atrophy and shape changes in the areas surrounding the affected region to construct interpretations of function (see Gilmour and Plomp, 2022 for a review). However, not all pathological conditions are localised, many affect the skeletal system in a diffuse manner, and others visibly affect the condition of the bone in a way that would inhibit measurements. For example, conditions, such as Paget's disease, are clinically associated with statistically significant levels of mobility impairments (Lyles et al., 1995). Assuredly, this would affect the amount, quality, and distribution of bone in a skeletal element. However, given the way that bone condition and shape is remarkably altered by remodelling in individuals with Paget's disease, it would often not be possible to evaluate cortical or trabecular bone losses in a traditional sense. New methods are needed to address these types of limitations if we wish to continue to investigate and assess disuse and impairment for all people in the past.

11.7 Conclusion

This chapter has presented a summary of the current state of research into disuse and impairment in bioarchaeology. We have incorporated both animal and clinical studies, which underpin the multiple skeletal examples provided. The foundation of this research lies in an explanation of bone adaptation, supported by the distribution of bone loss and directional asymmetry, which can be meaningful tools for assessing the repercussions of pathological conditions. In bioarchaeology, these have been used in studies of bed rest and frailty, asymmetric impairment, assistive devices and amputation, as well as recovery after pathology.

Through the discussion of numerous examples, ranging in time from fossil hominins to contemporary modern humans, we have illustrated the current state of research in this field and identified areas in which bioarchaeological investigations into injury, illness, adaptation, and care can expand. We encourage the use of comparative samples and multiple lines of evidence, including the examination of endosteal remodelling, to provide a more holistic understanding of the pathological condition as well as the recovery process. We envision a multi-method approach involving entheseal changes, cross-sectional geometry, histology, and trabecular bone microarchitecture playing an important role in understanding life after injury, particularly as a more thorough understanding of these adaptations emerge. Further, we caution that bioarchaeologists should be careful when making interpretations about impairment and suffering, as the severity of the skeletal pathology does not always directly correlate to the illness experience.

References

Acosta, M. A., Henderson, C. Y., & Cunha, E. (2017). The effect of terrain on entheseal changes in the lower limbs. *International Journal of Osteoarchaeology, 27*(5), 828–838.

Alexandre, C., & Vico, L. (2011). Pathophysiology of bone loss in disuse osteoporosis. *Joint, Bone, Spine, 78*(6), 572–576. https://doi.org/10.1016/j.jbspin.2011.04.007.

Arnstein, A. (1972). Regional osteoporosis. *Orthopedic Clinics of North America, 3,* 585–600.

Auerbach, B. M., & Ruff, C. B. (2006). Limb bone bilateral asymmetry: Variability and commonality among modern humans. *Journal of Human Evolution, 50*(2), 203–218. https://doi.org/10.1016/j.jhevol.2005.09.004.

Barrett, C. K., Guatelli-Steinberg, D., & Sciulli, P. W. (2012). Revisiting dental fluctuating asymmetry in neandertals and modern humans. *American Journal of Physical Anthropology, 149*(2), 193–204. https://doi.org/10.1002/ajpa.22107.

Becker, H. P., Rosenbaum, D., Kriese, T., Gerngro, H., & Claes, L. (1995). Gait asymmetry following successful surgical treatment of ankle fractures in young adults. *Clinical Orthopaedics and Related Research, 311,* 262–269.

Belcastro, M. G., & Mariotti, V. (2000). Morphological and biomechanical analysis of a skeleton from Roman Imperial necropolis of Casalecchio di Reno (Bologna, Italy, II-III c. A.D.). A possible case of crutch use. *Collegium Antropologicum, 24*(2), 529–539.

Berner, M., Pany-Kucera, D., Doneus, N., Sladek, V., Gamble, M., & Eggers, S. (2021). Challenging definitions and diagnostic approaches for ancient rare diseases: The case of poliomyelitis. *International Journal of Paleopathology, 33,* 113–127.

Biering-Sørensen, F., Bohr, H., & Schaadt, O. (1990). Longitudinal study of bone mineral content in the lumbar spine, the forearm and the lower extremities after spinal cord injury. *European Journal of Clinical Investigation, 20*(3), 330–335.

Biewener, A. A., & Bertram, J. E. A. (1993). Skeletal strain patterns in relation to exercise training during growth. *Journal of Experimental Biology, 185,* 51–69.

Binder, M., Eitler, J., Deutschmann, J., Ladstätter, S., Glaser, F., & Fiedler, D. (2016). Prosthetics in antiquity—An early medieval wearer of a foot prosthesis (6th century AD) from Hemmaberg/Austria. *International Journal of Paleopathology, 12*, 29–40.

Bloomfield, S. A. (2010). Disuse osteopenia. *Current Osteoporosis Reports, 8*, 91–97.

Burr, D. B. (1997). Muscle strength, bone mass, and age-related bone loss. *Journal of Bone and Mineral Research, 12*(10), 1547–1551.

Campanacho, V., & Santos, A. L. (2013). Comaprison of the entheseal changes of the os coxae of Portuguese males (19th-20th centuries) with known occupation. *International Journal of Osteoarchaeology, 23*(2), 229–236.

Castells-Navaro, L., Southwell-Wright, W., Manchester, K., & Buckberry, J. (2017). Interpretation of a probable case of Poliomyelitis in the Romano-British social context. *Archaeological Review from Cambridge, 32*(1), 34–52.

Ceroni, D., Martin, X. E., Delhumeau, C., Farpour-Lambert, N. J., De Coulon, G., Dubois-Ferrière, V., et al. (2013). Recovery of decreased bone mineral mass after lower-limb fractures in adolescents. *The Journal of Bone and Joint Surgery, 95*(11), 1037–1043.

Chan, G. K., & Duque, G. (2002). Age-related bone loss: Old bone, new facts. *Gerontology, 48*(2), 62–71.

Chevalier, T. (2019). Trauma in the upper limb of an Upper Paleolithic female from Caviglione cave (Liguria, Italy): Etiology and after-effects in bone biomechanical properties. *International Journal of Paleopathology, 24*, 94–107.

Churchill, S. E., & Formicola, V. (1997). A case of marked bilateral asymmetry in the upper limbs of an upper palaeolithic male from Barma Grande (Liguiria), Italy. *International Journal of Osteoarchaeology, 7*, 18–38.

Ciesielska, J. A., & Stark, R. J. (2020). Possible neurogenic disorder in a female buried in the monastic cemetery at Ghazali (ca. 670–1270 CE), northern Sudan. *International Journal of Osteoarchaeology, 30*(1), 33–42.

Coupaud, S., McLean, A. N., Purcell, M., Fraser, M. H., & Allan, D. B. (2015). Decreases in bone mineral density at cortical and trabecular sites in the tibia and femur during the first year of spinal cord injury. *Bone, 74*, 69–75.

Cowgill, L. W., Mednikova, M. B., Buzhilova, A. P., & Trinkaus, E. (2012). The Sunghir 3 Upper Paleolithic juvenile: Pathology versus persistence in the Paleolithic. *International Journal of Osteoarchaeology, 25*, 176–187.

Cunningham, H. C., West, D. W., Baehr, L. M., Tarke, F. D., Baar, K., & Bodine, S. C., et al. (2018). Age-dependent bone loss and recovery during hindlimb unloading and subsequent reloading in rats. *BMC Musculoskeletal Disorders, 19*(1), 223.

Darton, Y. (2010). Scapular stress fracture. A palaeopathological case consistent with crutch use. *International Journal of Osteoarchaeology, 20*(1), 113–121. https://doi.org/10.1002/oa.1002.

Duncan, P. W. (1994). Stroke disability. *Physical Therapy, 74*(5), 399–407.

Eastell, R., O'Neill, T., Hofbauer, L., Langdahl, B., Reid, I., Gold, D., et al. (2016). Postmenopausal osteoporosis. *Nature Reviews. Disease Primers, 2*, 1–16.

Epstein, S. (1972). Art, history and the crutch. *Clinical Orthopaedics and Related Research, 89*, 4–9.

Eser, P., Frotzler, A., Zehnder, Y., Wick, L., Knecht, H., Denoth, J., et al. (2004). Relationship between the duration of paralysis and bone structure: A pQCT study of spinal cord injured individuals. *Bone, 34*(5), 869–880.

Eyres, K. S., & Kanis, J. A. (1995). Bone loss after tibial fracture: Evaluated by dual-energy X-ray absorptiometry. *The Journal of Bone and Joint Surgery. British Volume, 77-B*, 473–478.

Farr, J., & Khosla, S. (2019). Cellular senescence in bone. *Bone, 121*, 121–133.

Frotzler, A., Berger, M., Knecht, H., & Eser, P. (2008). Bone steady-state is established at reduced bone strength after spinal cord injury: A longitudinal study using peripheral quantitative computed tomography (pQCT). *Bone, 43*(3), 549–555.

Garland, D. E., Stewart, C. A., Adkins, R. H., Hu, S. S., Rosen, C., Liotta, F. J., et al. (1992). Osteoporosis after spinal cord injury. *Journal of Orthopaedic Research, 10*(3), 371–378.

Giangregorio, L., & Blimkie, C. J. (2002). Skeletal adaptations to alterations in weight-bearing activity. *Sports Medicine, 32*(7), 459–476.

Giangregorio, L., & McCartney, N. (2006). Bone loss and muscle atrophy in spinal cord injury: Epidemiology, fracture prediction, and rehabilitation strategies. *Journal of Spinal Cord Medicine, 29*, 489–500.

Gilmour, R. J., Brickley, M., Jurriaans, E., & Prowse, T. (2019). Maintaining mobility after fracture: A biomechanical analysis of fracture consequences at the Roman site of Ancaster (UK) and Vagnari (Italy). *International Journal of Paleopathology, 24*, 119–129. https://doi.org/10.1016/j.ijpp.0401.

Gilmour, R. J., & Plomp, K. A. (2022). The changing shape of palaeopathology: The contribution of skeletal shape analyses to investigations of pathological conditions. *American Journal of Biological Anthropology, 178*(Suppl. 74), 151–180. https://doi.org/10.1002/ajpa.24475.

Guo, X. E., & Kim, C. H. (2002). Mechanical consequence of trabecular bone loss and its treatment: A three-dimensional model simulation. *Bone, 30*(2), 404–411.

Hart, N. H., Nimphius, S., Weber, J., Spiteri, T., Rantalainen, T., Dobbin, M., et al. (2016). Musculoskeletal asymmetry in football athletes: A product of limb function over time. *Medicine & Science in Sports & Exercise, 48*(7), 1379–1387.

Hawkey, D. E. (1998). Disability, compassion and the skeletal record: Using musculoskeletal stress markers (MSM) to construct an osteobiography from early New Mexico. *International Journal of Osteoarchaeology, 8,* 326–340.

Heaton, J. L., Pickering, T. R., Carlson, K. J., Crompton, R. H., Jashashvili, T., Beaudet, A., et al. (2019). The long limb bones of the StW 573 Australopithecus skeleton from Sterkfontein member 2: Descriptions and proportions. *Journal of Human Evolution, 133,* 167–197.

Heile, A. J., Pickering, T. R., Heaton, J. L., & Clarke, R. J. (2018). Bilateral asymmetry of the forearm bones as possible evidence of antemortem trauma in the StW 573 Australopithecus skeleton from Sterkfontein Member 2 (South Africa). *BioRxiv,* 486076. https://doi.org/10.1101/486076.

Henderson, C. Y., Salega, S., & Silva, A. M. (2018). Portuguese women's activity in the past: Comparing entheseal changes through time. *Annales Universitatis Apulensis Series Historica, 22*(1), 195–222.

Hernigou, P. (2014). Crutch art painting in the middle age as orthopaedic heritage (part I: The lepers, the poliomyelitis, the cripples). *International Orthopaedics, 38*(6), 1329–1335.

Hershkovitz, I., Edelson, G., Spiers, M., Arensburg, B., Nadel, D., & Levi, B. (1993). Ohalo II man—Unusual findings in the anterior rib cage and shoulder girdle of a 19000-year-old specimen. *International Journal of Osteoarchaeology, 3*(3), 177–188.

Holt, B. M., Fornaciari, G., & Formicola, V. (2002). Bone remodelling following a lower leg fracture in the 11,000-year-old hunter-gatherer from Vado all' Arancio (Italy). *International Journal of Osteoarchaeology, 12,* 402–406.

Jaworski, Z. F. G., Liskova-Kiar, M., & Uhthoff, H. K. (1980). Effect of long-term immobilisation on the pattern of bone loss in older dogs. *The Journal of Bone and Joint Surgery. British Volume, 62*(1), 104–110.

Jaworski, Z., & Uhthoff, H. K. (1986). Reversibility of nontraumatic disuse osteoporosis during its active phase. *Bone, 7*(6), 431–439.

Jee, W. S., & Ma, Y. (1999). Animal models of immobilization osteopenia. *Morphologie: Bulletin de l'Association Des Anatomistes, 83*(261), 25–34.

Jiang, S.-D., Dai, L.-Y., & Jiang, L.-S. (2006). Osteoporosis after spinal cord injury. *Osteoporosis International, 17*(2), 180–192.

Jones, G. (1969). Radiological appearances of disuse osteoporosis. *Clinical Radiology, 20,* 345–353.

Jørgensen, L., & Jacobsen, B. K. (2001). Functional status of the paretic arm affects the loss of bone mineral in the proximal humerus after stroke: A 1-year prospective study. *Calcified Tissue International, 68*(1), 11–15.

Kazakia, G. J., Tjong, W., Nirody, J. A., Burghardt, A. J., Carballido-Gamio, J., Patsch, J. M., et al. (2014). The influence of disuse on bone microstructure and mechanics assessed by HR-pQCT. *Bone, 63,* 132–140.

Knüsel, C. J., Chundun, Z. C., & Cardwell, P. (1992). Slipped proximal femoral epiphysis in a priest from the medieval period. *International Journal of Osteoarchaeology, 2,* 109–119.

Knüsel, C. J., & Goeggel, S. (1993). A cripple from the medieval Hospital of Sts James and Mary Magdalen, Chichester. *International Journal of Osteoarchaeology, 3,* 155–165.

Knüsel, C. J., Kemp, R. L., & Budd, P. (1995). Evidence for remedial medical treatment of a severe knee injury from the Fishergate Gilbertine monastery in the City of York. *Journal of Archaeological Science, 22,* 369–384.

Kozłowski, T., & Piontek, J. (2000). A case of atrophy of bones of the right lower limb of a skeleton from a medieval (12th-14th centuries) burial ground in Gruczno, Poland. *Journal of Paleopathology, 12*(1), 5–16.

Kulkarni, J., Adams, J., Thomas, E., & Silman, A. (1998). Association between amputation, arthritis and osteopenia in British male war veterans with major lower limb amputations. *Clinical Rehabilitation, 12*(4), 348–353. https://doi.org/10.1191/026921598672367610.

Lang, T., LeBlanc, A., Evans, H., & Lu, Y. (2006). Adaptation of the proximal femur to skeletal reloading after long-duration spaceflight. *Journal of Bone and Mineral Research, 21*(8), 1224–1230.

Lang, T., LeBlanc, A., Evans, H., Lu, Y., Genant, H., & Yu, A. (2004). Cortical and trabecular bone mineral loss from the spine and hip in long-duration spaceflight. *Journal of Bone and Mineral Research, 19*(6), 1006–1012.

Lau, R. Y., & Guo, X. (2011). A review on current osteoporosis research: With special focus on disuse bone loss. *Journal of Osteoporosis, 2011,* 1–6.

Lazenby, R. A. (1990). Continuing periosteal apposition I: Documentation, hypotheses, and interpretation. *American Journal of Physical Anthropology, 82*(4), 451–472.

Lazenby, R. A. (2002). Skeletal biology, functional asymmetry and the origins of "handedness". *Journal of Theoretical Biology, 218,* 129–138. https://doi.org/10.1006/yjtbi.3052.

Lazenby, R. A., & Pfeiffer, S. K. (1993). Effects of a nineteenth century below-knee amputation and prosthesis on femoral morphology. *International Journal of Osteoarchaeology*, 3(1), 19–28.

Leblanc, A. D., Schneider, V. S., Evans, H. J., Engelbretson, D. A., & Krebs, J. M. (1990). Bone mineral loss and recovery after 17 weeks of bed rest. *Journal of Bone and Mineral Research*, 5(8), 843–850.

Li, C., Laudier, D., & Schaffler, M. (2003). Remobilization restores cancellous bone mass but not microarchitecture after long term disuse in older adult dogs. In *Transactions of the 48th annual meeting of the Orthopaedic Research Society, New Orleans, LA*.

Li, X., Wagner, M., Wu, X., Tarasov, P., Zhang, Y., Schmidt, A., et al. (2013). Archaeological and palaeopathological study on the third/second century BC grave from Turfan, China: Individual health history and regional implications. *Quaternary International*, 290, 335–343.

Lieverse, A. R., Metcalf, M. A., Bazaliiskii, V. I., & Weber, A. W. (2008). Pronounced bilateral asymmetry of the complete upper extremity: A case from the early Neolithic Baikal, Siberia. *International Journal of Osteoarchaeology*, 18(3), 219–239. https://doi.org/10.1002/oa.935.

Loebl, W., & Nunn, J. (1997). Staffs as walking aids in ancient Egypt and Palestine. *Journal of the Royal Society of Medicine*, 90(8), 450–454.

Lovell, N. C. (2016). Tiptoeing through the rest of his life: A functional adaptation to a leg shortened by femoral neck fracture. *International Journal of Paleopathology*, 13, 91–95.

Lyles, K. W., Lammers, J. E., Shipp, K. M., Sherman, L., Pieper, C. F., Martinez, S., et al. (1995). Functional and mobility impairments associated with Paget's disease of bone. *Journal of the American Geriatrics Society*, 43(5), 502–506.

Mansukoski, L., & Sparacello, V. (2018). Smaller long bone cross-sectional size in people who died of tuberculosis: Insights on frailty factors from a 19th and early 20th century Finnish population. *International Journal of Paleopathology*, 20, 38–44. https://doi.org/10.1016/j.ijpp.2017.12.005.

Marchi, D., Sparacello, V. S., Holt, B. M., & Formicola, V. (2006). Biomechanical approach to the reconstruction of activity patterns in Neolithic Western Liguria, Italy. *American Journal of Physical Anthropology*, 131(4), 447–455. https://doi.org/10.1002/ajpa.20449.

Mariotti, V., Milella, M., Orsini, E., Trirè, A., Ruggeri, A., Fornaciari, G., et al. (2013). Osteobiography of a 19 th century elderly woman with pertrochanteric fracture and osteoporosis: A multidisciplinary approach. *Collegium Antropologicum*, 37(3), 985–994.

Mays, S. A. (1996). Healed limb amputations in human osteoarchaeology and their causes: A case study from Ipswich, UK. *International Journal of Osteoarchaeology*, 6(1), 101–113.

McGee, M. E., Maki, A. J., Johnson, S. E., Nelson, O. L., Robbins, C. T., & Donahue, S. W. (2008). Decreased bone turnover with balanced resorption and formation prevent cortical bone loss during disuse (hibernation) in grizzly bears (*Ursus arctos horribilis*). *Bone*, 42(2), 396–404. https://doi.org/10.1016/j.bone.2007.10.010.

Micarelli, I., Paine, R., Giostra, C., Tafuri, M. A., Profico, A., Boggioni, M., et al. (2018). Survival to amputation in pre-antibiotic era: A case study from a Longobard necropolis (6th-8th centuries AD). *Journal of Anthropological Sciences*, 96, 1–16.

Milella, M., Alves Cardoso, F., Assis, S., Perreard Lopreno, G., & Speith, N. (2015). Exploring the relationship between entheseal changes and physical activity: A multivariate study. *Ameican Journal of Physical Anthropology*, 156(2), 215–226.

Milella, M., Belcastro, M. G., & Zollikofer, C. P. E. (2012). The effect of age, sex, and physical activity on entheseal morphology in a contemporary Italian skeletal collection. *International Journal of Osteoarchaeology*, 27, 828–838.

Minaire, P. (1989). Immobilization osteoporosis: A review. *Clinical Rheumatology*, 8(2), 95–103.

Miszkiewicz, J. J., Rider, C., Kealy, S., Vrahnas, C., Sims, N. A., Vongsvivut, J., et al. (2020). Asymmetric midshaft femur remodeling in an adult male with left sided hip joint ankylosis, Metal Period Nagsabaran, Philippines. *International Journal of Paleopathology*, 31, 14–22.

Møller, A. P., & Swaddle, J. P. (1997). *Asymmetry, developmental stability and evolution*. UK: Oxford University Press.

Morbeck, M. E., Zihlman, A. L., Sumner, D. R., & Galloway, A. (1991). Poliomyelitis and skeletal asymmetry in Gombe chimpanzees. *Primates*, 32(1), 77–91.

Neri, R., & Lancellotti, L. (2004). Fractures of the lower limbs and their secondary skeletal adaptations: A 20th century example of pre-modern healing. *International Journal of Osteoarchaeology*, 14, 60–66.

Novak, M., Čavka, M., & Šlaus, M. (2014). Two cases of neurogenic paralysis in medieval skeletal samples from Croatia. *International Journal of Paleopathology*, 7, 25–32.

Osipov, B., Emami, A. J., & Christiansen, B. A. (2018). Systemic bone loss after fracture. *Clinical Reviews in Bone and Mineral Metabolism*, 16(4), 116–130.

Oxenham, M. F., Tilley, L., Matsumura, H., Nguyen, L. C., Nguyen, K. T., Nguyen, K. D., et al. (2009). Paralysis and severe disability requiring intensive care in Neolithic Asia. *Anthropological Science, 117*(2), 107–112. https://doi.org/10.1537/ase.081114.

Ozcivici, E., & Judex, S. (2014). Trabecular bone recovers from mechanical unloading primarily by restoring its mechanical function rather than its morphology. *Bone, 67,* 122–129.

Palmer, A. R., & Strobeck, C. (1986). Fluctuating asymmetry: Measurement, analysis, patterns. *Annual Review of Ecology and Systematics, 17,* 391–421.

Quek, S.-T., & Peh, W. C. G. (2002). Radiology of osteoporosis. *Seminars in Musculoskeletal Radiology, 6*(3), 197–206.

Reichel, H., Runge, H., & Bruchhaus, H. (1990). Die Seitendifferenz des Mineralgehaltes und der Breite am Radius und ihre Bedeutung für die Händigkeitsbestimmung an Skelettmaterial. *Zeitschrift für Morphologie und Anthropologie, 78*(2), 217–227.

Rhodes, J. A., & Knusel, C. J. (2005). Activity-related skeletal change in medieval humeri: Cross-sectional and architectural alterations. *American Journal of Physical Anthropology, 128*(3), 536–546. https://doi.org/10.1002/ajpa.20147.

Rittweger, J., & Felsenberg, D. (2009). Recovery of muscle atrophy and bone loss from 90 days bed rest: Results from a one-year follow-up. *Bone, 44*(2), 214–224. https://doi.org/10.1016/j.bone.2008.10.044.

Rittweger, J., Simunic, B., Bilancio, G., De Santo, N. G., Cirillo, M., Biolo, G., et al. (2009). Bone loss in the lower leg during 35 days of bed rest is predominantly from the cortical compartment. *Bone, 44*(4), 612–618.

Ruff, C. B. (2008). Biomechanical analyses of archaeological human skeletons. In M. A. Katzenberg, & S. R. Saunders (Eds.), *Biological anthropology of the human skeleton* (2nd ed., pp. 71–102). Hoboken: Wiley.

Ruff, C. B., & Jones, H. H. (1981). Bilateral asymmetry in cortical bone of the humerus and tibia—Sex and age factors. *Human Biology, 53*(1), 69–86.

Sánchez, N., Schweighofer, N., & Finley, J. M. (2021). Different biomechanical variables explain within-subjects versus between-subjects variance in step length asymmetry post-stroke. *IEEE Transactions on Neural Systems and Rehabilitation Engineering, 29,* 1188–1198.

Schäfer, M. L., Böttcher, J., Pfeil, A., Hansch, A., Malich, A., Maurer, M. H., et al. (2012). Comparison between amputation-induced demineralization and age-related bone loss using digital X-ray radiogrammetry. *Journal of Clinical Densitometry: Assessment of Skeletal Health, 15*(2), 135–145. https://doi.org/10.1016/j.jocd.2011.08.006.

Schlecht, S. H. (2012). Understanding entheses: Bridging the gap between clinical and anthropological perspectives. *The Anatomical Record, 295*(8), 1239–1251.

Schrader, S., & Buzon, M. (2017). Everyday life after collapse: A bioarchaeological examination of entheseal changes and accidental injury in postcolonial Nubia. *Bioarchaeology International, 1*(1–2), 19–34.

Schrenk, A., Gregoricka, L. A., Martin, D. L., & Potts, D. T. (2016). Differential diagnosis of a progressive neuromuscular disorder using bioarchaeological and biogeochemical evidence from a bronze age skeleton in the UAE. *International Journal of Paleopathology, 13,* 1–10.

Segal, G., Elbaz, A., Parsi, A., Heller, Z., Palmanovich, E., Nyska, M., et al. (2014). Clinical outcomes following ankle fracture: A cross-sectional observational study. *Journal of Foot and Ankle Research, 7*(1), 50.

Sevastikoglou, J. A., Eriksson, U., & Larsson, S.-E. (1969). Skeletal changes of the amputation stump and the femur on the amputated side: A clinical investigation. *Acta Orthopaedica Scandinavica, 40,* 624–633.

Sharrard, W. J. W. (1955). The distribution of the permanent paralysis in the lower limb in poliomyelitis. *The Journal of Bone and Joint Surgery. British Volume, 37-B*(4), 540–558. https://doi.org/10.1302/0301-620X.37B4.540.

Shaw, C. N. (2011). Is "hand preference" coded in the hominin skeleton? An in-vivo study of bilateral morphological variation. *Journal of Human Evolution, 61*(4), 480–487. https://doi.org/10.1016/j.jhevol.2011.06.004.

Sherk, V. D., Bemben, M. G., & Bemben, D. A. (2008). BMD and bone geometry in transtibial and transfemoral amputees. *Journal of Bone and Mineral Research, 23*(9), 1449–1457. https://doi.org/10.1359/jbmr.080402.

Sibonga, J. D., Evans, H. J., Sung, H. G., Spector, E. R., Lang, T. F., Oganov, V. S., et al. (2007). Recovery of spaceflight-induced bone loss: Bone mineral density after long-duration missions as fitted with an exponential function. *Bone, 41,* 973–978.

Sievänen, H. (2010). Immobilization and bone structure in humans. *Archives of Biochemistry and Biophysics, 503*(1), 146–152.

Skerry, T. M. (2008). The response of bone to mechanical loading and disuse: Fundamental principles and influences on osteoblast/osteocyte homeostasis. *Archives of Biochemistry and Biophysics, 473*(2), 117–123.

Sparacello, V. S., Roberts, C. A., Canci, A., Moggi-Cecchi, J., & Marchi, D. (2016). Insights on the paleoepidemiology of ancient tuberculosis from the structural analysis of postcranial remains from the Ligurian Neolithic (northwestern Italy). *International Journal of Paleopathology, 15,* 50–64.

Spikins, P., Needham, A., Tilley, L., & Hitchens, G. (2018). Calculated or caring? Neanderthal healthcare in social context. *World Archaeology, 50*(3), 384–403. https://doi.org/10.1080/00438243.2018.1433060.

Stirland, A. J. (1993). Asymmetry and activity-related change in the male humerus. *International Journal of Osteoarchaeology, 3*, 105–113.

Tabensky, A. D., Deluca, V., Briganti, E., Seeman, E., & Williams, J. (1996). Bone mass, areal, and volumetric bone density are equally accurate, sensitive, and specific surrogates of the breaking strength of the vertebral body: An in vitro study. *Journal of Bone and Mineral Research, 11*(12), 1981–1988.

Takata, S., & Yasui, N. (2001). Disuse Osteoporosis. *The Journal of Medical Investigation, 48*, 147–156.

Tesorieri, M. (2016). Differential diagnosis of pathologically induced upper and lower limb asymmetry in a burial from late medieval Ireland (CE 1439–1511). *International Journal of Paleopathology, 14*, 46–54.

Thomas, T., Vico, L., Skerry, T. M., Caulin, F., Lanyon, L. E., Alexandre, C., et al. (1996). Architectural modifications and cellular response during disuse-related bone loss in calcaneus of the sheep. *Journal of Applied Physiology, 80*(1), 198–202.

Thompson, A. R. (2012). Differential diagnosis of limb length discrepancy in a 19th century burial from Southwest Mississippi. *International Journal of Osteoarchaeology, 24*(4), 517–530. https://doi.org/10.1002/oa.2238.

Thurston, A. J. (2007). Pare and prosthetics: The early history of artificial limbs. *ANZ Journal of Surgery, 77*(12), 1114–1119. https://doi.org/10.1111/j.1445-2197.2007.04330.x.

Tilley, L. (2015). Theory and practice in the bioarchaeology of care. In *Bioarchaeology and social theory*. Cham: Springer.

Tilley, L., & Oxenham, M. F. (2011). Survival against the odds: Modeling the social implications of care provision to seriously disabled individuals. *International Journal of Paleopathology, 1*(1), 35–42. https://doi.org/10.1016/j.ijpp.2011.02.003.

Tilley, L., & Schrenk, A. (2017). New developments in the bioarchaeology of care. In *Bioarchaeology and social theory*. Cham: Springer.

Titorenko, V. (2018). Molecular and cellular mechanisms of aging and age-related disorders. *International Journal of Molecular Sciences, 19*(7), 2049.

Trinkaus, E. (2018). The palaeopathology of the Ohalo 2 Upper Paleolithic human remains: A reassessment of its appendicular robusticity, humeral asymmetry, shoulder degenerations, and costal lesion. *International Journal of Osteoarchaeology, 28*(2), 143–152.

Trinkaus, E., Churchill, S. E., & Ruff, C. B. (1994). Postcranial robusticity in homo. II: Humeral bilateral asymmetry and bone plasticity. *American Journal of Physical Anthropology, 93*, 1–34.

Trinkaus, E., Formicola, V., Svoboda, J., Hillson, S. W., & Holliday, T. W. (2001). Dolnı Věstonice 15: Pathology and persistence in the Pavlovian. *Journal of Archaeological Science, 28*(12), 1291–1308.

Uebelhart, D., Demiaux-Domenech, B., Roth, M., & Chantraine, A. (1995). Bone metabolism in spinal cord injured individuals and in others who have prolonged immobilisation. A review. *Spinal Cord, 33*(11), 669–673.

Uhthoff, H. K., & Jaworski, Z. F. G. (1978). Bone loss in response to long-term immobilization. *The Journal of Bone and Joint Surgery, 60-B*(3), 420–429.

Van Cant, M. (2018). Surviving amputations: A case of a late-medieval femoral amputation in the rural Community of Moorsel (Belgium). In *Trauma in medieval society* (pp. 180–214). Leiden: Brill.

Vico, L., Collet, P., Guignandon, A., Lafage-Proust, M.-H., Thomas, T., Rehailia, M., et al. (2000). Effects of long-term microgravity exposure on cancellous and cortical weight-bearing bones of cosmonauts. *The Lancet, 355*(9215), 1607–1611.

Wang, R., Thur, C. K., Gutierrez-Farewik, E. M., Wretenberg, P., & Broström, E. (2010). One year follow-up after operative ankle fractures: A prospective gait analysis study with a multi-segment foot model. *Gait & Posture, 31*(2), 234–240.

Weatherholt, A. M., & Warden, S. J. (2016). Tibial bone strength is enhanced in the jump leg of collegiate-level jumping athletes: A within-subject controlled cross-sectional study. *Calcified Tissue International, 98*(2), 129–139.

Weinreb, M., Rodan, G. A., & Thompson, D. D. (1989). Osteopenia in the immobilized rat hind limb is associated with increased bone resorption and decreased bone formation. *Bone, 10*(3), 187–194.

Więckowski, W. (2016). A case of foot amputation from the Wari imperial tomb at Castillo de Huarmey, Peru. *International Journal of Osteoarchaeology, 26*(6), 1058–1066.

Young, D. R., Niklowitz, W. J., Brown, R. J., & Jee, W. S. S. (1986). Immobilization-associated osteoporosis in primates. *Bone, 7*(2), 109–117.

Zaidi, Z. F. (2011). Body asymmetries: Incidence, etiology and clinical implications. *Australian Journal of Basic and Applied Sciences, 5*(9), 2157–2191.

Acting on what we have learned and moving forward with skeletal behaviour

Rebecca J. Gilmour[a], Kimberly A. Plomp[b], and Francisca Alves Cardoso[c,d]

[a]Department of Sociology & Anthropology, Mount Royal University, Calgary, AB, Canada, [b]School of Archaeology, University of the Philippines, Diliman, Quezon City, Manila, Philippines, [c]LABOH—Laboratory of Biological Anthropology and Human Osteology, CRIA—Center for Research in Anthropology, NOVA University of Lisbon—School of Social Sciences and Humanities (NOVA FCSH), Lisbon, Portugal, [d]Cranfield Forensic Institute, Cranfield University, Defence Academy of the United Kingdom, Shrivenham, United Kingdom

Humans, as with all primates, are highly social animals with complex behavioural repertoires. Thus, understanding human behaviour is a key aspect of all anthropological disciplines, including biological anthropology and archaeology. Fortunately for us, the way we act and what we 'do' is linked closely with our body's ability to perform actions, making the human skeleton a key piece of evidence that can help us investigate human behaviour. This means that we can look to human skeletons preserved in archaeological and palaeontological contexts to help us understand the multitude of different ways that humans and our ancestors behaved in the past. In this book, biological anthropologists outlined a variety of ways in which they attempt to read and interpret behaviour from the human skeleton. Some questions were motivated by a desire to understand human evolution, others more specific behaviours, occupations, or stressors, but they were all united by the aim to keep the biocultural link between humans as both biological and social animals in mind.

In this book, we aimed to fully consider not only the many ways that biological anthropologists infer behaviour from human skeletons, but also to carefully contemplate the limitations and challenges inherent in these interpretations. Interpreting behaviour from human remains can be a murky area of investigation, mainly because many overlapping skeletal changes can be indicative of very disparate activities, and both healthy and pathological skeletal variation often has multifactorial causes, such as genetic background, hormones, and diet. Donald Ortner perfectly encapsulated this when he wrote, "Reconstruction of behaviour from bioarchaeological data is a minefield littered with the wreckage of inadequate methodology and careless science" (Ortner, 1999, p. 107).

One of the first challenges we face when talking about 'behaviour' is the variety of terminology used. Words such as 'activity', 'occupation', and 'behaviour' are employed differently or synonymously by researchers looking to describe what humans 'do', with each term going in and out of favour over the years (Perréard Lopreno et al., 2013). Presently, all three terms persist in the literature in various ways. For example, studies involving individuals from documented collections that have associated occupational information are used to assess activity-related bone changes as they relate to specific occupations (Alves Cardoso and Henderson, 2010, 2013; Villotte et al., 2010). This 'occupational' information is also used to elucidate generalised patterns of behaviour in terms of workload intensity and repetitiveness, as well as particular technical gestures (e.g. manual and nonmanual work).

Although skeletal evidence has been used as indicative of 'occupations', going forward we must remember that we are assessing musculoskeletal movements that are an expression of the use of the body while undertaking a series and wide range of many different daily activities over the life course. Some of these activities may relate to professional occupations, others with hobbies and everyday life tasks. Differentiating any of these based on bone changes poses an extraordinary challenge that is magnified when studying human remains from archaeological contexts. When human remains are documented, and the ranges of activities performed by the individuals under study are inferred based on occupation at death, limitations also exist. Not only can a single occupation consist of an enormous array of different movements and actions, a person's occupation at death may not be representative of the occupations they had over their lifetime – people change professions, and there is missing data on additional activities such as hobbies and nonoccupational tasks (e.g. domestic duties and recreational sports). A further consideration is that the way occupational descriptions are conceptualised, and the duties they are thought to include, can be influenced by the cultural constructs of the researchers, curators, and even people in the past who described each role (Alves Cardoso and Henderson, 2013). Not only would one occupation include a wide range of activities, each with different biomechanical impacts on the musculoskeletal system, but a single occupation may also have different meanings or stages, each of which may reflect different duties. A shoemaker could be either an apprentice or a shop manager, and a housewife could be a maid, a servant, or a 'stay at home mom' (to use a modern term). For these reasons, the bone changes related to activities and those that are specifically 'occupational' are not necessarily synonymous because occupation only encapsulates one part of an individual's life.

Beyond the conceptual and interpretive problems that terminology poses, more specific criticisms of using skeletal indicators to infer behaviour in the past include insufficient integration of clinical information, an overeagerness to interpret activity over all other explanations, and the need for better contextualisation of interpretations. As stated by Jurmain (2011), it can be easy for biological anthropologists working with human remains to have an "over willingness to leap to simple behavioural conclusions" (p. 262). Jurmain further warns that "since many different activities, or combinations of activities, can produce similar bone involvement, it follows that skeletal data offer little hope of inferring *specific* behaviour from *specific* lesions" (p. 265). In regard to the multifactorial or complex origin of skeletal changes, as discussed by

Plomp in Chapter 7, *Behaviour and the bones of the thorax and spine*, pathological lesions of the spine, such as Schmorl's nodes or arthritis, have often been used to indicate specific activities, such as carrying heavy objects. But clinically, these conditions are well known to have multiple aetiologies that are unrelated to movement or activity, such as age and genetics. In a similar vein, Decrausaz and Laudicina, in Chapter 8, *Behaviour and the pelvis*, critically evaluate how a skeletal lesion in the pelvis, known as 'parturition scarring', has often been considered to indicate that a female has given birth. They discuss evidence from various sources that contradicts this notion and suggest instead that bioarchaeologists should use caution when inferring birth from pelvic lesions. Pathological lesions are not the only behavioural indicators that have complex causes. In Chapter 3, *Biosocial Complexity and the Skull*, White and Menédez illustrate the importance of untangling the complicated evolutionary, environmental, developmental, and biocultural factors that shape human skulls to provide insight into the evolution, history, and lifestyle of human populations.

Biological anthropologists have been criticised for interpreting morphology and/or skeletal lesions as evidence for activity and behaviour without clearly defined modern correlates to support their conclusions. Important volumes on behaviour in the skeleton (e.g. Jurmain, 2011; Larsen, 2015) have long called for 'substantiation using *known* analogs' (Jurmain, 2011, p. 265) and 'a science-based approach' (Larsen, 2015, p. 426) to interpret behaviour from the human skeleton. Earlier we discussed how biological anthropologists have used documented collections to investigate 'occupation', albeit not without significant limitations. There is still more work that can be done in this area, but we believe that biological anthropologists are, more and more, moving to incorporate clinical and documented data to test their hypotheses about activity and help interpret observations made from archaeological and fossil bone. By understanding this need for verified correlates to interpret behaviour from skeletal remains, especially in bioarchaeology and palaeoanthropology, many of the chapters in this book have woven clinical, cadaveric, animal, and/or documented skeletal sample data to help inform their hypothesis-driven questions throughout. For instance, in Chapter 2, *Bone biology and microscopic changes in response to behaviour*, DeMars et al. discuss the nature of bone responses at the microscopic level as they are understood in the clinical and biological literature and applied to archaeological remains. Gilmour et al. in Chapter 11, *Injury, disease, and recovery: Skeletal adaptations to immobility & impairment*, build on these strategies and combine clinical studies and animal models to understand how bone responds to unloading and disuse events, the evidence for which can then be quantified in archaeological assemblages to interpret treatment, convalescence, impairment, and even recovery from pathological conditions. Similarly, in Chapter 10, *Locomotion and the foot and ankle*, Tommy and Barak expertly integrate multidisciplinary lines of evidence, including experimental animal models, functional anatomy, and measurements of bone modelling and remodelling, to investigate the evolution of the hominin foot and ankle throughout our lineage to help us understand how the foot–ankle complex has changed as we evolved bipedalism. These chapters illustrate the wealth of information that can be gained when the question of behaviour is addressed using multiple methodologies and theoretical frameworks. The biocultural impact of behaviour on the human body is not simple, static, nor discrete, and so it stands to reason that developing a comprehensive and reliable understanding of the relationship between our skeleton and our behaviour would require complex, variable, and encompassing approaches, as was attempted throughout this book.

In addition to the efforts made by biological anthropologists to ground their data in comparative studies, they also adopt a great number of methodological approaches to understand behaviour. Many of these are introduced, discussed, and integrated throughout this book, including analysis of skeletal shape, trabecular bone microstructure, entheseal changes, and pathological lesions. Each method brings with it its own unique benefits and insights but also challenges and limitations. Since earlier books on behaviour were published, biological anthropologists have continued to develop and apply various approaches and techniques to better understand the effect of behaviour on the skeleton. Jurmain (2011) identified diaphyseal geometry as an area with great testable potential to provide insight into behaviour. Biomechanical approaches, including cross-sectional geometries, are adopted by many researchers seeking to understand behaviour as it can be interpreted through loading patterns. More recently, trabecular bone microarchitecture is integrated in behavioural studies as a more sensitive indicator of loading, providing biological anthropologists with another nondestructive and quantitative approach to behavioural investigations. The benefits and potential of these methods are perfectly illustrated in Chapter 8, *Tool use and the hand*, where Dunmore et al. explain how analyses of bone modelling/remodelling, cross-sectional geometries, and traces of soft tissues of the hand bones can help us understand how the hominin hand evolved the ability to make and use tools, which will also help us identify when these adaptations evolved in the hominin lineage.

Other approaches, such as the investigation of entheseal changes, have also undergone large-scale growth and re-envisioning over the last decade. Not only has the terminology used to refer to these features developed over time (e.g. musculoskeletal markers, enthesophytes, entheseal changes), but we also see the emergence of greater specificity in repeatable methods (e.g. Coimbra Method, Henderson et al., 2016). In later years the emphasis was placed on understanding that entheseal changes encompass any change to the normal visual appearance of the enthesis, and this may be multifactorial in origin and not related to activity and muscle use alone. Considering this, scholars now advocate for a better understanding of the biology and anatomy of entheses, and caution against using an oversimplified approach where any and all changes are seen as activity related. One of the most relevant advancements in this area of study was the acknowledgement that age is a major factor that needs to be considered and controlled for when assessing entheseal changes. The new methodological advancements reinforce the need to record all changes observed at entheseal sites, aiming to better our understanding of what may cause those changes and how each may relate to known causative factors such as the ageing process, underlying pathological conditions, and ultimately, activity. Hence, although entheseal changes are tested and continue to be used in activity assessments, there is also a growing awareness that bone changes at entheseal sites are closely linked to factors other than activity. Such is exemplified in Chapter 5, *Archery and the Arm*, as Jessica Ryan-Despraz explores archery from a kinesiological and an osteological perspective. Ryan-Despraz illustrates the growing awareness of an inclusive approach when assessing bone changes accredited to specific activities (archery). As they show, to most accurately understand and interpret behaviour, this approach needs to consider age, distinct entheseal morphology, and a better understanding of actions and muscle biomechanics linked to activities. This serves as an additional reminder to biological anthropologists that behavioural evidence in the skeleton continues to be murky and difficult to interpret; it is necessary to control for as many biological and contextual factors as possible prior to making interpretations from human remains.

Finally, this book was divided into chapters that focus on separate skeletal regions. However, in reality, the body and the skeleton does not perform in discrete units or sections, but rather as a highly integrated and complex system. A behaviour or activity that, initially, may seem to only involve one region of the body could actually influence bone development and features in other areas. As a broad and simplified example, a person who habitually lifts heavy objects will not only be subjecting their arms and shoulders to loads that may cause observable morphological changes in these regions, but these loads will also be applied through their axial skeleton and onto their lower limbs. This transference of force has the potential to cause spinal changes (e.g. Schmorl's nodes), and also increase the loads experienced in the lower limb, which could result in cross-sectional and trabecular bone microstructural changes. There is also research that demonstrates crosstalk among loaded and unloaded bones, in that bone modelling and remodelling can be stimulated in regions distant to the loaded area (Osipov et al., 2018; Sample et al., 2008). Essentially, forces applied to (or removed from) one region may lead to increased (or decreased) remodelling and bone formation, and therefore have structural implications throughout the entire skeletal system. Biological anthropologists should be encouraged to think about the body as an interactive system in which the actions affecting one region also have the potential to impact other regions.

The systemic nature of behaviour's effect on the skeleton is made clear in a number of chapters in this book. Despite the organisation of this book by skeletal region, the chapters are interlocking, leading the reader to explore how similar behaviours might manifest across the body. The intersection of activities across the body is nicely traced, for example, between Chapters 7, 8, and 9, from the thorax and spine, through the pelvis, and down to the lower limbs. Starting with Plomp's Chapter 7 on the spine and thorax, she begins to discuss the effects of corsetry. These themes are revisited in the pelvis by Decrausaz and Laudicina (Chapter 8) who go on to also discuss pelvic changes associated with horseback riding. These indicators of horseback riding activity are then pursued in greater detail and in the lower limb in Chapter 9 by Berthon et al. In each of these cases, the lines are blurred between chapters, helping researchers identify a starting point (perhaps in a single skeletal region) and encouraging them to continue exploring how similar behaviours may manifest across the skeleton. Biological anthropologists should therefore be inspired to consider the widespread effects of broad behaviours on the entire body, and not just in isolated elements. This overlap, as we flow from one skeletal area and chapter to another, acts to emphasise just how interconnected our bodies are.

The chapters in this book have shown how behaviour is more than just our physical motions, but also how we think, react, and interact. Researchers who pursue skeletal behaviour studies should remember the existence of biocultural connections between their bodies and the sociocultural milieus in which they exist. The authors in this book have all presented us with skeletal evidence, explaining how this evidence is used to inform the actual physical motions of people. As anthropologists, we often aim to go beyond description of physical morphology and seek their root sociocultural causes and explanations. Many of the chapters in this book delve into these lines of interpretation, using the physical evidence with an eye to how it informs social and cultural behaviours. Building on this understanding of verified responses in properly contextualised remains, it makes sense to continue to ask questions about how our development and evolution has influenced not only the types of behaviours we are investigating, but also how those behaviours can be preserved in the skeleton. Overall, we encourage readers to recognise the individual methods and themes in each chapter and pick them up piece by piece and reassemble them – step out of the box to use the diversity in approaches to develop new ways of thinking about behaviour as it is preserved in our bones.

References

Alves Cardoso, F., & Henderson, C. (2013). The categorisation of occupation in identified skeletal collections: A source of Bias? *International Journal of Osteoarchaeology, 23*(2), 186–196. https://doi.org/10.1002/oa.2285.

Alves Cardoso, F., & Henderson, C. Y. (2010). Enthesopathy formation in the Humerus: Data from known age-at-death and known occupation skeletal collections. *American Journal of Physical Anthropology, 141.* https://doi.org/10.1002/ajpa.21171, 550560.

Henderson, C. Y., Mariotti, V., PanyKucera, D., Villotte, S., & Wilczak, C. (2016). The new 'Coimbra method': A biologically appropriate method for recording specific features of fibrocartilaginous entheseal changes. *International Journal of Osteoarchaeology, 26*(5), 925–932.

Jurmain, R. (2011). *Stories from the skeleton: Behavioral reconstruction in human osteology.* Oxfordshire: Taylor & Francis.

Larsen, C. S. (2015). *Bioarchaeology: Interpreting behavior from the human skeleton.* Cambridge: Cambridge University Press.

Ortner, D. J. (1999). Review of the book bioarchaeology: Interpreting behavior from the human skeleton, by C.S. Larsen. *Journal of the Royal Anthropological Institute, 5*(1), 106–108.

Osipov, B., Emami, A. J., & Christiansen, B. A. (2018). Systemic bone loss after fracture. *Clinical Reviews in Bone and Mineral Metabolism, 16*(4), 116–130.

Perréard Lopreno, G., Alves Cardoso, F., Assis, S., Milella, M., & Speith, N. (2013). Categorization of occupation in documented skeletal collections: Its relevance for the interpretation of activity-related osseous changes. *International Journal of Osteoarchaeology, 23*(2), 175–185.

Sample, S. J., Behan, M., Smith, L., Oldenhoff, W. E., Markel, M. D., Kalscheur, V. L., et al. (2008). Functional adaptation to loading of a single bone is neuronally regulated and involves multiple bones. *Journal of Bone and Mineral Research, 23*(9), 1372–1381.

Villotte, S., Castex, D., Couallier, V., Dutour, O., Knüsel, C. J., & Henry-Gambier, D. (2010). Enthesopathies as occupational stress markers: Evidence from the upper limb. *American Journal of Physical Anthropology, 142*(2), 224–234. https://doi.org/10.1002/ajpa.21217.

Index

Note: Page numbers followed by *f* indicate figures, *t* indicate tables, and *b* indicate boxes.